DATE DUE

DEMCO 38-296

BEACHAM'S GUIDE TO
ENVIRONMENTAL ISSUES & SOURCES

BEACHAM'S GUIDE TO
ENVIRONMENTAL ISSUES & SOURCES

Volume I
Pages 1 - 652

BEACHAM PUBLISHING, INC.
WASHINGTON, D.C.

BEACHAM'S GUIDE TO ENVIRONMENTAL ISSUES AND SOURCES

Editor
Walton Beacham

Staff Editors
Charles J. Moseley
Don Raymond Marr

Book and Cover Design
Amanda Mott

Production
Deborah M. Beacham

Library of Congress
  Cataloging-in-Publication Data

Beacham's Guide to Environmental Issues and Sources/
edited by Walton Beacham.
  Washington, D.C.: Beacham Publishing, Inc.

  Includes bibliographical references and index

  1. Environmental sciences. 2. Environmental
sciences—Bibliography. I. Beacham, Walton 1943-. II.
Beacham Publishing.
  GE115.B43   1993   363.7—dc20   92-42027

  Assessment of issues and bibliography of
sources for environmental concerns of the 1980s and
1990s.

ISBN: 0-933833-31-8
Printed in the United States of America
First printing, May 1993

Printed on acid-free, recycled paper with vegetable inks

PREFACE

There were three goals for this reference work: to locate and compile a comprehensive database of sources related to environmental issues worldwide; to organize it by topic and by type of source so that researchers could easily identify materials useful to their area of study; and to identify and isolate environmental ideas so that students could use section heads and the index as a starting point for developing term paper topics, thesis statements, and report ideas.

Although this seemed like a straightforward task when we began the project, it was not. There are tens of thousands of environmental resources, many of which are not included in standard indexes. Of the sources that are commonly indexed, it is difficult to determine which ones relate to a particular topic. Even electronic indexes that allow you to search by key word often do not produce a satisfactory list of sources because of the wide disparity in cataloging terminology for many environmental ideas. If you are looking for sources on how to responsibly dispose of your old automobile tires, there is little chance that you'll find them by searching the words "car," "automobile," or "tire." The word "rubber" might lead to "rubber recycling programs" but most likely you'll have to start with "recycling" and work your way through dozens of subcategories.

Beacham's Guide to Environmental Issues and Sources has done that task for you. We've extracted resources from a variety of databases and catalogs and grouped them into meaningful and manageable categories. Within each category the sources are grouped by type and to some extent by their appropriateness to different levels of research. Technical books, for example, are separated from works directed at a general audience.

HOW THIS BOOK IS ORGANIZED

Because there is such an enormous amount of material about the environment covering widely diverse topics, our primary aim was to group related ideas into chapters. But, because many environmental issues are interconnected, some topics appear as subsections in several chapters. "Great Lakes," for example, appears as a full subsection in the chapters:

Great Lakes
    Coastal Zone Management
    Fishery Management
    Fresh Water Pollution
    Water Supply

and as a topic in the chapter Acid Rain: Lakes.

The organizational tradeoff was whether to keep all of the coastal zone management sources together, including the Great Lakes, or to keep the Great Lakes together. Recognizing the inconvenience of any reference work that forces researchers to consult multiple sections, we have facilitated cross referencing by giving "See also" page numbers at the beginning of each subsection. The index also groups topics and subsections, providing quick reference to disparate issues.

Each of the chapters is organized similarly. Sources that address multiple aspects of a topic are collected under the heading **Introductory Sources**. These provide an overview of the entire subject area and are sometimes quite comprehensive. They should not be considered as "beginning" or cursory materials. The remaining sources are collected into topical subgroups that make it easier to research specific ideas. The chapter on recycling, for example, contains topical subgroups on "tires," "oil," "plastic," and "beverage containers."

Within each of the topical categories, sources are grouped by type. Most categories contain the following:

Books (annotated)
Books (unannotated)
General interest periodicals (annotated)
General interest periodicals (unannotated)
Environmental and professional journals (annotated)
Environmental and professional journals (unannotated)
Law journals
Reports
Federal government reports
Conference proceedings

Where sources exist, additional subgroupings include

Technical books (annotated)
Technical books (unannotated)

Bibliographies
Reference
Databases (electronic)
Books for young adults
Videos

Because not every section contains bibliographies, reference, electronic databases, books for young adults, and videos, we have listed them on the contents page preceding each chapter so they will be easier to locate.

We have grouped resources by type to accommodate different kinds of research and library holdings. In addition, since some researchers are looking for general, non-technical, easy-to-locate materials, we have culled the entire chapter and selected representative works, grouped under the title **Short List of Sources for General Research**, found at the beginning of each chapter. All of the **Short List** citations appear again in the main body of sources, and researchers need not consult the **Short List** if they are consulting the subsections. In the same spirit of accommodating different kinds of libraries and research, we have separated general interest periodicals from environmental and professional journals, which tend to have a greater depth of coverage than the general interest periodicals. The bibliography is unusually strong in its representation of annotated journals and reports, usually included only in specialized bibliographies. We have also indicated if an article is the cover story or a column, which gives some indication as to the depth of coverage and type of reporting. Books containing illustrations, color illustrations, maps and charts are so indicated.

For a number of international issues, the best, and sometimes only, resources are studies conducted by organizations such as the World Bank or the United Nations, and published as reports to the organization. These are likely to be fairly technical and are probably not available in libraries with general holdings. Similarly, U.S. government reports are usually available only in regional federal repositories and some academic libraries. Conference proceedings, which tend to be of interest to specialists but which often contain unique information, are listed separately.

With rare exception citations are not repeated within a chapter (except in the **Short List**). Sources covering the topic **Food Supply: Nutrition** will not appear again under the topic **Food Supply: Health**. In light of our decision to avoid excessive duplication of sources within or across chapters, we have established a system of cross referencing. These "(See also. . .)" directions appear on the chapter contents pages and beneath the headings of each of the chapter subtopics. For example, the chapter on **Wildlife Conservation and Management** has the subtopic **Birds,** which contains sources related directly to bird conservation. Endangered birds are covered in the chapter on **Endangered Species**, so that the cross reference appears as:

Wildlife Conservation and Management: Birds
(See also *Endangered Species*) 1223

This permits researchers to study threatened birds separately, or to study all bird conservation by referring to both sections. "(See also . . .)" directs researchers to other sections of the book that specifically cover a similiar topic, and research is not complete without referencing both sections. "(Related topics in . . .)" indicates other chapters that contain associated but not directly related materials.

While the chapter titles and subsections, used in conjunction with the cross references, provide a compass to the issues, the key word index directs readers to specific sections. Our intention is not to index every entry that applies to a topic, but to direct researchers to the chapter(s) or section(s) where sources on that subject may be found. Often, a section contains multiple citations about the subject, and researchers should browse the entire section to locate all relevant material.

LOCATING SOURCES

In addition to citing sources, we have attempted to provide information for locating the source. For books we have usually been able to provide the Library of Congress call number.

> Alerstam, Thomas. *Bird migration*. New York: Cambridge University Press, 1990. 420 p.: ill., maps. (QL698.9 .A4413 1990)
>> Primarily examines the migratory habits, flight journeys, and the navigational capabilities of birds. Also addresses how climate and ecosystems influence the bird patterns and habits. Includes maps, sketches, tables, a bibliography, and an index.

Magazine and journal entries include page numbers, cited in one of two ways: by the range of page numbers

> Ackerman, Diane. "A reporter at large." *New Yorker* 66 (September 24, 1990): 61-62, 64-74, 76, 78, 80-82, 84-86, 88.

or by the first page number (164) followed by the number of pages (10).

Beissinger, Steven R.; and Bucher, Enrique H. "Can parrots be conserved through sustainable harvesting?" *BioScience* 42,3 (March 1992): 164(10).

For reports published by difficult-to-locate organizations, we have tried to provide an address. Often the report can be obtained for a nominal charge from the organization.

Nielsen, Leon. *Chemical immobilization in urban animal control work.* Milwaukee: Wisconsin Humane Society (4151 North Humboldt Avenue, Milwaukee 53212), 1982. 93 p.: ill. (QL62.5 .N54 1982)

Similarly, many federal government documents can be obtained from the Government Printing Office (G.P.O.) or from the National Technical Information Service. We have included agency reference numbers to these documents when available to facilitate ordering them.

Hendrey, George R.; et al. *Geological and hydrochemical sensitivity of the Eastern United States to acid precipitation.* Corvallis, Oreg.: Office of Research and Development, Environmental Protection Agency, 1980. Available from National Technical Information Service, Springfield, Va. 100 p. (EPA-600/3-80-024)

Renewable sources of energy. Paris: Organisation for Economic Co-operation and Development, 1987. Distributed by OECD Publications and Information Centre, Washington, D.C. 334 p.: ill. (TJ808 .R46 1987)

The addresses for these and other frequently cited organizations are provided in Appendix A.

We also provide addresses of video producers with each citation.

Acid rain: a North American challenge. Directed by Gary Toole. National Film Board of Canada (1251 Avenue of the Americas, New York, 10020): 1988. 16 min.

WHAT HAS BEEN INCLUDED AND EXCLUDED

The body of literature on the environment is enormous, and in order to make this book useful to students, librarians, and professional researchers we had to make decisions about what to exclude. Because information about the environment changes so rapidly, our first impulse was to include only current materials, and we initially began collecting magazine and journal articles published since 1988 and books published since 1985. But the longer we studied the issues the more we began to recognize that useful coverage of the issues sometimes went beyond those time frames. Some environmental concerns tend to be studied in cycles. Nuclear energy, for example, was the center of a great deal of attention, and subsequent research, in the 1970s and early 1980s. Then, with the nuclear industry put on hold as a result of cost factors, accidents, and federal regulation, research tapered off and the topic receded from public view. Then at the end of the decade, concern over global warming caused, in part, by power plant emissions, produced a resurgence of interest in nuclear power.

Consequently, we realized that our sources had to include periods of public concern and research that, on the surface, might not appear to be timely. Also, while some reports are as old as ten years, they may represent the only research available on a narrow topic, such as nutritional problems in the Seychelles islands. Some older works remain the definitive work on the subject or the book that brought an issue to social awareness, such as Rachael Carson's *The Sea Around Us* (1961), and we have included some of those seminal works. It is our intention to provide the latest information and to continually update and expand the material.

Because many libraries have newspaper indexes, we have excluded all daily newspaper articles except for feature stories in the *New York Times Magazine*. We have also excluded magazine and journal articles that are less than one page long. For federal government reports, we have included those issued by federal agencies, the General Accounting Office, and the Office of the Comptroller General, but with rare exception we have not included reports by Congressional committees or studies requested by members of Congress. We have collected and saved all the reports and one-page articles and may make them available in electronic format at some future date.

There are two additional criteria that affected what materials we included. Many works have been produced in foreign languages and have not been translated into English. Because we decided to include English sources only, there is an appearance that some areas of the world have been neglected by researchers. This may or may not be true. Second, we decided that all sources included must be available in the U.S. Many of the sources are widely available; others are more difficult to locate, but all are catalogued by, and should be available in the Library of Congress. Many of the reports, federal government reports, and conference proceedings may be ordered from the organization or agency that produced them.

Certain types of books were excluded because, although they relate to environmental concerns, they are not really research materials. Vegetarian cook books, photograph books of endangered animals, coffee table books of shorelines, field guides to birds and plants, and travel guides of the Rocky Mountains are examples of books that have been excluded. We have also excluded text books, operational manuals, and highly specialized technical books.

OVERVIEW AND ASSESSMENT OF ENVIRONMENTAL ISSUES

The material presented in Chapter 1 is taken from the Council on Environmental Quality's (CEQ) 1992 annual report. CEQ compiled data on twelve areas of the environment, then analyzed the data to produce an overview and assessment of the problems, the efforts from the government and private sector to solve them, and goals for future efforts.

The Council on Environmental Quality was created by the National Environmental Policy Act (NEPA) in 1969 to provide the president with an expert source of review of national policies, environmental problems, and trends, and to coordinate federal agency compliance with NEPA. Under NEPA, CEQ is responsible for assisting the president in preparing an annual environmental quality report to Congress, thus CEQ's annual report reflects the views of each administration. Therefore, this report is necessarily partisan and presents the views of the Bush administration. That many environmentalists might disagree with CEQ's evaluation of specific issues should not obscure the overall value of this report, especially in the areas of federal legislation and programs. It identifies and explains many aspects of the environment that will elucidate the issues in *Beacham's Guide to Environmental Issues and Sources*, and researchers are encouraged to refer to the CEQ assessment for clarification of laws, terms, and programs.

We appreciate CEQ's assistance in creating this chapter, especially the efforts of Dale Curtis and Seth Miller. We also appreciate assistance from the Environmental Protection Agency in reviewing the contents of the bibliographical chapters for completeness.

CHAPTER TITLES

(Full contents precedes each chapter)

WASTE

WILDLIFE AND ECOLOGY

NATURAL RESOURCES

APPENDICES

CONTENTS

VOLUME I

Contents

Contents

Contents

Contents

BEACHAM'S GUIDE TO

ENVIRONMENTAL ISSUES & SOURCES

CHAPTER 1

COUNCIL ON ENVIRONMENTAL QUALITY
1992 ENVIRONMENTAL ASSESSMENT

continued

POLICY AND GOVERNMENT PROGRAMS

EDUCATION

Public-private partnerships are the key to success in environmental education as well as in other environmental disciplines.

Conditions and Trends

Public interest in environmental education has fluctuated over past decades, but in the 1990s it received new impetus from the America 2000 Education Strategy developed by President Bush and the state governors. The strategy aims to accomplish excellence-in-education goals by the year 2000. Three of these goals—making U.S. students first in the world in science and mathematics, spurring student citizenship, and continuing education through adulthood—are key factors in environmental education.

Another boost came with passage of the National Environmental Education Act of 1990, which targets students in elementary and secondary schools, as well as post-secondary students, senior citizens, and the general public. The law calls for collaboration between the federal government and the media, private industry, museums, libraries, parks, and recreation facilities, and it recognizes the need for international cooperation.

To coordinate these efforts, the 1990 law required EPA to establish an Office of Environmental Education (OEE), which is now in place. The office will work to ensure that U.S. environmental education efforts remain well organized and consistent, regardless of future peaks and valleys in public interest.

Policies and Programs

Achieving the America 2000 goals of excellence in education will require the cooperation of individuals in schools and colleges, federal, state and local governments, professional and public interest groups, business, and industry. In 1992 the EPA Office of Environmental Education, acting as an advocate for the mission nationally and internationally, emphasized that environmental educators have a dual charge:

- Instill an environmental ethic in America's young people that will prepare them to deal responsibly with the environment throughout their lives.

- Raise the environmental awareness of adults as informed consumers in the global shift toward sustainable development and pollution prevention.

In 1992 OEE set goals to help federal agencies implement America 2000. A listing of these goals and examples of current efforts to achieve them follow. With the Clinton/Gore administration in place, it is not yet clear which of these programs will continue in their present form.

Goal 1. Expand communications with environmental experts.

A strong body of research and literature exists along with an experienced cadre of environmental educators. Federal agencies are establishing communications and advisory networks with educators to ensure that the nation has full benefit of their expertise.

National Environmental Education Advisory Council. The 1990 Environmental Education Act called for creation of an advisory council to make recommendations on implementing the act and to prepare a biennial report to Congress assessing the quality of environmental education in the nation and offering recommendations for improvement. EPA established an 11-member advisory council that represents the nation's geographic areas, minority groups, and a variety of occupations.

Federal Task Force on Environmental Education. NEEA also called for a network to coordinate environmental education efforts among federal agencies. Members of the Federal Task Force on Environmental Education include CEQ, EPA, the departments of the Interior, Agriculture, Defense, Education, Energy, State, and Health and Human Services, the Agency for International Development (USAID), National Aeronautics and Space Administration (NASA), National Oceanic and Atmospheric Administration (NOAA), National Science Foundation, Peace Corps, Tennessee Valley Authority, and U.S. Information Agency. Task Force subcommittees are developing specific areas for collaboration.

EPA Environmental Education Advisory Board. This board consists of 35 senior representatives from EPA headquarters, regional offices, and research laboratories. Six subcommittees are developing programs, policies, and procedures required by the 1990 act.

Pollution Prevention Advisory Group. This external advisory group makes recommendations to EPA on developing pollution prevention educational materials for students and on teacher training. The 20 members have expertise in education, publishing, communications, minority concerns, and the environment.

NEPA Education and Training Courses. The National Environmental Policy Act of 1969 (NEPA) established a national policy to promote environmental quality, as well as procedures to help federal agencies focus on that goal. Dozens of federal agencies as well as private entities and universities conduct training and educational courses on the NEPA environmental impact statement process, its requirements, and how it can be better used to integrate environmental values into the federal decisionmaking process to reduce conflict, litigation, and cost. CEQ has compiled a compendium that lists and describes NEPA training courses; copies can be obtained from the Council on Environmental Quality, Executive Office of the President.

Goal 2. Develop Partnerships to Increase Effectiveness and to Maximize Use of Limited Resources

Federal agencies are forming new coalitions and cooperative ventures with groups not traditionally involved in environmental education, such as industry, religious organizations, and senior citizen associations. By engaging all stakeholders in the process and by developing partnerships with grassroots organizations, federal agencies can ensure that programs are responsive to local needs.

Goal 3. Increase the number and skills of teachers who infuse environmental issues into their curricula.

Federal agencies are supporting the development of training programs and materials to ensure that all teachers, including those in nonscientific fields, have the opportunity to receive training and materials that will allow the infusion of environmental education into their curricula.

Teacher-Training Grants. In 1992 EPA awarded a $1.6 million grant to a consortium of academic institutions, corporations, and nonprofit organizations headed by the University of Michigan to develop and operate a national environmental education and training program. The program targets in-service teachers (K-12) as well as other educators and includes curriculum development and evaluation, teacher training, and information dissemination.

Wetlands Education. The Fish and Wildlife Service, in cooperation with the U.S. Geological Survey, has produced an educational poster for elementary and middle school teachers. Entitled, "Wetlands: Water, Wildlife, Plants and People," the poster features color aerial views of various types of wetlands and on the back has lesson plans and classroom activities. FWS has distributed 200,000 copies to staff, other federal agencies, private organizations, and educators throughout the United States. Sciencescope and Science for Children, publications which circulate to 90,000 teachers, featured the poster as a special pullout in their September 1992 issues. Copies are available from USGS.

Pollution Prevention. The EPA Pollution Prevention Task Force, an external advisory group, is developing a teacher-training plan and educational materials. With guidance from the task force and input from teachers and environmentalists, EPA has designed materials for grades K-12. The project complements the national program being developed by the University of Michigan consortium and may be incorporated into it.

Parks as Classrooms. Through this initiative, the National Park Service enters into partnerships with local schools to develop curricula and provide educational services for students who visit parks on field trips. In 1992 NPS supported 41 projects with funding of $780,000. Training courses emphasizing cultural, heritage, and environmental themes prepare interpreters to develop NPS educational programs. Other projects include curriculum development and an exchange-education project developed by the Everglades National Park involving two schools, one in Miami and the other in Chicago.

Heritage Program. The Bureau of Land Management (BLM) has designed a program to strengthen school children's sense of personal stewardship of America's cultural heritage. In 1992 BLM launched the program with an educational video, featuring the Teenage Mutant Ninja Turtles. Two articles on the program appeared in Science and Children, a professional magazine for science teachers, and BLM took part in seven partnership projects to publicize the program.

Goal 4. Encourage and support grassroots efforts.

Because innovative grassroots projects can be particularly responsive to local needs, federal agencies support environmental education efforts at the local level. In 1992 OEE awarded $2.4 million in grants to schools, states, and nonprofit organizations to support environmental education projects around the country. Individual grants were as high as $250,000, but the majority were $5,000 or less. Some projects focus on specific issues such as water, air, and solid waste management, while others emphasize general environmental education either locally or globally. Educational approaches range from community cleanup to computer games. EPA selects projects for their innovation and applicability in other communities, soliciting grant applications in the fall of each year by means of notices in the Federal Register.

Goal 5. Promote careers in the environmental field.

As the public becomes more aware of global environmental problems, federal agencies are increasing called on to support environmental research. The nation needs environmental scientists, engineers, and other professionals capable of developing preventive solutions to environmental problems. EPA promotes environmental careers and encourages participation of minority students in environmental degree programs through a number of programs, such as the following:

Internships and Fellowship. EPA administers programs that place college interns and in-service teachers in federal agencies active in environmental education.

Internships encourage college students to pursue careers in environmental science, and fellowships encourage teachers to take their environmental experience back into the classrooms, thereby educating students about the environment and promoting the pursuit of environmental careers.

NOAA administers the Dean John A. Knauss Marine Policy Fellowship that annually supports up to 25 graduate students in the Sea Grant College Program Network for one year of educational experience in Washington, D.C. Federal agencies or Congressional staffs host the fellows, who work on assignments related to the marine environment.

President's Environmental Youth Awards. Young people in all 50 states compete annually in this program which honors outstanding commitment to the environment. National award winners, selected by EPA regional offices, participate in a recognition program in Washington, D.C. Since 1989 President Bush has hosted ceremonies at the White House to honor the ten annual award winners.

National Environmental Education Awards. A new federal awards program creates the following awards to recognize individuals for their contributions to environmental education:

Theodore Roosevelt Award for an outstanding career in environmental education, teaching, or administration;

Henry David Thoreau Award for outstanding literature on the natural environment and pollution problems;

Rachel Carson Award for an exceptional print, film, or broadcast contribution to public education on environmental issues; and

Gifford Pinchot Award for an outstanding educational contribution to forestry and natural resource management.

President's Environment and Conservation Challenge Awards.

CEQ, working with private-sector partners, manages this program to recognize U.S. citizens and organizations for outstanding contributions to the nation's environmental quality. One of the four award categories is Education and Communications, in which Presidential medals are awarded for developing informational programs that inspire respect for the environment and raise the public's environmental awareness. For more information, see the Private Sector Initiatives section.

National Network for Environmental Management Studies. NNEMS provides fellowships or paid internships to students in exchange for written studies or reports on environmental science, policy, and management in areas of concern to EPA. Most recipients are students conducting master or doctoral research.

Goal 6. Provide educators and the public with access to appropriate environmental education materials.

A quantity of environmental education materials exists, but quality varies and finding appropriate materials can be difficult. To categorize and provide access to the best materials available, federal agencies have the following efforts underway:

Clearinghouse. To facilitate information exchange, EPA is developing an interactive computer system for environmental education materials produced by federal agencies and other organizations. EPA is testing the clearinghouse, designed for educators (K-12), during the 1992-1993 school year and plans to have it available to the public for the 1993-1994 school year.

Earth Notes. This quarterly periodical, published by the EPA Office of Environmental Education, contains innovative ideas from educators (K-6) about their first-hand experiences in bringing environmental education into the classroom. The periodical provides an open forum for the exchange of teaching ideas, comments, and brief essays concerning environmental education in the elementary grades.

Wee Deliver Newsletter. The U.S. Postal Service publishes this newsletter read by 1 million school children. The Postal Service and EPA developed a multi-page environmental insert on recycling and pollution prevention for the March 1992 issue, which was distributed to 7,000 elementary schools nationwide.

Goal 7. Educate the general public to make informed decisions.

Federal agencies are improving their ability to communicate complex scientific and technical information to the public. To prepare citizens to make informed decisions on the environment, federal agencies are sponsoring projects such as the following;

Media Strategy. EPA is developing a media strategy that federal agencies can use to reach the general public with information about actions that individuals can take to make a difference in the environment. The strategy enlists the help of private partners and the media, including press, television, and radio.

Marine Environmental Outreach. In 1992 the U.S. Coast Guard developed an outreach program for the general public and the marine industry. The campaign, scheduled for implementation in early 1993, will raise awareness of regional, national, and international marine pollution issues impacting U.S. waters. To improve relations with private groups, the Coast Guard will publish a Marine Environment Protection Strategy and a newsletter with updates on such topics as plastics in the marine environment, spill-reporting mechanisms, and the unintentional introduction of nonindigenous aquatic species into U.S. waters by ship ballast discharges.

Sea Grant Marine Advisory Service. The NOAA Sea Grant Network sponsors public education/outreach programs that utilize a network of extension professionals in every coastal state. The programs target teachers, state and local policymakers, industrial leaders, and the general public. Educational efforts promote appropriate strategies to protect and wisely use aquatic resources.

Mail That Mother Earth Can Love. The U.S. Postal Service produced a brochure by this title to explain what each person can do at work or home to help the mail and Mother Earth. The brochure contains the following tips for business mailers:

• Design recyclable mailings by using uncoated paper stocks, water-based inks, and envelopes with water-based adhesives and recyclable windows or no windows at all

• Support recycled-products industries by dealing with suppliers that use environmentally sound products, papers, and packaging

• Use environmentally friendly packing materials

• Print on both sides of paper

• Reuse and recycle packaging materials

• Target mailings

• Reduce the amount of undeliverable mail by keeping addresses current

EPA and the Postal Service joined forces to produce posters promoting the "New 3Rs—Reduce, Reuse, Recycle." In September 1992 the posters appeared in postal lobbies throughout the country, and a multilingual design was distributed to foreign postal administrations.

Goal 8. Reach beyond national boundaries.

Because environmental risks transcend political boundaries, international information exchanges and joint training ventures have become an environmental education priority, and federal agencies are supporting initiatives around the world.

Sport Fishing Education. Under provisions of the Federal Aid in Sport Fish Restoration Act, up to 10 percent of federal funds apportioned by the Fish and Wildlife Service to the states may be used for aquatic resources education. Such programs can include aquatic ecology, aquatic resources management, aquatic safety, conservation ethics, and fishing. The audience is U.S. fishermen and those from other nations fishing in U.S. waters.

Trilateral Memorandum of Understanding. The United States developed a Trilateral Memorandum of Understanding on Environmental Education in cooperation with the governments of Canada and Mexico to promote environmental education on the North American continent. Educational activities include information exchanges on educational policies, approaches, and materials; support and participation in seminars, workshops, and conferences; and trilateral initiatives involving the environmental education of youth.

Fulbright Fellowships. The U.S. Information Agency and EPA are developing Fulbright Fellowships for senior research scholars in environmental research and education from the Near Eastern, North African, and South Asian regions. Senior scholars will be eligible for 9-month to 1-year grants to further their environmental studies at an institution of higher education.

International Public-Private Partnerships. Federal agencies are supporting public-private initiatives to promote environmental education in various regions of the world. For example, EPA is sponsoring the Regional Environmental Center for Central and Eastern Europe, initiated in 1990 in Budapest, Hungary; the Caribbean Environment and Development Institute, initiated in 1992 in Puerto Rico; and the Environmental Education and Information Center, initiated in 1992 in Kiev, Ukraine.

The Department of the Interior also is assisting the government of Hungary in developing a free-market-based mineral development program. Assistance includes training in process and techniques for assessing the potential environmental effects of mineral development. The objective is to provide the basic tools for including consideration of environmental values in decisionmaking.

Federal Partnerships in Environmental Education

Federal agencies are forming partnerships with an array of public and private organizations.

Students Watching Over Our Planet Earth

SWOOPE is offered on the East Coast through a partnership between EPA and the Department of Energy. It teaches students in grades K-12 environmental science through hands-on experience.

Boy Scouts and Girl Scouts

EPA has signed memoranda of understanding (MOU) with the Boy Scouts and the Girl Scouts to reinforce environmental education.

Trail Boss

EPA, the departments of the Interior, Agriculture, and Defense, and the Boy Scouts of America developed this program to teach skills for training and leading volunteer crews involved in conservation projects.

Future Farmers of America

EPA and FFA have signed an MOU to coordinate environmental education materials and activities. FFA, which includes environmental issues in its primary mission of agricultural education, will collaborate with EPA on new approaches and materials for environmental education. EPA is developing a similar MOU with the National 4-H Council.

Partnerships with Environmental Groups

Federal agencies are working with a number of environmental groups on initiatives, such as the following:

North American Association for Environmental Education

EPA and NAAEE have signed an MOU to coordinate conferences, grant and internship programs, research, and meetings of the National Environmental Education Advisory Council.

Alliance for Environmental Education

AEE, a coalition of organizations, agencies, corporations, and institutions committed to public environmental education is working with EPA and Time/Warner on a mass media campaign to broadcast environmental messages to households nationwide.

National Environmental Education and Training Foundation (NEET)

CEQ oversees participation by federal agencies in this nonprofit educational foundation. The EPA Office of Environmental Education, assisted by NEET, is working on plans for partnerships with businesses.

Toronto Global Conference on Environmental Education

This conference, held at Toronto, Canada, on October 16, 1992, was a follow-up to actions taken on global environmental education at the U.N. Conference on Environment and Development (UNCED) held in June 1992. National Audubon

National Audubon Society

The Audubon Society, Dow Chemical, and EPA collaborated on a pilot environmental education program for seventh graders in the Great Lakes Region.

Partners Outdoors Action Group for Environmental Education

The Department of the Interior organized this group, whose members represent federal agencies such as Fish and Wildlife Service, EPA, TVA, and USDA Forest Service; environmental organizations, such as NAAEE and the Wildlife Habitat Enhancement Council; and businesses such as L.L. Bean and Times Mirror Magazines, Inc. The mission is to incorporate environmental education in outdoor recreational facilities and products.

Academy for Educational Development

The Fish and Wildlife Service (FWS) is working with AED and NAAEE to develop social marketing and educational methods to resolve human caused environmental problems on wildlife refuges.

Partnerships with Health Organizations

Public-private partnerships are increasing educational efforts to reduce health problems associated with environmental factors, such as lead poisoning. The EPA Office of Environmental Education is concentrating on educational programs in urban areas to target segments of the population that are not being reached, such as minority and multi-ethnic communities, senior citizens, and illiterate persons.

Partnerships with State and Local Governments

Many state and local governments have environmental education programs in place that can serve as models for other locales. OEE is collaborating with public agencies at all levels to disperse information, reinforce consistent messages, and minimize duplication of effort.

Partnerships with Academia

Liaisons with colleges and universities inform federal agencies on current research and innovative educational methods. OEE is using grant programs to fund environmental initiatives at institutions of higher education.

Partnerships with Senior Citizen Groups

Many senior citizens have the time to take part in environmental education programs of benefit to their communities. OEE is contacting senior citizens organizations, especially retired educators, to enlist their support in carrying out environmental education projects that reach all segments of society.

Environmental Education Grants

The following are examples of over 100 environmental education grants awarded by EPA in 1992 under the National Environmental Education Act:

• **Millions of Trees.** The Napa Valley Unified School District in California received a $16,835 grant to fund this project, in which students raise living Christmas trees and a variety of native plants that are later used to reforest local areas.

• **If Everyone Lived Like Me.** The Elachee Nature Science Center and Creative Museum in Georgia received an $87,500 grant to create an interactive video software program for school children and adults to educate consumers to reduce global pollution.

• **Waste Not, Want Not.** The Colo-Nesco Community School in Iowa received a $5,000 grant to help equip students to assess environmental issues and to foster international cooperation by addressing such issues.

• **Seed-to-Table Farm Stewardship Exhibit.** Maine Organic Farmers & Gardeners Association received a $5,000 grant to prepare a walk-through exhibit for display at public events in New England. Models, games, stories, photos, and performers demonstrate farming practices to protect natural ecosystems.

• **Building a Wetlands.** The Howard County Public School System in Maryland received a $22,440 grant to create a wetlands using recycled materials and to develop study areas and curricula for school and community use.

ECONOMICS

During much of the last century, Americans—both producers and consumers—depleted natural resources with little thought for the environmental damage they were causing. The nation continued to overlook environmental damages until polluted water and air began to threaten human health, and native species and ecosystems began to disappear. Then, in the last 30 years, the public began to see how human activities could deplete resources and damage the environment, harming the wellbeing of present and future generations. The nation responded by enacting legislation, implementing policies, and taking private initiatives to restore and protect the environment. Americans also recognized the need to provide for a healthy economy while protecting the environment. To meet this need, policymakers in 1992 developed policies and practices to harmonize the nation's economic and environmental goals.

Conditions and Trends

Despite the environmental progress recorded during the last generation, many producers and consumers continue to make choices that unnecessarily deplete or degrade natural resources. One reason is that often the costs, particularly the environmental costs, of using natural resources are not fully paid by the users. If the costs of using natural resources do not reflect full costs, producers and consumers have little incentive to use them efficiently. For example, if polluters do not have to pay the cost of using air or water resources, then they have no incentive to limit or clean up their emissions. The federal government has begun to implement a variety of policies to ensure that economic actors recognize the full cost of using natural resources and, accordingly, better manage these resources. These policies include penalties for exceeding pollution limits and fees for the use of natural resources.

Shifting Emphasis

Early federal policies to protect the environment generally relied on a command-and-control approach that typically specified an environmental goal, a method to achieve that goal, a deadline, and penalties for failure to comply. While these and other approaches have produced appreciable results, pollution control costs continue to increase. Under existing and planned regulations, EPA estimates that pollution control costs will increase by nearly 50 percent between 1991 and the year 2000 (from $107 billion to $148 billion in 1986 dollars, annualized at 7 percent). Environmental policymakers must address these costs as well as more complex problems such as pollution from dispersed sources (nonpoint sources) or from hazardous waste disposal—problems that are difficult or impossible to resolve with traditional approaches.

The federal government has begun to consider and apply market-based incentives and other economic tools to help reduce the cost and improve the effectiveness of environmental protection. For example, using a market-based approach, the government specifies an environmental goal but grants polluters more flexibility in meeting that goal by using whatever methods are most cost-effective for them. Federal agencies are applying market-based incentives to a range of issues, from reducing air pollution to managing fishery resources.

New Developments

U.S. policies are constantly adjusting to both domestic and international developments. A sampling follows of developments likely to shape U.S. environmental policies in the years to come.

Budget Deficit

During the last decade, the United States was a consumer nation rather than an investor nation. Large public debts accumulated during this period, and interest payments on the national debt increased correspondingly. Because financing a growing national debt threatens to affect the U.S. standard of living, the nation is directing more attention to reducing the annual federal budget deficit. As a result, the ability of the federal government to fund new or existing environmental initiatives could be limited. In combination with public demand for a healthy environment, fiscal constraints are likely to create pressure to provide cost-effective environmental protection through innovative policy approaches.

Competitiveness

The pressure to achieve environmental goals at the lowest possible cost also extends to the private sector. As U.S. firms deal with intense competition in domestic and foreign markets, they will seek to reduce the costs of environmental regulations. Techniques such as market-based incentives, negotiated rulemakings, innovative production practices, and voluntary partnerships promise to improve the U.S. standard of living and environmental quality by reducing the cost of achieving environmental goals. In turn, the development and application of cost-effective environmental policies will assist U.S. firms and workers in remaining competitive in world markets. Firms that succeed in developing environmentally sensitive products and cost-effective solutions to environmental problems may also be able to market their products and services in the United States and around the world.

Managing Natural Resources

Every industry uses natural resources to some extent to produce goods and services. In the nation's $5.7 trillion economy, agriculture, forestry, fisheries, and mining contributed nearly $128 billion to national income in 1991. Even with the need to address the budget deficit, the conservation of natural resources is essential to U.S. economic and personal wellbeing.

Budget deficits create pressure for the federal government to increase net revenues from development of natural resources in the public domain. One possible approach to these budgetary pressures would be for the federal government to consider reducing further or eliminating subsidies and establishing more appropriate fees for uses of public lands that deplete or degrade natural resources, such as mining, timber harvesting, grazing, and fishing. Under a market-based approach, resource users pay fees that better reflect depletion or damage costs. Such fees provide an incentive to use natural resources more efficiently.

Expenditures and Revenues

EPA estimates that U.S. spending for 1991 pollution abatement and control by both public and private sectors was $107 billion (annualized cost at 7 percent using 1986 constant dollars). From 1972 to 1991 pollution control costs quadrupled from $26 billion (0.9 percent of GDP) to $107 billion (2.2 percent of GDP). EPA estimates that these costs will continue to grow, reaching approximately $148 billion (2.6 percent of GDP) by the year 2000. More than 37 percent of these expenditures are for capital costs such as technology and equipment, while 53 percent are for operating costs.

In 1992 direct gross federal expenditures for natural resources and the environment totaled an estimated $20.2 billion, with the largest portion, $6.1 billion, allocated to pollution control and prevention. Real federal expenditures (1987 constant dollars) for natural resources and the environment peaked in 1980, but these expenditures have steadily increased by a total of $3 billion from 1987 to 1991.

In 1992 the federal government collected an estimated $7.6 billion in revenues from natural resources and environment-related sources such the following:

- Royalties and fees from leasing and extraction of oil, natural gas, and minerals ($2.8 billion)

- Fees from timber harvesting, grazing, and other land product uses ($1.6 billion)

- Taxes on chlorofluorocarbons ($662 million)

- Penalties and recoveries from environmental cleanup ($2 billion)

Pressure to reduce the budget deficit and address the costs of pollution control and managing the nation's natural resources will challenge policymakers to develop innovative, cost-effective ways to address these concerns.

National Economic Accounts

Nations measure the monetary value of goods and services from economic activity as an indicator of national wellbeing. One of these measures is gross domestic product (GDP)—the value of goods and services produced within the borders of a nation. Currently, however, accounting systems used to estimate GDP do not reflect depletion or degradation of the natural resources used to produce goods and services. For example, the United States could increase GDP with goods and services by overharvesting the national forests, but the GDP figures would not reflect the net depletion of the nation's timber resources and associated environmental damages. As a result, an increase in GDP from economic activities that deplete or degrade natural resources is not a good indicator of a nation's ability to derive income and other benefits from natural resources in the future. Economists within government agencies and in the private sector are studying how to count depletion or degradation of natural resources in national economic accounts.

International Considerations

With the end of the Cold War, nations are devoting increased attention to shared economic and environmental challenges. This development is essential since many environmental problems, such as acid rain, ozone depletion, and climate change can only be addressed effectively through global cooperation.

Nations have also begun to consider longer-term or sustainable approaches to economic development and environmental protection. Although sustainable development has no universally accepted definition, it may be considered as a way to provide equal or greater opportunities for future generations to enjoy economic and personal wellbeing. The United Nations Conference on Environment and Development (UNCED), held in Rio de Janeiro in June 1992, adopted several landmark agreements among nations to promote sustainable development, with an emphasis on market-based strategies. For a discussion of these issues and agreements, see the International Issues section.

Programs and Policies

In 1992 the federal government continued its efforts to integrate economic and environmental policies. Such efforts included initiatives aimed at providing better information and incorporating market-based incentives into environmental programs.

Environmental Accounting and Valuation

Current efforts in environmental accounting and valuation include the following initiatives:

Satellite Accounts for Natural Resources

The Bureau of Economic Analysis (BEA), within the Department of Commerce, is investigating the feasibility of developing satellite accounts for natural resources that will supplement the U.S. national economic accounts. Such accounts would record the depletion of natural resources. BEA plans to initially investigate the feasibility of satellite accounts for petroleum—a nonrenewable natural resource, and timber—a renewable natural resource.

In 1993 the United Nations Statistical Commission likely will adopt new international guidelines on economic accounting. The guidelines would lay the groundwork for developing satellite accounts in each country to record the depletion of natural resources. BEA has participated in developing the U.N. guidelines and expects to move U.S. accounts closer to the international guidelines in the future.

Chesapeake Bay Accounting

EPA is exploring methods to introduce environmental considerations into conventional economic accounting systems. For example, the Chesapeake Bay may be considered as a collection of natural assets providing goods and services valued by society. Because most of these natural assets are not bought and sold in the marketplace, they have no observable market prices to record their value and depreciation over time in existing domestic income accounts. In 1992 EPA supported a study in which researchers developed an environmental accounting system of the Chesapeake Bay to incorporate the value and depletion of its natural assets into domestic income accounts. The EPA Science Advisory Board is reviewing the study, and a final report is in progress.

Coastal Environmental Valuation

To meet the needs of state and local planners and coastal zone managers, the National Oceanic and Atmospheric Association (NOAA) established an Economics Group within its Coastal Ocean Program. This group initiated a project to assist with valuation of natural resources and environmental impacts for damage assessments and resource management purposes. NOAA also assembled a panel of experts to review and comment on the use of the contingent valuation method, which surveys public opinion as a technique for estimating values of natural resources in monetary terms. The panel's review of the contingent valuation method should be available in 1993.

Harnessing Market Forces

To harness market forces to benefit the environment, federal agencies are undertaking many initiatives, including the following:

Marketing Guidelines

In July 1992 the Federal Trade Commission (FTC) issued national guidelines to prevent misleading environmental marketing claims and provide consumers with the information necessary to choose products that minimize environmental damage. Through the FTC guidelines and educational efforts on environmental claims, consumers will have access to information to make more informed choices in the marketplace. For more information on the FTC guidelines, see the Pollution Prevention section.

Regional Emissions Trading

The South Coast Air Quality Management District (SCAQMD) in California is developing a marketable permits program to establish a Regional Clean Air Incentives Market (RECLAIM). The goals are to lower costs, increase flexibility, and attain state and federal air quality standards. The district will submit the program to EPA for review and approval as part of the California State Implementation Plan required by the Clean Air Act. EPA is assisting in developing state regulations to implement the program, with special attention to emissions quantification, monitoring methods, and enforcement provisions.

Mobile-Stationary Source Trading Program

Under this program, EPA provides guidance to states to clarify how certain requirements of the Clean Air Act Amendments of 1990 can be satisfied by trading emission reductions between mobile and stationary sources of air pollution. For example, a stationary pollution source such as a refinery might be allowed to meet or offset emission reduction requirements by purchasing and scrapping old, high-polluting cars—a mobile pollution source. Removing old, polluting cars from the highways, known as "cash for clunkers," could be used as a cost-effective way for stationary sources to meet certain emission reduction requirements or generate tradeable offsets to help meet new source review requirements.

Point-Nonpoint Source Trading

In 1992 EPA continued studying the potential of trading between point and nonpoint sources of water pollution. Under most scenarios being considered, regulated point sources could defer water treatment system upgrades, if they would pay for or arrange for equivalent or greater reductions in nonpoint source pollution within the same watershed. EPA has approved some use of point-nonpoint source trading involving nutrient pollutants in water bodies that currently have water quality problems. Programs have been developed for Cherry Creek Reservoir and Dillon Reservoir in Colorado and for Tar-Palmico River Basin in North Carolina. EPA and other agencies are evaluating the results of these programs.

Damage Assessments

By requiring the polluter to pay the costs of damages to natural resources, the federal government can provide an incentive to prevent future pollution. Under the Oil Pollution Act of 1990 and the Superfund Amendments and Reauthorization Act of 1986, NOAA and the Department of the Interior (DOI) are developing regulations to measure damages resulting from releases of oil and hazardous substances into the environment. NOAA and DOI assist government agencies designated as natural resource trustees to measure the ecological effects and economic damages of such accidents. Damage assessments enable government agencies to recover damages to natural resources from liable parties, and trustees may use the monies recovered to restore injured resources.

Incremental Cost Analysis

In 1992 the U.S. Army Corps of Engineers invested $330 million in fish and wildlife restoration projects and programs. To assure the most efficient allocation of funds, the Corps is developing and applying incremental cost analysis and other economic-ecological based approaches to improve the environmental planning and evaluation process. Incremental cost analysis is a tool for identifying the best option for habitat protection, mitigation, or restoration based on the quality and cost of obtaining additional habitat units.

Conservation Reserve Program

In 1992 the Department of Agriculture (USDA) used economic and environmental principles to evaluate bids submitted by farmers to retire highly erodible cropland under the voluntary Conservation Reserve Program (CRP). More than half of current CRP acres are located in the Great Plains. However, since passage of the Food, Agricultural, Conservation, and Trade Act of 1990, USDA has shifted new enrollments to the midwest and eastern states through a redesigned bid acceptance process and new eligibility criteria. These changes allow USDA to emphasize water quality improvement and select acres to obtain the greatest conservation and environmental benefits per dollar. Of the 1.5 million acres enrolled in CRP in 1991, about 31 percent were located in the Great Plains and mountain regions and more than 13 percent

(195,000 acres) were enrolled in Conservation Priority watersheds, including the Chesapeake Bay, Long Island Sound, and the Great Lakes. USDA also accepted more than 24,000 acres in high priority watersheds to improve water quality in coordination with the President's Water Quality Initiative.

Desert Tortoise Conservation Program

The Desert Tortoise Conservation Program of suburban Las Vegas, Nevada, is a good example of integrating environmental and economic considerations. In the 1980s conflict arose between homebuilders and conservationists over the decline of the endangered desert tortoise. After years of costly, protracted litigation, the Southern Nevada Homebuilders Association agreed to negotiate a tortoise habitat conservation plan with local governments, the Fish and Wildlife Service and Bureau of Land Management in DOI, and two private groups—the Environmental Defense Fund and The Nature Conservancy.

The conservation plan calls for at least 400,000 acres of prime desert tortoise habitat to be set aside in exchange for development rights to more than 22,000 acres of private land in urban areas of Las Vegas. As a condition of federal permits allowing development, builders agreed to contribute $2.5 million to establish a Desert Tortoise Conservation Center. For every acre developed within the permit boundary, builders contribute $550 to the fund for habitat acquisition and management. The federal government has acquired 160,000 acres of land for the tortoise. This ongoing effort provides a model for incorporating economic considerations to protect endangered species and their habitats.

Trade and the Environment

The intersection of environmental and trade concerns received unprecedented attention in 1992. Trade liberalization, generally associated with shifts in economic activities, can provide additional financial resources to countries and regions that could be used for environmental protection. At the same time, economic changes generated by open trade can have significant environmental consequences.

NAFTA

The year's landmark event in this area was the North American Free Trade Agreement (NAFTA) reached by the United States, Mexico, and Canada in October. NAFTA has become an international model for the integration of trade liberalization and environmental protection. In May 1991 the Bush administration pledged to make environmental concerns a key component of NAFTA and related negotiations. In early 1992 the U.S. trade representative published a review of NAFTA environmental issues, including recommendations for negotiators. As a result, NAFTA includes the following provisions:

• The countries agreed to explicit language affirming their commitment to promote sustainable development.

• NAFTA states that signatories should not lower health, safety, or environmental standards for the purpose of attracting investment, and it states the resolve of the countries to implement the agreement in a manner consistent with environmental protection.

• The countries will strive for congruence of various environmental standards, while affirming each country's right to choose the level of environmental protection that it considers appropriate.

• NAFTA dispute settlement panels are encouraged to call on environmental experts to provide advice on factual questions related to the environment.

On a parallel track, NAFTA countries agreed to establish a North American Commission on the Environment (NACE) comprised of the environment ministers from the three countries. NACE will provide a forum at which the countries can cooperate on issues of mutual concern, including the environmental effects of trade.

Also in 1992 the United States and Mexico initialed an agreement to establish a Joint Committee for the Protection and Improvement of the Environment. This agreement lays the foundation for cooperation in environmental training and enforcement on a country-wide basis.

The two countries developed and implemented a 1992-1994 Integrated Environmental Plan for the U.S.-Mexico Border Area, focusing on water quality improvement, air pollution monitoring, hazardous waste tracking, emergency response, enforcement cooperation, and pollution prevention. The U.S. Congress provided $208 million in fiscal 1993 for implementation of the plan, and the Mexican government is expected to provide $160 million.

UNCED

Trade and environment discussions fostered by UNCED reflected an international consensus that trade liberalization and stronger environmental protection can and must proceed in tandem. Principle 12 of the Rio Declaration reads:

States should cooperate to promote a supportive and open international economic system that would lead to economic growth and sustainable development in all countries, to better address the problems of environmental degradation. Trade policy measures for environmental purposes should not constitute a means of arbitrary or unjustifiable discrimination or a disguised restriction on international trade. Unilateral actions to deal with environmental challenges outside the jurisdiction of the importing country should be avoided.

OECD and the GATT

Among the leading international economic institutions are the Organization for Economic Cooperation and Development (OECD), involving 24 industrialized nations, and the General Agreement on Tariffs and Trade (GATT), which governs international trade. Among topics discussed by these fora are the use of trade restrictions for environmental purposes and the competitive effects of environmental regulation. For example, in 1991 a GATT panel ruled that the U.S. Marine Mammal Protection Act, which bans the import of tuna from nations whose fleets have an excessive rate of dolphin bycatch, violates international rules against discriminatory trade practices. Trade and environmental experts also debated the implications of proposed changes in the multilateral trading system.

Within OECD, the United States is taking a leading role in developing guiding principles for trade and environment. In doing so, the Bush administration has sought the views of nongovernmental groups concerned with trade and the environment.

Tradable Permits

In 1992 the Environmental Protection Agency (EPA) sponsored initiatives that use tradable permits to improve national efficiency in attaining air and water quality goals. Under a tradable or marketable permit system, EPA issues firms permits for allowable pollutant emissions. Firms may then choose a compliance strategy that is the most appropriate and cost-effective for their operation. If a firm with relatively low compliance costs can reduce its pollutant emissions below the allowable level, it can sell or trade its extra pollution allowance to others. In short, the system achieves an overall national pollution reduction goal while creating an economic incentive for firms to reduce pollution emissions using more efficient and cost-effective approaches. The success of market-based trading will depend on the ability of polluters to efficiently reduce their emissions, engage in trading, and meet the overall environmental goal.

In contrast, the traditional command-and-control approach, which forces every firm to reduce emissions by a certain percentage or use a specific technology, imposes widely varying costs on polluters. Traditional approaches alone provide little incentive or flexibility to minimize costs and comply with requirements to achieve the environmental goal.

Market-Based Environmental Incentives

The Environmental Protection Agency began to make use of market-based environmental incentives in 1976 and since then has added new programs that include the following:

Incentive Program	Date
Offset Program	1976
Offset Banking Program	1977
Bubble Program	1979
Netting Program	1980
Point Source Trading in Water	1981
Wetland Mitigation Banking	1981
Steel Industry Effluent Bubble in Water	1982
Lead in Gasoline Phasedown: Trading Program	1983
Point-Nonpoint Source Trading in Water	1984
Lead in Gasoline Phasedown: Banking Program	1985
Heavy Duty Truck Engine Emissions Averaging	1985
Emissions Trading Policy	1986
New-Source-Performance-Standards	1987
Compliance Bubble Policy	1987
Stack Height Emissions Averaging	1987
CFC Trading Program	1988
Extended Heavy Duty Truck Engine Emissions Averaging (Banking and Trading)	1990
Acid Rain SO_2 Allowance Trading Program	1991
Acid Rain Industrial Source Opt-in Program	1991
Acid Rain NOx Averaging Program	1991
Air Toxics Early Reductions Program	1991
Air Toxics Offsets Program	1991
Oxygenated Fuels: Averaging and Trading	1991
Reformulated Gasoline: Averaging and Trading	1991
Clean Fuels Fleet Credit Program	1991
Clean Fuels Vehicle Credit Program (California Pilot)	1991
State and Local Incentive Programs	1992
Economic Incentives Rule Expansion	1992*
Mobile-Stationary Source Trading Guidance	1992*
Air Toxics MACT** Averaging, such as HON***	1992*
Scrappage of Old Cars	1992
Point-Nonpoint Source Trading	1992*
Privatization of Wastewater Systems	1992*
Safer Pesticides Incentives	1992*

Streamlining Regulations of Premanufacture Notification	1992*
Water Conservation Partnership	1992*
Municipal Solid Waste Pricing	1992*
State Grants for Air Incentives	1992*

* New initiatives
** MACT = Maximum Available Control Technology
*** HON = Hazardous Organic NESHAP (National Emissions Standards for Hazardous Air Pollution)

PRIVATE SECTOR INITIATIVES

A growing number of private-sector organizations, including for-profit and non-profit groups, recognize the synergy between environmental protection and economic growth. In 1992 this ongoing shift in American culture was evident in new and expanded environmental initiatives.

Conditions and Trends

American businesses are finding that a strong commitment to environmental protection makes good business sense. The Business Council for Sustainable Development—a group of leaders from 48 major international corporations— concluded in its 1992 report, Changing Course, that environmental concerns are not just a cost of doing business but potentially a source of competitive advantage. In 1992 many U.S. companies implemented large-scale programs and experiments, formed advisory groups, and revised their strategic plans, recognizing that environmental protection can be good for the bottom line.

Environmental initiatives can benefit businesses in the following ways:

• Cost Savings. Companies can prevent pollution and save money by improving efficiency, conserving costly materials, and reducing pollution control costs.

• Greater Flexibility. Proactive environmental initiatives allow companies to define problems, set priorities, and undertake solutions. Voluntary initiatives allow companies to reduce future liability risks and influence the development of more practical, efficient, and effective regulations.

• Competitive Advantage. Companies are finding that environmental leadership keeps them ahead of firms that merely respond to or comply with current regulations.

• New Markets. Environmentally sound products, services, and technologies are spurring consumer demand and commercial opportunities. According to Department of Commerce estimates, the global market for environmental goods and services is in the range of $200 billion-300 billion per year and rising by 5-10 percent per year, with a growing proportion available to U.S. exporters.

• Increased Performance. Corporate environmental leadership promotes investor confidence, attracts high-caliber employees, reinforces employee commitment, and improves public image and dialogue.

Nonprofit groups can also benefit from the trend by gaining access to decisionmakers, helping businesses set priorities, and earning credibility for their technical expertise.

Environmental Partnerships

Many of the benefits to business described above are being achieved through strategic partnerships between some businesses and environmental groups that traditionally viewed each other as antagonists. Partnerships are a growing trend for several reasons. The environmental regulatory process serves a vital function, yet power relationships can influence the outcome, with general social costs ending up secondary to competing interests. In environmental partnerships, interest-based negotiations occur among individuals on the basis of what is best for the community, and solutions emerge that benefit the environment and the economy.

Policies and Programs

Agencies throughout the federal government are working to encourage voluntary private sector initiatives to protect the environment. Some of these programs, such as the Green Lights initiative of the Environmental Protection Agency (EPA) and the Take Pride in America program of the Department of the Interior, are described in other sections of this report.

In 1992 the Council on Environmental Quality continued to encourage private sector initiatives. Among these efforts were coordination of the President's Commission on Environmental Quality (PCEQ) and the President's Environment and Conservation Challenge Awards.

Biodiversity on Private Lands. This project demonstrates the feasibility of systemically integrating biodiversity conservation techniques into private land management. In consultation with biodiversity experts, several large commercial landowners have undertaken demonstration projects involving more than 60 sites that represent a variety of U.S. landscapes and locations. The project team will publish information materials in 1993.

Business/NGO/Government Forum. Participants in PCEQ initiatives are demonstrating that leaders with diverse perspectives can collaborate on environmental problems. To expand the use of this successful approach, PCEQ is recommending the establishment of a national forum to promote regular collaboration on environment and development issues by businesses, nongovernmental organizations (NGOs), and government agencies.

Energy Efficiency. Increased efficiency in the use of electricity and fuel can improve air quality, save money, reduce emissions of greenhouse gases, conserve resources, and strengthen the national economy. PCEQ is encouraging greater energy efficiency on a national scale through a market-based approach. The PCEQ energy team has published a national directory of utility demand-side management programs. The team also has enlisted more than 30 companies in a campaign to make voluntary, cost-effective investments in energy efficiency and to take other steps to trim energy use.

International Environmental Survey. To promote more effective efforts toward environmentally sustainable development worldwide, PCEQ conducted an international survey of nongovernmental organizations in 25 countries. To clarify the concerns and agendas of NGOs and businesses, the commission is comparing results with those of an environmental survey of businesses worldwide, conducted in 1991 by a private group.

Lead and Child Health. U.S. public health officials have found that lead poisoning from historic uses of lead is the foremost preventable environmental health threat to children in the United States. In cooperation with federal agencies, PCEQ has undertaken a media-based educational campaign directed at parents of young children. The campaign will identify sources of lead poisoning, harmful effects, and simple, effective ways to reduce exposure. It also will promote a federal telephone hotline and information clearinghouse, mandated for EPA implementation by the 1992 Housing Bill.

Promoting Partnerships. In 1992 PCEQ designed two projects to study and promote environmental partnerships. One project team sponsored case studies, analyses, and a workshop attended by grassroots leaders to determine what conditions make natural resource conservation partnerships successful. The other team studied how partnerships among businesses, NGOs, and governments in the United States and other countries contribute to more effective environmental management and government policies. Both teams prepared reports identifying methods and issues for further collaborative action.

Quality Environmental Management. This initiative epitomizes the entire PCEQ effort. Eleven major corporations are conducting voluntary demonstration projects at 12 facilities to apply the Total Quality Management (TQM) philosophy to the task of pollution prevention. Using quality improvement methods to eliminate inefficiencies that waste resources and cause pollution, this project is cutting emissions at targeted facilities and saving money at the same time. Goals include exploring the TQM/pollution prevention process, devising credible measures of progress, and identifying incentives or barriers to success. The project team is preparing a report for publication in 1993 entitled Total Quality Management: A Framework for Pollution Prevention.

Reducing Vehicle Emissions. PCEQ is demonstrating the viability of the "cash for clunkers" concept—generating air emission credits through accelerated retirement of older, heavily polluting automobiles. The commission is augmenting an existing cash-for-clunkers project involving a PCEQ member company in Delaware by devising a method for precise quantification of emission reductions. The project team is analyzing the cost- effectiveness of such programs for promotion among businesses and regulatory agencies.

U.S. Environmental Training Institute (USETI). With PCEQ support, the nonprofit U.S. Environmental Training Institute was founded in 1991 to train participants from developing countries at U.S. companies and institutions, thus facilitating the transfer of cleaner technologies. Eight corporate and two nonprofit PCEQ members are assisting USETI by sponsoring site visits, conducting training classes, and providing scholarships. Two PCEQ participant companies are working with USETI to conduct an industrial technology cooperative project in the Republic of Indonesia.

Water Education and Conservation. PCEQ has joined an ongoing, long-term project of the National Geographic Society to raise national awareness of threats to freshwater and actions needed to assure availability and quality of this resource in the future. Commission members are creating partnerships to enhance water education and conservation, share expertise on water-related technologies, promote water conservation and cleanup projects at industrial facilities, and encourage conservation by the general public.

Workplace Waste Reduction. This project has produced a comprehensive solid-waste reduction guide for an audience that ranges from corporate offices to small businesses. The guide describes how to inventory materials use and waste generation to encourage wider application of waste-reduction practices throughout the business and retail sectors.

President's Environment and Conservation Challenge Awards

In November 1992, for the second year, President Bush bestowed medals for excellence in the field of environmental conservation, just as Presidents long have honored Americans for excellence in the arts, humanities, sciences, and world affairs. The presidential medals for environmental achievement represent the nation's highest honor in this field. The President's Environment and Conservation Challenge Awards recognize citizens and organizations for outstanding contributions to the environmental quality of the nation.

In 1992 an independent selection committee judged applications and selected 9 presidential medal recipients and 13 citation recipients in the following categories:

- Environmental Quality Management
- Partnership
- Innovation
- Education and Communication.

Descriptions of the nine medal recipients follow.

Save Our Everglades. This program to preserve and restore Florida's unique wetlands ecosystem includes restoration of the Kissimmee River and protection of Lake Okeechobee, state Water Conservation Areas, Big Cypress Swamp, the Florida panther, and other endangered wildlife. The program has acquired 290,000 acres of land and constructed 36 wildlife highway crossings and bridges to protect endangered species and improve hydrology. It also helped pass landmark federal and state legislation. Partners include the State of Florida, the 22 conservation groups that form the Everglades Coalition, the U.S. Army Corps of Engineers, the National Park Service, and the U.S. Fish and Wildlife Service.

Think Earth. The Southern California Think Earth Environmental Education Consortium, composed of 13 organizations, developed and distributes an elementary school curriculum to teach children to conserve natural resources, reduce waste, and minimize pollution. Helping to reach one million students nationwide are the following consortium partners: Atlantic Richfield Company, City of Los Angeles, Education and Development Specialists, GTE California, Los Angeles Department of Water and Power, Metropolitan Water District of Southern California, Orange County Sanitation Districts, Sanitation Districts of Los Angeles County, South Coast Air Quality Management District, Southern California Edison, The Gas Company, Ventura County Air Pollution Control District, and the Ventura Regional Sanitation District.

Energy Conservation Collaboration. The once unlikely union of an electric utility, the New England Electric System of Westboro, Massachusetts, and an environmental group, the Conservation Law Foundation of New England, has resulted in two major achievements: development of a successful energy conservation program and regulatory approval for a utility earnings incentive. The partners designed a new facility, known as the "powerplant that conservation built." The design has reduced air pollutants and the need for new capacity, while saving money for consumers and stockholders.

IBM Environmental Management and Conservation. IBM Corporation has demonstrated that a large company can institutionalize and practice environmental protection and conservation through sound policy practices, executive leadership, and employee commitment. The company's record of accomplishments results from meeting or exceeding government regulations and, where none exist, from setting and adhering to its own stringent standards.

Boston Park Plaza Environmental Action. The Boston Park Plaza Hotel and Towers has undertaken an environmental campaign with 85 initiatives involving all departments of the hotel. The environmental actions of this family-owned and operated landmark property demonstrate that the competitive hospitality service industry can make changes that benefit the environment, while maintaining quality service and standards.

CFC Solvent Phaseout. In 1988 Northern Telecom, Inc., pledged and attained the complete phaseout of CFC-113 solvents from its 42 manufacturing and research operations worldwide. By the end of 1991 Northern Telecom, the first global telecommunications company to make such a pledge, had eliminated the use of these solvents in all of its operations. The phaseout reduced CFC solvent emissions into the atmosphere from 1,000 tons per year to zero in three years.

South Coast Recycled Auto Project (SCRAP). The Unocal Corporation sponsored SCRAP, an unprecedented effort to improve air quality in the Los Angeles Basin by scrapping heavily polluting, pre-1971 cars. In four months, the company purchased and crushed for recycling 8,376 old cars. SCRAP reduced air pollutants by 13 million pounds per year at a cost of 50 cents per pound.

People for Community Recovery. A grassroots environmental education and advocacy organization, People for Community Recovery, Inc., serves a predominantly low-income, African- American community in Chicago. PCR assesses environmental hazards in the community and mobilizes constituents to alleviate the hazards. Hazel Johnson, the group's founder and executive director, is a national leader in the grassroots environmental movement.

Chesapeake Bay Environmental Education. The Outdoor Environmental Education Program of the Chesapeake Bay Foundation provides field instruction in estuarine issues to students, teachers, and other people throughout the Bay watershed. The goal is to create a constituency that not only values and understands the bay but works to restore it. Hands-on water activities involve canoes, rowboats, skipjacks, powerboats, and work on a model low-input farm operated by the Foundation.

Financial Benefits of Corporate Environmentalism

Many companies are realizing significant results from their environmental initiatives. A few examples follow.

Recycling CFCs. Film Tec, a subsidiary of the Dow Chemical Company, installed new equipment to recover unused chlorofluorocarbons (CFCs), collect CFC vapors, and process CFCs for reuse. The firm has reduced CFC emissions 87 percent since 1986; production of recovery equipment has tripled; and Dow has saved $1.6 million.

Eliminating Emissions. The AT&T Columbus, Ohio, plant saved $210,000 by implementing changes in processes that eliminated emissions of perchloroethylene.

Reducing Toxics. Chevron saved $10 million in disposal costs and reduced hazardous waste 60 percent in the first three years of its Save Money and Reduce Toxics (SMART) program.

Cutting Disposal Costs. The Ford Assembly Plant in Lorain, Ohio, cut its paint sludge volume by 30 percent and its annual disposal costs by $56,000 in one year.

Reusing Wastes. The International Paper Company expects to save $100 million in disposal expenses between 1988 and 1995 by recycling and reusing manufacturing wastes.

Preventing Pollution. The 3M Pollution Prevention Pays program has resulted in cumulative worldwide savings from 1975 to 1990 of $537 million ($450 million from U.S. operations).

Using Chemical Wastes. A Monsanto plant in Texas found a use for chemical wastes that increased production by 20 percent and reduced 5 million pounds of discharge.

An Environmental Collaboration:
General Motors and the Environmental Defense Fund

A giant automaker and a nonprofit environmental group have teamed up to develop innovative strategies for reducing air pollution. The two organizations are focusing on expanding market-based incentives for businesses to find cost-effective ways to clean up pollution. After examining GM environmental programs, EDF will assess the most cost-effective actions the company can take. The two organizations are working on an accelerated auto-scrappage program to take older, heavily polluting cars off the road in California.

ENERGY

The quality of the environment has become closely linked to energy use. In 1992 the federal government expanded energy initiatives to improve environmental quality and energy security and to stimulate economic growth. The basis for the initiatives is the National Energy Strategy (NES) issued by the President in February 1991 as a blueprint for the energy future of the United States. During 1991 and 1992 the federal government introduced over 90 NES initiatives, including the following:

- Expanded energy efficiency and renewable energy programs

- Increased funding for advanced transportation technologies

- Expanded interagency efforts in the U.S. Global Change Research Program

- Formed partnerships among federal agencies, industries, and states to reduce energy and water use

The National Energy Strategy Act of 1992, enacted by the Congress and signed by President Bush in October 1992, allows implementation of a balanced, pro-environment, pro-growth national energy strategy.

Conditions and Trends

Energy Consumption

Residential and Commercial. This sector accounts for 36 percent of U.S. end-use energy consumption; 35 percent of which is currently in the form of electricity. As the number of electrically heated and cooled homes and buildings expands, the Department of Energy (DOE) expects residential and commercial demand for electricity to rise to over half the total end-use energy demand for this sector by the next century.

Industry. Industry accounts for another 36 percent of end-use energy consumption, relying on a mix of fuels. Industry uses 70 percent of the energy it consumes to provide heat and power for manufacturing. DOE expects industrial energy demand to grow steadily over the next two decades. During this period, however, energy use per dollar of economic output may decline as a result of energy efficiency improvements. Industry also will increase its generation of electricity and steam for its own use.

Transportation. The United States devotes 27 percent of its energy consumption to the transport of people and goods. Virtually all of this energy consists of petroleum products used to power automobiles, trucks, ships, airplanes, and trains. The transportation sector accounts for two-thirds of U.S. petroleum use. While the number of alternative-fueled vehicles will rise in the future, petroleum fuels likely will continue to dominate transportation energy use for the next 20 years.

Energy Production

U.S. domestic energy resources are extensive and diverse, with coal, oil, natural gas, and uranium found in significant quantities within U.S. borders. Unconventional sources, such as coal-bed methane, are also potential future energy sources. Renewable energy sources, such as geothermal, solar, and wind, are available and limited only by the cost-effectiveness of technologies to harness them.

The value of electricity to the U.S. economy continues to increase. By 2010 DOE estimates that 41 percent of the primary energy consumed in the United States will be used to generate electricity, up from 36 percent in 1992. The nation currently uses coal to generate 55 percent of the electricity it consumes; nuclear power supplies 20 percent; and the remainder is generated from oil, natural gas, and renewable energy.

Petroleum. Production of crude oil and refined products accounts for a fourth of U.S. annual energy production. Over 40 percent of the crude oil used in the United States is imported—much of it from politically volatile regions of the world such as the Persian Gulf, which holds two-thirds of known global oil reserves. DOE expects demand for petroleum to increase, even if petroleum prices increase significantly. U.S. production of crude oil, however, is on a steady decline, which will mean greater use of imports in the future to meet demand.

Natural Gas. Natural gas accounts for another fourth of U.S. annual energy production. Natural gas reserves are found in several areas of the United States, and supplies are transported via pipeline. DOE expects consumption of natural gas to increase considerably over the next 20 years, with the bulk of gas demand met by domestic supplies.

Coal. American coal reserves are the largest in the world and supply a fourth of U.S. energy consumption. Nearly nine-tenths of domestic coal use is for electricity generation at large power stations. DOE expects consumption of coal for electricity generation to grow as demand for electricity rises over the next two decades.

Nuclear. Nuclear power supplies 20 percent of all U.S. electricity generation. While eight plants were under construction in 1992, no new nuclear powerplants are on order, primarily because of regulatory uncertainty, costs, and concern over nuclear waste disposal and safety. Unless these issues are resolved, the contribution of nuclear power to national energy needs is unlikely to increase.

Renewables. Renewable energy sources, including hydropower, solar, biomass, municipal wastes, wind, and geothermal energy, provide less than 10 percent of U.S. annual energy production, with hydropower as the leading source. Although half of U.S. renewable energy goes to generate electricity, the nation also uses biofuels for transportation and solar energy to heat buildings and water. Over the next 20 years, DOE expects the use of renewable fuels to increase gradually to meet more than 11 percent of U.S. energy demand.

Environmental Impacts

Over the last two decades, the United States has reduced energy-related air pollutant emissions, primarily with controls mandated by the Clean Air Act. The Clean Air Act Amendments of 1990 promise further reductions.

Lead. Between 1970 and 1990, emissions of lead declined 97 percent as a direct result of its removal from gasoline.

Particulate matter and sulfur dioxide. Over the same period, nationwide emissions of particulate matter (PM) decreased by 59 percent and sulfur dioxide emissions decreased by 26 percent. Such decreases resulted primarily from stringent controls on industrial processes and on such fuel combustion sources as electric utilities and industrial, commercial, and residential boilers. A quarter of the PM decline and a third of the sulfur dioxide decline resulted from controls on coal-fired fuel combustion sources.

Nitrogen oxide. In conjunction with an 83-percent increase in electricity generation between 1970 and 1990, nitrogen oxide emissions from electric utilities increased by 67 percent.

Carbon monoxide and volatile organic compounds. A nationwide 41-percent decrease in carbon monoxide and 31-percent decrease in volatile organic compounds resulted primarily from controls on new motor vehicles.

Carbon dioxide. In 1988 U.S. carbon dioxide emissions totaled 1.3 billion metric tons, a 20-percent increase over 1970 and a reflection of the increase in population and energy use in that period. During the same period, the contribution of coal to total carbon dioxide emissions increased from 22 percent to 38 percent.

Policies and Programs

The National Energy Strategy Act of 1992 could change the American way of life by improving energy efficiency, ensuring the availability of energy supplies, and improving environ- mental quality. In 1992 the Bush administration implemented NES initiatives and expanded energy assistance to state and local governments. Federal agencies also implemented the President's Executive Order on Federal Energy Management. For a discussion of the Executive Order, see the Federal Facilities Management section. The following are highlights of progress made in 1992.

Superclean Coal and Liquid Fuels

During the fourth round of the Clean Coal Technology (CCT) Program in 1991, DOE added nine demonstration projects for co-funding with industry. Of 42 current CCT projects, 3 have been completed, 10 are in operation, 9 are under construction, and 20 are in the preconstruction design phase. The first four rounds of CCT represent a federal research and development (R&D) investment of $2 billion and an industry investment of $3 billion. The President's fiscal 1993 budget requested $500 million to fund the fifth round of the CCT program.

In 1992 DOE focused its coal R&D program on the super-clean high-efficiency power generation systems needed for coal to compete as an energy source under stringent environmental requirements scheduled to take effect after the year 2000. The DOE goal is to assist the private sector in developing and producing high-quality coal-derived liquid fuels competitive with oil priced at less than $30 per barrel.

Renewable Energy

The President requested fiscal 1993 funding for renewable energy R&D at $250 million, a 65-percent increase over funding in fiscal 1989, the first year of the Bush administration. Increased R&D will accelerate application of emerging renewable energy technologies, reduce costs, improve performance, and make these technologies more competitive in the marketplace.

National Renewable Energy Laboratory. In September 1991 the President designated the Solar Energy Research Institute in Golden, Colorado, as the National Renewable Energy Laboratory. By elevating this facility to national laboratory status, the Bush administration demonstrated a commitment to renewable energy technology.

Solar Thermal Power. In October 1991 DOE entered into a joint venture with a U.S. engine manufacturer to develop a system that uses solar energy to operate a generator capable of providing 5 kilowatts of electricity at an estimated cost of 15 cents per kilowatt hour. This system, scheduled for commercialization in five years, could replace remotely located diesel engines for electricity generation worldwide. The President requested $27 million for solar thermal R&D in the fiscal 1993 budget.

Wind Energy. DOE completed construction of a full-scale blade-and-materials fatigue testing facility at the Wind Energy Test Center at Boulder, Colorado. The wind energy industry uses the center to test improvements aimed at increasing the durability and life of wind turbines. Through R&D at the National Renewable Energy Laboratory, DOE demonstrated a 15- to 30-percent improvement in the performance of wind turbine airfoils compared with those currently available. With such advances in technology, wind energy is on the verge of becoming a competitive resource for electricity generation in many areas of the country. The President requested $22 million for wind energy R&D in the fiscal 1993 budget.

Alternative Fuels

The President's fiscal 1993 budget proposed $15 million to accelerate the acquisition of new vehicles that operate on fuels other than gasoline and the conversion of current vehicles. Funding would allow the purchase and conversion of an additional 5,000 alternative-fuel vehicles (AFVs). For more information on AFVs, see the Federal Facilities Management section.

U.S. Advanced Battery Consortium. In October 1991 the President announced a 4-year $260 million joint research venture with the nation's three largest automakers and others to develop a new generation of batteries for high-performance electric vehicles. These new batteries will be lightweight and capable of increasing significantly the range of electric vehicles. This government-industry effort to address a major technological challenge is strengthening the U.S. position in the international competition for electric vehicle development.

Hybrid Vehicles. Proposed 1993 funding of $20 million would create a new government-industry consortium for hybrid-vehicle technology, based on the model of the Advanced Battery Consortium. This new generation of vehicles would use batteries in combination with other systems, such as fuel cells or gas turbines.

Biofuels. Proposed 1993 funding for DOE biofuel R&D for transportation was $48 million, a 35-percent increase over fiscal 1992 and a 60-percent increase over 1991. In 1991 the departments of Energy and Agriculture signed a memorandum of understanding to coordinate federal efforts in the area of biomass research. The National Renewable Energy Laboratory signed a cooperative R&D agreement with a corn-ethanol producer to test a process that converts corn cellulose into ethanol. DOE signed a similar cooperative agreement with a major oil company to explore ethanol production from waste paper.

Energy Conservation and Efficiency

The President proposed increased funding for energy conservation R&D in the fiscal 1993 budget with a request of $330 million. This represents a doubling of DOE conservation R&D funding since 1989.

Industrial Energy Audits. In 1991 DOE added five Energy Analysis and Diagnostic Centers (EADCs) at U.S. universities, bringing the total of these centers that conduct energy audits to 18. To date, 3,600 EADC audits, conducted at a cost of $18 million, have resulted in $400 million in energy savings to industry. DOE has initiated discussions with industries, states, and electric and natural gas utilities on ways to improve industrial energy audit techniques. The Department is participating with the Electric Power Research Institute and several utilities in an audit program for selected large industrial firms. Industrial energy audits are a first step in determining where efficiency improvements are most desirable and cost-effective.

Energy-Efficient Buildings. DOE expanded support for R&D on a range of energy efficient building technologies. Working with the Department of Housing and Urban Development on a DOE-HUD Initiative for Energy Efficiency in Housing, DOE funded efforts to increase energy efficiency in HUD-aided housing and to help reduce utility costs of $2 billion per year. Areas of interest include home energy rating systems, energy-efficient mortgages, energy performance contracting in public housing, energy standards for manufactured housing, and guidelines for making Native American housing more energy efficient.

Integrated Resource Planning. DOE also increased funding for Integrated Resource Planning (IRP) to develop better analytical capabilities and to support state and local resource planning programs. The Department is evaluating 50 proposals from utilities and state regulators in response to a solicitation issued through the DOE Oak Ridge National Laboratory for IRP research.

Community Energy Systems. In 1992 HUD funded localities performing feasibility studies on preliminary designs for potential Community Energy Systems. Such systems provide reasonably priced energy for heating and cooling to businesses, industries, and residences. The goal is to foster community and economic development, maximize fuel efficiency, and reduce pollution.

Energy Production

During 1992 DOE received energy production proposals that included developing the Alaskan North Slope and sought Outer Continental Shelf leasing.

Alaskan North Slope Development Task Force. In 1992 the interagency Task Force on Alaskan North Slope Development began considering approaches to reduce regulatory and technical barriers to environmentally responsible development of northern Alaska. Among the barriers under analysis are restrictions on causeway and pipeline construction and difficulties in recovering heavy oil at low temperatures. Targeted fields have the potential to yield 1 billion barrels of oil.

Outer Continental Shelf Leasing. In February 1991 the Department of the Interior issued a draft proposal for a 5-year 1992-1997 leasing plan for the Outer Continental Shelf (OCS). The proposal reflects the President's call for a more targeted OCS leasing program fully supported by scientific studies and sensitive to the concerns and needs of local areas. The plan would increase efforts to gather information on environmental and natural resources and to respond to the views expressed by states, local governments, and the public.

National Technology Initiative

In February 1992 the federal government launched the National Technology Initiative (NTI), with participation by DOE, the Department of Commerce, the National Aeronautics and Space Administration (NASA), and the Department of Transportation. NTI promotes rapid commercialization of advanced technologies, including energy and environmental technologies. For more information, see the Technology section.

Highlights of the 1992 Energy Bill

At the height of the Persian Gulf war, with national security implications over U.S. reliance on foreign oil at their peak, the President unveiled a U.S. National Energy Strategy—more comprehensive and balanced than any that preceded it. Nineteen months later, after extensive negotiations with the Congress, President Bush signed the Comprehensive National Energy Policy Act into law to implement his energy strategy. The highlights of the 1992 Energy Bill, with its equal measures of energy conserved and energy produced, are described on the facing page.

• **Energy Efficiency.** New standards for improving energy efficiency in homes, offices, utilities, and factories.

• **Alternative Fuels.** Ensuring greater production and use of alternative fuels such as compressed natural gas, propane, ethanol and electricity for automotive transportation use. For example, in 1993, the federal automotive fleet will purchase at least 5,000 new alternative fuel vehicles; 10,000 in 1995; and by the year 2000, 75 percent of the federal fleet will be alternatively fueled.

• **Market-Based Sources of Electricity.** Major overhaul of the nation's electricity-producing law, known as the Public Utility Holding Company Act (PUCHA), to spur competition, reduce regulation, and create new market-based sources of electricity.

• **Nuclear Powerplants.** Streamline the federal licensing of nuclear powerplants to allow for one-stop permitting for construction and operation.

• **Strategic Petroleum Reserve.** Allow for more flexibility in the administration of the strategic petroleum reserve to avoid severe disruptions in the petroleum market that would cause unwarranted price increases.

• **Energy Technology.** Establish a research strategy for energy technology development and transfer.

• **Greenhouse Gas Emissions.** Directs the Energy Information Administration to establish a baseline inventory of greenhouse gas emissions and outlines the basis for a voluntary system for private sector reporting on emissions.

TECHNOLOGY

At an increasing rate, U.S. research organizations—public and private—are developing, implementing, and marketing technologies to fix environmental problems or achieve social objectives with the least possible impact on the environment.

A Growing Global Market

The market for environmental technologies is large and growing. The global market for environmental goods and services is in the estimated range of $200 billion to $300 billion per year and rising by 5-10 percent per year. An increasing share of this market is available to U.S. exporters. Expansion in this field will result from the following factors:

• New international environmental accords

• Increased government investment in environmental technology

• Market-based environmental programs such as the 1990 Clean Air Act Amendments that create economic incentives for pollution prevention

Public-Private Collaboration

The federal government funds just under half of all research and development (R&D) in the United States but performs only about 10 percent in its own laboratories. Industry conducts nearly 75 percent of the nation's R&D, and academia, about 10 percent. Recognizing the national economic benefits to be gained from increased public-private cooperation on technology R&D, the Congress enacted the Federal Technology Transfer Act of 1986. Under its provisions, the number of cooperative R&D agreements (CRADAs) between government labs and private companies increased from 460 in 1990 to 730 in 1991, with a total number to date of 1,100. A quarter of these CRADAs are with small businesses, and many involve technologies relevant to environmental protection.

Key Technologies

Recent surveys of national technology priorities reflect the need for cleaner, more efficient methods in fields ranging from manufacturing to transportation to agriculture. For example, the Twenty-first Annual Report of the Council on Environmental Quality reviewed current programs and policy reforms needed to spur development of technology to prevent pollution. A 1992 report of the World Resources Institute (WRI), a private, nonprofit environmental group, proposed a list of environmentally critical technologies with the following criteria:

• Their use brings about large, cost-effective reduction in environmental risk

• They embody a significant technical advance

• They are generic—applicable to broad sets of problems—and precompetitive—at a stage where R&D costs are too high for any one company to bring to market

• Their adoption involves a favorable ratio of social-to-private returns

The CEQ and WRI studies both noted that many technologies developed primarily for other purposes, such as supercomputers and computer-integrated manufacturing, may also have significant environmental uses.

Policies and Programs

Because technology is diffused—pervasively and rapidly—in modern society, no government can successfully centralize and direct its development. However, the U.S. Government supports and influences technology R&D through a variety of programs. Partnerships involving many federal departments, agencies, and private organizations create economies of scale and enhance innovation.

Federal Research and Development

The Office of Science and Technology Policy (OSTP) in the Executive Office of the President coordinates and integrates federally sponsored science and technology development through an interagency mechanism known as the Federal Coordinating Council for Science, Engineering, and Technology (FCCSET). Much of the research conducted through FCCSET is relevant to long-term environmental issues. In addition, several other agency-specific and interagency technology development projects are underway.

Committee on Industry and Technology (CIT). Coordinated by FCCSET, this committee supports research and assists private industry in the development of technology. CIT sponsors an interagency initiative in Advanced Manufacturing Technology (AMT), whose environmental component emphasizes development in the following areas:

- Environmentally benign manufacturing processes
- Waste minimization
- Advanced pollution monitoring and sensors
- Closed-loop recycling

Subcommittee on Environmental Technology (SET). Administered by the FCCSET Committee on Earth and Environmental Sciences, SET fosters interagency coordination for the research, development, and demonstration of solutions to environmental problems. SET coordinates the following efforts:

- Basic research for innovative technologies
- Applied research for solving environmental problems
- Demonstrations or pilot projects to foster rapid application of the products of new technologies

Private Enterprise/Government Interaction (PEGI). The FCCSET Committee on Earth and Environmental Sciences also administers the PEGI Task Group, which facilitates R&D interactions and collaborative efforts between the private sector and federal agencies, with an emphasis on global change issues. In 1992 PEGI collaborated with the Electric Power Research Institute, GEOSAT, Gas Research Institute, and American Petroleum Institute.

Strategic Environmental Research and Development Program (SERDP). In 1990 the Congress directed the Department of Defense (DOD) and the Department of Energy (DOE) to establish SERDP to identify DOD and DOE programs developed for defense purposes but possessing useful applications in the environmental field. For example, data gleaned from submarines and orbiting satellites could help scientists track long-term environmental changes. SERDP also conducts environmental research with private sector and government organizations.

NICE³ Program. Established in 1991 the National Industrial Competitiveness through Efficiency—Environment, Energy, and Economics program (NICE³) is a $2 million grant program jointly administered by the Department of Commerce (DOC), the Environmental Protection Agency (EPA), and DOE to fund development of generic, precompetitive technology to enhance industrial competitiveness through improved energy efficiency. Grants are awarded to state and industry consortia to develop energy efficient technologies with promise of commercial viability for U.S. industry. For example, a project was funded in Ohio to develop a new curing system for solvents in the manufacture of pressure-sensitive labels. This technology could result in an estimated savings of 604 billion British thermal units (Btu) per year, with an annual reduction of 7 million pounds of hazardous waste. For more information on NICE³, see the Pollution Prevention section.

Innovative Concepts Program (InnCon). Sponsored by DOE, EPA, and the Department of the Interior, InnCon provides seed money in amounts up to $20,000 to innovators who develop energy efficient technology and processes that increase industrial productivity. The program also supports innovators by publicizing their concepts and by introducing them to potential sponsors. The program, in its sixth year, has provided $11.5 million in funding for innovative environmental technology.

EPA Technology Innovation Office. This office promotes new cleanup methods for problems such as contaminated soils or groundwater. In 1991 EPA conducted 21 pilot and full-scale field demonstrations of new technologies prior to commercialization and scheduled 74 new technologies for demonstration. Among the techniques tested in 1992 were soil washing, chemical dechlorination, underground vacuum extraction, and the forcing of airstreams through water to evaporate volatile organic compounds. EPA also evaluated bioremediation—the use of microbes to break down organic contaminants—at demonstration cleanups across the country.

U.S. Bureau of Mines, Division of Environmental Technology. This office conducts research on innovative ways to achieve the following objectives:

• Reduce the volume and toxicity of mining and minerals-processing wastes

• Identify hazardous waste sites contaminated with heavy metals and other toxic constituents

• Mitigate the environmental impacts resulting from active and abandoned coal and other mining

Recent work has applied biotechnology to remediate toxic and hazardous waste sites created by mining, minerals processing, and related industries. These projects also provide a foundation for environmental regulatory analysis and development.

Marketing New Technologies

To complement its support for environmental technology R&D, the federal government also administers programs designed to accelerate the commercialization and deployment of new technologies.

National Technology Initiative. DOC, DOE, and EPA sponsor this initiative to promote public-private cooperation on technology development and commercialization. In 1992 the agencies cosponsored a series of ten conferences for entrepreneurs across the country to highlight opportunities for partnerships among federal agencies, universities, and the private sector on cooperative R&D, financing, and manufacturing. Five additional conferences are scheduled, and, in response to industry interest, DOE and EPA are organizing follow-on workshops.

Cooperative Research Centers/Engineering Research Centers. Funded by the National Science Foundation, this program has established a network of cross-disciplinary centers at major research universities, linking academia and industry. Currently, 23 centers are in operation at 21 universities in 16 states around the nation. The centers focus on environmental problems and technology solutions in fields such as microelectronics, telecommunications, biotechnology, energy resources and recovery, and design and manufacturing.

International Cooperation. As world leaders in science and technology, Americans are working with their counterparts in other nations to develop and share technological advances. In particular, the United States supports an active program of environmental technology cooperation with developing nations and those making the transition to market economies. Cooperation takes the form of information exchange, training and technical assistance, and financial assistance for selected projects. A sampling of current efforts follows.

United States-Asia Environmental Partnership (US-AEP). This coalition of U.S. and Asian businesses, government agencies, and community groups seeks to protect the environment and promote economic progress in the Asia-Pacific region. US-AEP is establishing centers in Hong Kong, Taiwan, the Philippines, and India. Additional sites are under consideration for Malaysia, Singapore, Thailand, Indonesia, and Korea. The US-AEP program has the following components:

- Technology cooperation
- Environmental fellowships and training
- Environmental infrastructure
- A regional biodiversity conservation network

Technology cooperation fosters opportunities through trade, investment, and business exchanges. The partnership is establishing environmental business centers in Asia to support trade and investment linkages. The Agency for International Development (USAID) administers U.S. efforts in the program with input from 20 other federal agencies.

U.S. Environmental Training Institute (USETI). This joint effort of U.S. industry and seven federal agencies provides training to professionals from public and private sectors of developing countries. In 1992, its first year of operation, the institute offered five courses at U.S. facilities on such topics as advanced landfill management, bioremediation, and restoration of nuclear sites. Attending the courses were environmental professionals from Thailand, India, Hong Kong, Morocco, South Korea, Taiwan, Mexico, Indonesia, the Philippines, and Puerto Rico.

Climate Country Studies. The United States has committed $25 million over two years to assist developing nations with needs assessments and reporting obligations under the United Nations Framework Convention on Climate Change. Country studies will facilitate future U.S. assistance to developing nations in taking actions to mitigate and adapt to climate change.

National Environmental Technology Applications Corporation (NETAC). Established in 1988 at the University of Pittsburgh's Applied Research Center with funding from EPA and the private sector, NETAC is accelerating the application of environmental technologies at home and abroad. NETAC has undertaken 11 cooperative technology projects in Eastern and Central Europe and the former Soviet Union, involving such organizations as USAID, the U.S. Information Agency, and the private, nonprofit World Wildlife Fund. In one example, NETAC conducted an independent evaluation of a biological remediation process, which the client was able to use to successfully market the technology.

Technology Transfer in Energy Efficiency

U.S. federal agencies, led by the Department of Energy (DOE), are promoting the transfer of energy efficient technology worldwide through efforts such as the following:

Committee on Renewable Energy, Commerce, and Trade (CORECT). This interagency committee focuses on market assessments for export of renewable energy technologies. Committee members represent DOE, the Department of Commerce (DOC), and the U.S. Agency for International Development (USAID). CORECT has supported photovoltaics in the Dominican Republic, wind systems in Guatemala, and geothermal energy in Honduras.

Technology Transfer Clearinghouse. An Environmental and Energy Efficient Technology Transfer Clearinghouse is being supported by USAID, DOE, and the Environmental Protection Agency (EPA). The clearinghouse provides information on pollution control, renewable energy, and energy efficient technologies. Pilot programs are in process in Mexico City and Vienna.

Assisting Deployment of Energy Practices and Technologies (ADEPT). This program helps nations to choose and apply energy technologies. Joint ventures, sponsored by DOE and host country governments, involve personnel from U.S. national laboratories, industrial groups, and host-country counterparts. ADEPT emphasizes indigenous fuel sources—biomass, geothermal, solar, and wind—and improved efficiency in fossil fuel use.

21st Century Program. Through the Americas' 21st Century Program, DOE is assisting the U.S. renewable energy industry to form joint ventures with nations of Latin America and the Caribbean.

Other DOE Programs. DOE also is participating in international programs to help other countries achieve the following objectives:

- Capture methane released in coal mining
- Develop small hydropower projects
- Conserve electricity through demand-side management.

AGRICULTURE

U.S. agriculture is in the midst of a major transition motivated by economic and environmental factors. These include the budget deficit, the expansion of international trade, water quality and quantity, topsoil erosion, and the compatibility of agricultural practices and the environment. In the midst of change, U.S. agricultural policies seek to balance several objectives:

- An abundance of food and fiber at reasonable prices
- Economic security for agricultural producers
- Conservation of natural resources

Conditions and Trends

Between 1982 and 1987 nonfederal cropland acreage increased by 1.4 million to 422.8 million acres, and nonfederal rangeland acreage decreased by 6 million to 401.7 million acres. Land-use shifts can be complex. For example, the net increase in cropland reflected the conversion of 16.2 million acres of cropland to other uses, while 17.6 million acres were converted from other uses to cropland.

The environmental consequences of these land-use shifts are also complex. Shifting land from other uses to crop production increases soil erosion and the risk of nutrient and pesticide contamination associated with current agricultural practices. Shifting land from crop production to other uses, on the other hand, increases environmental impacts from nonfarm uses, such as runoff of toxins from roadways and parking lots and vehicle emissions associated with urban development.

Since 1985 several hundred million acres of cropland have been idled as a result of conservation provisions in farm legislation. In 1987 alone 76 million acres out of a total cropland base of 321 million acres were idled under Farm Bill conservation and supply-control provisions. In 1992 a total of 54 million acres were idled out of the 343 million acres of crop acreage base. This figure includes acreage enrolled in the Conservation Reserve Program.

Policies and Programs

Federal farm legislation in 1985 and 1990 linked farm program benefits and the conservation of natural resources. The Department of Agriculture (USDA) conducts conservation programs to reduce soil erosion, protect wetlands, maintain and enhance wildlife habitat, and improve water quality, while achieving supply-control and income-support objectives in a cost-effective manner.

Conservation Compliance

The 1985 Farm Bill contained conservation compliance provisions that require farmers to develop and implement approved conservation plans on highly erodible cropland by 1995. Farmers who do not implement conservation plans lose USDA program benefits. As of July 1992 field personnel of the USDA Soil Conservation Service had helped 1.3 million farmers develop compliance plans that will reduce erosion on 135 million acres of highly erodible cropland. These plans call for 215,000 miles of terraces, 1.3 million acres of grassed waterways, 4,200 miles of erosion-control diversions, cropland contour farming on 25 million acres, and cropland conservation tillage on 100 million acres. As of July 1992 farmers had implemented conservation plans on over half the acres requiring conservation practices.

To bolster conservation compliance, USDA supports alliances among agricultural businesses, commodity groups, grower associations, and the farm news media. These alliances encourage conservation practices such as leaving crop residue on the soil surface during critical erosion periods. Farmers have chosen some form of crop residue management on 75 percent of the highly erodible land in compliance plans. Working with the alliances, USDA provides technical information and assistance through demonstrations, direct mail, brochures, workshops, advertisements, and features in the farm-oriented media.

Conservation Reserve Program

The Conservation Reserve Program (CRP) offers farmers an incentive to remove highly erodible cropland and other environmentally sensitive land, such as wetlands, from production for at least ten years. Farmers who enroll in the program receive ten annual rental payments for idling the land and cost-share help in establishing a cover of grass or trees. Producers who plant hardwood trees or construct wildlife corridors, windbreaks, or shelterbelts may request contracts up to 15 years in length.

As of August 1992 farmers had enrolled 36.5 million acres in the program, resulting in numerous environmental benefits. Average soil loss on CRP lands has declined to one-tenth of the former rate, from 21 to 2 tons per acre per year. Farmers are planting 2.4 million acres in trees; 31 million acres are in grass, predominantly in the prairie and Great Plains region; and another 2 million acres are in plantings beneficial to wildlife. In 1989 farmers entered 410,000 acres of farmed wetlands in the program. In the prairie pothole region of North and South Dakota, during one sign-up period alone, the program enrolled 150,000 acres of farmed wetlands. Under one program option, 13,000 acres of shallow-water wetlands are being restored. Eligibility has been extended to saline lands and to cropland that contributes to water quality degradation. Incentives encourage the planting of hardwood trees in woodland, windbreaks, shelterbelts, and wildlife areas. A total of 23 U.S. watersheds, including the Chesapeake Bay and the Great Lakes, are designated Conservation Priority Areas and, as such, receive preference in program enrollments.

In addition to reducing soil erosion, CRP reduces pesticide use on land enrolled in the program by an estimated 30,000 tons annually and fertilizer use by an estimated 2.4 million tons annually. Because CRP may create an incentive to increase the intensity of fertilizer and pesticide use on remaining cropland, however, the net reduction in fertilizer and pesticide use may be smaller than these estimates suggest.

Swampbuster

The 1985 Farm Bill took major steps to slow the loss of ecologically valuable wetlands in the United States. Producers who convert existing wetlands for crop production are disqualified from receiving USDA program benefits. The 1990 Farm Bill further refined wetlands provisions, making them easier to understand and enforce by changing the trigger for wetlands conversion from planting of an annual crop to the actual conversion of the wetlands. The legislation allows producers to mitigate wetlands loss through restoration of previously converted wetlands and also provides for graduated penalties. To date, the Soil Conservation Service has made 2 million determinations, delineating which areas are considered wetlands protected by the wetland conservation provisions of the 1985 Farm Bill.

Wetlands Reserve Program

The Wetlands Reserve Program (WRP), created by the 1990 Farm Bill, is similar to CRP in that it offers landowners financial incentives to restore and protect wetlands on their property. Landowners who grant long-term easements to restore and maintain wetlands receive easement payments, technical assistance, and up to 75 percent of costs for work approved to restore the wetlands. Landowners are then responsible for protecting and maintaining wetlands within the boundaries of the easement. WRP does not allow cropping or other alterations harmful to wetlands.

The Bush administration proposed full funding for the Wetlands Reserve Program at nearly $800 million through 1995 to restore and protect 1 million acres of wetlands. In fiscal 1992, with $46 million appropriated to enroll 50,000 acres, WRP began as a pilot program in nine states. During a sign-up held in July 1992, landowners offered 500,000 acres for enrollment in the pilot program.

Water Quality Incentives

The Agricultural Water Quality Incentives Program provides payments and technical assistance to landowners who reduce the movement of contaminants from cropland to surface and subsurface waters. As part of this pilot effort, USDA conducts research to improve conservation technology. The Department has selected 16 demonstration areas to show how economically sound agricultural practices can improve water quality. During fiscal 1991 a total of 1,679 producers used such practices on 163,000 acres of cropland. For example, better nutrient management reduced the application of nitrogen to cropland by 675,000 pounds and phosphorous by 216,000 pounds.

In addition, USDA has selected 74 areas for intensive action against specific nonpoint contamination problems. In these areas, farmers apply both existing and newly developed conservation practices to eliminate water quality problems, including those that involve nitrogen, phosphorous, and pesticide use. In 1991 a total of 8,606 producers installed conservation practices on 390,000 acres of cropland with a resulting reduction of 1.8 million pounds of nitrogen and 1.5 million pounds of phosphorous. Cooperating agencies also assist with the improvement of pasture management in a number of these areas.

Windbreaks and Shelterbelts

Trees planted in windbreaks and shelterbelts protect agricultural crops, livestock, and rural farmsteads. In addition to reducing wind erosion, these trees save energy and improve water quality. Assisted by federal, state, and local programs, farmers have established, to date, 200,000 miles of windbreaks and shelterbelts in the United States, with an annual increase of 3,500 miles.

Management System Evaluation Areas

A joint research initiative on Management System Evaluation Areas (MSEAs), begun in 1991 by the Agricultural Research Service and the Cooperative State Research Service, helps control and reduce the impacts of agricultural chemicals on water quality. The U.S. Geological Survey has contributed to the effort by conducting geohydrologic site characterizations and research on the transport of selected pesticides and nitrates into groundwater at five locations in the Midwest.

Agriculture in Concert with the Environment

Agriculture in Concert with the Environment (ACE) is a grant program sponsored by the U.S. Department of Agriculture and the Environmental Protection Agency. The program funds research, education, and demonstration projects to foster environmentally sustainable agriculture. In 1992 ACE awarded $2 million in grants for 27 projects with the following objectives:

- Prevent groundwater pollution from nutrients and pesticides
- Prevent surface water pollution from livestock production operations
- Promote nonchemical pest control

AIR QUALITY

As a world leader in air pollution control, the United States is moving toward air quality goals through the use of advanced technologies and innovative policies. However, in some parts of the nation, air pollutants continue to pose risks to human health and the environment. The Clean Air Act Amendments of 1990, a major achievement for air quality protection, were designed to achieve goals in a more flexible, cost-effective, market-based manner than was traditionally the case in the past.

Conditions and Trends

The United States made substantial progress in improving air quality nationwide during the 1980s. Monitors showed reductions in levels of six common pollutants for which the U.S. Environmental Protection Agency (EPA) has set health-based national ambient air quality standards.

The biggest success story from 1982-1991 was an 89-percent reduction in lead levels in the air, resulting primarily from the removal of lead from most gasoline. In addition, the gradual phase-in of cleaner automobiles and powerplants caused atmospheric levels of carbon monoxide to fall by 30 percent, nitrogen oxides by 6 percent, ozone by 8 percent, and sulfur dioxide by 20 percent. Levels of fine particulate matter (PM-10, otherwise known as dust and soot) have dropped 10 percent since the PM-10 standard was set in 1987.

Despite this progress, 86 million people live in U.S. counties where pollution levels in 1991 exceeded at least one national air quality standard, based upon data for a single year. Urban smog continues to be the most prevalent problem; 70 million people live in U.S. counties where 1991 pollution levels exceeded the standard for ozone, the primary constituent of smog. Ozone pollution is formed from hydrocarbon and nitrogen oxide emissions, which come from motor vehicles and stationary sources such as factories, powerplants, and certain consumer products. Ozone can cause respiratory problems in sensitive individuals exposed for extended periods and also can damage forests and crops.

Carbon monoxide and particulates also exceed federal limits in some areas. Moreover, relatively large annual releases of sulfur dioxide and nitrogen oxides contribute to acid rain, visibility degradation, and health risks for sensitive parts of the population. Toxic pollutants—those known or suspected to cause cancer or other serious health effects—are released into the air in many areas. The latest EPA Toxics Release Inventory shows a total of 2.2 billion pounds of air toxics released nationwide in 1990.

Global Atmospheric Pollution

All over the world, air pollutants cross international borders and rise miles above the Earth's surface. For example, chlorine concentrations in the upper atmosphere resulting from the release of chlorofluorocarbons (CFCs) and certain other chemicals have increased to a level that damages the Earth's ozone layer, which shields the planet from harmful doses of solar radiation. Another concern is possible climate change due to the accumulation of greenhouse gases that trap heat in the atmosphere. Emissions of greenhouse gases are still growing, but the rate of increase has slowed over the last 20 years. With direct and indirect controls established on greenhouse gas emissions in the last four years, emissions are expected to rise by as little as 1-2 percent during the 1990s.

Indoor Air Pollution

The population's general exposure to indoor air pollutants is increasing because of the following factors:

- Construction of more tightly sealed buildings
- Reduction of ventilation to save energy
- Use of synthetic building materials and furnishings
- Use of chemically formulated personal care products, pesticides, and house hold cleaners

Indoor air pollutants may include tobacco smoke, radon, volatile organic compounds, biological contaminants, combustion gases, respirable particulates, lead, formaldehyde, and asbestos. Scientific data is limited regarding exposures to and risks of indoor air pollution.

A focus of concern is radon, a naturally occurring gas that could be a leading cause of lung cancer in the United States. Although few homes have been tested (6 percent), federal and state environmental agencies have detected elevated radon levels in some areas of every state.

Policies and Programs

Current U.S. policy on air quality is embodied in the Clean Air Act Amendments of 1990 (CAAA), which contain several innovative, market-based mechanisms to help reduce pollution. Major international initiatives include the U.N. Framework Convention on Climate Change, the Montreal Protocol on Substances that Deplete the Ozone Layer, and the U.S./Canada Bilateral Agreement on Air Quality.

Clean Air Act Amendments

When fully implemented, CAAA will reduce air pollutant emissions by 56 billion pounds annually. The amendments mandated significant air quality improvements:

- Attainment of air quality standards nationwide by the year 2010

- Reductions in pollutants that cause acid rain, chiefly sulfur dioxide from coal-burning powerplants, by approximately 50 percent below 1980 levels by the year 2000

- Reductions of toxic air pollutants by at least 75 percent

- Cleaner cars, fuels, factories, and powerplants

- A phaseout of pollutants that threaten the stratospheric ozone layer

Since passage of CAAA, EPA has proposed or finalized rules and guidance documents that together will accomplish more than three-quarters of the emission reductions called for under the act. In an effort to prevent the kind of hostile litigation that has delayed environmental progress in the past, several rules were developed through regulatory negotiations and other forums in which industry, environmental groups, and state and local officials helped EPA develop regulatory proposals.

Market-Based Acid Rain Control

A prime example of the CAAA cost-effective, market-based methods is the acid rain control program, which employs an emission-allowance system to reduce annual nationwide sulfur dioxide emissions by up to 10 million tons per year, or roughly 50 percent below the 1980 level. Under this program, the federal government will allocate emission allowances to electric utilities whose powerplants emit sulfur dioxide. Nationwide emissions will be capped at the lower level and individual utilities must stay within their allowance limits. If, however, a utility can cut its emissions more than required, it can sell its extra allowances to another utility, or bank them for future use. This creates an economic incentive to the utility to maximize its emissions reductions at the lowest possible cost.

Initially, concerns with the market-based approach existed on two fronts. Some observers questioned whether a market actually would develop for the emission allowances, while others feared a lack of correspondence between the location of emission reductions and sensitive regions on the receiving end. Concerns over market development lessened substantially in 1992 when the Chicago Board of Trade established an allowance market; the first major transaction occurred in May 1992 when Wisconsin Power and Light sold 10,000 emission allowances to the Tennessee Valley Authority. Concerns over the distribution of pollution reductions have lessened because the cap on emissions will force all utilities to phase out older, dirtier equipment over time. EPA has instituted a monitoring system to identify potential problem areas.

The following is a summary of other CAAA rules that EPA has proposed or finalized to date.

Cleaner Gasoline

National limits on gasoline's volatility—its tendency to evaporate—already are reducing the release of hydrocarbons, which contribute to ozone smog. Starting in 1995 a new generation of cleaner, reformulated gasolines will reduce hydrocarbon and toxic emissions by at least 15 percent in the nine cities with the worst ozone pollution. A companion rule to increase the oxygen content of gasoline in 39 cities with carbon monoxide pollution will cut those emissions by an estimated 20 percent in 1993.

Cleaner Cars and Trucks

Tighter emissions limits for cars and light trucks will be phased in starting with 1994 models, cutting exhaust hydrocarbons by 30 percent and nitrogen oxides by 60 percent relative to current standards. A new cold-temperature standard is expected to reduce wintertime carbon monoxide emissions by 20 percent. Smoke from diesel trucks and buses will be cut by proposed particulate standards and by new limits on sulfur in diesel fuel.

Vehicle Inspection and Maintenance

High-tech vehicle inspection and proper maintenance could do more to improve urban air quality than any other single measure. Under a rule issued in 1992, EPA will require improved vehicle emissions inspection programs in cities having difficulty attaining federal standards for ambient levels of ozone and carbon monoxide. The rule also requires vehicle owners to tune-up or repair high-polluting vehicles.

Cash for Clunkers

An estimated 20 percent of all vehicles on the road create more than 60 percent of auto-related air pollution. A proposed "cash for clunkers" program would allow states and industries to buy older, higher-polluting cars and trucks, retire them permanently, and use the resulting pollutant reductions to help satisfy federal clean air requirements. Several companies and organizations have conducted pilot tests of the program in conjunction with state and federal officials in California, Illinois, and Delaware.

Fleet Vehicles

Beginning in 1998 in 22 cities, EPA will require new fleet vehicles, such as taxis and delivery vans, to meet tailpipe standards more stringent than those required for conventional vehicles. EPA guidelines will provide incentives for fleet owners to purchase ultra-clean vehicles fueled with natural gas, propane, pure alcohol, or electricity.

Offshore Oil Facilities

Offshore oil and gas platforms within 25 miles of a state's seaward boundary will be subject to the same air pollution control requirements as onshore facilities.

Incinerators and Landfills

New standards for municipal waste incinerators will result in a 90-percent reduction in emissions of metals, organic chemicals, and acid gases. Proposed standards for large, solid waste landfills, as well as hazardous waste treatment and disposal facilities, would cut toxic and smog-forming emissions. The landfill standards also would cut releases of methane, a greenhouse gas.

Permits

To facilitate enforcement of the amendments, states will develop comprehensive permit programs for air pollution sources.

Hazardous Organics Rule

New standards cover emissions from such diverse sources as storage tanks, transfer operations, and wastewater treatment plants. As a result, toxic air pollutants will fall by 1 million pounds per year, and volatile organic compound emissions by 2 billion pounds per year.

Air Quality in National Parks

The CAA amendments also cover air quality and visibility in pristine areas, such as national parks. Many national parks are affected by pollution haze that can degrade scenic vistas for park visitors. Building on a 1991 agreement to cut utility emissions near Grand Canyon National Park, the Grand Canyon National Visibility Transport Commission began an assessment in 1992 of visibility conditions at several additional parks and wilderness areas in the West. Recommendations on measures needed to protect these and other scenic resources will be presented to EPA in 1995. Special efforts have been made to involve the public in the commission's activities.

Stratospheric Ozone Layer

To slow and eventually reverse ozone-layer depletion, the United States signed the 1987 Montreal Protocol, amended in 1990 in London and again in 1992 in Copenhagen. The Protocol calls for rapid elimination of chemicals that are damaging the ozone layer. In 1989 the United States imposed an excise fee on the production of chlorofluorocarbons (CFCs); the fee was increased by the Energy Policy Act of 1992. In 1990 the Clean Air Act Amendments required ozone-depleting substances to be phased out by the year 2000.

In February 1992, acting on scientific evidence that ozone depletion is occurring more rapidly and widely than previously believed, President Bush accelerated the U.S. phaseout deadline for CFCs to the end of 1995, four years ahead of the original Montreal Protocol deadline. To accelerate progress in the near term, the President asked U.S. producers voluntarily to cut 1992 output to half the 1986 levels. In response to the President's call for swifter international action, protocol signatories meeting in Copenhagen in November 1992 agreed to accelerate the international CFC phaseout deadline to the end of 1995.

Currently the United States is ahead of other nations in phasing out CFCs and has a world lead in developing safer substitutes. The nation is also the leading contributor to funds to assist developing nations in reducing reliance on ozone-depleting substances.

Greenhouse Gases

In June 1992 the United States signed the U.N. Framework Convention on Climate Change, a binding treaty designed to promote continuing domestic and international actions to address climate change. The U.S. Senate ratified the convention in October 1992, making the United States the first industrialized nation to ratify the treaty. The U.S. government has established a national climate action plan that includes the following elements:

- Measures to reduce greenhouse gases directly and indirectly and to enhance greenhouse gas sinks, such as forests

- Development of new technologies to improve energy efficiency and reduce emissions

- The world's most extensive program of climate change research to better understand changes in the Earth's systems and develop appropriate responses

Actions included in the U.S. action plan will reduce greenhouse gas emissions by an estimated 7-11 percent below levels otherwise projected for the year 2000. For more information on U.S. climate change policy, see the International Issues section.

Indoor Air Pollution

Within the U.S. government, 20 federal agencies have responsibilities associated with indoor air quality, either through statutory responsibilities or because they are major property managers. An interagency Committee on Indoor Air Quality (CIAQ) coordinates these activities and includes members from EPA, the Consumer Product Safety Commission, the Department of Energy, the National Institute for Occupational Safety and Health, and the Occupational Safety and Health Administration.

The federal government administers indoor air programs under the authority contained in several statutes. For example, Title IV of the Superfund Amendments and Reauthorization Act (SARA) requires EPA to conduct research and disseminate information. The Federal Insecticide, Fungicide and Rodenticide Act (FIFRA) and the Toxic Substances Control Act (TSCA) authorize EPA to regulate products that adversely affect indoor air quality.

Among the indoor air pollutants being studied by federal agencies is radon. The U.S. Postal Service contracted with the DOE Oak Ridge National Laboratories to develop a sampling protocol to study radon in postal facilities. After conducting a sampling of postal buildings and evaluating new radon-detection technology, Oak Ridge scientists developed a sampling protocol and guidelines for the Postal Service radon program. The first phase of radon testing, which began in May 1992, focused on 40 postal facilities of different sizes throughout the nation.

Efforts by other federal agencies include research and dissemination of data on the following indoor air quality topics:

• Health effects
• Building design and control techniques
• Development of nonregulatory guidance on indoor air quality issues
• Chemical screening to prohibit or restrict potentially harmful products
• Radon testing
• Educational and outreach programs on health risks and prevention practices.

NAPAP: Research to Evaluate Policy

The Clean Air Act Amendments of 1990 reauthorized the National Acid Precipitation Assessment Program (NAPAP) to answer two questions:

• What are the costs, benefits, and effectiveness of Title IV of the Clean Air Act in reducing acid rain?

- What reductions in acid deposition are necessary to prevent adverse environmental effects?

NAPAP addresses these questions and identifies research gaps on an interagency basis by forming expert working groups. The program currently emphasizes the following topics:

- Aquatic and terrestrial ecosystems
- Human health
- Materials
- Visibility
- Monitoring and modeling of precursor emissions and changes in acid deposition in urban and non-urban areas

Cities with New Vehicle-Emissions Testing

Under a 1992 rule to enforce the Clean Air Act Amendments, the Environmental Protection Agency is requiring cities that have not met federal standards for ambient levels of ozone and carbon monoxide to improve inspection and maintenance (I/M) programs for vehicle emissions.

Basic I/M Programs. The following urban areas are required to initiate basic I/M programs:

Akron, OH	Grand Rapids, MI	Petersburg, VA
Ann Arbor, MI	Holland, MI	Port Arthur, TX
Aurora, IL	Huntington-Ashland	Port Huron, MI
Beaumont, TX	WV-KY-OH	Richmond, VA
Charleston, WV	Joliet, IL	Round Lake Beach, IL
Dayton, OH	Lewiston-Auburn, ME	Sheboygen, WI
Denton, TX	Lewiston, TX	Springfield, OH
Elgin, IL	Muskegon, MI	Texas City, TX
Galveston, TX	Parkersburg, WV-OH	Toledo,OH-MI

Enhanced I/M Programs. Enhanced programs, required in areas with higher pollution, involve more sophisticated testing. Several cities with basic testing in place, are required to adopt the stricter program, such as New York, Atlanta, Philadelphia, Boston, Houston, San Diego, and Washington. The following metropolitan statistical areas are required to start enhanced I/M programs:

Albany, NY Area   Jamestown, NY   Reading, PA
Altoona, PA   Johnstown, PA   Rochester, NY
Binghamton, NY   Lancaster, PA   Scranton, PA Area
Buffalo, NY   Manchester, NH   Sharon, PA
Burlington, VT   Niagara Falls, NY   State College, PA
Erie, PA   Orange County, NY   Syracuse, NY
Glen Falls, NY   Portland, ME   Tacoma, WA
Hagarstown, MD   Portsmouth-Dover-   Utica-Rome, NY
Harrisburg-Leban-   Rochester, NH-ME   Williamsport, PA
  on-Carlisle, PA   Poughkeepsie, NY   York, PA

Clean Air Act Business Opportunities

Complying with the Clean Air Act Amendments will be costly for some industries and businesses. However, the amendments also create business opportunities.

A study commissioned by the Environmental Protection Agency (EPA) projected that annual revenues to the air pollution control industry will grow by as much as $9 billion by 1996, with average annual growth of approximately 10 percent. Thousands of jobs may be created in the manufacture of air pollution control equipment. Enhanced vehicle inspection rules will generate jobs for inspectors, auto mechanics, and auto parts manufacturers. EPA also anticipates job shifting to such diverse fields as low-sulfur coal mining, chemical engineering for reformulated products, and manufacture of fuel additives and clean fuels such as natural gas and ethanol.

The Urban Heat Island Effect and Air Pollution

In urban areas, the mass of buildings, heat from cars and machinery, black asphalt of roads, and sparse tree cover combine to raise temperatures and slowly release heat back into the atmosphere. During the afternoon, city temperatures can be 6 degrees hotter than surrounding countryside. At night, city streets continue to radiate heat, keeping city temperatures 10 to 15 degrees warmer than the countryside. Studies have shown that this "urban heat island effect" aggravates air pollution and increases peak energy demand. The following strategies can help reverse the impact of the urban heat island effect, and in so doing, cut energy bills and reduce air pollution:

- Select light colors for building exteriors and roofs to reflect rather than absorb sunlight

- Use concrete or light-colored aggregate in asphalt for roads

- Plant trees next to buildings to shade roofs, and shrubs to shade walls and windows

- Shade exterior central air conditioning units to improve efficiency

- Include natural shading in building energy codes

- Plant trees to shade parking lots

The Department of Energy, Environmental Protection Agency, USDA Forest Service, and American Forestry Association are cosponsoring a 2-year study to evaluate urban heat island effects and shade-related issues in 12 demonstration cities.

HAZARDOUS AND SOLID WASTES

The United States has some of the world's most stringent laws regulating the handling and disposal of hazardous and solid wastes. These laws have spawned an array of complex regulations, some of which place a heavy burden on those regulated, especially small businesses and communities. In 1992 President Bush issued a directive to federal agencies to review and revise existing regulatory programs, and Congress began considering reforms that would achieve environmental protection with less burdensome regulatory control.

Conditions and Trends

Until 1976, when Congress passed the Resource Conservation and Recovery Act (RCRA), federal agencies had little authority to regulate hazardous and solid wastes. RCRA instituted a cradle-to-grave management system for hazardous wastes and developed disposal criteria for solid wastes, including municipal solid wastes. Over the past decade, RCRA, with its provisions for state regulation, has been joined by other federal environmental laws that authorize the Environmental Protection Agency (EPA) and the states to undertake programs to prevent pollution, manage hazardous wastes, clean up hazardous waste sites, ensure the safety of underground storage tanks containing petroleum or hazardous substances, and address the growing volume of municipal solid waste to ensure its safe management.

By current EPA estimates, the United States spends $35 billion a year on the following activities:

• Managing and disposing of hazardous waste and cleaning up contamination from such wastes at hazardous waste management facilities

• Collecting, managing, and disposing of municipal and other solid wastes

• Managing the underground storage tank improvement and cleanup program

EPA estimates that these expenditures will rise significantly by the year 2000—faster than any other category of environmental regulation. For example, with no changes in current law, the costs of the Superfund hazardous waste site cleanup program could rise from approximately $2.5 billion in 1992 to $7 billion in the year 2000.

In 1992, acting upon the Presidential directive to review and revise existing programs, the EPA Office of Solid Waste and Emergency Response (OSWER) proposed a package of reforms. To avoid regulatory relief at the expense of environmental protection, OSWER proposed the following reforms that would balance regulatory control with environmental risk to protect human health, the environment, and the economy:

- RCRA Reform Initiative. Regulatory reforms would reduce cost to industry by tailoring controls to risk and by removing certain administrative requirements.

- Superfund Revitalization. Reforms would improve the effectiveness, efficiency, and equity of the Superfund program, administered by EPA under the 1980 Comprehensive Environmental Response, Compensation, and Liability Act (CERCLA or Superfund) and the 1986 Superfund Amendments and Reauthorization Act (SARA).

- OSWER Reforms. The EPA Office of Solid Waste and Emergency Response would improve its responsiveness.

Municipal Solid Waste

Regulatory reforms would promote cost-saving pollution prevention in such areas as municipal solid waste. EPA estimates that in 1990 Americans produced approximately 196 million tons of trash (4.3 pounds per person per day), double the amount produced in 1960. In the absence of additional actions to reduce trash, EPA estimates that the amount generated in the United States could reach 222 million tons (4.5 pounds per person per day) by the year 2000.

Meanwhile, the number of landfills open to receive wastes declined from 20,000 in 1978 to 6,000 in 1986 as many landfills and open dumps reached capacity or were closed because of inappropriate design or operation. Many of the remaining landfills are expected to reach their capacity during the coming decade. Public opposition to the siting of new waste management facilities, expressed by "not in my backyard" sentiments, makes it difficult to site new landfills, even though they are safer.

The national trend is to reduce the amount and toxicity of wastes through such approaches as source reduction and recycling. EPA estimates that the rate at which materials in the municipal solid waste stream were diverted for recycling or composting rose from 7 percent in 1960 (6 million tons) to 17 percent (33 million tons) in 1990. Between 1985 and 1990 this amount doubled.

The nation can address local solid waste disposal problems by reducing the amount and toxicity of waste generated or, using techniques such as recycling and composting, by reducing the amount of waste needing disposal. Federal standards help protect human health and the environment, but implementation of solid waste programs remains largely a state and local matter. In 1988 EPA set a national goal of a 25-percent reduction by the end of 1992 in the amount of solid waste requiring disposal, using source reduction and recycling. Data for 1990 showed a national reduction in wastes of 21-23 percent, primarily through recycling. A sampling of current source reduction and recycling efforts follows.

Variable-Rate and Full-Cost Pricing. In 1992 EPA encouraged local and municipal governments to charge disposal fees that vary by the weight or volume of waste generated by a household or that reflect both the direct costs of the service and indirect environmental costs. The market signals sent by variable-rate and full-cost pricing create an incentive for households to reduce their waste stream and increase their recycling.

As an example, 39 percent of single-family-home garbage customers in Seattle in 1988 subscribed to have two or more cans picked up each week at a cost higher than single-can service. That same year, the city initiated curbside collection of recyclables and yard waste. The following year, the city increased the rate for multiple cans from $5 to $9 per month and lowered rates for homes that recycle. By 1991 customers subscribing for pickup of two or more cans had fallen to 11 percent, while those subscribing for pickup of less than one can had increased from 1 percent to 25 percent. Between 1987 and 1991, largely because of the revised variable-rate structure and initiatives on source reduction, recycling, and composting, the tonnage of residential solid waste disposed by the Seattle solid waste utility declined by 26 percent.

Federal Waste Reduction and Recycling. In response to the President's executive order of October 1991, requiring the U.S. government to promote waste reduction and recycling, 50 federal agencies named federal recycling coordinators and 36 submitted status reports on affirmative procurement programs to purchase products made of recycled materials. For more information, see the Federal Facilities Management section.

State and Local Recycling. Since 1988 local governments have increased the number of curbside-collection recycling programs by a factor of four. These programs now serve 65 million people. To assist state and local governments, federal agencies sponsor recycling initiatives, such as the following:

- An EPA Federal Procurement Hotline

- A procurement publication by the General Services Administration

- An EPA solid waste information clearinghouse

- EPA publications such as Markets for Scrap Tires, Recycle Today, and guides on composting and recycling materials

Recycling on National Parks. Now in its third year, a program cosponsored by the National Park Service, Dow Chemical, and Huntsman Chemical is among the first and largest recreation recycling efforts. To date, 600,000 pounds of material in six national parks have been recycled through the program. In 1992 the National Mall in Washington, D.C., became the seventh participating national park. The 160 bins to be set up on the Mall will collect hundreds of tons of material per year. Dow and

Huntsman are underwriting program costs, including the design and installation of recycling bins, collection of recyclables, informational exhibits, and technical support. Start-up costs for the Mall project are $250,000 for trucks and daily collection. In 1992 alone, Dow and Huntsman spent $500,000 on establishing new recycling in the seven national parks that together attract 23 million visitors each year.

Paper and Wood Recycling. The Forest Products and Harvesting Research Program of the USDA Forest Service has embarked on a national recycling effort to reduce by half the paper and wood in the municipal waste stream. Only 25 percent of the wood and paper discarded annually is recycled, while waste paper accounts for 40 percent of the volume in landfills. Scientists are seeking new technologies to make more efficient use of recovered paper, paperboard, and wood waste. They are working to improve the quality of recyclable material from paper and to develop alternative, nonpaper products from recycled wood fiber. They also are evaluating the economics of technologies for using recycled wood fiber and the economic and environmental impacts of expanded recycling.

Waste to Energy. In 1990, of all U.S. municipal solid wastes, 16 percent were combusted. Currently, 150 municipal waste-to-energy plants are in operation, combusting waste and converting 15 percent of the nation's solid waste into energy. In February 1991 EPA issued regulations for waste-to-energy plants, controlling air emissions of organics, heavy metals, and acid gases such as hydrochloric acid. Regulations ensure the maintenance of proper combustion conditions. Although combustion products—residue or ash—may be high in concentrations of heavy metals, they can be safely handled in landfills that meet the new national standards. The National Energy Strategy projects that waste-to-energy plants in the year 2010 will produce seven times the amount of power that they currently generate.

Landfills. In 1991 EPA issued regulations to improve the safety of municipal solid waste landfills to become effective in October 1993. New management standards cover the following categories of concern: location restrictions, operating requirements, design standards, groundwater monitoring and corrective action, closure and post-closure care, and owner/operator financial responsibility. Regulations cover facilities throughout their operating life and during their post-closure care period for 30 years. The rule may create an incentive for increasing source reduction and recycling nationwide because of the increased costs of disposal.

Interstate Transport of Solid Waste

In recent years, transport of municipal and other solid wastes across state boundaries became a contentious issue as half the states enacted new restrictions or requirements on imports of out-of-state solid waste. The commerce clause of the U.S. Constitution, which prohibits states from restricting interstate commerce without the consent of the Congress, has been the basis for court challenges to laws impeding the interstate transport of wastes. Recent sessions of the U.S. Congress have considered legislation without resolving the issue.

Hazardous Waste

U.S. environmental laws regulate the generation, transport, storage, treatment, and disposal of hazardous wastes. A summary follows on relevant provisions of RCRA; the Hazardous and Solid Waste Amendments of 1984 (RCRA amendments); CERCLA; SARA, including the Emergency Planning and Community Right-to-Know Act; the Clean Air Act Amendments of 1990; and the Hazardous Materials Transportation Uniform Safety Act of 1990.

RCRA and the RCRA Amendments. RCRA and the RCRA amendments regulate the generation, transportation, storage, treatment, and final disposal of hazardous and solid wastes. RCRA regulates solid wastes produced by 200,000 generators. Today many of these wastes are banned from land disposal unless treated to meet EPA standards, which require best available treatment technology—chemical or biological treatment or incineration—to reduce toxicity and stabilize residues.

Policies and Programs

As the volume of hazardous and solid waste increased in conjunction with national growth, government agencies—federal, state, and local—have responded with a number of programs, many in partnership with private-sector groups.

• **Land Disposal Restriction.** The EPA Land Disposal Restriction Program has resulted in 42.5 million tons of hazardous wastes being treated each year prior to final disposal. Of this total, provisions of the Safe Drinking Water Act regulate 34 million tons of treated wastes pumped into the ground in underground injection wells. The remaining 8.5 million tons are disposed of in or just below the land surface in RCRA-regulated landfills, surface impoundments, and land application units.

• **Corrective Actions.** EPA can require corrective actions to address all releases of hazardous waste or hazardous constituents from any facility requiring a RCRA hazardous waste permit. Of 4,600 hazardous waste facilities subject to corrective actions, EPA estimates 3,600 will require some form of cleanup. To date, EPA and states authorized to run the corrective action program have conducted 3,000 initial assessments. Of this total, 800 facilities are pursuing corrective action under consent orders or permit requirements, and 135 facilities are conducting cleanup activities.

- **RCRA Reform.** In 1992 EPA began a major RCRA Reform Initiative that could save businesses and communities billions of dollars annually. RCRA reform has the following objectives:

- Redirect RCRA towards higher risk wastes and cleanup activities

- Expedite the pace of corrective action cleanups at RCRA facilities

- Tailor management standards to environmental risks posed by waste management

- Streamline the RCRA permit system to facilitate cleanups and hazardous waste recycling and also to address closed facilities

- **Underground Storage Tanks.** Chemicals that escape from underground tanks containing petroleum or other hazardous substances can contaminate drinking water supplies. Fumes from these tanks can pose health hazards, including fires and explosions. The RCRA amendments require tank registration, leak detection, leak prevention, and corrective action. From 1990 through August 1992, states and private parties confirmed 167,000 releases, initiated 114,000 cleanups, and completed 45,000 cleanups. When a responsible party for a tank cannot be found to oversee corrective action, EPA authorizes 49 states, 5 U.S. territories, and the District of Columbia to use the Leaking Underground Storage Tank Trust Fund for cleanups. As of June 1992 the Trust Fund had collected $810 million for cleanup actions.

- **Superfund.** CERCLA created the Superfund program to respond to a release or threatened release of hazardous substances, pollutants, and contaminants stemming from accidents or uncontrolled hazardous waste sites. Under the law, those responsible for the contamination are required to conduct the cleanup. Where enforcement is not successful, the federal government can clean up a site using the CERCLA Trust Fund (Superfund), which is supported by excise taxes on feedstock chemicals and petroleum and by a more broadly based corporate environmental tax. If the Superfund program conducts the cleanup, the government can take court action against responsible parties to recover costs with, in some cases, as much as treble damages.

To accelerate responses, EPA seeks to have responsible parties begin the remedial design before entry of a consent decree. In fiscal 1991 parties liable for contamination at nonfederal Superfund sites agreed to perform a record 63 percent of new cleanup work; in 1992 that figure rose to a record 72 percent. The total value of responsible party commitments to conduct site study and cleanups at Superfund sites has exceeded $1 billion for each of the past three years. Since Superfund was created in 1980, EPA has obtained $7 billion in site work commitments from responsible parties, with three-quarters of this amount obtained since 1988. Examples of recent Superfund accomplishments follow.

• **Site Evaluations.** Through November 1992 EPA had identified 37,592 potentially hazardous waste sites across the nation. Of these, 93 percent have undergone a preliminary assessment to determine the need for further action.

• **National Priorities List.** EPA maintains the Superfund National Priorities List (NPL), which identifies the nation's most seriously contaminated hazardous waste sites eligible for permanent Superfund cleanup. As of mid-November 1992 a total of 1,252 sites had been listed, on which work was underway at 95 percent and cleanup construction was in process or complete at 41 percent. In fiscal 1992 permanent cleanup construction was completed at 86 sites—a total equalling more than the number completed in the first 11 years of Superfund.

• **Environmental Progress.** To date, at 3,000 sites, the Superfund program has treated, isolated, neutralized, or removed from the environment 13 million cubic yards of contaminated soil and solid wastes; 1 billion gallons of liquid waste; 6 billion gallons of contaminated groundwater; and 316 million gallons of polluted surface water. These actions have reduced potential risks to the 23 million people who live within four miles of a Superfund site, although such risks are difficult to quantify and depend on many variables.

• **Superfund Accelerated Cleanup Model.** Because the number of abandoned and uncontrolled hazardous waste sites far exceeds original estimates, EPA, in fiscal 1992, developed the Superfund Accelerated Cleanup Model. This management approach, which is being tested in 27 pilot projects nationwide, has the following features:

• Single-site assessments save time by eliminating redundant activities, resulting in earlier enforcement search and notification;

• Regional management decision teams prioritize the work load by developing standardized cleanup methods for similar sites and promote innovative cleanup technology;

• Early action mechanisms address immediate threats to public health and safety in a timely manner; and

• Long-term action mechanisms help the public understand the extended time commitment required to complete cleanup of a Superfund site, once immediate threats are removed.

• **Remedy Selection.** Cleanup of the average Superfund site currently takes 7 to 10 years from listing to completion. EPA compresses each phase of remediation by qualifying fewer, better focused alternatives. Remedy-selection guidelines provide standard approaches for similar types of sites, such as municipal landfills and PCB-contaminated sites.

• **Cleanup Targets.** Having completed cleanup at nearly 150 sites as of November 1992, the Superfund program moved toward a target of 200 completed sites by the end of fiscal 1993 and 650 by the end of fiscal 2000.

• **Public Involvement.** EPA increasingly involves communities near Superfund sites in cleanup decisions. New measures of cleanup success, based on protecting public health and the environment, initiate uniform outreach at the start and completion of each remedial action. Such measures enable the public to identify accomplishments within the Superfund cleanup process.

• **Innovative Technologies.** Federal agencies are supporting the development of new cleanup technologies. Examples follow.

• **EPA Technology Innovation Office.** This office, created in March 1990, promotes development and use of new treatments for cleanup problems such as decontaminating soils and groundwater. Promising new treatments clean up contamination more quickly, more effectively, and less expensively. The EPA Superfund Innovative Technology Evaluation Program measures the effectiveness of new techniques for destroying, immobilizing, or reducing hazardous waste and contaminated materials. EPA has completed 41 full-scale field demonstrations of new technologies prior to commercialization and 14 laboratory and pilot-scale studies; and 91 new technologies are set for future evaluation. Projects include soil washing, chemical dechlorination, underground vacuum extraction, and forcing of airstreams through water to evaporate volatile organic compounds. EPA is evaluating bioremediation—the use of microbes to break down organic contaminants—at demonstration cleanups across the country. Through fiscal 1991 EPA had selected 210 new technologies for use in remedial actions at NPL sites.

• **U.S. Bureau of Mines, Environmental Technology Division.** This division is undertaking similar research in developing new technologies for immobilizing or reducing hazardous waste and contaminated materials resulting from mining and mineral processing. Most of the contaminates are inorganic. For example, techniques have been developed to recover lead from scrap batteries and to treat lead-contaminated soil at battery breaker superfund sites.

Chemical Emergencies

The Environmental Protection Agency (EPA) sponsors a cooperative program to ensure that local communities are prepared for accidents involving hazardous materials, to help reduce the number of accidents, and to assist in mitigating effects on public health and the environment should they occur. Participants include state, tribal, and local governments, industries, labor groups, environmental groups, and other stakeholders. The following laws and initiatives support state and local right-to-know and response planning:

Emergency Planning and Community Right-To-Know Act. Part of the Superfund Amendments and Reauthorization Act of 1986 (SARA), this law has two components:

• **Right-To-Know.** The Community Right-to-Know program involves reporting the presence and release of hazardous chemicals in communities, maintenance of the Toxics Release Inventory by EPA, and other measures to expand public access to such information.

• **Emergency Response Preparedness.** The 50 states and 6 U.S. territories have state emergency response commissions that oversee 4,000 local emergency planning committees. Local committees receive data on hazardous materials produced, used, or stored in their community; provide information to the public; and develop, update, and test emergency plans for responding to accidental releases of chemicals.

• **Clean Air Act Amendments of 1990.** The CAA amendments require federal agencies to develop means of preventing and planning for chemical accidents that might discharge hazardous materials into the air.

• **Hazardous Materials Transportation Uniform Safety Act of 1990.** This act establishes standards and provides for planning and training.

• **Transportation Standards.** The Department of Transportation (DOT) imposes standards for use by states and Native Americans in setting and enforcing routes for transportation of certain hazardous materials. In 1992 the DOT Research and Special Programs Administration (RSPA) issued a rule, establishing a national registration program for certain shippers and carriers of hazardous materials and a collection system to fund a national emergency response planning and training grant program.

• **Planning and Training Grants.** The act authorizes EPA to administer an assistance program that will award $12.8 million annually during 1993-1998 to enhance preparedness and response capabilities of states and tribes.

• **Federal Response Plan.** In 1992 EPA completed an interagency Federal Response Plan, with guidelines for responding to significant and catastrophic national emergencies, including earthquakes and hurricanes. EPA has lead responsibility for hazardous materials responses under the plan, which has the following components:

• **National Response Team.** Key to activating the plan, this team is chaired by an EPA official and includes representatives of the U.S. Coast Guard, the departments of the Interior, Defense, Energy, Agriculture, Commerce, Health and Human Services, Justice, Labor, Transportation, and State, the Nuclear Regulatory Commission, and the Federal Emergency Management Agency.

- **National Oil and Hazardous Substance Pollution Contingency Plan.** This plan, developed by EPA in coordination with members of the National Response Team, would assist in a national response to a chemical emergency.

- **EPA Emergency Operations Center.** This center would coordinate and monitor the federal response.

- **EPA National Incident Coordination Team.** Comprised of senior level officials from each of the major EPA pollution control offices and branches, this team would staff the Emergency Operations Center.

- **International Right-to-Know.** The United States has been successful in using the community right-to-know program to provide a foundation for international chemical accident programs such as the following:

- **Organization for Economic Cooperation and Development.** OECD has developed a program called Guiding Principles for Chemical Accident Prevention, Preparedness, and Response;

- **United Nations Environment Program.** UNEP sponsors a program for developing countries called Awareness and Preparedness for an Emergency at the Local Level.

- **United Nations Conference on Environment and Development.** At UNCED in Rio de Janeiro in 1992, the United States championed inclusion of community right-to-know policies and toxic release inventories in the Agenda 21 list of actions that the 180 countries present agreed by consensus should be standards for all governments. The principle of community right-to-know also was included in the UNCED Rio Declaration.

Hazardous Materials Transportation

The Hazardous Materials Transportation Uniform Safety Act of 1990 imposed standards for use by states and Native Americans in setting and enforcing routes for transportation of certain hazardous materials. In July 1992 the DOT Research and Special Programs Administration (RSPA) issued a rule establishing a national registration program for certain shippers and carriers of hazardous materials and a collection system to fund a national emergency response training and planning grant program. In addition to these standards, RSPA developed the following regulations to coordinate U.S. domestic requirements with various international agreements:

- **Packaging Standards.** Several rules issued in 1990 to conform with United Nations recommendations on hazardous materials packaging became effective in 1992. The U.N. recommendations provide performance-oriented standards that encourage innovation and facilitate international commerce.

• **Marine Pollutants.** In 1991 the U.S. Senate ratified Annex III of the International Convention for the Prevention of Pollution from Ships, which mandated changes in the list of materials identified as marine pollutants by the International Maritime Organization. Currently the DOT Research and Special Programs Administration is developing regulations to implement the treaty obligations.

Marine Transportation and Oil Pollution

The Oil Pollution Act of 1990 (OPA) strengthened the authority of the federal government to prevent oil spills in marine and freshwater environments. It provided for the cleanup of spills and compensation for public and private damages, without preempting states rights of recovery. DOT administers the law in cooperation with EPA.

Pollution Response. DOT regulations proposed in 1992 would require tank vessels and marine transportation facilities to develop oil spill response plans to be in place by February 1993 and operating by August 1993. Vessels transporting oil and hazardous bulk cargo would be required to carry pollution discharge removal equipment. Response plans also would be required for other kinds of transportation facilities, such as pipelines, rails, and highways.

In response to OPA, the Coast Guard placed pollution response equipment in each of its 49 marine safety units located in ports throughout the United States and purchased enhanced response equipment for 19 sites.

As of September 1992 a new Coast Guard Atlantic Strike Team became fully operational for both oil and hazardous substance spill response. The National Strike Force Coordination Center in Elizabeth City, North Carolina, administers Atlantic, Gulf, and Pacific Strike teams. The center trains federal on-scene coordinators in pollution response and provides technical assistance, equipment, and other resources in the event of a spill.

Liability. In response to OPA, the Coast Guard created the National Pollution Funds Center in February 1991 to address funding, liability, and damage claims arising from discharges and threatened discharges of oil. In August 1992 DOT issued an interim rule establishing procedures for filing of claims for removal costs or damages resulting from the discharge of oil.

Double-Hull Tankers. Under OPA, new tank vessels operating in U.S. waters must be constructed with double hulls, and existing single-hull vessels must be phased out between 1995 and 2015. In August 1992 DOT issued an interim rule providing guidance on double-hull construction.

DOT also sponsored a 1991 National Academy of Sciences study on alternative tank vessel design, which found the double hull to be an effective design and recommended further study of the mid-deck design. In addition, the Coast Guard has participated in efforts by the International Maritime Organization to evaluate alternatives to double-hull construction and is preparing a report on the subject to the Congress.

Nonregulatory Initiatives. States have considerable leeway under OPA in regulating the maritime industry, which often results in inconsistent federal and state requirements regarding pollution prevention, preparedness, liability, and response. To address this problem, the Coast Guard has begun a program to improve coordination with states and to seek a consistent marine environmental protection strategy. The Coast Guard will encourage and support regional initiatives that bring citizen groups and marine industry associations together to foster long-term environmental compliance.

Maritime Oil Pollution. In March 1992 the United States became the first nation to ratify the International Convention on Oil Pollution Preparedness, Response and Cooperation (OPRC), negotiated under the auspices of the International Maritime Organization (IMO). The convention—proposed by President Bush in June 1989 and completed 17 months later—protects the marine environment from oil spills and calls for planning, reporting, technology sharing, and cooperation. OPRC requires parties to establish national systems for preparedness and response, a system the United States established through the 1990 Oil Pollution Act. The U.S. Coast Guard is working with IMO to refine OPRC implementation plans and to bring existing bilateral contingency plans with Canada, Japan, Mexico, Bermuda, and Russia into conformity with the convention.

Transboundary Movements of Hazardous Wastes

Regulating exports of hazardous and other wastes is an emerging trend in the international community. In 1990 the United States signed the Basel Convention on the Control of Transboundary Movements of Hazardous Wastes and Their Disposal. In August 1992 the Senate gave consent to the U.S. ratification of the Convention, and Congress must now enact enabling legislation. Proposed federal legislation to implement the Basel Convention would prohibit the export and import of hazardous and municipal wastes and residues from the incineration of municipal wastes, except in cases where the United States has a bilateral agreement with another country ensuring that wastes will be handled in an environmentally sound manner.

WATER QUALITY AND SUPPLY

Streams and rivers, lakes and ponds, and groundwater aquifers supply water for U.S. homes and businesses. In addition, the nation's aquatic resources support irrigation, electric power, navigation, fisheries, wildlife habitat, and recreation. Once managed as separate entities according to political boundaries, water resources increasingly are managed as interconnected, essential parts of large aquatic ecosystems.

Conditions and Trends

Water Quantity

In 1992 the nation's potable freshwater needs were met by withdrawals from rivers, streams, lakes, reservoirs, and groundwater aquifers. On a national scale, water quantity was ample, with 1.4 billion gallons per day of renewable water supplies withdrawn in the lower 48 states, up from offstream withdrawals of 408 billion gallons per day in 1990.

A different picture emerges, however, at the regional level, because of an uneven natural distribution of the resource in relation to regional and seasonal water demand. Rising demands for limited supplies in certain parts of the nation, particularly the West, are matters of public concern on a number of fronts—health, environmental, economic, recreational, and industrial. Changes in engineering, management, or public behavior can alleviate some of this concern.

The natural availability of water in any geographic area is determined by hydrologic conditions, which reflect the continual circulation of water from the sea to the atmosphere to the land and back again. The amount of rain and snow is the primary factor in the availability of surface water resources. Average annual precipitation in the United States is 30 inches per year; however, the range varies from a few tenths of an inch per year in the desert areas of the Southwest to 400 inches per year at some locations in Hawaii.

• **Surface Water.** Streamflows are the standard measure of water quantity, whether in-stream or off-stream and whether for recreation, irrigation, or public water supplies.

• **National Water Conditions Report.** Six-year streamflow trend data (1986-1991) issued by the U.S. Geological Survey for 12 hydrologic areas illustrate the variation in natural distribution of water supplies across the nation. Based on this data, 1991 monthly streamflow is above the long-term median in four basins, at the median in five basins, and below the median in three basins.

• **Augmenting Supplies.** The following methods can increase the efficiency of surface-water use for potable water supply:

- Manmade surface-storage reservoirs
- Water conservation
- Improved water system management
- Reuse of water of impaired quality or water recycling
- Reduction of evaporation from the soil, surface-water bodies, and plants
- Greater utilization of groundwater aquifers through conjunctive use of groundwater and surface water

Although surface-storage reservoirs can augment natural water supplies, construction is limited by such factors as cost, competing land uses, engineering problems, and environmental concerns associated with land inundation and diversion or removal of water from downstream aquatic systems.

• **Users.** Currently, irrigation use accounts for the greatest share of freshwater withdrawals, at 40 percent, with industrial use accounting for 6 percent of that total. Thermoelectric powerplants use the greatest share of both fresh and saline withdrawals combined, at 48 percent. Because this water is used for cooling purposes, most of it is returned to the source. Another major nonconsumptive use of managed water systems is the generation of hydropower.

Nine states each withdraw greater than 10 billion gallons per day of surface water for combined offstream uses, constituting 42 percent of the total surface-water withdrawals in the country. The largest withdrawals occur in California, Illinois, New York, and Texas. Public water supply use continues to increase, reflecting population growth as well as an increase in per capita water use.

• **Groundwater.** Although surface-water diversions continue to exceed groundwater pumping by volume, the nation has increased its use of groundwater for all purposes, except thermoelectric power generation. Groundwater provides drinking water supplies for 40-50 percent of the U.S. population. It also supplies 40 percent of average annual base flow to streams.

• **Use.** California, Idaho, Kansas, Nebraska, Texas, Florida, and Arizona—states where irrigated agriculture is common—use the most groundwater. While groundwater is currently an alternative to surface-water supplies, particularly during drought, it may in the future become the preferred source for water.

• **Overuse.** In water-deficient areas, such as Arizona, large volumes of groundwater continue to be withdrawn to supply agricultural and municipal needs. Because of limited supplies, such withdrawals cannot be sustained indefinitely. In Florida, groundwater development has redistributed natural flow patterns in the aquifers, resulting in sinkholes, saltwater intrusion, and land surface subsidence. Groundwater mining in California's San Joaquin Valley has resulted in sediment compaction and land subsidence. Such development trends suggest the need to conserve existing groundwater supplies if the nation is to meet future water needs.

• **Floods and Droughts.** Floods are natural and recurring hydrologic events that perform useful ecological functions, yet for people, they remain one of the most destructive natural hazards in the United States. Many areas of the country have flood-control facilities through programs of the Army Corps of Engineers, but many other areas susceptible to flooding remain unprotected. At the other extreme, droughts can severely reduce water availability and cause significant environmental and economic impacts. Adverse effects of a drought on water supplies depend on the amount of water stored or available from the preceding year, the water demands relative to average flow, soil conditions, and the natural flow during a drought period. The multi-year droughts of the late 1920s and 1930s resulted in extensive regional impacts. For the past several years, numerous river basins in the western states have experienced drought conditions.

Water Quality

Despite decades of research and regulation, the federal government still lacks a comprehensive assessment of the quality of the nation's waters. Efforts to acquire this information depend, in part, on the states, which report every two years to EPA, which in turn reports to the Congress on the quality of the water in each state. Results are published in the EPA National Water Quality Inventory, also known as the 305(b) report. According to the 1990 National Water Quality Inventory, the states assessed about a third of total U.S. river miles, half of lake acres, and three-quarters of estuarine square miles, which represented a substantial increase over areas assessed for the previous 1988 water inventory.

Of assessed water resources, about two-thirds met federal water quality standards and achieved state-designated uses such as fishing, swimming, and drinking. One-third of the assessed waters did not fully meet designated uses. The major remaining impairment to water quality was wet-weather runoff from such sources as farmlands, city storm sewers, construction sites, and mines. Pollutants resulting from wet-weather runoff included siltation, pathogens, toxic chemicals, and excess nutrients that produce low dissolved oxygen levels capable of suffocating fish and contaminating groundwater. Loss of wetlands also can contribute to water quality problems.

Surface Water Impairment. The 1990 National Water Quality Inventory indicated that agricultural sources were by far the leading source of contaminants, contributing to 60 percent of the impaired stream miles and 57 percent of the impaired lake acres in the assessment. Agricultural pollutants included siltation, nutrients (fertilizers and animal wastes), and pesticides. For rivers and streams, sewage treatment plants followed agriculture as a source of contaminants, contributing to 16 percent of the impaired stream miles. For lakes, hydrologic/habitat modification contributed to 40 percent of impaired acres. Such impairments are not additive, since more than one source can contribute to impaired water quality.

Groundwater Impairment. For many years, land surface and subsurface disposal of wastes was considered safe and convenient. Only recently did researchers discover that natural processes have a limited capacity to convert contaminants into harmless substances before they reach groundwater. One EPA study estimates that more than half of the nation's land area has geologic factors that would allow groundwater contamination. The EPA 1992 National Pesticide Survey estimates that 10.4 percent of community water supply wells and 4.2 percent of rural domestic wells contain detectable levels of one or more pesticides. The study also estimates that 1 percent (68,500) of all U.S. drinking water wells exceed the EPA health-based limits on contaminants. Elevated levels of nitrates also have been detected in groundwater. These and other groundwater contaminants, such as organic and inorganic chemicals, radionuclides, and microorganisms may cause adverse health, social, environmental, and economic impacts. Among these impacts are the health risks of exposure to contaminants and expenditures such as groundwater purification systems. Because groundwater provides base flow to streams, the potential for adverse impacts on surface-water quality also exists, especially under conditions where dilution is minimal.

Policies and Programs

Among the federal agencies with water-related responsibilities, the U.S. Geological Survey (USGS) in the Department of the Interior (DOI) provides the hydrologic information needed to manage the nation's water resources. EPA administers water pollution control and safe drinking water programs. The Army Corps of Engineers (COE) and EPA regulate water issues such as wetlands filling and ocean dumping. Several DOI bureaus conduct water quality programs on public lands, and the Bureau of Reclamation and COE oversee a vast system of levees, dams, and reservoirs for flood control, irrigation, hydropower, navigation, and environmental protection. In addition, many states assume principal administrative roles in managing water resources through delegated programs.

Water Quantity

Although the federal government administers a significant portion of the nation's water storage and conveyance facilities, water allocation and administration rests principally with the states. The Army Corps of Engineers and the Bureau of Reclamation provide federal assistance to the states for water quantity programs.

Drought Assistance. In 1992 the governors of 14 states joined 4 tribal officials in requesting temporary drought assistance under the Reclamation States Emergency

Drought Relief Act of 1991. The Bureau of Reclamation maintains Water Conservation and Advisory Centers in regional offices and in Washington, D.C., headquarters to coordinate water conservation programs with the states. In 1992 the bureau took the following measures to lessen the impacts of the western drought:

• Administered $25 million designated for use in meeting emergency needs for areas stricken with drought

• Initiated seven studies requested by water districts to assess the potential to improve efficiency of water-delivery schedules

• Facilitated water banking and transfers

• Inventoried available water inactive in reservoir storage as a potential emergency water supply

• Identified studies needed on drought mitigation and water conservation to maintain crop production and protect the environment

• Continued climate change research to evaluate probable impacts of drought on reservoir system management in the western United States.

Water and Power. The Army Corps of Engineers manages over 600 water management projects nationwide, and the Bureau of Reclamation, which manages the majority of projects in the West, operates more than 350 reservoirs. These projects manage water resources for irrigation, flood control, hydroelectric power, navigation, municipal and industrial use, fish and wildlife purposes, and recreation. Bureau of Reclamation projects, with a storage capacity of 125 million acre-feet of water, deliver water to 28 million people and 10 million acres of land each year. A total of 52 hydropower plants generate 60 billion kilowatt hours of electricity each year, making the bureau the eleventh largest producer of electric power in the nation.

Water Transfers. Federal water transfers among willing buyers and sellers, in accordance with state law, represent one way to respond to changes in demand for western water supplies without building new storage facilities. The Bureau of Reclamation facilitates trades such as the following, in partnership with the states and other interested parties:

• **Snake River Drainage.** The Bureau of Reclamation and the State of Idaho jointly established three water banks in the Snake River Drainage. Water users with reservoir rights can "deposit" surplus water in the banks, thus making it available to others.

• **Central Valley Project.** On the Friant Unit of the Central Valley Project, water transfers occurred between districts as part of a conjunctive-use program, which manages groundwater and surface water for a common purpose. Southern California water users propose expanding the conjunctive use of Central Valley Project water, and studies are underway to further increase water-use efficiency.

• **Columbia River Basin.** The Bureau of Reclamation traded flood-control storage space with the Army Corps of Engineers to aid salmon migration.

• **Indian Water Rights Settlements.** Water settlements have resolved conflicts between Native American tribes and other persons over title to western water. The Bush administration supported inclusion of leasing in Indian water rights settlements. Examples include Fort Hall, Montana, and Jicarilla, New Mexico. With clear title, Native Americans can lease water to their neighbors, putting water to use or maintaining existing uses. Water leasing precludes the need for new storage construction and allows Native Americans a return for the use of their assets.

Water Quality

The goals of the Clean Water Act are ambitious—fishable, swimmable rivers throughout the nation and zero discharge of pollutants into U.S. waters. The act requires all municipal sewage and industrial wastewater to be treated to reduce or remove pollutants before being discharged into waterways, and it provides federal grants and capitalization of state revolving loan funds to help communities build sewage treatment plants. EPA, in cooperation with the states, establishes limits on the amounts of specific pollutants that may be discharged by municipal sewage treatment plants and industrial facilities, based on available technologies and economic costs of compliance.

National Pollution Discharge Elimination System. Under the Clean Water Act, EPA or approved states administer the National Pollutant Discharge Elimination System (NPDES). EPA or approved states issue permits that establish effluent limits for all municipal and industrial dischargers. The federal government currently authorizes 38 states and one territory to operate the NPDES permit program. In addition to technology-based limits, EPA may develop limits based on water quality criteria where technology-based controls are not stringent enough to make waters safe for such uses as fishing, swimming, and drinking. For industrial dischargers, EPA has established stringent standards to control up to 126 toxic pollutants. Currently, about 50 major industries comply with these standards, which are based on the best available technology that is economically achievable.

EMAP. EPA is developing the Environmental Monitoring and Assessment Program (EMAP) to monitor and assess the ecological health of major ecosystems in an integrated, systematic manner. EMAP will monitor and assess surface waters, forests, near-coastal waters, wetlands, agricultural lands, arid lands, and the Great Lakes. The program will operate at regional and national scales, over a period of decades, to evaluate the extent and condition of entire ecological resources. A 4-year resampling cycle will allow approximately 800 lakes and 800 stream sites to be evaluated annually.

Improving Water Quality in New York Reservoirs. In July 1992 New York City officials and 400 farmers in the Catskill Mountains and lower Hudson River Valley agreed to voluntary measures designed to reduce water pollution from agricultural practices. If successful, these measures could save the City of New York from having to spend $4 billion on a new filtration system to clean river water. The city's six Catskill and Delaware River reservoirs, which provide drinking water to more than 8 million people, are being contaminated by manure, insecticides, fertilizers, and road runoff from farming areas.

Under the agreement, the city will pay $3.4 million to the state's Soil Conservation Committee that will in turn provide technical assistance to the farmers through Cornell University and other state and local agricultural and environmental groups. Farmers contend that successful implementation of voluntary measures could achieve up to 80-percent compliance with water quality requirements.

Groundwater Quality Protection. To protect groundwater, EPA is implementing a Ground Water Strategy for the 1990s that emphasizes pollution prevention. The strategy draws upon federal environmental laws that control solid and hazardous wastes, pesticides, surface waters, underground storage tanks, and waste cleanup, as well as drinking water. As part of a new Comprehensive State Ground Water Protection Program, the states will integrate all federal and state programs relating to groundwater. EPA has provided $12.2 million to the states to implement the program.

Safe Drinking Water Program. EPA sets standards for drinking water quality and requirements for treatment under the Safe Drinking Water Act (SDWA). Federal standards control both manmade and naturally occurring contaminants, and the Public Water Supply Supervision Program, authorized by SDWA, supervises compliance. In most cases, states have the primary responsibility for oversight and enforcement. EPA supports states through grants and technical assistance and, if necessary, enforces SDWA regulations.

National Water Quality Assessment. The U.S. Geological Survey administers the National Water Quality Assessment Program (NAWQA) to describe the status and trends in the quality of the nation's surface waters and groundwater. Within a 10-year period, NAWQA plans to study the populations served by public water supplies and conduct investigations in 60 study units that will represent 70 percent of the water Americans use. Personnel will use physical, chemical, and biological measures to investigate the occurrence of pesticides, nutrients, and sediments. In 1992 the Geological Survey had NAWQA field investigations underway in 20 study units distributed across the nation. Study units consist of watersheds that range in size from 1,200 to 60,000 square miles. Each study unit has a liaison committee with representatives from federal, state, and local agencies, universities, and the private sector.

Endangered Species Protection. Approximately a third of all federally listed threatened and endangered species rely on aquatic ecosystems. In July 1992 EPA and the U.S. Fish and Wildlife Service signed a Memorandum of Understanding (MOU) linking Clean Water Act standards to species protection. EPA agreed to an assessment and consultative process to determine whether CWA standards interfere with provisions of the Endangered Species Act. Where adverse impacts are found, EPA could move to alter CWAstandards to mitigate impacts.

President's Water Quality Initiative. The U.S. Department of Agriculture has undertaken a major research initiative called the Management System Evaluation Areas (MSEA). The emphasis is on nonpoint-source contamination of ground and surface water by agricultural chemicals. Research results from five study locations in the Midwest will help farmers control and reduce the impact of agrichemicals on water resources. The U.S. Geological Survey is supporting the initiative by conducting geohydrologic studies and research on the fate and transport into groundwater of the pollutants atrazine, carbofuran, alachlor, and nitrate.

National Irrigation Water Quality Program. The Department of the Interior administers this program to investigate irrigation-induced water quality problems in the western states. Reconnaissance investigations check for contamination by trace elements and toxic substances including selenium, mercury, arsenic, and pesticides. In 1992 DOI completed ten investigations, during which waterfowl deformities and reproductive problems were found at several locations. The Department has published detailed study reports for the Kendrick Reclamation Area in Wyoming and the Stillwater Wildlife Management Area in Nevada.

Nonpoint Source Pollution. Despite progress in improving water quality impacts from industrial and municipal dischargers (point sources), the nation faces significant water quality challenges. The nation's major remaining water pollution problems are caused by less obvious and more widespread sources of pollution—nonpoint sources that affect both surface water and groundwater. These relatively uncontrolled sources of pollution may contribute more to water quality degradation than point sources. Leading nonpoint sources of pollution include agricultural and urban runoff, forestry practices, and hydromodification. In 1992 EPA awarded 52.5 million to the states to assist in the development of nonpoint management programs. In addition, a 1991 survey by the National Association of State Foresters indicated that 32 states have forestry nonpoint source pollution control programs that emphasize prevention through the use of best management practices (BMPs). In 18 states conducting BMP compliance surveys, compliance ranged from 79 percent for streamside management to 98 percent for forest-site preparation.

Combined Sewer Overflow Initiative. Combined sewer overflows (CSOs) occur where sanitary and storm sewers are interconnected. During rainstorms, these systems become overloaded, bypass treatment works, and discharge as much as 90 percent of the pollutants they contain. CSO discharges may contain high levels of suspended solids, bacteria, heavy metals, floatables, nutrients, oxygen-demanding organic compounds, oil and grease and other pollutants. Discharges of these pollutants often exceed state water quality standards. In 1989 EPA issued the National Combined Sewer Overflow Strategy, reaffirming that CSOs are point sources subject to National Pollution Discharge Elimination System (NPDES) permit requirements. NPDES permits are based on technology-based and water-quality-based requirements of the Clean Water Act. In 1991 EPA launched an initiative targeting those CSO systems causing the most severe water quality problems. EPA will issue new or revised permits to CSO systems lacking or having inadequate discharge permits and plans enforcement actions against systems in violation of permit limits.

Mexican Border Initiative. In February 1992 the United States and Mexico adopted a multimedia plan to address environmental problems along their common border. The plan includes construction of international wastewater treatment facilities, provision of safe drinking water, conducting joint monitoring and enforcement, and providing technical assistance and training to Mexico. The two nations have begun technical assistance and training; they plan joint monitoring in 1993 along the Rio Grande River and propose monitoring of groundwater for the Nogales and El Paso areas. Construction continues on the Tijuana international wastewater facility as do negotiations toward international agreements for wastewater facilities to serve Nogales and the area of Mexicali/Calexico. In 1993 numerous U.S. agencies will coordinate efforts to provide over $85 million in grants and loans to Mexican communities or colonias across the border from Texas and New Mexico.

Key Points of the 1992 Water Bill

In signing the Reclamation Projects Authorization and Adjustment Act of 1992, known as the 1992 Water Bill, President Bush and the Congress enacted the most extensive western water policy legislation in decades. The legislation affects more than 24 water projects in 17 western states. Provisions include the following:

• **Central Valley Project**—expands the purpose of the California Central Valley Project, which distributes 20 percent of the state's developed water supply, from primarily irrigation to include municipal use and the restoration and protection of fish and wildlife habitat.

• **Central Utah Project**—authorizes $992 million for one of the last giant federal water projects under construction.

• **Glen Canyon Dam**—protects the Grand Canyon from environmental damage caused by fluctuating water releases from Glen Canyon Dam.

• **San Gabriel Valley**—authorizes the federal portion of funding for design, planning, and construction of a multipurpose facility to improve water quality and store water in the San Gabriel Valley groundwater basin.

• **Buffalo Bill Dam**—authorizes an $80-million expansion of the Buffalo Bill Dam and Reservoir in Wyoming, including construction of recreation facilities.

• **Salton Sea Area**—authorizes a $10-million saline water project in the Salton Sea area of California.

• **Mono Lake Basin**—initiates a project to develop 120,000 acre-feet/year of reclaimed water in Southern California to offset water diversions from the environmentally sensitive Mono Lake Basin.

• **Western Water Review**—calls for a comprehensive federal review of western water resource problems and of programs administered by the U.S. Geological Survey in the Department of the Interior

Monitoring Water Quality

To provide water quality information and to coordinate the federal, state, local, and private entities involved, the federal government has formed an Intergovernmental Task Force on Monitoring Water Quality (ITFM). Eight federal and eight state agencies joined to form the 3-year task force, which is chaired by the Environmental Protection Agency and cochaired by the U.S. Geologic Survey. Four task groups are addressing the following topics:

- A nationwide institutional framework to achieve more efficient and effective water resource monitoring

- Environmental indicators

- Comparable data collection methods

- Data management and more effective information sharing

The task force is recommending a nationwide coordinated monitoring strategy with national overviews and regional implementation. The strategy will link individual efforts into a comprehensive nationwide effort to support effective decisionmaking.

Geographic Targeting and Watershed Management

Geographic targeting focuses not on political boundaries, such as federal regions, states, counties, or municipalities, but on ecological units—estuaries, river basins, corridors, or the critical habitat of a species or group of species. Watershed management is an example of this approach.

In April 1991 EPA identified watershed management as a tool for dealing with problems affecting watersheds selected for special attention. Watershed management honors the interconnectedness of aquatic systems by managing human impact on those systems in an integrated fashion. The objective is to align existing water quality management programs to complement and strengthen the efforts of others, including other federal agencies, state and local governments, and private citizens. This approach shifts the emphasis from particular pollutants or pollution sources to a process that begins with the questions, "What is affecting this watershed? Can the resource and its stressors be managed differently?" In 1992 EPA was involved in watershed projects in the following locations:

- Anacostia River, Maryland, and the District of Columbia
- Bear Creek Floatway, Alabama
- Canaan Valley, West Virginia
- Elkhorn Slough, California
- Merrimack River, Massachusetts, and New Hampshire
- Pequea and Mill Creeks, Pennsylvania
- Tomki River, California

TRANSPORTATION

The mobility of people and goods is essential to national wellbeing and competitiveness in the world marketplace. The vast U.S. transportation system supports the nation's dynamism and freedom. At the same time, transportation consumes much energy and space and produces a large portion of the nation's pollution problems. The transportation corridor and infrastructure can contribute to habitat stress and ecosystem disruption, and transportation accidents involving hazardous cargoes can have major adverse impacts on human health and the environment.

Conditions and Trends

The U.S. transportation network is an ever-changing complex of public and private roads, airports, transit systems, railroads, ports, waterways, and pipelines. In the 1990s the nation asks transportation planners to ensure that the system grows and operates in a manner compatible with the environment.

During this century and especially since the 1950s, the U.S. transportation system has come to be dominated by petroleum-powered automobiles and trucks, especially for personal travel. As part of this shift to motor vehicle travel, other significant changes have occurred. For example, in 1992 the Department of Transportation (DOT) issued a report on commuting, which indicated that from 1980 to 1990, the number of vehicles on the road grew 17.4 percent, almost twice as fast as the 9.7 percent increase in the number of workers. The number of drive-alone commuters increased faster than any other mode of travel, while carpooling declined, and most carpools involved only two people. Commuting by transit declined slightly, and those working at home increased slightly.

Trends in highway travel and fuel consumption diverged in recent years. In 1990 highway travel in the United States totaled 2.2 billion vehicle miles, up 2.4 percent since 1989, while highway use of motor fuel was 131.6 billion gallons, a slight decrease from 1989. The shift is attributed in part to a steady rise in the average fuel economy of automobiles.

The most severe problem affecting America's highways is congestion. The Federal Highway Administration reported that 65 percent of peak-hour travel on urban interstate highways was congested in 1987, up from 49 percent in 1981; on rural highways, that figure rose from 4 percent to 17 percent; and truck traffic alone increased by 44 percent. Traffic congestion, which wastes billions of dollars in lost time, money, and fuel, also is a major factor in causing smog.

The predominant mode in terms of volume for intercity freight is rail, followed by waterways, highway, pipeline, and air. U.S. freight railroads have substantial excess capacity and in the last decade have invested heavily in advanced technology to improve performance. Railroads are more fuel efficient than trucks and barges and, even while carrying more total freight, emit far fewer air pollutants. And while all transportation modes pose serious risks from accidents involving hazardous materials, the accident rate of railroads has declined by about two-thirds and that of trucks by about one-third since 1980.

Oil tanker safety has improved markedly in recent years, possibly because of stricter federal and state rules adopted since the 1989 Exxon Valdez spill. According to a private study conducted for the Department of Energy, tankers in 1991 spilled 55,000 gallons of oil or petroleum products in U.S. waters, the lowest level in 14 years. The study found that no major spills occurred in the first six months of 1992 and that tanker operators were taking increased safety precautions.

Policies and Programs

In 1992 DOT worked to bring the U.S. transportation system into greater harmony with the environment, especially to reduce transportation's contribution to air pollution. Recent legislation, including the 1991 Intermodal Surface Transportation Efficiency Act (ISTEA), the 1990 Clean Air Act Amendments, and the Oil Pollution Act of 1990, contain numerous provisions that will change the way the nation moves people and goods.

Surface Transportation

The 1991 Intermodal Surface Transportation Efficiency Act established new sources of funding for environmental initiatives, increased flexibility in the major funding categories, and set more rigorous planning requirements. While ISTEA will result in large new investments in highway and transit programs, scarcity of land and traffic congestion suggest the need for emphasis on maximizing the efficiency of the system, not just expanding capacity.

Transportation and the Section 404 Program. The Intermodal Surface Transportation Efficiency Act of 1991 (ISTEA) will provide billions of dollars for improvements to the nation's transportation infrastructure. To ensure that these projects receive appropriate, expeditious environmental review, the secretary of Transportation, the EPA administrator, and the assistant secretary of the Army for Civil Works signed a joint memorandum to senior regional managers for each agency. The memo emphasized the need for improved coordination, innovative and cost-effective approaches, and to the extent practical, integration of section 404 alternative analyses with those conducted under the National Environmental Policy Act. It also included 13 specific action items or commitments and provided headquarters contacts to resolve issues that cannot be addressed quickly in the field.

Environmental Initiatives. ISTEA dedicated funds to mitigate congestion, improve air quality, and enhance transportation. In fiscal 1992 the law authorized $809 million for an air quality program and $353 million for transportation enhancement, which includes historic preservation, greenways, bicycle paths, and efforts to protect scenery and water quality. DOT worked closely with CEQ, EPA, and the Department of Justice to set environmental priorities and to select environmental indicators and, in April 1992, issued guidelines for use of the new funding categories.

Funding Flexibility. In the major infrastructure projects of the surface transportation program, ISTEA provided greater spending flexibility, which can allow for an increase in environmental benefits. For example, during fiscal 1992 the act obligated $243 million to mass transit projects from traditional highway revenue sources. This amount was in addition to $3.8 billion available in fiscal 1992 transit funds.

Transportation Planning. In late 1992 DOT drafted funding provisions for transportation planning at the metropolitan and state levels. Integrated transportation, land-use, and environmental planning yields transportation projects that prevent or minimize such adverse environmental impacts as air and water pollution, habitat fragmentation, and disruption of communities.

In related efforts, the DOT Federal Highway Administration (FHWA) and the Environmental Protection Agency (EPA) signed a memorandum of understanding that provides a framework for enhanced coordination in 13 key environmental areas. The Clean Air Project of the National Association of Regional Councils, funded jointly by EPA and DOT, provided technical guidance to metropolitan planning organizations on clean air issues, including guidance on best modeling practices.

DOT also continued efforts to streamline processing of wetlands dredge and fill permits for transportation projects. DOT, the Army Corps of Engineers, and EPA signed a joint memorandum to field staffs aimed at improving coordination on transportation projects involving complex wetlands issues.

Intermodalism. The degree to which the various parts of a transportation system, such as highways, airports, and railroads, work together in a harmonious whole—intermodalism—is one of the factors determining the effectiveness and efficiency of the nation's transportation system. In response to ISTEA, the Department of Transportation formed an Office of Intermodalism to maintain and distribute intermodal transportation data, coordinate federal research on the subject, and oversee grants to states to develop model transportation plans.

Auto Fuel Efficiency. In 1992 the costs and benefits of proposed major increases in corporate average fuel economy (CAFE) standards were debated in the Congress during deliberations on national energy legislation and also in the 1992 presidential campaign. Some lawmakers proposed increasing the CAFE standard for passenger cars from the present 27.5 miles per gallon to 40 miles per gallon by the year 2000; presidential candidate Bill Clinton proposed reaching 45 miles per gallon by 2010.

In response to a request by DOT, the National Academy of Sciences published a report in April 1992 addressing potential fuel economy levels for new cars and light trucks through model year 2006. The study identified several technologies that would lead to improved fuel economy but also noted other factors that should be considered:

- Tighter tailpipe emission standards mandated by the 1990 Clean Air Act Amendments

- Consumer preferences for larger vehicles, especially among senior citizens and farmers

- Potentially increased costs per vehicle

- Impacts on employment and international competitiveness

Bicycling and Walking

Bicycling and walking are two environmentally benign, low-cost, healthy ways to get around. In 1993 DOT will submit a National Bicycling and Walking Study to the Congress, outlining how federal, state, and local governments, private organizations, and individuals can work together to increase bicycling and walking. The report includes new data on bicycling and walking and examines such issues as safety and health benefits.

Railroads

The efficiency of rail transportation suggests that it may gain increased prominence in the transportation debates of the 1990s. In recent years, in addition to passage of ISTEA, the primary issues affecting railroads and the environment have involved advanced technologies and wetlands protection.

High Speed Rail. Magnetically levitated (maglev) trains are an energy-efficient, high-speed mode of ground transportation. In 1990 the DOT Federal Railroad Administration (FRA) formed the interagency National Maglev Initiative in cooperation with the U.S. Army Corps of Engineers, the Department of Energy, and EPA. The National Maglev Initiative is working with the private sector and state governments to assess the potential of these high speed trains. The goal is to improve intercity transportation in the twenty-first century through development of a commercially viable maglev system. Further research will determine the technological, economic, environmental, and safety issues associated with maglev, as well as its potential role in the United States.

The first commercial maglev system in the world is being planned in the Orlando, Florida, area. The DOT Federal Highway Administration, DOT Federal Railroad Administration, and the Florida Department of Transportation jointly are preparing an environmental assessment for this project, the first assessment of its kind. FRA and the Texas High Speed Rail Authority are preparing an environmental impact statement on proposed high-speed rail projects between Dallas/Ft. Worth and Houston, and between Dallas/Ft. Worth and San Antonio.

Historic Preservation. Along Amtrak's Northeast Corridor, restored railroad stations in New Haven, Connecticut, and Washington, D.C., received recognition as winners of the 1992 National Historic Preservation Awards for rehabilitation of public buildings. With FRA financial and technical assistance, the two stations were updated to meet current demands while preserving historic elements. The award is administered by the Advisory Council on Historic Preservation.

Wetlands Protection. In 1992 DOT became a partner in Coastal America, an intergovernmental initiative to enhance the coastal environment. In particular, DOT is participating in a project involving the EPA, Department of Commerce, Army Corps of Engineers, and State of Connecticut to restore coastal marshes along the Amtrak railroad line. The project will take advantage of ISTEA provisions that allow federal-aid highway funds to be used to support wetlands conservation efforts. Other states may also use this expanded funding flexibility to establish wetlands conservation banks and wetlands conservation plans. For more information on Coastal America, see the Coasts and Oceans section.

Airport Noise

The Airport Noise and Capacity Act of 1990 set national aviation noise policy and provided for a transition to a quieter aircraft fleet. In September 1991 the DOT Federal Aviation Administration (FAA) issued rules to phase out noisier aircraft by December 31, 1999, using an interim compliance schedule. In 1992 FAA received the first set of annual compliance reports, which indicated that the percentage of the noisier Stage 2 airplanes in the national fleet fell below 50 percent for the first time. Implementation of other regulations required by the Airport Noise Act has also begun.

America's Treeways: Planting Trees Along Highways

As part of President Bush's America the Beautiful tree-planting initiative, a coalition of public and private organizations is planting thousands of trees along U.S. highways. Cosponsors include the DOT Federal Highway Administration, the USDA Forest Service, the DOI Take Pride in America program, the National Tree Trust (a private foundation), the National Association of State Foresters, and state departments of transportation. The program relies on volunteers to plant and care for trees donated by forest product companies.

In 1992 volunteers planted 500,000 trees on highway rights-of-way primarily in Ohio, with Michigan, Texas and Virginia undertaking programs. Trees planted along U.S. highways benefit the environment in several ways:

- Enhance visual beauty and roadside scenery

- Create habitats for wildlife and plants

- Improve air quality

- Reduce soil erosion by controlling water runoff

- Provide windbreaks and shade

Planting trees promotes the spirit of volunteerism across the country, educates children about the environment, and builds partnerships among groups and individuals concerned with the preservation of nature.

POLLUTION PREVENTION

It is the national policy of the United States that pollution should be prevented or reduced at the source. The Pollution Prevention Act of 1990 gives the force of law to the common sense notion that very often, the best, most economically efficient way of reducing the impact of society's waste on the environment is to make less of that waste in the first place. Pollution prevention means changing the way the nation produces and consumes goods and services so that fewer pollutants are generated, and as a result, fewer pollutants are released into the environment.

Conditions and Trends

Federal initiatives, spurred on by the Pollution Prevention Act, have contributed to recent increases in the nationwide acceptance of the principles of pollution prevention.

State Legislation

In addition to federal legislation, 27 states now have pollution prevention laws. Most of these laws, enacted since 1987, target hazardous and solid wastes, and one state, Iowa, includes air emissions. Many follow the example of the federal Pollution Prevention Act and include a waste management hierarchy with the preferred option—pollution prevention or source reduction—at the top, followed next by environmentally sound recycling and finally by waste treatment and disposal. Some of the laws extend technical and financial assistance to waste generators to speed pollution prevention efforts, and others require waste generators to formulate facility-wide pollution prevention plans. Such plans help the generators analyze their waste streams and identify opportunities for pollution prevention.

Toxics Release Inventory

The 1986 Emergency Planning and Community Right-to-Know Act requires many sources of toxic emissions and transfers to report the amount of such releases and transfers publicly. Public concern about toxic chemicals has created a parallel incentive for companies to cut emissions and transfers voluntarily. Pursuant to the Community Right-to-Know Act, the Environmental Protection Agency (EPA) maintains a Toxics Release Inventory (TRI).

In 1990 releases and transfers of toxic chemicals measured by TRI showed an 11-percent decrease (600 million pounds) to 4.8 billions pounds of toxic chemicals released into the environment or transferred. Releases into the air totaled 2.2 billion pounds, a 14-percent decrease from the 1989 level of 2.5 billion pounds. Releases to land decreased from 454 million pounds in 1989 to 440 million pounds in 1990. Releases into lakes, rivers, and other bodies of water increased by 2 percent, or 4 million pounds.

Municipal Solid Waste

Americans generated 4.3 pounds of municipal solid waste per person per day in 1990, an 8-percent increase over the 1988 level. EPA estimates that the amount of waste generated will increase through the end of the decade but at a slower rate, reaching 4.5 pounds per person per day by the year 2000. Efforts to promote source reduction, production of more durable products, and backyard composting could help to further slow or even reverse this trend.

Policies and Programs

The federal government promotes pollution prevention through regulatory and enforcement actions, cooperation with state and local governments and private sector groups, pollution prevention grants, and encouragement of voluntary, private efforts.

Regulation and Enforcement

Federal agencies, such as the Council on Environmental Quality (CEQ), EPA, and the Federal Trade Commission (FTC) are shaping regulatory and enforcement actions to promote pollution prevention.

National Environmental Policy Act. NEPA includes among its objectives promotion of efforts to prevent or eliminate damage to the environment. Accordingly, CEQ provides guidance to federal agencies on incorporating pollution prevention into the design and execution of projects. Federal pollution prevention efforts, where appropriate, are a part of Environmental Impact Statements (EIS) and Environmental Assessments (EA). Whether federal agencies manage public lands, issue regulations, or purchase office supplies, the CEQ guidance helps ensure that federal actions and federal funds serve as a catalyst for national efforts to reduce pollution at its source.

Supplemental Environmental Projects. EPA encourages pollution prevention through the inclusion of Supplemental Environmental Projects (SEPs) in settlement agreements resolving violations of environmental statutes. As part of such an agreement, a defendant conducts pollution prevention or other projects that reduce risks to human health and the environment, beyond what is required by law. To date, EPA has negotiated 164 SEPs in settlements under the Emergency Planning and Community Right-to-Know Act, Resource Conservation and Recovery Act (RCRA), Clean Water Act (CWA), Clean Air Act (CAA), and Toxic Substances Control Act (TSCA). Facilities generally commit to spending $1-5 for every penalty dollar reduced as part of the settlement. Annual reductions in toxic chemicals used or released by defendants have ranged from 8,500 pounds to 1.5 million pounds per project. The goal of this EPA program is to encourage SEPs that eliminate or significantly reduce pollutants or decrease natural resources usage, while showing little or no increase in net costs to the offending facility.

Source Reduction Review. The Pollution Prevention Act requires EPA to review its own regulations to determine their potential for source reduction. EPA established the Source Reduction Review Program to ensure early review of key regulations mandated by CAA, CWA, and RCRA. In developing industrial standards, EPA is considering the use of source reduction technologies, along with market-based approaches and command-and-control options, as the basis for regulatory standards.
Government-Business Cooperation

The federal government cooperates with private businesses on pollution prevention through a number of initiatives.

President's Commission on Environmental Quality. In cooperation with the President's commission (PCEQ), 11 major corporations have undertaken voluntary demonstration projects at 12 facilities to test the integration of total quality management (TQM) and pollution prevention. Using quality improvement methods to eliminate inefficiencies that waste resources and cause pollution, the project seeks to cut emissions at targeted facilities and save money at the same time. PCEQ is preparing a report, Total Quality Management: A Framework for Pollution Prevention, to present quality environmental management (QEM) incentives and barriers, management processes, and measures of success to guide the development of new QEM programs.

Integrated Resource Planning. Under the President's National Energy Strategy, the Department of Energy (DOE) supports Integrated Resource Planning (IRP) by state and federal agencies as well as by individual utility companies. Expanding the traditional electric power planning process that typically considers only utility-gene-rated energy options for meeting electricity demand, IRP adds conservation, load management, and independent power purchases. By considering alternative sources of energy services, utility companies and utility regulators can avoid the pollution that would result from only building more powerplants and generating more electricity. By the year 2000 DOE estimates that 40 states will have implemented IRP programs, and planning decisions for 90 percent of the nation's generating capacity will be made with pollution prevention in mind.

Green Lights. EPA sponsors this voluntary program to encourage corporations and state and local governments to install energy efficient lighting in their facilities. Participants reduce energy consumption by replacing existing lighting with technology that delivers the same or better quality lighting while saving electricity, preventing pollution, and saving money on electricity bills. By October 1992 a total of 651 facilities had committed 2.9 billion square feet of office space to the program—more office space than exists in New York City, Los Angeles, Chicago, Houston, Dallas, San Francisco, and Washington D.C., combined. Once enrolled, participants survey their facilities to see where lighting efficiency can be improved and commit to making 90 percent of the identified cost-effective lighting upgrades within five years.

In 1992 Green Lights resulted in an average reduction in lighting electricity demand of 53 percent, total energy savings of 100.5 million kilowatt hours, and a $6.8 million reduction in energy bills. These upgrades achieved annual emission reductions of 133.5 million pounds of carbon dioxide, 1.2 million pounds of sulfur dioxide, and 482,000 pounds of nitrogen oxides. When fully implemented, the project has the potential to prevent several thousand times these quantities of air emissions. Green Lights maintains a Hotline number, (202) 775-6650, to provide information on the program.

Energy Star. A major obstacle to marketing energy efficient equipment is the lack of reliable information on whether the equipment will perform as promised. A new EPA product-labeling program identifies proven energy efficient products with a distinctive Energy Star logo. Consumers will find the Energy Star seal of approval on commercial and residential products such as computers, showerheads, and room air conditioners beginning in the summer of 1993.

Golden Carrot. EPA and a consortium of electric utility companies are working on an initiative called the Golden Carrot to overcome another barrier to developing the energy efficiency market—the time lag between development and commercialization of energy efficient products. Participating utilities have established a $30 million prize for the first company to market a nonCFC refrigerator that is at least 25 percent more energy efficient than 1993 standards. By the year 2000 such refrigerators could reduce electricity consumption by 3 billion kilowatt hours per year, saving consumers $240 million on their annual electric bills. The Golden Carrot competition could help the nation implement the U.N. Framework Convention on Climate Change by preventing 600,000 metric tons of carbon emissions each year. It also could help meet U.S. obligations under the Montreal Protocol to phase out CFC production.

Future Golden Carrots could bring to the market high efficiency clothes washers and dryers, water and space heaters, and central air conditioning systems. Golden Carrot also is inspiring other nations to promote energy efficiency. In Thailand, public utilities have developed a Golden Mango program to bring high-efficiency refrigerators to its market.

NICE[3]. Short for National Industrial Competitiveness through Efficiency: Energy, Environment, and Economics, NICE[3] is a state grant program jointly administered by

EPA, and the departments of Commerce and Energy. Grants support innovative technologies in waste reduction and energy efficiency that can improve industrial competitiveness. In 1991 the following grants helped save energy, reduce hazardous wastes, and generate cost savings.

• **Methanol Recovery.** A grant helped the Peroxygen Chemical Division of FMC Corporation in Texas develop a method to use steam distillation to recover spent methanol from its hydrogen peroxide purification process, thus reducing the facility's energy consumption by 300 billion BTU, its hazardous waste generation by 14.6 million pounds, and its operating costs by $11.9 million per year.

• **Reclaimed Cleaning Water.** A grant helped a PPG Industries facility in Ohio design and install a combined ultrafiltration/reverse osmosis unit to reclaim cleaning water used in the production of water-based primers. The unit will save the company $205,000 per year while reducing hazardous wastes by 3.1 million pounds and cutting energy consumption by 190 billion BTU.

• **Reduced Emissions of Volatile Organic Compounds.** A grant to the Niagara Mohawk Power Corporation and Carrier Corporation in New York helped demonstrate how reducing the emission of volatile organic compounds can lead to energy efficiency improvements. If adopted nationwide, the program has the potential to save 30 trillion BTU and avoid the generation of 126 million pounds of hazardous waste each year.

• **33/50 Program.** The EPA 33/50 program derives its name from its goals—a reduction of 33 percent by 1992 and of 50 percent by 1995 in environmental releases and off-site transfers of 17 priority pollutants. The baseline year for reductions is 1988, when releases and transfers of these chemicals totaled 1.4 billion pounds, as measured by the Toxics Release Inventory (TRI). By October 1992 a total of 937 companies had volunteered to reduce their releases and transfers by 347 million pounds—25 percent below the 1988 baseline. See following page.

• **National Science Foundation Research Grants.** In 1992 NSF and the Council for Chemical Research, Inc., initiated a program to fund research aimed at reducing manufacturing pollution at its source by designing new chemical synthesis and processing methods. In 1992 the program funded ten joint university-industry research programs and anticipates rapid expansion in 1993. Research projects seek feedstock substitutions, alternative synthetic and separation procedures, better catalysts, and process modifications to reduce resource and energy consumption and minimize the production of byproducts and wastes. Industrial participation in the design and evaluation of the projects ensures the applicability of research to real-world manufacturing problems and speeds technology transfer from the laboratory to the manufacturing plant.

• **Dye Industry Initiative.** EPA and the Ecological and Toxicological Association of Dyestuffs Manufacturers (ETAD) have developed a pollution prevention program for the dye industry. In October 1991 a task force, with representatives from ETAD, individual companies, and EPA, issued a pollution prevention manual for the dye industry. The manual identifies opportunities for pollution prevention and recycling for all operations related to dye manufacturing, from purchasing materials to packaging the product. To monitor progress, ETAD is using a survey that collects information on waste reduction activities in the industry.

• **Design for the Environment.** In 1992 EPA launched Design for the Environment (DFE) to apply pollution prevention principles to the design of chemical processes and products. By disseminating information on the comparative risk and performance of chemicals, DFE helps industries make informed, environmentally responsible design choices. The program includes the following components:

• **Chemical Design Grants Program.** DFE is funding research at six universities to minimize or eliminate toxic substances in the synthesis of chemicals.

• **National Center for Pollution Prevention.** This consortium, located at the University of Michigan, is developing academic curricula on chemical design for use in business, engineering, and natural resources programs.

• **Demonstration Projects.** In the Printing Industry Pilot Project, EPA is bringing together printers, printing trade associations, and chemical and equipment suppliers to investigate alternative chemicals and processes to prevent pollution.

• **Insurance Incentive.** This pilot project is designed to work with the insurance industry to investigate whether rates can be adjusted to reflect the reductions in environmental liability associated with pollution prevention projects.

• **Full Cost Accounting.** As part of the program, EPA is preparing a workshop on how companies can account more accurately for the costs of their polluting activities.

• **Recycled Products Trade Fair.** In June 1992 the Council on Federal Recycling and Procurement Policy hosted a trade fair to increase the use of recycled-content products by the federal government. The theme, "Closing the Loop," stressed that unless materials recovered from the waste stream are turned into products that someone can use, they have not really been recycled. The fair brought together the makers of recycled-content products with one of the country's largest consumers—the federal government. Recycled content product manufacturers demonstrated the availability and quality of their products, while government officials explained the process of selling those products to the government.

Department of Defense Purchasing. As major purchasers and users of hazardous materials, the armed services have the potential to reduce pollution in their operations. In 1992 the Army, Air Force, Navy, and Marines recorded pollution prevention accomplishments in systems acquisition, material substitution, process improvement, and improved material management. For more information, see the Department of Defense section.

• **Postal Service Initiatives.** The pollution prevention program of the U.S. Postal Service includes source reduction and recycling. For a discussion of recycling efforts, see the Federal Facilities Management section.

• **Source Reduction.** Efforts include the replacement of wooden pallets by longer-life, more cost-effective plastic pallets. The cost per trip for plastic pallets is about half that of the wooden models. Changing to a thinner, yet equally strong cardboard for Express Mail packaging has reduced bulk by 15 percent, while also reducing shipping costs, warehousing, and post-consumer costs of disposal. A lighter, more economical Express Mail envelope, with no loss in service, security, or customer confidence, saves 170 tons per year—a 15-percent reduction in bulk. A change to Priority Mail flat-rate envelopes made of a lighter weight, 100-percent recycled (25 percent post-consumer waste) stock saves 251 tons of cardboard annually.

• **Third Class Mail Study.** A 1990 study indicated that 8.2 percent of third-class mail was undeliverable and thrown away. In response, the Postal Service has worked with mailers to improve the quality of their address lists through National Change of Address (NCOA), Address Change Service (ACS), Operation MAIL, and Address List Correction. As a result, bulk business mail has been reduced by 1.4 billion pieces in the last two years.

• **Global Mailstream Study.** The Universal Postal Union accepted a proposal by the U.S. Postal Service to study how postal services worldwide can be leaders in preventing pollution. A workgroup, composed of postal administrations from Sweden, Japan, China, the United States, and the former Soviet Union will survey 175 postal administrations on ways to prevent pollution from materials entering the global mailstream.

Voluntary Initiatives

Through awards programs, the federal government recognizes outstanding voluntary initiatives to prevent pollution. In May 1992 the EPA Administrator's Awards recognized 17 organizations for outstanding achievements in pollution prevention. Winners were selected from 840 applicants, and awards were made in the following categories: environmental, nonprofit, and community organization; small and large business; local governments; state government; federal government; and educational institution. Recipients included the following:

• A Chemical Hazards Prevention Program by Inform, Inc.
• Green Star Program by the Alaska Center for the Environment

- A Volatile Organic Compound Reduction Program by Mead Packaging
- A new single-coat paint developed by the Navy
- A pollution prevention planning model developed by the Air Force

Guidelines for Environmental Marketing

Consumers can help prevent pollution by considering the environmental impacts of the products they purchase. In July 1992 the Federal Trade Commission, with input from EPA, issued the following national guidelines to address application of section 5 of the Federal Trade Commission Act to environmental advertising and marketing practices:

- Environmental claims must be substantiated by reliable scientific evidence

- Qualifications and disclosures should be sufficiently clear and prominent on a product to prevent deception

- Environmental claims should make clear whether they apply to the product the package, or a component of either

- Environmental claims should not overstate the environmental attribute or benefit

- A claim comparing the environmental attributes of one product with those of another should make the bases for the comparison clear and should be substantiated.

The guidelines cover the following environmental claims being made by advertisers for products in the U.S. marketplace:

General Environmental Benefit. Unless substantiated, claims of general benefit, such as "environmentally friendly" or "environmentally safe," should be avoided.

Degradable, Biodegradable, and Photodegradable. Claims should be substantiated by evidence that the product will completely decompose into elements found in nature within a reasonable time after consumers dispose of it in the customary way.

Compostable. Claims should be substantiated by evidence that all materials in the product or package will break down into usable compost, such as soil-conditioning material and mulch, in an appropriate composting program or facility or in a home compost pile or device.

Recyclable. A product or package should not be marketed as recyclable unless it can be collected, separated, or recovered from the solid waste stream for use in the form of raw materials in the manufacture of a new product or package.

Recycled Content. Materials should have been recovered or diverted from the solid waste stream, either during manufacture or after consumer use.

Source Reduction. Claims that a product or package has been reduced or is lower in weight, volume, or toxicity should be qualified to avoid deception about the amount of reduction and the basis for any comparison asserted.

Refillable. Claims should provide for collection and return of the package for refill or for later refill of the package by consumers with a product subsequently sold in another package.

Ozone Safe and Ozone Friendly. A product should not be adver- tised as ozone safe, ozone friendly, or as not containing CFCs, if the product contains any ozone-depleting chemical.

While the guidelines themselves do not have the force of law, failure on the part of companies to follow them can result in the Federal Trade Commission bringing suit against offending companies. Between July 1991 and July 1992, FTC announced consent agreements in eight such enforcement actions. In three cases, products containing CFCs or other ozone-depleting chemicals claimed to be ozone-friendly. Four cases involved unsubstantiated claims of biodegradability of plastic trash bags and disposable diapers. The eighth case dealt with a manufacturer's improper claim that its product was pesticide-free. The settlements in these cases prevent the companies from using misleading claims in the future.

The Commission is seeking public comments on whether to modify the guidelines at the end of a 3-year period. In the meantime, interested parties may petition FTC to amend the guidelines.

BIODIVERSITY

Biodiversity has been defined as the variety and variability of life—the diversity of genes, species, and ecosystems—but it is the interactions between these living components that are of primary concern. Diversity is essential for full ecological functioning. For example, genetic diversity better enables a species to survive changing conditions. Variation among species is the key to such basic ecological concepts as the food web. The rich variety of plant and animal species that exists on Earth is ultimately the source of much of the world's food, clothing, and shelter. Biological systems continually recycle air, water, and land. Humanity disrupts biodiversity at the risk of limiting its options and quality of life both now and in the future.

Conditions and Trends

Along with other nations, the United States lacks full and accurate information on the status and distribution of biodiversity components and the functioning of ecosystems. Most information available focuses on particular species, but a recognition is emerging regarding the value of data on ecosystems, communities, habitat, and genetic diversity. Even for species of current concern, such as those that are declining or near extinction or have special value for commerce or recreation, information is incomplete, and for thousands of species and plant and animal communities, existing status and trend information is anecdotal and sketchy. Although numerous sources of information are available, the following elements are lacking:

- A coordinated biodiversity data reporting or assessment effort

- An overall scientific consensus on which statistics are most useful as biodiversity indicators and which procedures are most useful in comparing and managing data

The diverse global gene pool possesses the potential for vast economic value for humanity. Food products from wild species include fisheries, which in 1989 provided 100 million tons of food worldwide. Domesticated species account for 12-32 percent of gross domestic product (GDP) in countries around the world. In addition to providing timber, natural fibers, gums, and oils, nature is a major source of medicinal products for traditional medicine practiced by 80 percent of the people in developing countries. Diverse biota also fuel the development of modern pharmaceuticals. One quarter of all prescriptions dispensed in the United States contain active ingredients extracted from plants, and 3,000 antibiotics, including penicillin and tetracycline, are derived from microorganisms. Biodiversity also attracts recreation and tourism to unique sites throughout the nation and the world.

Policies and Programs

In 1992 federal agencies, the Congress, the states, and private organizations joined in efforts to sustain domestic and global biodiversity. A sampling of policies and programs follows.

National Biodiversity Center

In June 1992 the Bush administration announced its intent to establish a domestic center for biodiversity information. Interagency efforts are underway to develop such a center, whose basic functions would include the following:

Improving Access to Information. The center would serve as a mechanism to locate and provide access to biodiversity information in public and private institutions.

Assessing Biodiversity Information. The center would examine the current state of scientific information on biodiversity to identify gaps.

Improving and Standardizing Methodologies. The center would promote standard methods for data collection, reporting, and management, thus improving the comparability and utility of information.

Convention on Biological Diversity

In 1992 two years of intensive international negotiations concluded with a convention on biological diversity signed by 153 nations at the United Nations Conference on Environment and Development (UNCED), held in Rio de Janeiro, Brazil. The convention requires measures to identify biological resources in decline, reasons for such declines, and on-site and off-site measures to address the declines.

Although the United States participated actively in negotiations and hoped to be a signatory, the convention included several provisions, unrelated to the conservation sections of the treaty, that were unacceptable to the United States. These provisions concerned intellectual property rights, access to genetic information used in biotechnology, and international funding for biodiversity activities. As a result, the United States declined to sign the treaty in its present form. At the Rio meeting, however, President Bush stated that U.S. efforts to conserve domestic biodiversity will exceed the requirements of the treaty.

The United States conducts several programs for worldwide biodiversity research and conservation through the U.S. Agency for International Development (USAID) and the international forestry programs of the USDA Forest Service. The Biodiversity Support Program, a joint venture by USAID, the World Wildlife Fund, The Nature Conservancy, and the World Resources Institute, provides expertise on biodiversity around the world. A USAID/National Science Foundation collaborative program supports research and education in developing countries on potential threats to ecosystems and species.

Inventory and Research Proposal

At UNCED, the United States offered a proposal to foster international scientific cooperation on biodiversity surveys, inventories, and data management, with the goal of increasing the availability, comparability, and quality of information on biological diversity. The United States plans to host a meeting in early 1993 of international experts to consider topics such as data-gathering and reporting standards, approaches to conducting inventories and surveys, data management techniques, and institutional issues related to national biodiversity surveys.

Gap Analysis

Fish and Wildlife Service researchers use satellite imagery, habitat data, and geographic information systems to compare the distribution of species and plant/animal communities with existing land use and land management patterns. From this analysis, they can begin to assess current protection and identify areas in need of greater protection. In 1992 federal, state, and private organizations worked on gap analysis projects in 22 states. Nationwide coverage is planned.

Endangered Species Protection

Biodiversity conservation includes efforts to preserve the diversity of species and the ecosystems on which they depend. Foremost among such U.S. efforts is the Endangered Species Act (ESA), whose initial stated purpose is "to provide a means whereby the ecosystems upon which endangered species and threatened species depend may be conserved." Despite this broad mandate, most implementation efforts focus on protecting individual species in danger of extinction (endangered) or likely to become so in the foreseeable future (threatened).

The federal agencies that implement ESA—Fish and Wildlife Service (FWS) for terrestrial and some aquatic species and National Marine Fisheries Service (NMFS) for marine species and anadromous fisheries—have begun to increase the use of regional and multi-species, habitat-based approaches. For example, FWS emphasizes additions to threatened and endangered species lists that cover multiple species in a given region, so that conservation efforts can address their various needs more efficiently. FWS also supports multi-species recovery actions.

In a landmark decision to accelerate ESA implementation, FWS in December 1992 reached a legal settlement with environmental groups by agreeing to propose for listing within four years the 400 highest-ranked candidate species. In addition the service consented to expedite final consideration of 900 species that it believes may be worthy of protection but for which it does not have definitive scientific information. This settlement will create additional momentum for multi-species, habitat-based conservation approaches.

ESA allows the taking of listed species—defining "take" to include "harass, harm, pursue, hunt, shoot, wound, kill, trap, capture, or collect, or attempt to engage"—if the activity is incidental to an otherwise legal activity and after approval of a habitat conservation plan demonstrating that the taking will not appreciably reduce the likelihood of survival or recovery. Such plans often cover a significant portion of the habitat for a species and thus have the potential to provide protection for a range of species and plant/animal communities. As of October 1992 FWS and NMFS had issued 20 incidental-take permits and 17 permit amendments. FWS staff had provided pre-application technical assistance on 52 habitat conservation plans.

In 1991 the BLM director invoked a section of ESA that allows a cabinet-level committee to authorize a project that would jeopardize a species. This group, known as the Endangered Species Committee, convened in late 1991 and in May 1992 reached a decision to allow some of a number of disputed timber sales to proceed. (For more information, see the Forestry section.)

Natural Community Conservation Planning

In 1992 the State of California continued an innovative program of Natural Community Conservation Planning (NCCP) to anticipate and prevent controversies resulting from listing of species under federal or state endangered species laws. The goal is to protect entire ecosystems while allowing compatible economic development. The program works through voluntary, permanent, enforceable conservation strategies that address specific habitats and species, setting out protected areas and areas appropriate for development.

A pilot program is underway to address coastal sage scrub, a habitat type supporting several potentially endangered or threatened species, including the California gnatcatcher. Private landowners have agreed to forestall development on 141,000 acres, some of which contains coastal sage scrub, for an 18-month period, while final conservation plans are developed. During this period, public land-use and regulatory agencies will increase their discretionary review of projects on 1.1 million acres, containing 210,000 acres or 53 percent of the identified sage scrub habitat. Development of the conservation plans involves landowners, local governments, and state and federal wildlife agencies.

California Bioregions

The State of California, in cooperation with federal agencies, began a ground-breaking effort in 1991 to reorient existing environmental and natural resource management efforts around natural ecological regions or bioregions. The first major activity of the effort was a report released in 1992 on the Sierra bioregion. The Sierra Report, based on a broad public outreach effort, contained 18 recommendations for increasing public understanding of the effects of human activities on the bioregion. The effort involved local communities in decisionmaking and recognized the linkage between environmental and economic concerns.

Every Species Counts

The USDA Forest Service has adopted landscape and ecosystem approaches to managing habitat for 230 threatened or endangered species. In 1992 the agency surveyed 1 million acres for threatened, endangered, and sensitive species (a Forest Service designation). The agency is preparing species management guides for the 50 states, in conjunction with such groups as The Nature Conservancy, which assist with inventory, data management, and development of conservation strategies. Examples of activities during 1992 follow.

- Release of two captive-bred California condors and two Andean condors into historic condor habitat in California

- Placement of 450 nest boxes for the hurricane-ravaged red-cockaded woodpecker in South Carolina national forests, which contributed to a 9-percent increase in woodpecker populations

- Careful use of fire to restore specialized forest ecosystems, such as southern pine stands that host red-cockaded woodpeckers, tallgrass prairie in Illinois, and mountain golden heather in North Carolina

- Development in the Southwest of management guidelines for the northern goshawk, adopting an ecosystem approach that benefits other species as well.

Bring Back the Natives!

Non-indigenous (exotic) species have invaded or been introduced to many areas of the country, where they often disturb natural ecosystems. In 1992 the Forest Service, BLM, and the National Fish and Wildlife Foundation (a Congressionally chartered nonprofit organization) cooperated on 18 Bring Back the Natives projects. This national effort attempts to restore the health of entire stream systems and their native aquatic species, including fish, mollusks, and plants. On public lands, restoration of entire watersheds can make a critical difference in the recovery and conservation of vulnerable species.

Pacific Salmon and Steelhead Conservation

In 1992 the USDA Forest Service initiated a strategy to provide an ecological approach to managing salmon and steelhead habitat in western national forests. The strategy will identify and implement measures that contribute to the sustained natural production of salmon and steelhead stocks, possibly precluding the need for listing under the Endangered Species Act. The Forest Service is coordinating development and implementation of the strategy with other fish management agencies and conservation groups.

Biodiversity and NEPA

In 1992 CEQ, with support from EPA and the departments of the Interior, Agriculture, Defense, and Transportation, completed a series of conferences on ways to incorporate biodiversity and ecosystem management into analyses prepared under the National Environmental Policy Act (NEPA).

Biodiversity on Private Lands

Since the majority of lands in the United States are in private ownership, any national strategy to conserve biodiversity must find ways to encourage conservation actions by private landowners. In 1992 the President's Commission on Environmental Quality (PCEQ) undertook an initiative to develop best practices for integrating biodiversity considerations into private land management decisionmaking. The initiative involves case studies of 16 demonstration projects proposed by private firms, ranging from 100 acres on a suburban Philadelphia corporate campus to 280,000 acres of a forested and mixed-use watershed on the coast of Washington state. A PCEQ report will present case studies and best-practice recommendations along with outreach materials that offer guidance for private landowners.

International Trade in Endangered Species

The eighth conference of the parties to the Convention on International Trade in Endangered Species (CITES) took place in Japan in March 1992. Delegates agreed to continue the African elephant's listing on Appendix I, which bans all trade in ivory. They discussed the listing of commodity species such as blue fin tuna and several tropical timber species, including varieties of mahogany. Although blue fin tuna was not listed, several varieties of mahogany were, and delegates agreed to reevaluate the scientific criteria for CITES listing.

In 1992 the Congress, with the support of the Bush administration, enacted new legislation to facilitate U.S. implementation of decisions and resolutions under CITES with regard to exotic birds. The new legislation authorizes regulation of imports of exotic wild birds, whose populations may be adversely affected by trade. It also authorizes the Secretary of the Interior to support wild bird management efforts in other nations.

Caribbean Regional Treaty

In 1992 the Bush administration sent to the Senate for ratification the Protocol Concerning Specially Protected Areas and Wildlife. This protocol builds upon general obligations outlined in the Convention for the Protection and Development of the Marine Environment of the Wider Caribbean Region, which calls upon signatories to establish protected areas to conserve rare or fragile ecosystems as well as habitats for threatened and endangered species.

Neotropical Migratory Birds

Partners in Flight/Aves de las Americas is a cooperative program that involves state, federal, and international agencies and conservation organizations. It is the first integrated program for research, monitoring, and habitat management for migratory birds—neotropical migrants—that breed in North America and migrate to Latin and South America during the northern winter. In 1992 the program sponsored a national workshop on the status of such birds as the indigo bunting, scarlet tanager, and many species of warblers and the effect of land-use practices on their long-term survival. Initiatives focused on bird population management, habitat monitoring, habitat improvement, training, and public education.

Marine Sanctuaries

The NOAA National Marine Sanctuary Program aims to conserve and protect habitats and associated biological communities necessary to sustain living marine resources. It also provides opportunities for research. In 1992 Monterey Bay off the central California coast and the Florida Garden Banks off the Texas-Louisiana border were added to the National Marine Sanctuary system. These additions brought the number of marine sanctuaries to 13 and more than doubled the size of the system.

Estuarine Research Reserves

The goal of the National Estuarine Research Reserves Program is to establish and manage a system of estuarine research areas representative of the various U.S. biogeographical regions and estuarine types. In 1992 the Ashepoo-Combahee-Edisto (ACE) Basin and North Inlet/Winyah Bay reserves were designated in South Carolina, bringing the total number of reserves to 21.

Research Natural Areas

Research natural areas (RNA) are areas of forest, range, aquatic, geologic, alpine, shrubland, and grassland vegetation set aside for nonmanipulative research, study, monitoring, and education. In 1992 the Forest Service reviewed 300 sites for addition to the system. Established in 1927, the RNA program now encompasses 290 sites and 250,000 acres.

Biosphere Reserves

Biosphere reserves provide a framework for regional and international cooperation in biodiversity conservation and research. The U.N./U.S. Man and the Biosphere Program (MAB) has initiated interdisciplinary research on ecosystem sustainability in biosphere reserves in New Jersey, Virginia, and Florida. In 1992 UNESCO designated The Land Between the Lakes, a demonstration area for integrated management in Kentucky and Tennessee administered by the Tennessee Valley Authority, as the forty-seventh U.S. biosphere reserve. A Biosphere Reserves Integrated Monitoring Program (BRIM), launched under the auspices of EuroMAB, is promoting cooperation among European and North American biosphere reserves. Among the first BRIM efforts is documentation of biodiversity in the reserves.

National Park Service

NPS has launched a national program to promote public awareness of biodiversity management issues and the value of biodiversity to human societies. In 1992 efforts included a comprehensive manual for interpreters and a range of media used in educational programs at many national parks and monuments.

Forest Service

The Forest Service Natural Resource Conservation Education program supports lifelong learning about natural resources and ecosystems, their interrelationships, conservation, use, management, and value to society. In 1992, the program's first full year, the Forest Service committed $2.5 million to 150 education projects, including the following: development of curricula for elementary and secondary schools; use of national forests and grasslands as living classrooms; support for individual teachers; events promoting environmental awareness; and development of local and statewide strategic plans for conservation education. Through partnerships with other federal, state, and local agencies, schools, and private industry, the Forest Service was able to double funding for biodiversity education.

Bureau of Land Management

BLM, along with other federal agencies, held the first session of a field training course that applies principles of ecosystem management to field activities to conserve biodiversity.

Biodiversity Symposium

In July 1992 the USDA Forest Service held an international symposium, "Biodiversity in Managed Landscapes: Theory and Practice," in Sacramento, California. Public and private organizations, including a number of federal agencies, were cosponsors. The symposium, attended by 400 people from 12 countries, had the following objectives:

- Document case examples of bio-diversity theory and concepts applied at differing scales, such as local, national, and global

- Examine policies that affect biodiversity conservation

- Emphasize that the conservation of biodiversity requires more than a network of protected areas and preserves

- Provide a forum for dialogue and interchange that will lead to increased understanding and cooperation

U.S. Biodiversity: Trends in Key Species

Biodiversity monitoring data is available for a number of key species. Species data often is collected on an annual basis, whereas changes in ecosystem status generally are not. Recent CEQ annual reports addressed changes in ecosystem diversity, such as loss of tallgrass prairie, old-growth forest, and wetlands. Additional critical biodiversity issues include concerns over the effect of the widespread use of hatchery fish on the genetic diversity of wild fish.

Endangered Species

As of December 1992 the Fish and Wildlife Service (FWS) has classified 755 U.S. species of plants and animals as threatened or endangered. The Threatened and Endangered Species Lists, including foreign species, have a total of 1,284 species, and 3,964 species are candidates. The most recent (1990) assessment of recovery status for listed species revealed 38 percent declining, 10 percent improving, 31 percent stable, 2 percent extinct, and 19 percent of unknown status.

Ducks and Geese

Although estimated numbers of ducks migrating south in 1992 remained stable or showed an increase over recent years, the numbers are below long-term averages. By population estimates, 1992 was the sixth lowest year on record, even though breeding populations of three duck species showed a significant increase over 1991 figures, and populations were relatively stable for ten principal species. Estimates of goose populations, based on less rigorous information than for ducks, indicate that despite recent declines, most goose populations are at or above normal levels. An exception is Atlantic flyway Canada geese, which declined to their lowest level in 23 years.

Raptors

Fall 1991 and spring 1992 surveys of raptors showed large declines in counts of sharp-shinned hawks, one of the most abundant of the eastern raptors. The cause of the decline is unknown. Osprey, Cooper's hawk, merlin, and peregrine falcon—several of which are thought to have been affected by the now-banned pesticide DDT—apparently are recovering.

Pacific Salmon

In late 1991 the American Fisheries Society (AFS) classified 214 native, naturally spawning stocks of Pacific salmon, steelhead, and sea-run cutthroat trout as being at high risk of extinction (101 stocks), at moderate risk of extinction (58), or of special concern (54). One stock was already listed under federal and State of California endangered species laws, and during 1992, three Columbia River stocks were listed. AFS indicated that 18 of the classified stocks may already be extinct, and that 106 other major West Coast salmonid populations have also become extinct.

Linking Ecosystems and Biodiversity

In June 1992 the Bureau of Land Management (BLM) in the Department of the Interior and the USDA Forest Service jointly adopted a philosophy of ecosystem management that includes the following elements:

- Managing lands and waters to restore and sustain biodiversity and productivity of renewable natural resources

- Developing conservation partnerships and improving the participation of citizens in public resource decisionmaking

- Bringing the scientific community into stronger partnerships with land managers

Ecological land management is based on the following principles:

- Protecting or restoring the integrity of soils, air, water, biodiversity, and ecological processes

- Meeting the basic needs of people and communities who depend upon the land for food, fuel, shelter, livelihood, recreation, and spiritual renewal

- Promoting efficiency by ensuring cost-effective production of natural resources

- Working to meet the needs of a variety of interests, regions, and generations in land-use decisionmaking

In 1992 the departments of Agriculture and the Interior issued America's Biodiversity Strategy: Actions to Conserve Species and Habitat, which discusses the framework for managing species and protected areas that has evolved over the past century. Public land managers, scientists, and the public increasingly take an ecosystem approach to specific situations found in the various parts of the country. For example, BLM has developed a proposal for managing 2.5 million acres of federal land in western Oregon that incorporates biodiversity and sustainable development. An environmental impact statement released in August 1992 contains an analysis of the proposal.

FISHERIES AND MARINE MAMMALS

Marine fisheries are an invaluable national asset. Commercial and recreational fishing and associated enterprises contribute $24 billion annually to the U.S. economy. The U.S. Exclusive Economic Zone (EEZ)—at 2 million square miles, the largest of any nation—supports diverse and abundant living marine resources that provide rich economic as well as other quality-of-life benefits.

Conditions and Trends

In 1991 commercial fishermen landed 9.4 billion pounds of fish and shellfish in U.S. waters for a catch worth $3.3 billion at dockside. U.S. exports of edible domestic fishery products totaled a record $3 billion. Alaskan pollack was the leading species in terms of quantity, and shrimp was the most economically valuable species. In the 1991 report, Our Living Oceans: The First Annual Report on the Status of U.S. Living Marine Resources, the National Marine Fisheries Service documented a recent decline in marine species, including fish, shellfish, marine mammals, and sea turtles. The cause is a combination of fishing-related pressures and habitat loss or degradation. Among U.S. fishery stocks, 52.6 percent are fully or over-utilized, and only 13.4 percent are under-utilized. For 34.1 percent of fishery stocks in U.S. waters, too little information is available to make any determination regarding their status.

Low Populations

Populations of many fish with recreational and commercial value are at historically low levels. Populations of New England groundfish, the most valuable component of the region's fishery, are down to 45 percent of their estimated long-term potential. Many Pacific salmon, a symbol of the nation's rich natural heritage, are near extinction. The American Fisheries Society reports that at least 106 stocks of west coast salmonids are extinct and 214 others are at risk. Four stocks are presently listed as threatened or endangered under the Endangered Species Act (ESA). Present numbers of Atlantic bluefin tuna, which can have a value of $94 per pound or $67,000 for a single large fish, are only 10 percent of what they were 20 years ago.

Declines such as these can have a profound effect on local economies and regional ecosystems. For Massachusetts alone, recent all-time low landings of groundfish translate into annual losses of 88 million pounds or $193 million and 8,000 jobs, when compared to the estimated long-term potential of these species. For all of New England, groundfish losses represent 137 million pounds or $350 million and 14,000 jobs.

The major factors contributing to declines in marine species include overharvesting, bycatch waste, habitat degradation, and, in the case of marine mammals, fishery/marine mammal interactions.

Intensive fishing operations affect more than the abundance of targeted stocks. The functioning and balance of the marine ecosystem can be profoundly impacted. Overharvest of New England groundfish has resulted in steady declines in the abundance of the principal groundfish and flounder species. These declines have been accompanied by steady increases in the numbers of skates and dogfish that inhabit the same ecological niche but have no present commercial value.

Removal of large quantities of historically dominant and commercially valuable groundfish has shifted the relative abundance of different species in the ecosystem. The feasibility of rebuilding groundfish and flounder stocks will depend upon the extent to which these species have been displaced by the now dominant skates and dogfish.

Marine Mammals

Of some 100 species of marine mammals found worldwide, 58 inhabit U.S. waters, and of these, 13 are listed as threatened or endangered under ESA. The status of most marine mammal stocks is uncertain because of a lack of information. The U.S. Marine Mammal Commission, established by the Marine Mammal Protection Act of 1972, administers a research program to fill existing information gaps. The commission also reviews federal policies and actions for their impact on marine mammals.

In 1992 the U.S. Marine Mammal Commission reported the following species in most urgent need of protection:

Northern Right Whale. The most endangered marine mammal in U.S. waters, this species is also the world's most endangered large whale. The largest known population, estimated at 350 animals, occurs in coastal waters off the east coasts of Canada and the United States. After being devastated by commercial whaling in the last century, populations now represent 5 percent of estimated pre-commercial exploitation levels. Today, entanglement in fishing gear and collisions with ships may be the principal human-related causes of mortality and injury. In the eastern Pacific Ocean, only seven sightings of right whales have been made in the last 25 years, indicating critically low numbers.

• **Hawaiian Monk Seal.** The most endangered seal in U.S. waters may number as few as 1,500 animals that inhabit the remote northwestern Hawaiian Islands. Interactions with the Hawaiian swordfish industry, which in recent years expanded from 50 to 150 vessels, may be contributing to declines in births and beach counts.

• **West Indian Manatee.** Occurring in the coastal waters and rivers of Florida and Georgia, the long-term survival of this population of 1,800 animals is in doubt. Known deaths between 1988 and 1991 exceeded 550, of which 150 were caused by water craft. Habitat degradation resulting from coastal development may pose the more serious long-term threat to this species.

• **Steller Sea Lion.** Declines in the number of Steller sea lions, throughout their range but particularly in Alaska, led to their being listed in 1990 as threatened under the Endangered Species Act. Fisheries interactions appear to be a contributing factor. Declines of 90 percent have been recorded for major rookeries over the past 30 years, with the greatest losses occurring in the last decade. Likely causes of declines include incidental take by trawl fisheries with over 20,000 sea lions taken between 1966 and 1988, commercial exploitation of sea lion prey species (principally pollack), and shooting by fishermen attempting to protect their gear or catch.

• **California Sea Otter.** Numbering around 1,900 animals, the remaining population of sea otters along the central California coast continues at risk and is listed as threatened under ESA. Incidental takes in gillnets, a contributing factor in the decline, were stopped by state actions that prohibit the use of gillnets in sea otter habitat; numbers appear to be increasing.

Policies and Programs

The federal agencies with major responsibilities for living marine resources, including anadromous fishes—those that travel between fresh and salt water—are the National Marine Fisheries Service (NMFS) and the U.S. Fish and Wildlife Service (FWS). The U.S. Coast Guard shares enforcement responsibilities at sea.

National Marine Fisheries Service

The National Marine Fisheries Service has primary responsibility for stewardship of the nation's living marine resources, including most marine mammals. A part of the National Oceanic and Atmospheric Administration (NOAA) in the Department of Commerce, NMFS oversees the economic welfare of U.S. fisheries as well as the biological health of the marine life and habitat on which these fisheries depend.

Fishery Management Plans. The Magnuson Fishery Conservation and Management Act provides for the conservation and management of all fishery resources within the U.S. Exclusive Economic Zone. The act created eight Regional Fishery Management Councils, whose members represent NMFS, the fishing industry, and affected states. The Fish and Wildlife Service is a nonvoting member of each council, and the Coast Guard advises on enforceability of proposed regulations. The councils are charged with preparing Fishery Management Plans (FMPs) and in 1992 NMFS reported 32 FMPs in place.

Strategic Plan. NMFS has adopted a 5-year strategic plan that reflects a commitment to addressing the problem of declining stocks of marine species. The plan includes the following priority items:

• Rebuild overfished marine fisheries

• Maintain currently productive fisheries

• Advance fishery forecasts and ecosystem models

• Integrate conservation of protected species and fisheries management

• Improve seafood safety

• Protect living marine resources habitats

• Improve the effectiveness of international fisheries relationships

• Reduce impediments to U.S. aquaculture

Controlling Harvest. Until recently, efforts to control total harvests depended on regulations that decrease overall fishing efficiency: trip limits, size limits, gear restrictions, quotas, and seasonal and area closures. NMFS and the management councils are now considering limiting access to some fisheries. Historically, U.S. fisheries have been managed as open-access resources in which anyone may fish. Under this system, as new fisheries opened and fishing technology improved, fishermen received substantial returns, encouraging more fishermen to enter the fishery. As a result, total harvests approached or exceeded the maximum that many fisheries could support, leading to increased competition among fishermen as each received a smaller share of the declining resource. Fishermen responded by investing in more efficient and more costly technology, which compelled them to seek larger catches to remain economically viable.

Under controlled access systems, eligible fishermen would be guaranteed rights to a predetermined percentage of the total allowable catch, thereby eliminating competition among fishermen for the same resource. Schemes range from a simple moratorium on additional entries to a fishery, to an Individual Transferable Quota (ITQ) system. ITQ allows fishermen to own quotas in a fishery and to trade, sell, or buy them. In 1992 NMFS used controlled access to manage seven different U.S. fisheries and proposes its use in four additional ones.

Habitat Restoration. For Pacific Northwest salmon stocks, habitat degradation rather than overharvesting is the primary issue. The American Fisheries Society has identified 101 stocks at high risk of extinction, 58 at moderate risk, and 54 stocks of special concern. Four are listed under ESA as threatened or endangered. These populations have been most seriously impacted by human activities occurring in and along the rivers in which the fish spawn. Hydropower generation, logging, mining, agriculture, and urban growth block the passage of migrating fish, alter or destroy spawning habitats, and degrade water quality.

Restoring degraded habitats is complicated by the number of parties with a direct stake in such efforts. Maintaining adequate river water quantity and quality for anadromous fish requires coordination at the regional level, with the involvement of federal, state, and local governments, water and power authorities, the timber industry, farmers, developers, private landowners, conservation groups, native American tribes, and recreational and commercial fishermen.

Under the Fish and Wildlife Coordination Act, NMFS comments on federal proposals that could affect U.S. waters essential to living marine resources. These include waters within the EEZ, under state jurisdiction, and throughout the migratory and spawning range of anadromous fish. NMFS provides advice and recommendations to agencies on potential adverse affects of proposed actions on fish habitat.

Protected Species. NMFS manages protection programs for marine mammals, with the exception of polar bears, sea otters, manatees, walrus, and dugongs, which come under the jurisdiction of the Fish and Wildlife Service. All marine mammals receive protection under the Marine Mammal Protection Act, and some receive further protection under ESA. In addition, the International Whaling Commission (IWC) oversees the management and conservation of large whales, most of which are listed as threatened or endangered. NMFS and the Marine Mammal Commission provide technical advice to the U.S. Commissioner to the IWC, who is also the NOAA administrator. Under the Endangered Species Act, NMFS also manages six species of sea turtles federally listed as threatened or endangered.

International Aspects

The migrations of commercial species such as yellowfin tuna, anadromous Atlantic and Pacific salmon, and Central Bering Sea pollack, have international implications. The conservation and management of such fish and of many marine mammals require high seas fishery management that neither the United States nor any other nation can address alone.

Atlantic Salmon. At the 1992 annual meeting of the North Atlantic Salmon Conservation Organization (NASCO), the United States announced a long-term policy to curtail commercial fisheries that target migrating Atlantic salmon in the high seas. The United States will continue to recognize the legitimate rights of subsistence users and personal-use fisheries and the fact that interceptions of salmon cannot be entirely eliminated. NASCO approved a Protocol to the Convention on Conservation of Salmon in the North Atlantic, calling on other nations to prohibit the fishing of Atlantic salmon stocks beyond national jurisdictions.

Tuna-Dolphin Conservation. The United States also is participating in international efforts to reduce the incidental take of marine mammals in high seas fisheries. In June 1992 the United States and nine other nations adopted a multilateral Dolphin Conservation Program for the Eastern Tropical Pacific Ocean to reduce the incidental mortality of dolphins in the tuna fishery to the lowest possible level. Over the last three decades, purse-seine tuna fishery operations have killed large numbers of dolphins, where the two species swim together. The agreement establishes a strict schedule to reduce mortality to 5,000 dolphins by 1999, which is one-tenth of 1 percent of the estimated dolphin population in these waters. The conservation program includes a research component to develop techniques for catching yellowfin tuna without harming dolphins and provides detailed procedures to ensure that mortality limits are not exceeded. In October President Bush signed into law the International Dolphin Conservation Act of 1992, amending the Marine Mammal Protection Act to require a dolphin-safe U.S. tuna market by June 1994.

High Seas Driftnet Fishing. This highly destructive fishing technique results in large incidental takes of marine mammals, seabirds, and other forms of life. In keeping with a U.S.-led resolution of the U.N. General Assembly, a moratorium on driftnet fishing went into effect on January 1, 1991.

The High Seas Driftnet Fisheries Enforcement Act of 1992 reflects the U.S. commitment to translate international commitments into domestic action. The Driftnet Act extends the domestic prohibition on high seas driftnet fishing and prohibits the import of all fish and fish products and sport fishing equipment from any country engaged in high seas driftnet fishing beginning January 1, 1993. The law provides for denial of port privileges and navigation in U.S. waters for vessels engaged in the prohibited fishing practice. The act includes measures to establish an international regime for the management of the high seas fishery in the Central Bering Sea.

U.N. Conference on Environment and Development. UNCED addressed high seas fishing issues in *Agenda 21*, the document intended to provide guidance for setting a global course to sustainable development. In the Agenda 21 chapter on oceans, the United States supported language to ensure that populations of marine species are maintained at healthy levels, that populations of marine mammals are protected, and that fishing gear minimizes the incidental catch of nontarget species. Following UNCED, the U.N. General Assembly adopted a resolution endorsing an intergovernmental conference on high seas fishing issues to be held in late 1993. The United States is cooperating with the U.N Food and Agriculture Organization (FAO) and the Organization of Economic Cooperation and Development (OECD) to address conservation as well as trade questions pertaining to high seas fishing.

Whaling Moratorium. The United States has supported an international moratorium on commercial whaling since inception of the ban in 1985 under the auspices of the International Whaling Commission (IWC). Because several nations other than the United States desire an easing of the ban, the commission, at its annual meeting in 1992, moved toward adopting a sustainable, scientifically based procedure for calculating whale quotas. IWC parties agreed that until this procedure is adopted and in force, the moratorium on commercial whaling should continue, a stance supported by the United States.

Sea Turtles. Commercial shrimp-fishing operations in the Caribbean region often result in the incidental capture and drowning of threatened and endangered species of sea turtles. In response, during the past three years, the United States has implemented legislation prohibiting Caribbean shrimp imports unless those countries adopt specific measures to reduce incidental capture of sea turtles.

For example, shrimp fishermen in the United States are required by law to equip their nets with a piece of gear known as a "turtle excluder device" (TED), which reduces the incidental capture of sea turtles by more than 97 percent. TED usage is considered essential for the survival of endangered sea turtle populations in the wider Caribbean. Caribbean nations that do not undertake comparable steps to install TEDs and reduce incidental turtle takes may face shrimp import embargoes. The United States offers technical assistance to help affected countries incorporate TED technology into their shrimp fleets.

Causes of Declines in Marine Species

In waters managed by the United States, recent declines in marine species are attributable to the following causes:

• **Overharvesting.** The removal of fish in such quantities that populations are unable to replace themselves depletes fisheries stocks. Increased domestic fishing operations are a major cause of declines in U.S. fishery stocks. Contributing factors are increased efficiency in fishing technology, overcapitalization of the fishing industry, unlimited access to fisheries, and noncompliance with fishery management regulations.

• **Bycatch Waste.** Inadvertent capture of nontarget and undersized marine species occurs in most fisheries. Fishermen return bycatch to the water but the animals rarely survive. Impacts are significant especially when bycatch would have commercial or recreational value at larger sizes. Shrimp fishermen discard nine pounds of potentially valuable juvenile fish for every pound of shrimp harvested. Fisherman also discard economically valuable species to comply with quotas or other conservation regulations. In New England, 90 percent of yellowtail flounder and cod catches consist of noncommercial sizes lost as potential future catch. Mortality of marine mammals and endangered species such as sea turtles trapped or entangled by fishermen is also a serious bycatch problem.

• **Habitat Degradation.** The loss or degradation of habitat critical to various life stages, particularly reproductive and juvenile stages, of marine life threatens the future viability of U.S. fisheries. Among the nation's valuable commercial and recreational species, 75 percent depend on coastal estuaries where impacts of development and pollution may render habitats inadequate. Fully 90 percent of the

Gulf coast fishery is critically dependent on estuaries and coastal wetlands. Lack of access to spawning habitat and physical alteration of these sites have profound implications for anadromous salmon stocks. Hydroelectric dams, freshwater diversions for irrigation, and habitat degradation from logging and development have led to the listing of four west coast salmon stocks under the Endangered Species Act; several other stocks face a similar fate.

• **Interactions between Fisheries and Marine Mammals/Sea Turtles.** Fishing operations can directly disturb marine mammals and sea turtles by harassing, injuring, or killing them. These animals may become trapped or entangled by fishing gear, leading to injury or death as well as to damaged fishing equipment. In 1989 the National Marine Fisheries Service initiated a fishery-wide observer program to collect information on the incidental take of marine mammals in commercial fisheries. Indirect interactions may occur as marine mammals compete for the same fish and shellfish resources as commercial or recreational fishermen. Along the North Atlantic coast, harbor seals are known to remove bait from lobsterpots and eat pen-raised salmon. In the commercial fisheries for pollack and other groundfish in Alaska, fisheries interactions are a contributing factor in the decline of threatened Steller sea lion populations. In addition to being caught in nets and shot by fishermen, the sea lions may be facing a decline in food supplies in the fisheries.

COASTS AND OCEANS

Coasts and oceans contribute significantly to the nation's economy, and to the quality of life for Americans. The coastal zone deserves protection for a variety of reasons:

- Aesthetic and recreational offerings

- Critical ecological functions such as protection of water quality and protection from flood and storm damage

- Economic activities such as fishing, tourism, and commerce

Conditions and Trends

The vast reaches of the oceans are relatively clean and healthy primarily because of their enormous volume and capacity for dilution. However, the edges of the oceans—the coastal waters—are subject to the disruptive impacts of human endeavors. These include the consumption of resources, disposal of wastes, and alteration of shores for social and economic benefits. A 1990 assessment by the Environmental Protection Agency (EPA) of 26,693 estuarine coastline miles and 4,320 ocean coastline miles revealed that only 56 percent of estuarine waters fully support designated uses such as swimming and fishing, while 89 percent of coastal ocean waters fully support designated uses. Only 2 percent of Great Lakes coasts assessed in the study support designated uses.

Population Pressures

As defined by the National Oceanic and Atmospheric Administration (NOAA), the U.S. coastal zone is comprised of the 666 coastal counties, cities, boroughs, or other county equivalents within 50 miles of ocean, bay, sound, or Great Lakes coasts. Excluding Alaska, U.S. coastal counties support nearly half of the nation's population while accounting for only 11 percent of the land mass. NOAA studies anticipate a 15-percent increase in coastal population over the next two decades, concentrated in California, Florida, and Texas. Coastal population density increased from 248 people per square mile in 1960 to 341 in 1988, more than four times the national average.

Multiple Use

Varied interests use coasts, oceans, and their associated resources in a multitude of ways:

- Commercial fishing
- Marine transportation
- Oil, gas, and other extractive activities
- Recreation
- Residential development
- Federal, state, and local government operations
- Activities of Native American tribes and of the general public

The multiple demands placed on the coastal zone, combined with increasing population density, reflect the zone's high economic value and contribution to the nation's quality of life. For example, coastal recreation alone represents significant expenditures. Recreational fishermen spent an estimated $6.2 billion in 1985, and recreational boaters spent $17.1 billion in 1989. An estimated 94 million people participate annually in coastal boating and fishing. At the same time, the multitude of uses concentrates pressure on coastal resources and processes.

Coastal Pollution

Pollution stress results from activities in and adjacent to coastal waters. Oil spills and waste discharges from boats, industrial facilities, and municipal wastewater treatment plants are direct sources of pollution. Activities occurring throughout watersheds that drain to coastal waters provide indirect sources of pollution in the form of point-source discharges into rivers. Nonpoint sources such as runoff from agricultural or urban areas carry nutrients, chemical pesticides and herbicides, and increased sediment loads resulting from development.

Coastal pollution has its greatest impacts on recreational use, fish and shellfish, and wildlife habitat values. Residents and recreationists in many coastal areas continue to experience beach closures because of unsafe levels of fecal coliform bacteria or marine debris. Other problems include fish and shellfish consumption advisories, shellfish bed closures, and diminished aesthetic quality. The viability of 75 percent of the nation's commercial fisheries depends on clean and functioning estuaries. Pollution and physical alteration may render coastal habitats incapable of providing elements critical to the life cycles of many fish and shellfish.

Coastal Hazards

As coastal development increases, so do potential threats to life, property, and natural resources, resulting from coastal hazards—natural and human-induced.

Flooding, high winds, tsunamis, coastal erosion, sea level rise, and storm surges associated with hurricanes and typhoons are all of concern to coastal planners, managers, and residents. The need to protect natural features such as wetlands, mangroves, dunes, and coastal barrier islands is reinforced each time the nation experiences the destructive forces of hurricanes such as Hugo (South Carolina, 1989), Andrew (south Florida and Louisiana, 1992), and Iniki (Kauai, Hawaii, 1992).

Improvements in the ability to predict the severity and landfall of coastal storms have resulted in far fewer injuries and deaths resulting from these events. However, even as storm-related mortality decreases, the costs of storm damage to heavily developed and rapidly developing coastal areas increase. Damages from Hurricane Andrew could reach $30 billion, making it the most expensive natural disaster in U.S. history.

Oceans

The world's oceans are under stress, with their productive and regenerative capacity threatened by pollution and over-utilization of marine resources. In contrast to coastal regions, the open sea remains relatively clean. Lead, synthetic organic compounds, and artificial radionuclides are widely detectable but at low levels. The oil slicks and litter common along sea lanes remain, for the most part, of minor consequence to communities of organisms living in the open-ocean areas.

Policies and Programs

Several federal agencies share responsibility for coastal and ocean issues. NOAA (Department of Commerce) and the Fish and Wildlife Service (Department of the Interior) are responsible for stewardship of living marine resources. NOAA, EPA, FWS, the Minerals Management Service (Department of the Interior), the National Science Foundation (NSF), and the Navy are all active in marine-related scientific research. Navigational safety and national security are covered by NOAA, the Army Corps of Engineers, the Navy, and the Coast Guard, while offshore minerals development is primarily the responsibility of MMS. EPA is the lead agency responsible for the regulation of pollution sources, and NOAA and the Coast Guard also address coastal marine pollution.

Planning and Management

Programs to manage the nation's coasts and oceans are mandated by federal laws, including the Coastal Zone Management Act Amendments of 1990 (CZMA) and the Clean Water Act (CWA).

Coastal Zone Management

The Coastal Zone Management Program, authorized by CZMA, is administered primarily by NOAA. Through a federal-state partnership, states with federally approved Coastal Zone Management (CZM) plans are eligible for federal grants and technical assistance to manage their coastal resources. Presently 29 of the 35 U.S. coastal states and territories have approved CZM programs that cover 94 percent of the U.S. coastline (84,117 miles). Texas, Ohio, Minnesota, and Georgia are in the process of developing CZM programs, and presently, the Great Lakes states of Illinois and Indiana do not participate in the program. In 1992 the majority of participating states prepared assessments and strategies to address priority coastal enhancement in such areas as natural hazards, shoreline access, wetlands, marine debris, ocean resources, special area management planning, government facility siting, and cumulative and secondary impacts of development.

Nonpoint Source Pollution

The CZMA Amendments of 1990 require participating coastal states to develop Coastal Nonpoint Pollution Control Programs subject to approval by NOAA and EPA. Nonpoint source pollution is now believed to contribute more to water quality degradation than point sources. Nonpoint pollution programs will need to focus on the following sources of water pollution that impair or threaten U.S. coastal waters:

- Agricultural runoff
- Urban runoff
- Silvicultural (forest management) runoff
- Marinas
- Recreational boating
- Hydromodification such as dams, levees, and shoreline erosion

Geographic Initiatives

In response to the multiple demands placed on coastal resources, coastal managers increasingly base data collection and management programs on ecologically determined geographic units such as drainage basins, estuaries, or ecosystems. The geographic approach considers the full range of factors that contribute to resource degradation. Such an approach requires the collaboration of public and private groups to balance multiple values.

For example, the EPA National Estuary Program is a partnership established by section 320 of the Clean Water Act that identifies nationally significant estuaries threatened by pollution, development, or overuse. Comprehensive Conservation and Management Plans (CCMPs) identify management options designed to restore targeted estuaries. Federal, state, regional, and local authorities as well as affected industries, academic institutions, and citizens take part in developing CCMPs. The program currently has 21 estuaries, 4 of which were added in 1992. Following approval in 1991 of the first CCMP, which covered Puget Sound, Washington, EPA approved a CCMP for Buzzard's Bay, Massachusetts, in 1992.

Other EPA geographic initiatives, such as the Chesapeake Bay Program, Great Lakes Program, and Gulf of Mexico Program target coastal resources. The Near Coastal Waters Program helps EPA identify priority coastal areas and integrate protective efforts by addressing gaps in coastal management.

National Marine Sanctuaries

The Marine Protection, Research, and Sanctuaries Act of 1972 provides for the protective management of unique marine recreational, ecological, historical, research, educational, and aesthetic resources through designation of National Marine Sanctuaries (NMS). Managed by NOAA, these select areas are the marine equivalent of national parks. In 1992 the United States doubled the area protected by this program with establishment of four new sanctuaries: the Flower Garden Banks sanctuary in the Gulf of Mexico, the Monterey Bay sanctuary off the coast of California, the Stellwagen Bank sanctuary off the coast of Massachusetts, and the Hawaiian Islands Humpback Whale sanctuary off the Hawaiian Islands. The Monterey Bay sanctuary, which encompasses 4,024 square nautical miles, is second only to Australia's Great Barrier Reef as the largest marine protected area in the world. The total area now protected under the National Marine Sanctuaries Program is over 9,000 square nautical miles.

Estuarine Research Reserves

NOAA protects 425,000 estuarine acres through the National Estuarine Research Reserves System established by CZMA. In 1992 a total of 144,000 acres was added to the system with final designation of the ACE-Basin Reserve (134,710 acres) and the North Inlet/ Winyah Bay Reserve (9,000 acres), in South Carolina.

National Wildlife Refuges

Coastal refuges make up 77 million acres, or 86 percent, of the 91 million acres in the National Wildlife Refuge System administered by the Fish and Wildlife Service.

Coastal Barrier System

The Coastal Barrier Resources Act restricts federal spending and assistance for development in the Coastal Barrier Resources System. The goal is to reduce damage to these valuable and ecologically sensitive barrier islands. The system includes 560 units and 1.27 million acres.

Restoration

Along with efforts to prevent and control degradation of the coastal environment, a number of federal programs emphasize restoration of damaged and depleted resources. For example, the Coastal Wetlands Planning, Protection, and Restoration Act of 1990 (CWPPRA) provides grants to coastal states or territories for acquisition of coastal lands or waters and for restoration, enhancement, or management of coastal wetlands ecosystems. States are required to maintain acquired coastal lands or waters, providing long-term protection of the property itself, its water quality, and its dependent fish and wildlife. FWS coordinates the process through which states and other applicants submit annual proposals for CWPPRA grant funding. To date, under the program, states have acquired, protected, or restored 69,500 acres of coastal wetlands.

NOAA protects and restores habitats of living marine resources as part of its trustee responsibilities under various laws. The National Marine Fisheries Service administers the NOAA Restoration Center, which develops and directs national expertise in identifying and evaluating restoration alternatives, restoring injured trust resources, and research and development in the area of restoration methodologies.

Science and Research

While NOAA conducts most federal science programs on the coastal ocean, other agencies, particularly within the Department of the Interior, contribute as well. Half of the research conducted by the Fish and Wildlife Service concerns coastal areas and species. The Minerals Management Service and the U.S. Geological Survey conduct coastal-related research as well.

Coastal Ocean Program

The NOAA Coastal Ocean Program focuses scientific expertise on coastal ocean and Great Lakes issues. The program seeks to improve understanding and predictive capabilities regarding coastal ocean pollution and habitat degradation, fisheries productivity, and protection of life and property in coastal areas.

North Pacific Marine Science Organization

A Convention for the North Pacific Marine Science Organization (PICES) entered into force in March 1992. Canada, Japan, and the United States are inaugural members, with China and Russia expected to become members soon. PICES has a broad mandate to promote and coordinate marine and related research, with an emphasis on the ocean's interactions with land and atmosphere; its role in global weather and climate change; flora, fauna, and ecosystems; uses and resources; and the impacts of human activity.

EPA Office of Research and Development

This office devotes its efforts to understanding the cumulative impacts of multiple stresses on marine and estuarine environments. Such efforts enable EPA water programs to move beyond their traditional chemical and effluent focus to address ecological exposure and biological responses. New EPA initiatives on habitat and biodiversity have implications for the marine environment.

Minerals Management Service Environmental Studies

Through an Environmental Studies Program, MMS conducts research on fundamental physical, chemical, and biological processes; ecosystem function; and abundance and distribution of living marine resources. The agency applies this information to decisions on management of offshore oil, gas, and mineral resources.

Interagency Collaboration

Within the President's Office of Science and Technology Policy (OSTP), the federal interagency Committee on Earth and Environmental Sciences (CEES) has a Subcommittee on U.S. Coastal Ocean Science (SUSCOS). The subcommittee coordinates the work of several agencies to obtain a predictive understanding of the processes associated with the coastal environment. Members represent the departments of the Interior, Agriculture, Commerce, Defense, Energy, State, and Transportation along with CEQ, EPA, NASA, and the National Science Foundation. The subcommittee provides long-range, coordinated planning for the federal coastal ocean science effort.

Monitoring and Assessment

The NOAA Strategic Assessment Program organizes and synthesizes existing information on U.S. coastal and ocean characteristics. NOAA also is responsible for the Large Marine Ecosystem (LME) Initiative that is developing strategies for monitoring the health of coastal ocean ecosystems around the world.

At the watershed level, the EPA National Estuary Program (NEP) develops guidance for use by individual estuary programs in monitoring the effectiveness of estuary management. The report, Managing Troubled Waters, issued by the National Research Council, guides development of monitoring plans for the 21 NEP estuaries and provides a model for other programs.

Coastal America

In 1991 the federal government established an interagency partnership to restore, preserve, and protect the living coastal heritage of the United States. The President's Council on Environmental Quality (CEQ) coordinates the partnership, whose members include agencies with stewardship responsibilities for coastal resources. In 1992 the Environmental Protection Agency (EPA) and the departments of the Interior, Army, and Commerce were joined by the departments of Agriculture, Air Force, Housing and Urban Development, Navy, and Transportation.

Coastal America partners focus federal and nonfederal legislative authorities and funding on action-oriented projects that address habitat loss, contaminated sediments, and nonpoint-source pollution.

In 1992 Coastal America sponsored projects at the local, regional, and national levels:

• **Local.** Partners initiated shovel-in-the-ground projects in each of the seven Coastal America regions, with participation by at least three federal agencies, a state/local government agency, and a nongovernmental organization. A monitoring component assesses success or failure of each effort, and an education/outreach component involves the public. Projects underway are removing obstructions to fish migrations, constructing wetlands, reducing agricultural nonpoint-source pollution from fertilizers, and stabilizing eroding shorelines.

• **Regional.** Implementation teams have a list of priority projects for which they are seeking partnerships. The New England region developed a strategy to restore coastal embayments cut off from tidal flow by transportation facilities, and Coastal America signed a resolution with the State of Connecticut to help restore its coastal salt marshes.

• **National.** The partnership has issued *Coastal Wetlands of the Continental United States: The Fragile Fringe* and a coastal video catalog. In preparation are a Reporters Guide to Ocean and Coastal Issues and a Coastal Reference and Retrieval System to provide a single coastal information system or clearinghouse for education/information materials in CD-ROM format.

WETLANDS

Wetlands are critical ecological systems. An estimated 80 percent of the nation's coastal fisheries and one-third of the nation's endangered species depend on wetlands for spawning, nursery areas, and food sources. Wetlands are home to millions of waterfowl and other birds, plants, mammals, and reptiles. Wetlands also perform hydrologic functions. Serving as recharge areas, they help protect the quantity and quality of the nation's groundwater. Unaltered wetlands in a flood plain can reduce flood peaks by 80 percent, and their natural water filtration and sediment control capabilities help maintain water quality.

Wetlands are also vital to commercial and recreational sectors of the economy, such as the sports fishing and waterfowl hunting industries. In addition, the diversity of plant and animal life in wetlands make them a valuable resource for nonconsumptive fish and wildlife-related recreation. Wetlands provide educational and research opportunities and a variety of historical and archaeological values.

Conditions and Trends

More than half of the wetlands that existed in what is now the lower 48 states have been converted to other uses; several states have lost 80 percent or more of their wetlands. Estimated annual wetlands losses between the mid-1950s and the mid-1970s stand at 458,000 acres. Between the mid-1970s and the mid-1980s, estimated losses ranged from 120,000 to 290,000 acres per year. Loss estimates for nonfederal rural wetlands in the 1982-1991 period showed a further reduction. According to the Natural Resources Inventory issued by the Soil Conservation Service, annual wetlands loss rates in nonfederal rural areas outside Alaska fell from 131,000 acres during 1982-1987 to 108,000 acres during 1987-1992. The reductions are attributable to enforcement of the Swampbuster provisions of the 1985 and 1990 farm bills.

Because loss figures can be confusing, federal agencies—for purposes of inventory—consider lands cleared of native wetlands vegetation as converted to other uses and often do not count them as wetlands. Many of these converted acres, particularly in agricultural areas, may still exhibit wetlands characteristics. If upon discontinuation of cropping, these lands exhibit wetlands soil and hydrologic characteristics and again support wetlands vegetation, they may be considered wetlands for regulatory purposes.

Policies and Programs

In 1992 the U.S. government continued efforts to restore wetlands and reform federal wetlands programs. Federal policy focused largely on implementing the President's August 1991 comprehensive policy statement on wetlands.

Wetlands Restoration

Federal agencies restore wetlands on their own lands as well as providing technical and financial assistance to state and private landowners. A 1992 interagency report recorded the restoration and enhancement of 282,000 acres of federal wetlands from fiscal 1989 through fiscal 1992. A half million acres of wetlands on nonfederal lands also were restored, primarily with federal technical assistance.

Forest Service. Wetlands restoration, which is common on the national forests, can be a simple matter, as the following example demonstrates. During road construction on the Cibola National Forest in New Mexico, workmen uncovered a road culvert that had been installed below ground. The culvert had lowered the water table in an adjacent 3-acre meadow, altering the hydrology of the site and eliminating wetlands plant species. By installing an elbow on the inlet of the culvert, Forest Service staff were able to reestablish the original drainage, and the higher water table promoted the return of wetlands species to the meadow. Cost to the Forest Service was $75, and labor contributed by the Wild Turkey Foundation came to $120.

Bureau of Land Management. BLM developed a national strategy to restore wetlands-associated wildlife habitat on 20 million acres through cooperative public-private efforts. For example, in Idaho, BLM worked with Ducks Unlimited and the Idaho Department of Fish and Game to construct 11 miles of fence to enhance waterfowl nesting and breeding habitat along the South Fork and Henry's Fork of the Snake River. The Thousand Spring/ Chilly Slough area was selected by the North American Wetlands Conservation Council to receive a $125,000 grant to acquire adjacent wetlands habitat, and The Nature Conservancy transferred to the State of Idaho 18 acres of headwater spring crucial to the management and protection of this key waterfowl habitat area.

Conservation Reserve Program. Through the Conservation Reserve Program (CRP), designed to protect highly erodible cropland, the Department of Agriculture protects wetlands by means of 10-year contracts and provides assistance for wetlands restoration efforts. To date, 400,000 acres of wetlands have been enrolled in CRP. Upon expiration of these 10-year contracts, current CRP wetlands could be returned to active cropping, or they could qualify for enrollment in the Agricultural Wetlands Reserve.

North American Waterfowl Management Plan

The North American Waterfowl Management Plan provides an umbrella for activities of many federal agencies and for federal partnerships with public and private groups. Partners include the governments of the United States, Canada, and Mexico, 50 states, 10 provinces, and more than 250 conservation groups. Between 1986 and 1991, more than 250 wetlands conservation partnerships implemented habitat protection, restoration, and enhancement projects benefiting over 1 million acres of wetlands and associated habitats. Estimated 1992 accomplishments include 60,000 acres restored, 140,000 acres enhanced, and 210,000 acres protected.

Partners for Wildlife

The Fish and Wildlife Service provides technical and financial assistance to private landowners, through a Partners for Wildlife program. To date, the agency, working with private partners, has helped restore 177,000 acres through voluntary agreements with 9,000 landowners. In 1992 alone FWS helped restore 37,400 acres through voluntary agreements with 2,000 landowners.

Managing Riparian Areas

Given their relatively small acreages, riparian values and benefits are considerable. In the arid western United States, riparian areas are among the most productive ecosystems on public lands, but, over the years, many of these areas have deteriorated. Since riparian benefits cannot be fully realized under degraded conditions, public concern is growing over the condition of these areas, particularly in the western states.

Forest Service. The National Forest System contains 5.6 million acres of riparian ecosystems along 359,000 miles of streams, rivers, and shorelines. In 1990 only 42 percent of these riparian areas met standards set by forest plans. Based on inventories of over 400,000 acres of riparian ecosystems, the Forest Service developed a number of restoration plans for the future. The goal of the agency's Riparian Strategy is to complete forest-wide inventories of riparian conditions and ecological health by the year 1995. In 1991 the Forest Service issued the report, Riparian Management, A Leadership Challenge, with guidelines for restoring riparian areas and wetlands throughout the National Forest System and for assuring that healthy riparian areas are not degraded by human activities. The goal is to meet standards set by national forest plans on 75 percent of unsatisfactory riparian areas by the year 2000.

Over the past three years, the Forest Service has restored 22,000 acres of riparian areas on the Lassen, Plumas, Six Rivers, Cleveland, and Mendocino forests in the Southwest. Restoration techniques include meadow rehabilitation, channel and bank stabilization, road obliteration, mine rehabilitation, and gully restoration. The following groups have assisted with these efforts: California Department of Fish and Game, California Department of Forestry, California Conservation Corps, Youth Conservation Corps, Los Angeles County Parks and Recreation Department, local landowners, and national forest permit holders.

Bureau of Land Management. Riparian areas constitute 23.7 million acres or 8.8 percent of the 270 million acres of BLM lands. In an effort to restore, enhance, and protect these areas, BLM has undertaken a Riparian-Wetlands Initiative for the 1990s. With a goal of restoring 75 percent of BLM riparian areas by 1997, BLM has implemented 680 new on-site projects. Among these are new fence construction to improve grazing management, tree planting, and the use of prescribed fire. A total of 530 maintenance projects are upgrading deteriorated fences, water developments, and habitat-improvement structures. The bureau is conducting inventories on thousands of miles of riparian streams to determine their condition and potential for recovery. Examples of cooperative efforts follow.

• **Marys River.** Ranchers, local residents, and conservation groups are working together to improve riparian management on the Marys River in Nevada. As a part of this effort, 54 miles of stream critical to the threatened Lahontan cutthroat trout were acquired through a land exchange between BLM and private owners.

• **Trout Creek.** The Trout Creek area of Southeast Oregon is a 190,000-acre watershed with about 175 miles of stream. Due to poor livestock management practices, much of the watershed was in degraded condition. BLM, local ranchers, Oregon Trout, the Oregon Environmental Council, the Izaac Walton League of America, and others developed a rotational grazing system to give the degraded areas periodic rest from the grazing of 2,000 head of cattle.

• **Cedar Creek.** Cedar Creek in Northern California is an 8 1/2 mile stream on private land surrounded by a 60,000-acre watershed managed primarily by BLM. BLM, landowners and ranchers, the Alturas Riparian Steering Committee, The Nature Conservancy, and others are cooperating to protect and enhance upland and riparian-wetlands areas. Techniques include improved grazing management, prescribed burning to control sagebrush and juniper, and planting of willows, cottonwoods, aspen, and sedges.

What Is a Riparian Area?

The green area immediately adjacent to streams, rivers, and lakes—known as a riparian area—is identified by the presence of vegetation that requires large amounts of free or unbound water.

Healthy riparian areas can provide the following benefits:

• Produce more forage than uplands, resulting in higher livestock-weaning weights

• Shelter livestock during weather extremes

• Reduce flood velocities and bank erosion minimizing property loss

- Stabilize streambanks with dense vegetation that reduces damage from animal trampling, ice scouring, and erosive flood waters

- Increase late-summer streamflows for irrigation, stockwater, and fisheries by recharging underground aquifers and providing bank water storage

- Filter sediment, protecting water quality, prolonging irrigation pump life, and reducing siltation of ponds and irrigation ditches

- Improve wildlife habitat by providing food, water, and cover

- Improve fisheries by providing food, cool water, and cover

- Provide recreation sites for picnics and camping

Coastal America Saltmarsh Restoration

The Intermodal Surface Transportation Efficiency Act (ISTEA) allows federal-aid highway funds to be used for wetlands mitigation and conservation. In July 1992 the interagency Coastal America partnership, as one of its projects, signed agreements to restore ecologically valuable saltmarshes along the Connecticut coastline. These marshes have been cut off from tidal flow since rail and road corridors were developed along the coastline in the early twentieth century. The Coastal America partnership involves the Connecticutdepartments of Transportation and Environmental Protection, U.S. Environmental Protection Agency, Army Corps of Engineers, and departments of Transportation and Commerce. The agencies agreed to identify appropriate candidate sites and cooperate on restoration activities. For more information on Coastal America, see the Coasts and Oceans section.

Aquatic Ecosystems Restoration Strategy

A committee of the National Research Council (NRC) recommended in 1992 that a national aquatic ecosystems restoration strategy be developed for the nation, to accomplish, over the next 20 years, restoration of the following:

- 2 million acres of lakes

- 400,000 miles of river-riparian systems

- 10 million acres of wetlands, largely through reconverting crop and pastureland and modifying or removing water-control structures. The committee recommended this as a net gain—that is, 10 million acres over and above losses of wetlands from other activities.

Coastal Wetlands Conservation

The Coastal Wetlands Planning, Protection, and Restoration Act of 1990 created a National Coastal Wetlands Conservation Grants Program for state coastal conservation efforts. In 1992 a total of 13 projects received $5.7 million in federal funding under this grants program.

Wetlands Delineation Manual. In 1992 EPA and the departments of the Interior, Agriculture, and Army continued work on revising the 1989 Federal Manual for Identifying and Delineating Jurisdictional Wetlands. Prior to 1989 federal agencies had no common delineation method; some used different techniques in different areas of the country. For example, the 1987 Wetlands Delineation Manual of the Army Corps of Engineers (COE) was not binding upon COE staff conducting delineations, and thus its use was inconsistent. An interagency manual prepared in 1989 became the subject of intense controversy and led, in August 1991, to a proposed revision. In response to that proposal, federal agencies received over 76,000 comments from the public. In 1992 deliberations continued on these comments and on options for proceeding. In the meantime, the Congress directed the Army Corps of Engineers to resume using its 1987 manual for implementation of the section 404 Clean Water Act regulatory program. The Department of Agriculture continued to rely on the National Food Security Act Manual to implement Swampbuster and other wetlands provisions of the 1985 and 1990 farm bills.

Nonregulatory Programs. The President's policy statement included elements to strengthen acquisition and other nonregulatory programs that protect wetlands, in many cases through better coordination of existing programs.

• **National Restoration Program.** President Bush called for a government-wide wetlands restoration and creation program on federal lands. The Interagency Committee on Wetlands Restoration and Creation, created to fulfill this mandate, completed its report, A National Program For Wetlands Restoration and Creation, in August 1992. The committee recommended long- and short-term objectives, consistent with a recent National Research Council (NRC) report on restoration of aquatic ecosystems. The committee provided criteria for project selection and made the following recommendations:

- Restoration on 1.3 million acres by 1995 (200,000 acres of federal and 1.1 million acres of nonfederal lands) as a step toward achieving a net gain of 10 million acres of wetlands on federal and nonfederal land by 2010.

- Voluntary federally assisted measures on nonfederal lands, since approximately 90 percent of restorable wetlands are on state and private lands

- Initial focus on wetlands restoration rather than creation, since restoration has a higher likelihood of success

- Uniform definitions and reporting standards

- A limited-term federal-state-private National Wetlands Restoration Council to coordinate restoration, enhancement, and creation activities. The NRC report also recommended a single coordinating entity.

• **Inventory.** Because many federal programs collect information on wetlands, the objectives of the following programs ensure efficient and comparable inventory data.

- A long-term effort is underway to assess the feasibility of coordinating and integrating the National Wetlands Inventory, conducted by the Fish and Wildlife Service, with the USDA National Resources Inventory, which includes resources in addition to wetlands, uses different methods, and covers only rural, nonfederal lands.

- In 1992 the interagency Federal Geographic Data Committee recommended updating and distributing a report on federal coastal wetlands mapping programs prepared by the Fish and Wildlife Service and the National Oceanic and Atmospheric Administration. The report describes the technologies used in federal mapping efforts.

- In 1992 the federal agencies undertaking inventory efforts (USDA, DOI, NOAA, and EPA) agreed to use a common wetlands classification/ identification scheme—the Cowardin system—to provide consistency in describing wetlands.

• **Research.** The Federal Coordinating Committee on Science, Engineering, and Technology (FCCSET) established a Subcommittee on Wetlands Research to define research responsibilities and identify areas where future research could support wetlands programs and coordinate wetlands research. The subcommittee produced an inventory of federal wetlands research efforts.

• **Regulatory Streamlining.** The third element of the President's wetlands plan was to improve and streamline the section 404 regulatory program of the Clean Water Act. In pursuit of this goal, federal agencies undertook the following activities in 1992:

• **Project Management.** COE issued a Regulatory Guidance Letter (RGL) clarifying its decisionmaking role for the evaluation of permits in the section 404 program. The guidance is aimed at improving coordination between federal agencies and applicants in requesting information, providing comments, and developing permit conditions.

• **Permit Review.** In late summer 1992 the federal government revised interagency agreements among COE, EPA, and the National Marine Fisheries Service. Developed under section 404(q) of the Clean Water Act, these agreements allow for orderly resolution of disputes between the Corps, which issues section 404 permits, and the principal agencies that review permits. Procedures governing elevation of permit decisions for higher level review are a principal component of the agreements. The revisions differ from past section 404(q) interagency agreements by limiting elevation of specific permit cases to those applications affecting aquatic resources of national significance.

• **Regulation of Prior Converted Cropland.** The Army Corps of Engineers and EPA proposed regulations that would incorporate into section 404 regulations existing Corps guidance providing that "prior converted" croplands are not within the scope of the section 404 program. Prior converted croplands are areas that were converted mostly through drainage prior to passage of Swampbuster provisions in the 1985 farm bill.

• **Outreach and Public Information.** In an effort to clarify the federal wetlands program and improve communication with the agriculture community, the Department of Agriculture, EPA, and COE developed six fact sheets on issues relating to wetlands and agriculture. Additional brochures and other materials are being developed. To increase private stewardship of wetlands, EPA expanded its Wetlands Hotline (1-800-832-7828), which distributed publications, to include information on voluntary public and private assistance programs available to private landowners.

• **Alaskan Wetlands.** In October 1992 the Bush administration proposed a rule change, using the process known as sequencing to remove the requirement that section 404 permits in Alaska be subject to review. Sequencing requires applicants first to avoid all negative impacts; second, to minimize unavoidable impacts; and finally, to compensate for remaining impacts with mitigation measures—usually restoration or creation of wetlands—preferably on or near the site. In place of sequencing, applicants would need only to demonstrate that they had minimized impacts. In proposing the rule, EPA requested comments on approaches—including retention of sequencing—for protecting areas defined as high-value wetlands by the state of Alaska.

Agricultural Wetlands Reserves

The Agricultural Wetlands Reserve Program is a new, permanent-easement program created by the 1990 Farm Bill. The program, administered by the Department of Agriculture, is authorized to purchase permanent wetlands easements on up to 1 million acres of farmland over five years and to pay up to 75 percent of restoration costs. While budget requests were in line with the 1 million-acre target, the Congress began with $46 million for a 50,000-acre pilot program in fiscal 1992. During the initial sign-up period, U.S. farmers attempted to enroll a total of nearly 500,000 acres of wetlands in conservation easements in the following nine states:

State	Acres Offered
California	85,000
Iowa	45,000
Louisiana	119,000
Minnesota	33,000
Missouri	29,000
Mississippi	115,000
New York	2,000
North Carolina	25,000
Wisconsin	13,000
Total	**466,000**

Wetlands Acquisitions

In 1992 federal acquisitions of key wetlands continued under a variety of programs. Federal agencies purchased a total of nearly 133,000 acres, using various funding sources led by the Land and Water Conservation Fund for 49,196 acres and the Migratory Bird Conservation Fund for 83,746 acres. Examples of wetlands acquisitions follow.

Agency	Project	Location	Acreage
NOAA	ACE Basin NERR	North Carolina	3,408
NOAA	North Carolina NERR	North Carolina	45
NOAA	Sapelo Island NERR	Georgia	206
FWS	Morgan Brake NWR	Mississippi	3,897
FWS	Grasslands WMA	California	6,129
FWS	Deep Fork NWR	Oklahoma	6,805
FWS	Bogue Chitto NWR	Louisiana	2,476
FWS	Waterfowl Production Areas	North Central States	63,159

NOAA = National Oceanic and Atmospheric Administration, U.S Department of Commerce
NERR = National Estuarine Research Reserve
FWS = Fish and Wildlife Service, U.S. Department of the Interior
NWR = National Wildlife Refuge
WMA = Wildlife Management Area

INTERNATIONAL ENVIRONMENTAL ISSUES

In 1992 the United Nations Conference on Environment and Development (UNCED), held in Rio de Janeiro in June, made front-page news of the linkages between environmental quality and economic development. Sustainable development —defined as managing the Earth's resources in a way that ensures their long-term quality and abundance—became the goal toward which nations committed themselves at home and at the "Earth Summit."

Conditions and Trends

During the past year, policymakers and the general public received a stream of scientific reports on the global nature of many environmental problems. Among the transboundary issues of greatest concern were biodiversity, forests, global climate change, oceans, and ozone-layer depletion.

Biodiversity

The biological wealth that scientists call biodiversity encompasses the Earth's variety of distinct species, the genetic variability within them, and the variety of ecosystems they inhabit. At the present time, no one has full and accurate information on the status and distribution of biodiversity components or on the functioning of ecosystems, but current evidence offers reason for concern.

Leading authority E.O. Wilson of Harvard University estimates that the current worldwide rate of extinctions exceeds 1,000 times the natural rate. Each of these losses from the world's natural heritage represents the potential loss of ecological, economic, and aesthetic benefits. The loss of biodiversity can result from a number of factors, particularly habitat loss and fragmentation, over-exploitation of species, introduction of intrusive species into an ecosystem, pollution, climate change, and industrial agriculture and forestry. These factors are not confined to tropical regions but affect temperate ecosystems as well. For a discussion of conservation efforts, see the Biodiversity section.

Forests

Forests that provide services and products of ecological and economic value locally, nationally, regionally, and globally are under stress around the world. According to the 1990 assessment of the U.N. Food and Agriculture Organization, tropical forests are disappearing at a rate of 17 million hectares (42 million acres) per year.

Responding to population pressures and poverty, many nations provide subsidies and incentives for forest conversion to farmland and pasture. Other causes include destructive logging practices; overharvest of commercial forests; subsidized timber sales; large-scale development projects; and lack of clearly defined property rights, leading to the "tragedy of the commons" problem.

Temperate and boreal forests, although expanding in some areas, are substantially degraded in others primarily because of air pollution, inefficient harvesting, and lack of reforestation. In some parts of Eastern Europe, forest dieback caused by air pollution is as high as 50 percent. For a discussion of conservation efforts, see the Forestry section.

Global Climate Change

Certain atmospheric gases—including carbon dioxide, methane, and nitrous oxide—trap solar rays that would otherwise be radiated back into space. Without this "greenhouse effect," Earth would be uninhabitably cold. But many scientists believe rising emissions of greenhouse gases from human activity could lead to "global warming" and negative changes in sea level, frequency and intensity of storms, precipitation patterns, and incidence of droughts.

Greenhouse gases are emitted by various sources and stored in biological or chemical "sinks." Sources and sinks include the burning of fossil fuels such as coal and oil, industrial processes, changes in forest cover and land use, certain agricultural practices, and biodegradation of wastes. Greenhouse gas emissions occur in all nations; for example, the National Academy of Sciences estimated in 1983 that about 25 percent of worldwide carbon dioxide emissions from fossil fuel combustion arises in the United States, 25 percent in Western Europe and Japan, 25 percent from the former Soviet Union and Eastern bloc, and 25 percent from developing nations. Industrialized country shares are declining relative to those of developing countries.

The Intergovernmental Panel on Climate Change (IPCC), created under the auspices of the World Meteorological Organization (WMO) and the United Nations Environment Program (UNEP), issued its first assessment in 1990 on the science, impacts, and possible responses to climate change. The report projected a global average temperature rise of 0.2 to 0.5 degrees Centigrade per decade through the next century, if no actions are taken to limit net greenhouse gas emissions. Global temperature increases at that rate would be greater than any experienced in human history.

However, the IPCC report also noted uncertainties that hamper scientists in making accurate predictions about the timing, magnitude, and regional distribution of climate change. Key unknowns include the way carbon cycles through natural systems and the function of clouds, oceans, and biota as natural feedback mechanisms. In 1992 IPCC issued a brief report that included the following supplementary findings.

• Global average surface temperature is sensitive to atmospheric concentrations of carbon dioxide or its equivalent.

• Atmospheric concentrations of greenhouse gases have risen substantially over pre-industrial era levels.

• Global mean temperatures have risen 0.3 to 0.6 degrees Centigrade over the last century, but no definitive association can be made with human activity, nor will scientists be able to verify such a hypothesis for a decade or more.

• Chlorofluorocarbons (CFCs) and sulfate aerosols may offset greenhouse warming.

• Current assumptions used to compare the global warming potential of greenhouse gases may be inaccurate.

Oceans

The productive and regenerative capacities of the world's oceans are also threatened by pollution and over-utilization. Coastal regions, as sites with concentrated natural resources and human development, are particularly at risk. Habitats are being lost irretrievably to the construction of harbors and industrial installations, the development of tourist facilities and mariculture, and the growth of settlements and cities. Although difficult to quantify, destruction of beaches, coral reefs, and wetlands, as well as increased erosion of the shore, are evident all over the world.

In contrast to the coasts, the open sea remains relatively clean, although trends for certain species are of concern. Lead, synthetic organic compounds, and artificial radionuclides are widely detectable at very low levels. Small oil slicks and litter common along sea lanes are unsightly but, for the most part, of minor consequence to communities of organisms living in open-ocean areas. For discussions of efforts to conserve the world's ocean resources, see the Coasts and Oceans section and the Fisheries and Marine Mammals section.

Ozone-Layer Depletion

Significant findings emerged over the last year on the condition of the stratospheric ozone layer, which protects the Earth's surface from harmful levels of solar radiation. Early in the year, UNEP released its latest International Ozone Assessment, and findings were issued from the second U.S. Airborne Arctic Stratospheric Expedition. Both studies presented evidence that ozone depletion may occur more rapidly and widely than previously suspected, leading to concerns about potential increases in incidence of skin cancer, cataracts, and ecological damage. In the fall, the U.S. government announced additional troubling news: the widest ever ozone depletion over Antarctica.

Concern also emerged over a substance previously not recognized as an ozone--depleting chemical. The 1992 UNEP assessment identified methyl bromide as potentially having a high ability to deplete stratospheric ozone, although it noted a need for additional, intensive research. For many agricultural uses of methyl bromide, safe alternatives have yet to be identified.

Policies and Programs

In 1992 worldwide progress in protecting the ozone layer, endangered species, and regional air and water quality continued under various international agreements and U.S. initiatives. Most significantly, two years of preparations for the United Nations Conference on Environment and Development (UNCED) culminated in adoption of new international policies, embodied in The Rio Declaration, Agenda 21, and A Statement of Forest Principles. At UNCED, newly negotiated conventions on climate change and biodiversity were opened for signature.

The following update on major international environmental issues is presented in alphabetical order.

Antarctica. In October 1991 the United States and most other Antarctic Treaty Parties signed a protocol on environmental protection in Antarctica with provisions on the conservation of flora and fauna, waste disposal, marine pollution, environmental impact assessment, and protected areas. The protocol prohibits the mining of mineral resources for at least 50 years. Pursuant to the protocol, Antarctic Treaty parties convened a meeting of experts on environmental monitoring in Buenos Aires in June 1992.

The Arctic. A number of meetings took place concerning the Arctic Environmental Protection Strategy (AEPS), adopted in 1991 by the United States and seven other Arctic Rim nations. Topics included conservation of fauna and flora and response to environmental emergencies. AEPS participants have established the Arctic Monitoring and Assessment Program, with a secretariat located in Norway.

Canada. In March 1992 the United States and Canada released the first progress report under the 1991 U.S.-Canada Air Quality Accord. Initial emphasis has been on acid rain and visibility. The two nations established a bilateral Air Quality Committee and increased the exchange of experts and information on North American air quality matters.

The two countries continued to work on Great Lakes protection, implementing a Great Lakes Toxic Air Pollution Deposition-Monitoring Program. In October 1992 the Convention on Environmental Impact Assessment in a Transboundary Context was presented to the U.S.-Canada International Joint Commission, and negotiations continued on guidelines for implementation.

Canada joined the United States and Mexico in agreeing to establish a North American Commission on the Environment in parallel to the development of the North American Free Trade Agreement (see the NAFTA discussion in the Economics section). In September 1992 EPA, Environment Canada, and the Mexican Secretariat of Social Development signed a memorandum of understanding on environmental education.

Climate Change National Action Plan. The United States presented initial elements of its national action plan at negotiating sessions for the U.N. Framework Convention on Climate Change in 1991 and added further measures in May 1992. These actions will limit net greenhouse gas emissions in the year 2000 to 6-11 percent below baseline projections, a limited increase of 1-6 percent above 1990 levels. The U.S. national action plan, published for public comment in the Federal Register (December 8, 1992) included several dozen measures, a sampling of which follows.

- Creating stronger standards and incentives for energy efficiency in the industrial, commercial, housing, and transportation sectors

- Accelerating the development and production of vehicles that operate on clean fuels such as natural gas, electric batteries, and biofuels

- Capturing methane from landfills, mines, and agricultural sources

- Research and development of high-performance aircraft engines, high-speed railroads, intelligent highway systems, industrial waste reduction and recycling systems, efficient coal technologies, safer nuclear reactors, and solar and wind energy technologies

- Research and planning for adaptation to climate change, focusing on coastal and freshwater resources, forests, and agriculture

Because net greenhouse gas emissions are rising most rapidly in developing and newly industrialized nations, the United States pledged $25 million in fiscal 1993-1994 to assist such countries in conducting climate change country studies. Such studies would emphasize the following topics:

- Vulnerability of developing countries to climate change

- Inventories of sources and sinks of greenhouse gases

- Technological options for limiting emissions and adapting to climate change

- Feasibility of specific climate-related projects

The United States hosted a Country Studies Workshop in September 1992 in Berkeley, California, and continued to fund over half the world's scientific research on climate change.

The U.S. Global Change Research Program is designed to reduce key scientific uncertainties and to develop more reliable scientific predictions upon which sound responses to global change can be based. Results from this work are available to the international scientific community.

An Inter-American Institute for Global Change Research was established in June 1992 under U.S. leadership. The institute is a network of centers engaged in basic research on global change processes of special interest to the Americas. The institute could be the first of several research centers worldwide to examine global change processes at the regional level.

Convention on International Trade in Endangered Species. As a party to the Convention on International Trade in Endangered Species of Fauna and Flora (CITES), the United States participated in the eighth conference of the parties in March 1992 in Kyoto, Japan. Among decisions reached by the conference, parties agreed to continue listing the African elephant and ban all trade in ivory. For the first time, CITES discussed the possible listing of commodity species such as blue fin tuna and several tropical timbers and listed several varieties of mahogany. Delegates decided to reevaluate scientific criteria upon which listing actions are based. The treaty is evolving to deal with a future in which trade issues and species conservation will become increasingly intertwined. The United States will host the ninth conference of CITES parties in late 1994.

In 1992 Congress enacted legislation that will facilitate U.S. implementation of CITES resolutions on exotic birds. The bill regulates imports of exotic birds whose populations may be adversely affected by trade. It also authorizes assistance to wild bird management efforts in other nations.

Eastern and Central Europe. Building on existing bilateral accords and using funds provided under the Support for Eastern European Democracy Act of 1989, the United States continued to assist countries in the region with development of environmental legislation and institutions. The United States also provided technical assistance and financial support for activities such as energy efficiency centers in Prague and Warsaw.

Enterprise for the Americas Initiative. The United States continued to implement the Enterprise for the Americas Initiative (EAI), launched by President Bush in 1991. The agreement aims to expand free trade throughout the Americas, stimulate economic reform and investment liberalization, and ease debt burdens. The debt-reduction plan includes innovative debt-for-nature swaps as a mechanism to support environmental protection and conservation.

Under the plan, the United States and eligible nations will establish the terms for reduction of certain official debt (PL-480 debt); payment of interest on the remaining debt will be made in local currency and deposited into trust funds supporting environmental projects in the host country. Local nongovernmental organizations will play a prominent role in planning trust-fund expenditures through membership on an oversight board. Participating countries qualify for the program by implementing macroeconomic and trade policies conducive to free-market development. In February

1992 the United States signed an EAI environmental framework agreement with Chile, and similar agreements are in effect with Jamaica and Bolivia.

Global Environment Facility. The Global Environment Facility (GEF) was established in 1990 as a 3-year pilot project to provide financial and other assistance to developing countries for innovative environmental projects that provide global benefits. The World Bank, the United Nations Development Program (UNDP), and UNEP serve as joint implementing agencies for the facility. GEF funds can be used for projects in the areas of global warming, ozone-layer protection, protection of biodiversity, and international waters. At UNCED, the facility was named the interim financial mechanism for both the Framework Convention on Climate Change and the Biodiversity Convention. Four rounds or tranches of GEF projects have been submitted to participating countries for approval.

Latin America and the Caribbean. In 1992 the Protocol Concerning Specially Protected Areas and Wildlife in the Caribbean Region (the SPAW Protocol) went to the U.S. Senate for its advice and consent to ratification. The protocol builds on existing international obligations to establish specially protected areas to preserve rare or fragile ecosystems, especially the habitats of threatened or endangered species.

Multilateral Development Banks. In 1992 the United States continued its efforts to encourage the World Bank and other multilateral development banks to integrate environmental considerations into lending practices, as required by the Pelosi Amendment. For example, prior to action by the World Bank Board, the U.S. executive director scrutinizes the potential environmental impacts of proposed bank projects and withholds support for those with significant potential impacts unless an environmental impact assessment or comprehensive summary of alternatives has been made available to affected parties, nongovernmental organizations, and World Bank executive directors at least 120 days in advance of a board vote. This U.S. policy has resulted in greater use of environmental impact assessments and public review of proposed projects in developing countries.

The World Bank developed new policies in 1992 for agriculture, water resource management, and energy efficiency and conservation. Reforms, undertaken at U.S. urging, strengthened linkages between environment and development goals in lending programs. For example, in October 1992 the World Bank completed a series of studies on energy conservation in the United States and in newly industrialized Asian countries, adopting a policy to make greater use of integrated resource planning—a tool that illuminates alternatives to powerplant construction. Policy shifts resulted in increased consultation with affected groups and nongovernmental organizations. The United States urged regional development banks to undertake similar reforms.

Formal discussions regarding a tenth replenishment of funding for the International Development Association (IDA) continued through most of 1992. U.S. negotiators demanded that IDA activities include full public participation and address environmental considerations.

North American Free Trade Agreement (NAFTA). The intersection of environmental and trade concerns received unprecedented attention in 1992. Trade liberalization, generally associated with increases in national incomes, provides financial resources that can be devoted to environmental protection. At the same time, the economic growth generated by open trade can have significant environmental consequences. For a discussion of NAFTA, the landmark event of 1992 in the area of trade and the environment, see the Economics section.

Ozone-Layer Depletion. In August 1992 amendments to the Montreal Protocol on Substances that Deplete the Ozone Layer entered into force for the United States and 19 other countries. Parties to the Montreal Protocol and its 1990 amendments have agreed to phase out the production of chlorofluorocarbons (CFCs) and other ozone-depleting substances by the end of the century.

In February 1992, in response to scientific evidence indicating that ozone depletion may be occurring more rapidly and widely than previously believed, President Bush took the following actions:

- Accelerated the U.S. phaseout deadline to the end of 1995—four years ahead of the Montreal Protocol deadline

- Called on other nations to ratify the Montreal Protocol and amendments and accelerate phaseout deadlines

- Asked U.S. producers to speed progress in the near term by cutting 1992 output to half of 1986 levels, even though the protocol allowed producers to be at 80 percent

By early 1992 U.S. production of CFCs had dropped more than 40 percent below 1986 levels largely as a result of innovative market-based mechanisms such as fees and tradable allowances.

Parties to the Montreal Protocol created the Interim Multilateral Fund in 1990 to assist developing countries undergoing the transition to CFC-replacement technologies. Total contributions to this voluntary 3-year fund rose from $160 million in 1991 to $200 million in 1992 and likely will increase substantially in 1993.

Technology Cooperation. The United States supports technology cooperation with developing countries and countries with economies in transition to help them find environmentally sound paths to economic growth and development. Technology cooperation takes the following forms:

- **Information Exchange.** A new Environmental and Energy Efficient Technology Transfer Clearinghouse, sponsored jointly by USAID, EPA, and the Department of Energy (DOE) will provide information on pollution control, renewable energy, and energy-efficient technologies. Pilot programs have been established in Mexico City and at the U.N. Industrial Development Organization (UNIDO) in Vienna.

• **Training and Technical Assistance.** The Environmental Training Institute shares U.S. environmental advances with developing countries by providing training courses to foreign executives. At UNCED, President Bush proposed an Environmental Technology Cooperation Corps to engage the U.S. private sector with counterparts abroad in the diffusion of environmentally sound management and technology.

• **Capacity Building.** The United States has established an Energy Efficiency Center in Russia to complement similar centers previously established in Eastern Europe. The purpose of the centers is to reduce energy-related wastes by increasing indigenous expertise and resources devoted to energy efficiency.

• **Technology Transfer and Development.** The United States promotes technology transfer and development through efforts that include the following DOE programs:

  • **ADEPT.** The ADEPT program (Assisting Deployment of Energy Practices and Technology) encourages joint ventures between U.S. national laboratories and industrial groups working in integrated project teams with developing country counterparts.

  • **America's 21st Century.** Through this program, DOE assists the U.S. renewable energy industry in forming joint ventures in Latin America and the Caribbean. U.S.-Asia Environmental Partnership. In January 1992 President Bush announced the United States-Asia Environmental Partnership (US-AEP), involving U.S. and Asian businesses, governments, and community groups in a program that includes technology cooperation, environmental fellowships and training, and conservation of regional biodiversity. More than 25 Asian countries are eligible to participate in the partnership, coordinated by a working group co-chaired by USAID and the Department of Commerce and including 17 other federal agencies.

The Earth Summit

The United Nations Conference on Environment and Development (UNCED), also known as the Earth Summit, brought together 6,000 delegates from over 170 countries for discussions and negotiations on such topics as climate change, biodiversity, deforestation, desertification, oceans, technology cooperation, and economic relations between developed and developing nations. More than 150 heads of state, including President Bush, attended UNCED, committing their nations to greater cooperation on many of the most pressing challenges facing the world.

Held June 3-14, 1992, in Rio de Janeiro, Brazil, the Earth Summit produced three principal documents and two framework conventions. Summaries follow on these documents, the Forests for the Future Initiative, the two conventions opened for signature at UNCED, and the U.S. plan for UNCED follow-up.

Rio Declaration. This non-binding statement sets forth 27 basic principles that will guide environmental protection and economic development worldwide. U.S. negotiators emphasized the principles of public participation in decisionmaking, public access to information, environmental assessments, and market-based economic and environmental strategies.

Agenda 21. This 580-page action plan charts a course toward global sustainable development. Its 40 chapters propose specific approaches to issues such as the following:

Finance. The result of intense negotiations, this chapter reflected a balance between the need for developing countries to mobilize internal resources for sustainable development and their need for additional assistance from industrial nations. The United States stressed realistic economic policies, the private sector's role in providing financial resources, and innovative methods of financing environmental protection, such as debt-for-nature swaps.

Forests. Providing a common approach for countries to integrate national and international actions in the conservation and sustainable development of forests, this chapter calls for action in four areas: multiple roles of forests; protection, conservation, and sustainable management; promoting efficient utilization; and planning, assessment, and monitoring. The United States was especially active in forest issues at UNCED (see below).

Hazardous and Solid Wastes. These chapters address key concepts such as waste prevention and minimization and call for ratification of the 1990 Basel Convention on the Control of Transboundary Movements of Hazardous Wastes and their Disposal.

International Institutions. This chapter calls for a number of reforms:

- Stronger interagency coordination in the U.N. system

- Enhanced roles for the U.N. Environment Program (UNEP), the U.N. Development Program (UNDP), and other U.N. agencies

- Preparation and comparison of national plans for sustainable development

- An enhanced role for nongovernmental organizations in U.N. deliberations and programs

- Establishment of a Commission on Sustainable Development by the U.N. General Assembly. As a senior body of the Economic and Social Council (ECOSOC), such a commission would provide a forum for discussing national and multilateral experiences in implementing UNCED recommendations

• **Oceans.** The oceans chapter includes comprehensive strategies addressing marine pollution from land-based activities and conservation of marine life. The United States successfully promoted the use of economic incentives to limit industrial and agricultural activities that pollute the seas. The United States also obtained language to ensure that populations of marine species are maintained at healthy levels, populations of endangered marine mammals are protected, and fishing gear is used that minimizes the incidental catch of nontarget species.

• **Technology Cooperation.** This chapter reflects U.S. efforts to highlight the concepts that make technology cooperation happen, such as protecting intellectual property rights, developing information networks, building indigenous technological capacity and regional research centers, and making effective use of technological assessment.

• **Toxic Chemicals.** This chapter proposes a structure for international work in chemical risk assessment and management that is likely to produce concrete results, given the prior momentum for international cooperation in this area. Reflecting lessons learned through the U.S. Community Right-to-Know Program (see "Toxics Release Inventory" in the Pollution Prevention section), the chapter encourages nations to establish similar programs. It also invites appropriate international organizations to develop guidance for such programs, and encourages industry to adopt voluntary right-to-know programs in the absence of government requirements.

Statement of Forest Principles. At the initiative of President Bush, heads of state at the 1990 summit of seven industrial nations (the G-7) announced the goal of negotiating a legally binding, international agreement on forests. Many developing countries were reluctant to rush into an agreement that might limit use of their sovereign, revenue-producing forest resources. A consensus emerged to negotiate a "non-binding authoritative statement of principles for . . . the management, conservation and sustainable development of all types of forests." After two years of negotiations, a statement of forest principles was adopted at UNCED, calling on nations to ensure that forests are managed to meet the needs of present and future generations. It addresses the role of international cooperation, including financial and technical assistance, trade, and science. The United States and other G-7 nations continue to support the goal of a full-fledged forest convention.

Forests for the Future. On June 1, 1992, to accelerate momentum for international forest conservation, President Bush proposed the Forests for the Future initiative, aimed at ending net worldwide forest loss within a decade. The President called upon aid-giving nations to double worldwide assistance for forest conservation from $1.4 billion to $2.7 billion. Leading by example, he pledged to double U.S. bilateral forest assistance—an increase of $150 million annually. To match international commitment with domestic action, the President announced that the United States will end clearcutting as a standard timber harvest practice on federal forestlands, as part of a new ecosystem approach to the management of forests. The President's America the

Beautiful initiative is already working to plant and maintain an additional one billion trees per year in the United States.

Climate Change Convention. At UNCED, the United States and more than 150 other countries concluded 15 months of negotiations by adopting the U.N. Framework Convention on Climate Change. The convention establishes the foundation for an international response to climate change with provisions such as the following:

- Nations must mitigate climate change through the design and implementation of national strategies that limit emissions of all greenhouse gases—the goal is to return emissions to the 1990 level by the year 2000—and enhance natural sinks of greenhouse gases such as forests.

- The treaty accommodates variations in the political and economic circumstances of nations, specifically avoiding uniform emissions targets and timetables for only one greenhouse gas, as proposed by some nations.

- Parties are encouraged to take account of climate change in developing economic, social, and environmental policies.

- Industrialized nations will provide assistance to developing countries, both in collecting data on net greenhouse gas emissions and in limiting the rate of growth in emissions.

- The treaty defines a mechanism for providing financial aid to developing countries for certain incremental costs of projects that produce global environmental benefits.

- Nations will raise awareness of the causes and implications of climate change and associated response policies through public education and training.

- Nations will improve their capacity to observe, model, and understand the global climate system through a continuing program of internationally coordinated research.

The convention enters into force upon ratification by 50 countries. At UNCED, President Bush supported a "prompt start" in treaty implementation, including early discussions by industrialized countries of national action plans to control greenhouse gases (see "Climate Change National Action Plan" in this section). On October 13, 1992, at the President's urging, the U.S. Senate voted unanimously to ratify the treaty, making the United States the first industrialized nation, and the fourth overall, to do so.

Biodiversity Convention. As a pioneer in domestic biodiversity conservation, the United States worked over the past several years for an international convention to reduce worldwide loss of biodiversity. Unfortunately the complexity of the issues, combined with international pressure to have a final text before UNCED, resulted in a convention unacceptable to the United States. At UNCED, President Bush stressed that the United States voluntarily would exceed the conservation-related provisions of the convention, but he rejected as unwise the convention's provisions on intellectual property rights, government access to genetic information in biotechnology, and financial assistance.

The U.S. commitment to biodiversity conservation is consistent with provisions of *Agenda 21.* In response to the call made at UNCED for increased research, the United States proposed a plan to foster international cooperation on biodiversity surveys, inventories, and data organization. The United States plans to host a meeting of international experts in 1993 to consider such topics as data gathering and reporting standards, approaches to inventories, surveys, and data management, and institutional issues. Efforts are underway to develop a National Biodiversity Center to coordinate and make such information available in the United States.

UNCED Follow-Up in the United States. In his remarks at Rio, President Bush emphasized the need for follow-through on commitments made there. In August 1992, under a cabinet-level policy coordinating group, the Department of State and the Council on Environmental Quality began to review action items in *Agenda 21,* comparing them to existing U.S. programs. This review will help identify priorities for further action, facilitate work on a U.S. sustainable development strategy, and provide the framework for U.S. participation in the work of the U.N. Commission on Sustainable Development.

PUBLIC LANDS

The U.S. government manages natural resources on federal lands to provide for the needs of present and future generations. A policy of sustainable multiple use provides range, timber, minerals, watersheds, fish and wildlife habitat, wilderness, and recreation, while maintaining scenic and cultural values. Protected from un- regulated development, public lands support resource development that sustains the long-term productivity of soils, waters, plants, and animals. Some federal lands and natural resources are managed for special uses, such as parks, military training, grazing, endangered species, and outdoor recreation. For example, on lands managed by the departments of the Interior and Agriculture, 95 million acres (15 percent) are designated wilderness.

Conditions and Trends

In 1992 public land management reflected the nation's demand for a balance between environmental and economic values. Federal land managers worked to meet the public desire to protect natural landscapes while providing resources for economic growth. Increased populations in the western states are influencing federal land management, which no longer occurs in isolated settings but in close proximity to residential areas. Today federal land managers coordinate actions closely with states, municipalities, and the private sector.

Policies and Programs

Public Rangelands

Together the Bureau of Land Management (BLM) and the USDA Forest Service manage a quarter of a billion acres of land available for use by livestock. Grazing fees for federal pastures remain a controversial issue.

Bureau of Land Management. BLM manages 170 million acres of rangelands, some in poor condition because of past livestock overgrazing, agricultural and commercial misuse, and off-road vehicle abuse. Fifty years ago, a third of BLM rangelands were in poor condition with only 16 percent in good to excellent condition. In 1991 rangelands in poor condition were down by half, while rangelands in good to excellent condition had increased to 36 percent, with an overall 87 percent in stable or improving condition. BLM is committed to increasing rangelands in good to excellent condition to 40 percent and to reducing those in poor condition to 10 percent. The bureau also has set a goal for establishing environmental balance by 1997 on 75 percent of riparian-wetlands areas, which can be impacted by grazing. For more information, see the Wetlands section.

BLM improves rangeland conditions by means of grazing management systems, protective fencing, water development in the uplands, plantings, and reseeding. Examples follow.

Incentive-Based Grazing Fees. Grazing fees for the use of federal lands are a contentious issue addressed annually by legislation. In 1992 the fee was $1.92 per animal-unit month (one cow and calf per month). In 1991 BLM collected $19 million in grazing fees. A BLM task force has developed a proposal that would give ranchers an incentive to improve leased federal lands. Incentive-based grazing fees could improve range management without significantly increasing the program budget. Although participation in incentive programs would be optional, ranchers who chose not to adopt sound management practices that maintain or improve rangeland would pay significantly higher fees.

Coordinated Interdisciplinary Resource Monitoring (CRIM). BLM is consolidating rangeland monitoring in CRIM, a unified data-collection effort for all of its programs. CRIM provides the information needed to evaluate management and to select best management practices.

Desired Plant Community (DPC). A vegetation management strategy called DPC achieves plant communities that best meet the overall needs of the public, wildlife, and livestock. Such communities, at the same time, protect water, soil, vegetation, and air.

Forest Service. Of the 191 million acres managed by the USDA Forest Service, 100 million acres are available for use by domestic livestock. A new Change-on-the-Range philosophy advocates ecology-based range management to accomplish resource stewardship and provide healthy rangeland ecosystems. In 1992 the Forest Service implemented a new information reporting system to measure the health of rangelands. A sampling of other projects follows.

Grazing Fees. The Forest Service administered 9,798 grazing allotments in 33 states in 1992. Grazing fees varied from $1.92 to $3.92 per animal-unit month. Fees collected from public grazing on national forests—excluding national grasslands—totaled $9.7 million in 1991 and $8.8 million in 1992.

Video Documentary. In 1992 the Forest Service and the American Farm Bureau produced and distributed a video documentary featuring on-the-ground ecological range management.

Noxious Weeds. A 1992 Noxious Weed Workshop, designed for range managers, had as cosponsors: the Forest Service, BLM, Western Weed Coordinating Committee, Inter-mountain Noxious Weed Advisory Council, Nevada State Department of

Agriculture, Montana State Department of Agriculture, and Montana Weed Control Association.

Livestock and Big Game. A 1992 Livestock-Big Game Workshop addressed conflicts that can arise when managing livestock and big game species on national forests and neighboring private lands. Cosponsors were the Forest Service, National Cattlemen's Association, Public Lands Council, International Association of Fish and Wildlife Agencies, and the Wildlife Management Institute.

Energy and Minerals

In fiscal 1991 and 1992 onshore and offshore energy minerals leases provided $4.3 billion in revenue to the U.S. Treasury, the Historic Preservation Fund, the Land and Water Conservation Fund, Indian tribes, and the states. Leasable minerals include oil and gas, coal, geothermal energy, phosphate, and, to a limited extent, hardrock minerals. BLM administers federal lands and interests that contain the nation's onshore mineral resources, and the Minerals Management Service manages the mineral leasing and development on the outer continental shelf.

Bureau of Land Management. BLM fosters development of onshore minerals to achieve sustained yield, multiple use, and conservation of natural resources. Federal lands managed by a number of agencies and subject to surface and subsurface mineral development total 732 million acres, of which 272 million acres are surface and subsurface rights on BLM lands. The remaining 460 million acres are acquired lands, lands that have been patented with minerals reserved to the United States, and public lands on which minerals are administered jointly by BLM and other agencies, such as the National Park Service and the Forest Service.

In fiscal 1991 non-energy leasable mineral development on federal lands provided $83 million in lease revenue. Of the $31 billion contributed to the gross national product by nonfuel minerals produced in the United States, federal lands in the West figured prominently. These lands hold a large volume of potentially minable resources—gold, copper, and other hardrock minerals. Policy debate continues over the environmental and economic aspects of hardrock mining.

Minerals Management Service. The nation's outer continental shelf (OCS) encompasses 1.2 billion acres. A portion of revenues derived from offshore oil and gas leases are designated to help preserve marine habitats, including biodiversity and historic and cultural resources. The OCS Program provides 80 percent of revenues credited to the Land and Water Conservation Fund (L&WCF). OCS revenues supplied the fund with $843.8 million in fiscal 1990; $885 million in fiscal 1991; and $888 million in fiscal 1992.

Watchable Wildlife

Federal land agencies have formed Watchable Wildlife partnerships with groups such as Defenders of Wildlife, National Audubon Society, National Wildlife Federation, and the Izaac Walton League. Begun in 1988 the program is a nationwide effort to increase wildlife viewing opportunities, provide information on the needs of wildlife, and promote wildlife conservation. BLM maintains 159 viewing sites in 11 western states, and the Forest Service maintains 165 viewing areas with plans for an additional 220. Accomplishments for 1992 included the following:

• **Viewing Areas.** In 1992 a National Watchable Wildlife Conference gathered and disseminated data on tools for creating viewing areas nationwide. In Colorado, BLM, the Forest Service, Colorado Division of Wildlife, Bighorn Sheep Society, and the City of Georgetown established the Big Horn Sheep Viewing Area, 30 miles west of Denver.

• **Viewing Guides.** Watchable Wildlife cooperators have published 12 state viewing guides and are planning guides for most of the other states. In 1992 the National Park Service, National Park Foundation, and National Fish and Wildlife Foundation prepared a viewing guide for Watchable Wildlife in the National Parks, and the Forest Service launched a similar program, Eyes on Wildlife, for the National Forest System.

Recreational Fisheries

In May 1992 the Forest Service and BLM cosponsored a meeting of government fisheries managers to promote a joint policy to improve the quality and quantity of recreational fishing on public lands. A sampling of projects follows.

Fish and Wildlife 2000. Since 1985 BLM has received $11.8 million in federal appropriations, matched by $18 million from private sources, to complete 1,200 wildlife and fisheries projects. In 1992 BLM issued a comprehensive plan for managing its wildlife and fisheries resources.

California Water 2002. As part of its environmental program, the Bureau of Reclamation sponsors the following initiatives in California:

• Contracting with the state to purchase and install salmon spawning gravels in the upper Sacramento River

• Improving the water supply for the Nimbus Fish Hatchery

• Contracting for a firm water supply to wildlife refuges in the Central Valley

- Initiating a demonstration cost-sharing program with nonfederal diverters to remedy the fish-screen effects of pump and direct-water diversion facilities on the Sacramento and San Joaquin rivers and in the Delta

Pathway to Fishing. In 1992 five federal agencies and two private groups signed a memorandum of understanding to work together to advance angler skills, ethics, and stewardship of fisheries resources on federal lands and waters. Cooperating in the effort are the USDA Forest Service; four DOI agencies—Bureau of Reclamation, Fish and Wildlife Service, National Park Service, and Bureau of Land Management; Berkley, Inc.; and In-Fishermen, Inc. The cooperators have produced an instructional kit called Pathways to Fishing that introduces the public to fishing, fish biology, and aquatic ecology. Kits are available from the U.S. Government Printing Office.

Fish and Wildlife Habitats

The Fish and Wildlife Service manages 92 million acres of national wildlife refuges, administers the Endangered Species Act, provides consultations on water development and water quality under the Fish and Wildlife Coordination Act and section 404 of the Clean Water Act, operates fish hatcheries, and provides grants to states under the Federal Aid in Sport Fish Restoration and Federal Aid in Wildlife Restoration acts. FWS is conducting a management review of the National Wildlife Refuge System to meet the nation's needs and interests through 2003, the hundredth anniversary of America's first national wildlife refuge. Refuges 2003, a combined management plan and environmental impact statement, will chart the course of the system for the next decade, considering facility and land needs for an increasing population. To encourage public involvement, FWS held 31 meetings around the country and received thousands of public comments.

Celebrate Wildflowers

In 1992 the Forest Service organized Celebrate Wildflowers, an annual, national event to emphasize wildflower appreciation by agency personnel and the public. Underscoring the event is an intensified effort by the Forest Service to inventory rare plants and implement conservation and recovery strategies. This objective is reflected in the number of Forest Service botanists, up from 32 in 1989 to 104 in 1992.

America's Great Outdoors

The President's recreation initiative, America's Great Outdoors, promotes interpretation and education, improved facilities, and acquisition of new recreational lands.

Forest Service. The national forests are the most visited federal lands, with 597.6 million visits per year. In 1992 the Forest Service improved recreational access for visitors, including those with disabilities. The agency expanded interpretive programs for cultural resources and joined interagency partnerships to encourage domestic and international tourism on federal lands. A visitor day equals 12 visitor hours by one or more persons, and in 1992 national forests and grasslands recorded 279 million recreation visitor days, a 6-percent increase over 1990.

National Park Service. In August 1992 President Bush signed a law establishing the Marsh-Billings National Historical Park, the first national park in Vermont. The 550-acre hillside site includes a mansion that was the boyhood home of George Perkins Marsh, who in 1864 wrote Man and Nature, a book that inspired the early conservation movement in America.

In 1992 land acquisition for national parks continued, but acquisition alone cannot solve seasonal or site-specific crowding; Yellowstone and Yosemite likely will remain among the nation's favorites, with summer months the most popular visitation times. In 1991 the 361 units of the National Park System recorded 268 million recreation visits, up from 154 million visits to 284 units in 1971. Although entrance to most national parks is not subject to specified capacity limits, some rivers and backcountry areas are restricted to permit holders to avoid overcrowding and natural resource degradation. The National Park Service, in conjunction with public and private organizations, has introduced measures to minimize traffic congestion, including the operation of visitor transportation systems in 12 park areas. Private concessionaires and others outside the parks also provide transportation services for visitors.

Private Concessions on Public Lands. In 1992 to improve management of the private concerns that operate concessions such as hotels, restaurants, and ski resorts on public lands, a Concessions Management Task Force, with members from DOI land management agencies, the USDA Forest Service, and the Army Corps of Engineers, issued a draft report. The task force recommended maintaining financial consistency among agencies, increasing competition by setting appropriate fees and durations of agreements, and integrating concessions management into federal planning.

Wilderness

Since 1964 the Congress has set aside 95 million acres of wilderness in 548 units of the National Wilderness Preservation System, largely by designating existing roadless areas on national forests and Department of the Interior lands, rather than by adding new lands to the federal estate.

Bureau of Land Management. In 1992 BLM completed a 15-year wilderness review mandated by the Federal Land Policy and Management Act of 1976. The review covered 26 million acres of lands in 750 wilderness study areas in nine western states.

In response, the secretary of the Interior recommended for wilderness designation 330 units of BLM lands in the nine western states for a total of 10 million acres, and the President transmitted the following recommendations to the Congress:

- California, 62 wilderness areas with 2.3 million acres

- Idaho, 27 wilderness areas with 1 million acres

- Nevada, 52 wilderness areas with 1.9 million acres

- New Mexico, 23 wilderness areas with .5 million acres

- Oregon, 49 wilderness areas with 1.3 million acres

- Utah, 69 wilderness areas with 1.9 million acres

Forest Service. The USDA Forest Service manages 33.6 million acres of wilderness in 382 units. In December 1991 the President signed the Chattahoochee Forest Protection Act, adding 25,840 acres of national forests in Georgia to the National Wilderness Preservation System. In June 1992 the President signed the Los Padres Condor Range and River Protection Act, adding 400,400 acres of national forests in California to the system. The Congress considered but did not pass legislation that would have created another 1.8 million acres of wilderness.

Wild and Scenic Rivers

The National Wild and Scenic River System consists of 151 designated rivers with a total of 10,410 protected river miles. The Forest Service manages 96 of these rivers with a total of 4,316 miles, and BLM manages 32 rivers in five states for 1,991 miles. In fiscal 1992 Congress designated 26 new rivers on national forests for a total of 914 miles in the following states:

- Michigan, 14 rivers on the Hiawatha, Ottawa, and Huron-Manistee national forests

- Arkansas, 8 rivers on the Ozark and Ouachita national forests

- California, 3 coastal streams on the Los Padres National Forest

- Pennsylvania, 85 miles of the Allegheny River on the Allegheny National Forest

Congress considered but did not pass bills that would have added 112 miles of the lower Salmon River in Idaho, along with sections of the Merced River in California, the Rio Grande in New Mexico, and the Black Canyon of the Gunnison River in Colorado.

Special Recreation Areas

The National Forest System contains 42 congressionally designated recreation areas, encompassing 7 million acres and including 18 national recreation areas, 5 national scenic areas, 4 national monuments, and 15 other recreation areas. In 1992 the Congress added the following lands on the Chattahoochee National Forest in Georgia:

• Coosa Bald National Scenic Area, a 7,100-acre scenic area that features the southern Appalachian "bald" vegetative type

• Springer Mountain National Recreation Area, a 23,330-acre recreation area that marks the southern terminus of the Appalachian Trail

America the Beautiful

The President's America the Beautiful (ATB) initiative calls for enhanced recreational opportunities on public lands. After revitalizing the Land and Water Conservation Fund (L&WCF) as a mechanism for acquiring new recreational lands, the President proposed the purchase of 174,991 acres of special land in 25 states. L&WCF grants funded state and local government acquisitions of parklands and development of outdoor recreation facilities. From its inception in 1965 through 1991, the L&WCF assisted 36,000 projects with $5.5 million; in fiscal 1992 alone, the fund awarded 250 grants totaling $15 million for basic recreation development and 35 grants for acquisition of 8,493 acres of park and open-space lands. ATB has initiated cost-sharing partnerships with states, local governments, and private entities to protect and preserve natural and cultural resources. Examples follow.

Federal Land Acquisition. As of September 1992 the Forest Service had invested $107.5 million in acquiring 166,781 acres of new recreational lands. In fiscal 1992 BLM acquired 189,426 acres of land for protection and recreation, FWS acquired 89,750 acres, and NPS acquired 153,675 acres for a total of 432,851 acres at a cost of $229 million, including the following:

• **Balcones Canyonland National Wildlife Refuge.** This Austin, Texas, refuge provides habitat for seven federally listed endangered species and other rare and potentially threatened species, including the black-capped vireo and the golden-cheeked warbler.

• **Fox River Trail.** After 11 years of efforts by 11 local jurisdictions and $3 million in seven acquisition and development grants, the Fox River Trail in suburban Chicago is now a 70-mile long scenic, riverside path.

• **Pompey's Pillar.** In 1992 BLM acquired a 366-acre landmark property on the Lewis and Clark National Historic Trail in Montana. The site features a large rock outcropping that bears the nation's only physical evidence of the Lewis and Clark Expedition—the carved signature of Captain William Clark. The acquisition also includes 200 acres of wetlands.

• **Stillwater National Wildlife Refuge.** This refuge is a staging area for spring and fall migrations of shorebirds, swans, geese, and bald eagles and a duck breeding site. To acquire additional wetlands and to dilute elevated levels of harmful trace metals in the water, the Fish and Wildlife Service, The Nature Conservancy, and the State of Nevada signed a cooperative agreement to purchase water rights and land from willing sellers to sustain 25,000 acres of wetlands habitat. In 1992 FWS prepared a draft environmental impact statement for acquiring water for wetlands and initiated socio-economic and hydrological studies.

Legacy '99 Corrective Actions. DOI is improving the infrastructure on 450 million acres of national parks, wildlife refuges, and other public lands. The goal is to clean up hazardous materials, increase maintenance, rehabilitate facilities, and inspect 383 dams managed by DOI. In 1992 the dam safety program recorded 194 inspections with corrective actions taken on 12 dams. BLM also completed the first-phase assessment of 47 sites potentially contaminated by hazardous materials and conducted a follow-up study on a classified high-risk site to determine the best method of cleanup.

Recreation and Resource Protection. BLM sponsored 330 challenge cost-share projects to protect natural resources at recreational sites, and the Fish and Wildlife Service sponsored 211, including the following:

• **Byways System.** Since its inception in 1989, the National Byways System has established 50 back-country byways—roads that extend through remote scenic corridors—with 2,450 miles in 11 states. BLM and private organizations such as Farmers Insurance and American Isuzu fund interpretive kiosks at entry points and cooperate on improvement and maintenance.

• **Off-Highway Vehicle Project.** BLM signed a memorandum of understanding with the Motorcycle Industry Council and the Specialty Vehicle Institute of America to protect public lands by promoting safe and responsible off-highway vehicle (OHV) recreation. The private groups, which serve as a clearinghouse for resource information and as a link between BLM and riding clubs, have published Right Rider, a series of four brochures for the general public. BLM provides expertise on OHV management and involves riders in planning, developing, and maintaining trails.

Partnerships

Partnerships between public and private agencies maximize the effectiveness of federal funds and encourage private sector involvement in preserving natural and cultural resources. Independent cost-sharing grant programs or in-kind grants from private individuals, groups, or organizations support public land management.

Habitat Improvement. The Forest Service administers a Challenge Cost-Share Program, which has increased from 57 partnerships in 1986 to 2,500 partnerships in 1992. In 1992 partners turned $12 million in federal funds into $31.4 million in habitat improvements. BLM enlists partners such as the Global ReLEAF program, sponsored by the American Forestry Association, which in 1992 assisted public land restoration in Nevada, New Mexico, Wyoming, and Arizona. In the Nevada project, 150 volunteers planted aspen, cottonwood, willow, and other shrubs along the Marys River to restore habitat for the Lahonton cutthroat trout. Cosponsoring the project were BLM, Nevada Mining Association, Nevada Chapter of Trout Unlimited, Fish and Wildlife Foundation, and Global ReLEAF.

Take Pride in America. This national public awareness campaign encourages responsibility and pride in America's natural, historic, and cultural resources. The program relies on volunteers and in-kind contributions from private sources on efforts such as the following:

• **America's Treeways.** Federal, state, and private groups are cosponsoring this program, in which volunteers have planted hundreds of thousands of trees donated by forest industries along U.S. highways.

• **U.S. Virgin Islands Cleanup.** Sponsored by a coalition of community groups, this project enlists volunteers of all ages to adopt an area—beach, park, or roadway—and accept responsibility for its cleanup.

• **National River Cleanup Week.** A new event, sponsored in 1992 by a coalition of federal agencies and national conservation groups, resulted in the cleanup of 6,000 river miles by 20,000 volunteers.

Mining on the National Forests

The Forest Service manages lands with mineral deposits to ensure that surface development is conducted in an environmentally acceptable manner and that lands are restored to a productive condition. The mining industry assists by funding environmental studies and mitigation, such as fish, wildlife, and recreation enhancement projects. Estimated industry contributions for environmental studies in 1992 were $5 million. The Forest Service minerals program covers both leasable and locatable minerals. In 1992 a total of 12 million acres were under lease, mostly for oil and gas. The Forest Service reviews proposals for leasing, informs BLM which lands can be made available, and approves surface-use operating and reclamation plans after leases are issued to ensure that mining is conducted in an environmentally sensitive manner. The Forest Service reports the following mining statistics for 1992:

- Production of 12 million barrels of oil and 210 billion cubic feet of gas from 2,300 wells on 12 million acres of leased land, disturbing only 11,500 acres (0.1 percent) of the land leased.

- Opening, on the Inyo National Forest in California, of the first geothermal powerplant on National Forest System (NFS) lands.

- Decline in land leased for oil and gas, from 35 million acres in 1983 to 12 million acres in 1992, reflecting the affect of economic conditions on the industry.

- Production from the largest surface coal mine in the world, a mine on NFS lands in Wyoming, of 3 percent of all coal mined in the nation.

- Over 5.5 million tons of phosphates, with a market value of $137.5 million, produced from leases on NFS lands in Idaho.

Locatable minerals, available for development under the Mining Law of 1872, include gold, silver, copper, zinc, and other minerals. The Forest Service regulates locatable mineral operations to mitigate environmental effects to surface resources and to ensure reclamation. Producing mines total 1,200 out of 500,000 mining claims on 10 million acres of NFS lands, primarily in the West. Only 44,000 acres (0.4 percent of total land under claim) are disturbed as a result of ongoing locatable mineral operations, while 100,000 acres are in various stages of reclamation.

Protecting American Battlefields

To promote public stewardship of historic battlefields, the American Battlefield Protection Program, a public-private partnership begun in 1990, identifies sites in critical need of protection.

Civil War Legacy 2000. The Civil War Trust leads a fundraising campaign to raise $200 million by the year 2000 to assist federal, state, local, and private efforts to acquire and protect historic battlefields.

Shenandoah Valley Study. The National Park Service completed a study of 15 battlefields in the Shenandoah Valley of Virginia in response to the Civil War Sites Study Act of 1990. The NPS Battlefield Protection Program provided technical assistance to local organizations involved in the study, which assessed needs and developed preservation strategies.

Priority Civil War Sites. In 1992 NPS entered cooperative agreements with organizations at priority Civil War sites to assist with interpretation and preservation. Cooperators include local governments, property owners, the Civil War Battlefield Foundation, National Trust for Historic Preservation, Civil War Society, and Association for the Preservation of Civil War Sites.

FORESTRY

Forests are a source of ecological services and economic goods—from soil and watershed protection to timber and fuelwood to carbon sequestration, wildlife habitat, and recreation. The ecology and economics of the forest cannot be separated, since forest values are at once local, national, and global in scope. In 1992, in the midst of heightened domestic and international debate about forests, the U.S. government adopted a philosophy of ecosystem management for federal lands and led the way in calling for global forest conservation.

Conditions and Trends

The call to conserve and sustain forests is urgent because many types are under stress around the world. Forest conversions on private lands are a trend both at home and abroad.

U.S. Forests

Forests occupy about a third of the nation's land area, with forest types varying from sparse scrub forests of the arid interior West to the highly productive forests of the Pacific Coast and the South. The majority of U.S. forests, about 525 million acres, are in state and private ownership, with the balance in federal forests. Most eastern forests are in state or private ownership, and western forests are divided between the federal government and other owners.

Americans reforested 2.9 million acres of public and private forestland in fiscal 1991, near record highs, and a comparable acreage was regenerated naturally. On managed acres, forest growth exceeded harvests by 37 percent, up threefold over the 1920s, but the nation continues to lose half a million acres of private forestlands per year because of conversions to such uses as urban expansion and agricultural crops. At this rate, total U.S. forests could decrease from 731 million acres in 1992 to 703 million acres in 2040, a 4-percent decline.

Reforestation can improve damaged ecosystems and help protect watersheds, soil, and crops. The number of trees planted in 1991 was the seventh highest of any year in U.S. history. Of acres planted in trees, 82 percent were on private lands, and of these, 39 percent had nonindustrial private owners.

U.S. forests, on the whole, are healthy, but problems arise, especially on managed forests, in response to stresses such as insects and drought. In some managed forests, past harvesting practices and fire suppression have resulted in stands poorly adapted to the site, prone to insect and disease problems, and stressed by drought. Potential for catastrophic wildfire is high because of the accumulation of dead and dying vegetation.

Throughout the nation, forest pests are causing serious damage. The European gypsy moth defoliated 4 million acres in 14 states in 1991, affecting the health of oaks and associated species. The southern pine beetle reached epidemic levels in 7 states and threatened timber, wilderness values, and habitat for the endangered red-cockaded woodpecker. The Douglas-fir tussock moth and western spruce budworm were at elevated levels in the Pacific Northwest. Root diseases and dwarf mistletoe in the West, fusiform rust in the South, and dogwood anthracnose in the East and South are causing concern.

International Forestry

According to a 1990 estimate by the U.N. Food and Agriculture Organization, tropical forests are disappearing at a rate of 17 million hectares (42 million acres) a year. Responding to population pressures and widespread poverty, many nations provide incentives for forest conversion to farmland and pasture. Other causes include destructive logging practices, overharvesting of commercial forests, subsidized timber sales, and large-scale development projects such as hydropower and mining.

Temperate and boreal forests, although expanding in some areas, are substantially degraded in other areas because of factors affecting forest health that include air pollution, inefficient harvesting, and lack of reforestation. In some parts of Eastern Europe, forest dieback caused primarily by air pollution is as high as 50 percent. In 1992, however, the general condition of European forests showed improvement over recent years.

Policies and Programs

Federal forestry initiatives, both domestic and international, are conducted primarily by the Forest Service in the Department of Agriculture, the Department of the Interior (DOI), the Department of Defense (DOD), and the U.S. Agency for International Development (USAID).

U.S. Forestry

In 1992 the DOI Bureau of Land Management and the USDA Forest Service announced complementary forestry initiatives based on ecosystem management.

• **National Forests.** In response to new scientific information and mounting public opposition, the Forest Service announced in June 1992 that it will phase out clearcutting as a standard practice on national forests. As part of its new ecosystem management philosophy, the agency will make greater use of other harvesting methods to provide more visually pleasing and ecologically sensitive forest management. The Forest Service will allow clearcutting only in situations where it best accomplishes specific objectives of the forest plan.

• **Reforestation.** In 1991 the national forests set a reforestation record with 503,000 acres reforested by planting, seeding, or natural regeneration. The Forest Service planted over 50 species of trees, many of genetically improved stock, to maintain vegetation diversity. Sites included timber sale harvest areas and areas devastated by natural catastrophes such as fires, insects and disease, and windstorms.

• **Harvesting.** National forests contain 47 percent of the nation's standing softwood sawtimber inventory, but, in addition to timber, these forests possess wilderness, wildlife, watershed, and recreation values. Recent Forest Service planning documents identified recreation as the fastest growing use of national forests in the 1990s. The Forest Service conducts harvesting, where permitted, in an environmentally sensitive manner to maintain long-term health and productivity. In fiscal 1991 national forests contributed 19 percent of the nation's total softwood sawtimber volume used for lumber, up from 13 percent in fiscal 1986.

In fiscal 1991 the Forest Service sold 6.4 billion board feet of timber, down a billion board feet from recent years. The decrease in volume reflected, in part, management concerns relating to biodiversity and the maintenance of old-growth forests and other habitat for endangered species. As of August 31, 1992, with 80 percent of the year's data in, national forest timber harvests had produced gross revenues to the national treasury of $985 million and net revenues of $326.7 million.

• **Forest Fire Behavior.** To combat catastrophic forest fires, the Intermountain Fire Science Laboratory in Missoula, Montana, developed a prediction system for extreme fire behavior that resulted in more effective use of fire suppression to protect human life, property, and natural resources. The lab is a unit of the Intermountain Forest Experiment Station in Ogden, Utah.

• **Interior Forestlands.** DOI manages 125 million acres of forestland, with objectives that reflect the different missions of the National Park Service (NPS), the Fish and Wildlife Service (FWS), the Bureau of Land Management (BLM), and the Bureau of Indian Affairs (BIA).

• **National Park Service.** National parks contain 37 million acres of forestland, managed solely for preservation with no commercial timber operations.

• **Fish and Wildlife Service.** National wildlife refuges contain 18 million acres of forestland managed for a diversity of indigenous wildlife populations. FWS allows limited commercial forestry operations, consistent with habitat objectives.

• **Bureau of Land Management.** BLM manages 48 million acres of forestland in 12 western states and Alaska. The BLM Total Forest Management policy, adopted in June

1992, provides an ecosystem approach to timber harvesting. In addition to timber, TFM objectives include water quality, soil conservation, fish and wildlife habitat, old growth, aesthetics, and recreation. Clearcutting will no longer be the management tool of choice, except where it is the most environmentally effective method to achieve specific management objectives, such as rehabilitating lands affected by fires, windstorms, insects and disease, and other natural disasters. TFM is reflected in the draft 10-year Resource Management Plan for western Oregon that BLM issued in August 1992 and in guidelines for managing BLM forestlands in western states and Alaska.

In 1991 the sale of BLM forest products, including lumber, fuelwood, ferns, and mushrooms, brought $145 million into the national treasury. Timber harvesting is restricted to about half of BLM forestlands because of wilderness designation, special study areas, sites that can be easily damaged, and sites with special wildlife and recreational values. In 1992 BLM planted 35,000 acres with 17.5 million seedlings and initiated an inventory of its forestlands in Alaska, in response to interest in expanding timber harvesting.

• **Bureau of Indian Affairs.** BIA manages 16 million acres of forestland in 23 states for the use and benefit of Native American owners. The objective is to perpetuate economic forest products and employment opportunities as well as ecological, environmental, and biological forest values. In November 1990 President Bush signed the National Indian Forest Resources Act, culminating a joint effort by Indian tribes, the Intertribal Timber Council, the BIA Division of Forestry, and the Congress to develop a comprehensive statute for managing tribal forests.

• **America the Beautiful Forestry.** To help meet the President's goal to plant and maintain an additional billion trees per year in urban and rural areas in this decade, the Forest Service and state foresters have established a cost-sharing Stewardship Incentive Program with the following components:

- Tree planting

- Stabilizing eroded lands

- Protecting riparian areas and wetlands

- Improving wildlife and fisheries habitat

- Enhancing forest recreation

- Establishing and renovating windbreaks and hedgerows

In 1991 the program helped landowners develop 13,596 forest stewardship plans to enhance the management of natural resources on 1.4 million acres of private forestland. In 1992 an additional 20,239 plans improved management of another 2.5

million acres. All 50 states have formed State Forest Stewardship Coordinating Committees to assist state foresters with the program. The USDA Forest Service, Soil Conservation Service, Extension Service, and Agricultural Stabilization and Conservation Service work with state foresters to provide technical assistance.

• **Forest Legacy.** The Forestry Title of the 1990 Farm Bill directed the Secretary of Agriculture to establish a Forest Legacy Program to identify environmentally valuable forestlands threatened by conversion to nonforest uses such as residential and commercial development and to protect them through voluntary conservation easements. The program protects both traditional uses of private lands and public forest values for future generations. In 1992 the Forest Service finalized guidelines to implement the program in five lead states—New York, New Hampshire, Vermont, Maine, and Washington. The fiscal 1992 appropriations bill added Massachusetts.

• **Forest Health.** In response to a serious loss of trees from insects and disease, federal agencies have accelerated forest health programs.

• **Forest Health Monitoring Program.** In 1992 state forest resource agencies worked with the Forest Service, BLM, and EPA to monitor long-term trends in health and productivity of U.S. forest ecosystems. A monitoring network now includes the New England states and parts of the mid-Atlantic, Southeast, and West. In 1992 these efforts focused on the condition of sugar maple trees through the U.S.-Canada North American Maple Project. When fully implemented, the Forest Health Monitoring Program will provide regional and national data on the health of all U.S. forestlands.

• **Accelerated Treatments.** In 1992 in response to tree mortality in the Blue Mountains of Oregon and Washington, the Forest Service directed $51.5 million to national forests in the area to salvage dead and dying trees, prevent and suppress insect and disease infestations, and conduct research on pest control techniques. Similar initiatives are planned for eastern Washington, California, and the Southeast.

• **Imitating Nature.** BLM is accelerating the harvest of dead and dying trees and, with the USDA Forest Service, is evaluating tree density control by imitating natural processes of forest development, such as thinning, planting diverse species mixes, and using fire.

• **Asian Gypsy Moth.** This forest pest, native to eastern Russia, was detected in Oregon and Washington in 1991. In April and May of 1992 the USDA Animal and Plant Health Inspection Service, the Forest Service, and the states of Oregon and Washington cooperated on eradication treatments on 124,000 acres, using a microorganism that kills the moth. A gypsy moth trapping network will monitor the success of the biocontrol treatments.

• **European Gypsy Moth.** Using biotechnology, scientists at the Center for Biological Control of Northeastern Forest Insects and Diseases in Hamden, Connecticut, developed a new strain of virus that kills the European gypsy moth. Simplified procedures for commercial production will allow its widespread use as an economical, environmentally benign control agent.

• **Pest Risk of Imports.** In response to a proposal by the timber industry to import logs from Russia, New Zealand, and Chile, the Forest Service assessed larch logs from Russia and identified several serious pest risks. The Animal and Plant Health Inspection Service used the assessments to formulate measures for excluding exotic pests from U.S. forests.

• **Hurricane Reparations.** In the aftermath of two major hurricanes in 1992, the Forest Service supplied tree-planting funds to Florida, Louisiana, and Hawaii. The agency also neared completion of reforestation efforts in South Carolina following Hurricane Hugo in 1990.

• **Pacific Yew and Cancer Treatment.** The National Cancer Institute (NCI) describes taxol, a substance found concentrated in the bark of Pacific yew trees, as the most promising anti-cancer agent of the last decade. However, large amounts of yew bark—up to 60 pounds from three mature trees—are needed to produce enough taxol for the treatment of one patient. Another complicating factor is the habitat of the yew—primarily federal old-growth forests slated for increased preservation in Oregon, Washington, Idaho, and California. Taxol may be approved for commercial use as early as 1993, when the demand likely will increase.

Federal agencies are cooperating to supply the taxol needed for research and treatment in an environmentally responsible manner. Recent actions included the following:

• The Forest Service implemented a cooperative agreement with the Bristol-Myers Squibb Company to ensure that bark is supplied for research, that the Pacific yew resource is managed to ensure sustainability, and that local people are involved in bark harvest;

• The Food and Drug Administration, BLM, and the Forest Service collaborated on preparation of an environmental impact statement covering a proposed 5-year program to collect Pacific yew bark for research, clinical trials, and treatment.

• BLM and the Forest Service each issued 1992 draft guidelines on managing the Pacific yew to help ensure a sustainable supply of yew for the medical community and for the continued existence of Pacific yew in the forest ecosystem.

The Forest Service also took the following initiatives:

• Inventoried national forests most likely to contain Pacific yew

- Initiated research on the silviculture, ecology, and management of Pacific yew

- Worked with Congressional committees for passage of the Pacific Yew Act of 1992, which authorizes the Forest Service to negotiate yew-bark sales and use receipts to pay for the program

- Collected 825,000 pounds of bark in 1991 on 10 national forests in Idaho, Montana, Oregon, and Washington and, as of August 1992, collected 646,000 pounds of bark for continued clinical trials

International Forestry

In 1992 public and private efforts were underway to promote sustainable forest management abroad, especially in developing countries. Examples follow.

USDA Forest Service. The 1990 Farm Bill authorized the secretary of Agriculture to create, within the Forest Service, an Office of International Forestry led by a deputy chief. Established in June 1991, the Office of International Forestry coordinates programs in the following areas:

- Tropical forestry technical assistance and training
- Cooperative work with USAID
- International scientific and technical exchanges
- Cooperative research with other countries
- Support to international organizations responsible for global forestry issues
- Operation of the Institute of Tropical Forestry in Puerto Rico as an international institution

The Office of International Forestry administers the following programs:

- **Forestry Support Program.** In 1992 the Forest Service provided forestry technical assistance and trainers in support of 18 USAID programs, including projects in Indonesia, Costa Rica, Nicaragua, Ghana, Uganda, Panama, Haiti, Guinea, and the Sahel.

- **Tropical Forestry Program.** The Forest Service held workshops in 1992 on agroforestry and remote sensing in East Africa, Latin America, and Panama. Forest Service staff provided training and technical advice to Brazil on wildfire suppression and helped establish "Twinning Relationships" between U.S. national forests and forestry organizations in Mali, West Africa, and Costa Rica.

- **Disaster Assistance Support Program.** The Forest Service provides emergency response, preparedness, prevention, and mitigation assistance for global disasters, including catastrophic forest fires.

• **Cooperative Research.** In 1992 Forest Service scientists participated in 11 cooperative research programs in four countries—Pakistan, Poland, the Republic of China (Taiwan), and Yugoslavia. In addition, they provided forest pest management assistance to Africa and China.

Forestry Carbon Dioxide Offset Schemes. Several private U.S. firms are experimenting with forestry projects in developing countries to offset carbon dioxide emissions from U.S. operations. For example, in August 1992 New England Power Company and Innoprise, a major Malaysian forest products company, announced a 3-year pilot project to make logging practices less damaging to the rainforest in Malaysia. The project is part of a previously announced plan by the utility to lower or offset its greenhouse gas emissions 45 percent by the year 2000. By reducing the number of surrounding trees destroyed during harvesting and by enhancing forest regeneration, the foresters hope to sustain trees capable of removing 300,000-600,000 tons of carbon dioxide from the atmosphere over the life of the project—trees that otherwise would have been removed from land that would not have been reforested. The project will be monitored by the Rain Forest Alliance, a New York-based environmental group. If successful, reduced impact logging techniques could be applied to tropical hardwood rain forests in the Philippines, Indonesia, New Guinea, and other developing nations.

Old-growth Forests and the Northern Spotted Owl

In 1992 efforts to protect old-growth forests in the Pacific Northwest by reducing federal timber harvests continued to raise concern over economic impacts on small, timber-dependent communities. These forests are the habitat of the endangered northern spotted owl, and the majority of old growth remaining in the region is located on federal lands managed by the USDA Forest Service and the DOI Bureau of Land Management.

Conservation Strategy. In January 1992 the Forest Service filed a final environmental impact statement (EIS) on managing old-growth habitat, and in March the Department of Agriculture adopted the Forest Service Conservation Strategy for the northern spotted owl, prepared pursuant to an EIS. The strategy combines old-growth preservation with methods to sustain or create old-growth characteristics for the future. Implementation of the strategy and timber sales in suitable spotted owl habitat have been stayed by federal courts pending the outcome of litigation.

Endangered Species Committee. In response to a petition from BLM, the Secretary of the Interior convened the Endangered Species Committee to consider 44 BLM timber sales in Oregon that the Fish and Wildlife Service (FWS) determined would jeopardize the existence of the northern spotted owl. As required by the Endangered Species Act (ESA), the committee consists of the secretaries of the Interior, Army, and Agriculture, administrators of the National Oceanic and Atmospheric Administration and Environmental Protection Agency, chairman of the Council on Economic Advisors, and a representative from any affected state, in this case, Oregon.

The Endangered Species Act allows the committee to exempt actions such as timber sales from restrictions of the act, if such actions meet the following criteria: absence of reasonable and prudent alternative courses of action; regional or national significance, defined on a case-by-case basis; benefits that outweigh those of alternatives; and identified mitigation measures.

In May 1992 the committee exempted 13 of 44 proposed BLM sales, denied exemptions for the other 31, and required, as mitigation for the exempted sales, that BLM implement a recovery plan for the northern spotted owl as expeditiously as possible to serve as the basis for the next BLM 10-year plan. The 13 sales encompass 1,742 acres of forestland with a timber volume of 88 million board feet. Court action has blocked the 13 exempted sales.

Recovery Plan for the Northern Spotted Owl. In January 1992 FWS designated critical habitat for the northern spotted owl, and in May 1992 the secretary of the Interior issued a draft recovery plan for the owl, as required by ESA. The recovery plan includes the following elements:

• A network of 196 Designated Conservation Areas (DCAs) on federal lands to be managed for owl habitat, encompassing 7.5 million acres, 2.1 million acres of which are within national parks and wilderness areas.

• Guidelines for forest management and other activities on DCAs, prohibiting commercial timber harvest in owl habitat.

• Management recommendations for federal lands outside DCAs, initially adding 410,000 protected acres.

• A set of standards for judging when the northern spotted owl has reached recovery and thus no longer needs ESA protection.

• Recommendations for contributions from nonfederal forestlands to support northern spotted owl populations.

• A monitoring and research program to seek information on the owl and its habitat and to develop and test techniques for creating and maintaining owl habitat while allowing forest management.

• Mechanisms to implement the recovery plan, providing oversight and coordination.

Alternative Plan. At the same time that DOI prepared the draft recovery plan, it also prepared an alternative plan to protect owl habitat at a lower cost to the region's economy. The alternative plan would create a network of 75 DCAs encompassing 2.8 million acres of nonreserved lands and 1.5 million acres of reserved lands. Habitat in other parts of the range would not be managed with owl needs as the primary consideration. Implementation of the alternative plan would require congressional approval, since it does not meet all the goals of the Endangered Species Act.

Resource Management Plans. Court actions also have halted BLM timber sales in western Oregon pending completion of Resource Management Plans (RMPs) for six districts. In 1992 the Bureau prepared draft RMPs and draft environmental impact statements. Based on an ecosystem approach to management, the plans will determine old-growth timber harvests for the next decade.

Forests for the Future

In preparation for the June 1992 U.N. Conference on Environment and Development (UNCED), President Bush launched the Forests for the Future initiative, calling on the donor community to double world assistance for forests to $2.7 billion. He pledged to increase annual U.S. bilateral forest assistance by $150 million and endorsed a pledge of $20 million to the Brazil Rainforest Pilot Program, an additional $2 million for the International Tropical Timber Organization, and reform of the World Bank forestry policy.

At UNCED, President Bush and 130 other heads of state signed a "non-legally binding authoritative statement of principles for a global consensus on the management, conservation, and sustainable development of all types of forests." President Bush first called for a world forestry agreement at the 1990 G-7 summit in Houston, and the United States and other G-7 countries support the UNCED statement of principles as a first step toward a binding forest convention.

At the Rio Conference, participating nations accepted the report, Conservation and Development of Forests, by the UNCED Secretary General on conditions of world forests and threats facing them. They also adopted Agenda 21, which includes a common approach for countries to use in integrating national actions and international cooperation with forest conservation and sustainable development.

NATIONAL ENVIRONMENTAL POLICY ACT

The National Environmental Policy Act (NEPA) is the nation's environmental magna carta. With passage of NEPA in 1969, the United States adopted a national policy to "encourage productive and enjoyable harmony between man and the environment" and to direct federal agencies, to the fullest extent possible, to interpret and administer U.S. policies and laws in accordance with this policy. NEPA established the Council on Environmental Quality (CEQ) to advise the President and assist federal agencies with compliance, and the act mandated procedural requirements to fulfill its substantive goals.

Conditions and Trends

To help ensure that consideration of environmental values is systematic in federal decisionmaking, NEPA section 102(2)(C) requires preparation of a "detailed statement" for "major federal actions significantly affecting the quality of the human environment." The "detailed statements" have come to be called environmental impact statements (EIS), and the process underlying the preparation of those statements by federal agencies has come to be called the environmental impact assessment or NEPA process.

CEQ has promulgated regulations to implement the procedural provisions of NEPA (40 CFR Parts 1500-1508). The primary purpose of these regulations is to ensure that federal agencies consider the substantive goals of NEPA in the course of decision-making and that the public and environmental agencies are involved in that process.

Over the years, some observers have come to associate NEPA implementation more with documentation of environmental analysis than with the substance of the analysis itself. However as CEQ regulations state:

> Ultimately, of course, it is not better documents but better decisions that count. NEPA's purpose is not to generate paperwork—even excellent paperwork—but to foster excellent action.

The purpose of NEPA then is protection of the environment, not production of paperwork. Many federal agencies are striving to better integrate NEPA goals with their own policies and actions. To recognize and encourage such efforts, CEQ developed a Federal Environmental Quality Awards Program to showcase model NEPA compliance in two categories:

• Exemplary Action Award, based on a specific example of an environmental impact analysis

- Exemplary Agency Award, presented to the agency that best demonstrates a consistently high standard of performance in the NEPA process

CEQ asked each federal department to nominate one exemplary action and one agency within that department. The Environmental Protection Agency (EPA) also nominated agencies in both categories based on its experience in reviewing environmental impact analyses. Award recipients will be recognized in future CEQ annual reports.

Policies and Programs

NEPA Regional Conferences

During 1991 and 1992 CEQ hosted five regional conferences that focused on ecological issues in NEPA analyses. The conferences, cosponsored by EPA with assistance from the Department of Defense and the U.S. Fish and Wildlife Service, were held in Denver, Atlanta, Boston, Chicago, and Anchorage. They afforded CEQ the opportunity to meet with NEPA practitioners in the field to discuss methods for implementing national policies that relate to conserving biodiversity. Experts in biodiversity conservation spoke at each conference, and NEPA practitioners presented papers and case studies on methods of incorporating ecological principles into the NEPA process. Methods include geographic information systems (GIS) and data available through The Nature Conservancy's Natural Heritage Program. In addition, CEQ devoted a segment of each conference to the purpose, theory, and practice of preparing environmental assessments. After reviewing information gained from the conferences, CEQ will issue a report on biodiversity in NEPA analyses. CEQ intends to issue guidance to the federal agencies on NEPA consolidation in the EA process.

Environmental Assessment Survey

In 1992 CEQ surveyed federal agencies to determine whether EAs are facilitating effective NEPA compliance. The survey, which covered 45,000 EAs prepared during the year, also showed that while the EA process often facilitates mitigation, many agencies do not use EAs as originally envisioned by the regulations in three significant respects:

- Agencies rarely use an EA to determine whether an EIS is necessary

- Agencies prepare EAs that are often quite lengthy and correspondingly costly

- Agencies appear to rely heavily on mitigation measures to justify EAs and FONSIs

The survey results suggest that some agencies have avoided preparing EISs based on the erroneous perception that preparation of EAs does not require public involvement. However, the results also indicate that increased reliance on EAs may be attributed in part to early NEPA integration in agency planning, which results in mitigation measures being incorporated into a project site and design, with adverse environmental effects consequently reduced or eliminated. The survey results are available in their entirety from CEQ.

Cumulative Impact Analysis

While most federal managers recognize the need to assess the cumulative impacts of agency actions, EISs are often limited in scope. As a result, EISs frequently miss the big-picture effects of individual proposals, and litigation ensues. While a consensus exists on the need to evaluate cumulative effects, federal agencies do not all agree on the methodology or conceptual approach to that analysis.

To promote dialogue on this issue, CEQ hosted workshops to develop methodological approaches to cumulative impact analysis. The Council is collecting case studies on how agencies conduct analyses, and a handbook in preparation will incorporate different agency approaches. CEQ also is working with the Canadian Federal Environmental Assessment Review Office to share information regarding approaches to cumulative impact analysis.

International Activities

CEQ is involved in several interagency efforts regarding environmental impact assessment (EIA) in the international context.

Protocol on Environmental Protection. CEQ helped to develop legislation for the Protocol on Environmental Protection signed in 1991 by the United States and other Antarctic Treaty parties. The Council also worked with the National Science Foundation to draft EIA regulations to implement the Protocol.

Pelosi Amendment. CEQ is involved in an interagency process to ensure that, pursuant to the Pelosi Amendment (22 U.S.C. 262m), the U.S. executive director of each multilateral development bank abstains from voting on loan proposals that have a significant impact on the environment unless the proposal is accompanied by an environmental assessment or comprehensive environmental summary received at least 120 days before the board votes on the proposal.

Convention on Environmental Impact Assessment in a Transboundary Context. In February 1992 the United States signed the Convention on Environmental Impact Assessment in a Transboundary Context, which was negotiated under the auspices of the Economic Commission for Europe. CEQ is developing guidelines to implement the Convention for projects with significant adverse transboundary environmental impacts.

Technical Assistance. Other countries frequently ask CEQ to assist them in formulating an EIA process. In 1992 CEQ staff met with delegations from Russia, Spain, Cameroon, Malawi, Egypt, Korea, China, Japan, France, and Italy, among others, to discuss the NEPA process and its role in federal decisionmaking. In addition, a CEQ attorney assisted officials from the Turkish Ministry of the Environment in drafting EIA procedures.

NEPA Training

In 1992 CEQ undertook a NEPA training initiative to ensure that quality NEPA education and training are available across the country. As part of the initiative, CEQ compiled a compendium of NEPA courses currently offered by federal agencies, universities, and private entities. Copies are available from CEQ. A sampling of courses in which CEQ took part follows.

Justice Department Course. In February 1992 the Legal Education Institute of the Department of Justice (DOJ), assisted by CEQ, presented a NEPA course designed for government lawyers in Washington, D.C. The course is being offered again in January 1993 in Seattle, Washington.

University Short Courses. In November 1992 CEQ cosponsored an annual week-long NEPA course at the Duke University School of the Environment. The course will be repeated in March 1993.

American Bar Association Course. CEQ and DOJ lawyers helped organize a NEPA course sponsored by the American Law Institute of the American Bar Association in Washington, D.C., in November 1992.

Federal Leadership Training. During the year, CEQ participated in leadership-training courses for federal agency personnel, including courses sponsored by the Federal Executive Institute and the Office of Personnel Management Executive Seminar Centers.

Agency Implementation

In 1992 agency implementation of NEPA regulations produced the following developments.

Emergency Alternatives. CEQ regulations provide for alternative compliance arrangements in the event of emergency circumstances.

Department of Energy. In July 1991 the Department of Energy (DOE) and CEQ reached an agreement regarding alternative NEPA arrangements for the immediate drawdown of Par Pond at the closed Savannah River nuclear weapons facility. CEQ concurred with DOD that the weakened state of the Par Pond Dam constituted an emergency under CEQ regulations and that drawdown of the pond was necessary to avert a potentially damaging and life-threatening flood in the case of dam failure. Alternative NEPA arrangements included preparation of a special assessment and implementation of specific mitigation measures.

Agency for International Development. When the Mount Pinatubo volcanic eruption in the Philippines was followed by the rainy season, the Agency for International Development consulted with CEQ regarding EIA procedures for disaster assistance. Consultation with CEQ also took place for the shipment of emergency supplies of fuel and food to the Bosnia area.

Department of Defense. Hurricane Andrew, which devastated south Florida in August 1992, virtually destroyed Homestead Air Force Base. Seeking to relocate some essential missions, the Department of Defense (DOD) sought CEQ assistance in ensuring that temporary relocation of activities were conducted in compliance with NEPA. In consultation with CEQ, DOD committed to preparing EAs for each relocation, with public involvement in preparation of documents to the extent practicable. DOD agreed to contact CEQ immediately for additional consultations on appropriate alternative NEPA arrangements, if an EA indicated that significant environmental impacts might occur as a result of relocation.

Agency NEPA Regulations. In 1992 after consultations with CEQ, DOE, the USDA Forest Service, and the Bureau of Land Management published new agency NEPA regulations. CEQ also consulted with the following agencies in the process of publishing new NEPA regulations or revising existing ones: the Office of Surface Mining, Bureau of Mines, Bureau of Indian Affairs, Indian Health Service in the Department of Health and Human Services, and National Indian Gaming Commission.

Agency Workshops on NEPA Implementation. In addition to regional conferences, CEQ also participated in the following NEPA workshops:

Federal Agency Workshops. CEQ took part in workshops on NEPA implementation issues sponsored by federal agencies, including EPA, DOE, the Department of the Army and Army Corps of Engineers, the Minerals Management Service, National Park Service, and Fish and Wildlife Service. Conferences on Native Americans and the environment included the First National Tribal Conference on Environmental Management, held in Cherokee, North Carolina, and the annual Bureau of Indian Affairs NEPA workshop.

Environmental Impact Assessment Conferences. CEQ took part in conferences on environmental impact assessments sponsored by EPA Regions III and IV, the

Advisory Committee on Intergovernmental Relations, and the American Association of Law Schools Environmental Law Committee.

NEPA Case Law

In 1992 NEPA case law included decisions by the United States Supreme Court and by federal courts of appeal and district courts.

Old Growth Timber Sales

Robertson v. Seattle Audubon Society, 112 S.Ct. 1407 (1992).

The case concerned challenges to the policy of the USDA Forest Service and the DOI Bureau of Land Management (BLM) to allow the harvesting and sale of timber from old-growth forests in the Pacific Northwest. The Portland Audubon Society originally brought suit against BLM in 1987, and the Seattle Audubon Society challenged Forest Service decisions in 1989. Congress, attempting to resolve the issues raised in these cases, passed section 318 of the Department of the Interior and Related Agencies Appropriations Act of 1990, which went into effect on October 23, 1989.

Section 318 mandated a specified volume of timber sales, established a timber management plan for national forest and BLM lands in Oregon and Washington, and prohibited sales in identified spotted owl habitat areas. It also declared that management of areas according to subsections (b)(3) and (b)(5) of this section on the 13 national forests in Oregon and Washington and BLM lands in western Oregon known to contain northern spotted owls is adequate consideration for the purpose of meeting the statutory requirements that are the basis for the consolidated cases captioned *Seattle Audubon Society et al. v. F. Dale Robertson*, Civil No. 89-160; *Washington Contract Loggers Assoc. et al. v. F. Dale Robertson*, Civil No. 89-99 (order granting preliminary injunction); and the case *Portland Audubon Society et al. v. Manuel Lujan, Jr.*, Civil No. 87-1160-FR. Pub. L. No. 101-121, 103 Stat. 701, 754-50, § 318(b)(6)(A). Based on section 318, the district court in Seattle Audubon vacated its preliminary injunction, and the district court in Portland Audubon granted the government's motion to dismiss.

On appeal, plaintiffs in the consolidated cases claimed that section 318 violates the separation of powers doctrine and was thus unconstitutional. The Ninth Circuit Court of Appeals agreed, holding that the Congress had not repealed or amended the environmental laws underlying the litigation but rather had endeavored to instruct the federal courts to reach a particular result in specified pending cases (Seattle Audubon Society v. Robertson, 914 F.2d 1311 (9th Cir. 1990)).

The United States Supreme Court, however, reversed the court of appeals and concluded that section 318 compelled changes in the law, not findings or results under old law. The citations to specific lawsuits, said the Court, "served only to identify the five 'statutory requirements that are the basis for' those cases . . ." To the extent that the applicable language affected the adjudication of those cases, it did so by modifying the provisions at issue. Further, the Court noted that the Congress may amend substantive law in an appropriations statute as long as it does so clearly and found that congressional intent in section 318 was "not only clear, but express."

Justiciability: Finality and Standing

Foundation on Economic Trends v. Lyng, 943 F.2d 79 (D.C. Cir. 1991)

Plaintiffs sought an injunction and declaratory judgment against the Department of Agriculture (USDA) for failure to prepare an EIS for the USDA "germplasm preservation program." On the merits, the district court granted summary judgment for defendants on the ground that plaintiffs had failed to identify a proposal for which an EIS should have been prepared under NEPA.

Before reaching the merits, the Court of Appeals for the District of Columbia Circuit questioned plaintiffs' standing under the recent Supreme Court decision in *Lujan v. National Wildlife Federation,* 110 S.Ct. 3177 (1990). First, the court noted that when a party seeks judicial review of agency action under the Administrative Procedure Act (APA), as in this case, plaintiffs must identify the agency action affecting their interests and must demonstrate that the interest sought to be protected is arguably within the zone of interests to be protected by the statute.

Next, the court explained that plaintiffs in the instant case claimed informational standing, that is, injury based on the absence of information contained in an EIS. The court concluded, however, that under the Lujan decision, an agency's refusal to prepare an EIS was not a particular agency action that could be the source of injuries sufficient to confer standing. Plaintiffs in NEPA cases must point to an action at least arguably triggering the agency's obligation to prepare an EIS.

Based on this analysis, the court held that the germplasm preservation program for which plaintiffs sought the preparation of an EIS was not an identifiable action or event. "Plaintiffs fail to target their complaint to a particular proposal for federal action and fail to identify any revisions or changes taken by any of the defendants in the germplasm program that would trigger the obligation to prepare an environmental impact statement under NEPA." The court vacated the judgment in the lower court and remanded the case with instructions to dismiss the complaint for lack of jurisdiction.

Public Citizen v. Office of the U.S. Trade Representative, No. 92-5010, 1992 (D.C. Cir. Aug. 7, 1992) (available on WESTLAW, CTADC database, WESTLAW No. 186568)

Plaintiffs claimed that the President and the U.S. Trade Representative were required to prepare EISs for negotiation of a North American Free Trade Agreement (NAFTA) and the Uruguay Round of multilateral trade negotiations under the General Agreement on Tariffs and Trade. The district court dismissed the suit on the ground that plaintiffs lacked standing to pursue their claims, and it also noted ripeness problems.

The Court of Appeals for the District of Columbia Circuit concluded that plaintiffs had failed to identify a final agency action within the meaning of the Administrative Procedure Act (APA). The basis for judicial review of NEPA claims is final agency action under the APA. In the absence of a final agreement in either the NAFTA or Uruguay Round negotiations, and the possibility that no final agreements would result, the court concluded that no final agency action under the APA existed, and the district court therefore did not have jurisdiction. The Court of Appeals, accordingly, affirmed the dismissal.

Following the President's announcement of a NAFTA agreement on September 12, 1992, Public Citizens filed a new action to compel preparation of an EIS only with respect to the NAFTA, *Public Citizens v. USTR*, No. 92-2102 (filed September 15, 1992). Public Citizens filed a motion for summary judgment on October 15, 1992.

Natural Resources Defense Council v. Lujan, 768 F.Supp. 870 (D.D.C. 1991)

National and local Alaskan environmental organizations challenged a legislative EIS (LEIS) prepared by defendants in 1987, concerning the future management of the Arctic National Wildlife Refuge (ANWR) and subsequently filed a motion for a preliminary injunction to compel defendants to circulate as a draft supplemental LEIS a report issued by defendants in 1991. In a motion to dismiss, defendants argued, among other things, that plaintiffs lacked standing under NEPA, that the NEPA claims plaintiffs sought to raise were not justiciable under the separation of powers doctrine, and that a statutory conflict between the Alaska National Interest Lands Conservation Act (ANILCA) and NEPA prevented defendants from revising the 1987 LEIS. The court held that plaintiffs had standing to challenge the adequacy of the LEIS under NEPA and issued a declaratory judgment that defendants had violated NEPA by not issuing the 1991 report as a supplemental LEIS.

Before reaching the standing issue, the court noted that earlier litigation between the same parties on procedural aspects of the issuance of the 1987 LEIS had resulted in a determination that the LEIS did not fall within the modified procedures for legislative proposals found in the CEQ regulations. Those regulations exempt legislative EISs from the requirement that an EIS be prepared in two stages, permitting an agency to transmit a single LEIS to the Congress, other agencies, and the public for review and comment. This modified procedure is not available, however, where "[t]he proposal results from a study process required by statute." The ruling in the earlier case, according to the court, had consequences in the instant case because it meant that the 1987 LEIS was subject to the same requirements as any EIS.

With respect to standing, plaintiffs alleged that the failure of the 1987 LEIS to comply with NEPA, ANILCA, and the APA made the LEIS so inadequate that the Congress would not have the full information it needed to make a decision on the future use of ANWR; if an uninformed Congress acted to allow oil exploration in ANWR, it would adversely affect plaintiffs who use and enjoy the area. In addition, plaintiffs alleged that the deficiencies in the 1987 LEIS adversely affected their public participation rights because the data and analyses in it were flawed.

The court found plaintiffs to have alleged that the inadequacies of the 1987 LEIS created "a risk that serious environmental impacts will be overlooked" and concluded that this sufficed to establish injury in fact. Since plaintiffs had alleged that they used and enjoyed the area, they had met both the requirement of geographical nexus and the requirement that their interests be within the NEPA zone of interests:

> There is no doubt that "recreational use and aesthetic enjoyment" are among the sorts of interests that NEPA was specifically meant to protect.

Plaintiffs' participatory rights were also found to be within the zone of interests protected by NEPA:

> If a decision subject to NEPA . . . is made without the information that NEPA seeks to put before the decisionmaker, the harm that NEPA seeks to prevent occurs.

On the separation of powers argument, the court ruled that the exercise by the Congress of its legislative powers would not nullify any declaratory relief the court might grant. A legislative EIS is not simply a legislative aid but rather serves in part to ensure that the public has an opportunity to participate meaningfully in decisionmaking at the administrative and legislative levels. Accordingly, the court held that the requirement of an adequate EIS for legislative proposals can be enforced by a private right of action.

The court also rejected defendants' argument that ANILCA overrides or repeals NEPA, thus precluding the revision or supplementation of the 1987 LEIS. ANILCA by its terms does not repeal NEPA, and repeal by implication is disfavored. In the context of NEPA, which requires federal agencies to comply with its mandate to "the fullest extent possible," courts have been especially reluctant to hold that another statute overrules it.

As to the merits of plaintiffs' claim regarding whether the 1991 report should be circulated as a supplement to the 1987 LEIS, the court first reiterated the finding of the earlier litigation that the 1987 LEIS fell into the exception to the exemption in the CEQ regulations for legislative EISs. Addressing defendants' argument based on the conclusion that a legislative EIS must be prepared in two stages did not mean that it must also be supplemented, the court then noted that "[a]n agency's duty under NEPA is a continuing one." So long as the decisionmaking process is still pending, the dissemination of relevant environmental information will be promoted by supplementation.

Having determined that the 1987 LEIS was subject to supplementation, the court found that the information contained in the 1991 report was "significant" as that term is defined in the CEQ regulations, although it was not environmental information and no change occurred in assessed impacts on the environment. Citing earlier NEPA cases, the court held that supplementation is required where new information so alters a project's character that a new "hard look" at the environmental consequences is necessary, even where the change is not strictly environmental. The court noted that defendants considered the new information in the 1991 report to be important inasmuch as it included new geologic as well as economic data and supported the

1987 LEIS recommendation of full oil leasing in ANWR's coastal plain. Because the information contained in the 1991 report was significant, the court held that defendants' failure to issue and circulate that report as a supplemental LEIS was a violation of NEPA. The United States has filed a notice of appeal.

Decision to Prepare an EIS

Roanoke River Basin Association v. Hudson, 940 F.2d 58 (4th Cir. 1991), cert. denied, 112 S.Ct. 1164 (1992)

The Army Corps of Engineers issued a permit to the City of Virginia Beach to construct a water-intake structure and pipeline to divert water from Lake Gaston—part of the Roanoke River system—to Virginia Beach's water supply. Prior to issuing the permit, the Corps prepared an environmental assessment on this proposal and concluded that, with mitigation, the environmental impacts would not be significant and the preparation of an EIS was not required. Plaintiffs challenged the Army Corps decision not to prepare an EIS in light of the possible impact that the pipeline might have on the striped bass population of the Roanoke River.

Affirming a lower court's finding in favor of the Army Corps, the Court of Appeals for the Fourth Circuit held that if a mitigation condition eliminates all significant environmental effects, no EIS is required. The only significant impact alleged was the impact on striped bass, and the Army Corps determination that a mitigation condition would eliminate any adverse impact was supported by the record. The court also noted that the existence of a disagreement between the Army Corps and other federal agencies as to the need for an EIS is not by itself grounds for a court to require an EIS.

Supplementing an EIS

Coker v. Skidmore, 941 F.2d 1306 (5th Cir. 1991)

In this decision, the Court of Appeals for the Fifth Circuit overturned the district court's determination that a supplemental EIS had to be prepared prior to construction of a 5.4-mile levee on the Yazoo River in Mississippi. The levee construction was part of the Yazoo River Basin Flood Control Project, for which the Corps of Engineers prepared a programmatic EIS in 1975. The District Court found that the 1975 EIS was outdated and thus had to be supplemented before the levee construction could proceed.

The court of appeals overturned the district court's ruling, because the lower court had not evaluated the case using the CEQ and Corps NEPA standards for supplementation. The CEQ regulations (40 CFR § 1502.9(c)) state that an agency must supplement an EIS if the agency makes substantial changes in the proposed action relevant to environmental concerns, or if significant new circumstances or information relevant to environmental concerns and bearing on the proposed action or its impacts arise. The district court explicitly found no evidence that construction of the levee would have a significant environmental effect. The court noted that an EIS need not be supplemented "whenever new information concerning a project comes to light...or when portions of it become out of date."

Functional Equivalent Doctrine

Western Nebraska Resources Council v. EPA, 943 F.2d 867 (8th Cir. 1991)

In relevant part, the Court of Appeals for the Eighth Circuit found that EPA need not comply with NEPA prior to its actions under the Safe Drinking Water Act (SDWA). Following a long line of cases dealing with EPA administration of pollution control laws, the court found that the SDWA is the "functional equivalent" of the NEPA process and thus, adherence to NEPA formal requirements is not necessary.

Environmental Assessments

Sierra Club v. James D. Watkins, No. 88-3519 (34 E.R.C. 2057, D.D.C. Dec. 9, 1991; available on LEXIS, Genfed library, Dist file, LEXIS No. 17482)

This case dealt with NEPA compliance for the importation of spent nuclear fuel rods from Taiwan to the United States. The court found the EA prepared by DOE to be inadequate, despite finding that plaintiffs had failed to demonstrate that the proposed action would have a significant environmental impact. The court upheld the DOE FONSI for the project.

The court premised what it acknowledged to be an unusual opinion on a discussion of the purpose and function of EAs. While acknowledging that one function is to determine whether the proposed action will have significant environmental impacts and thus require preparation of an EIS, the court relied on that portion of the CEQ regulation which states that an EA serves to aid an agency's compliance with NEPA, where no EIS is necessary (40 CFR §1508.9(a)(2)). The court found further support for its determination that EAs have an independent function in section 102(2)(E) of NEPA, which requires agencies to "study, develop and describe appropriate alternatives to recommended courses of action in any proposal which involves unresolved conflicts concerning alternative uses of available resources."

The court noted that the examination of alternatives was bounded by the "rule of reason" and that the level of analysis should be concomitant with the severity of the impacts. In this instance, however, the court found both the agency's choice of alternatives and analysis of cumulative risks of radiation exposure to be inadequate. The court also ruled that DOE should have considered risks associated with a worst case scenario, without discussing the 1986 amendment to the CEQ regulations related to that issue.

Natural Resources Defense Council v. Duvall, 777 F.Supp. 1533 (E.D.Cal. 1991)

In this case, the district court found that the Bureau of Reclamation FONSI for proposed regulations to implement provisions of the Reclamation Reform Act could not be sustained. The court first discussed the differences it perceived between the standards for review of the adequacy of an EIS and review of a FONSI determination. It decided not to follow the United States Supreme Court choice of the arbitrary and capricious standard in *Marsh v. Oregon Natural Resources Council*, 490 U.S. 360 (1989), which concerned a decision to prepare a supplemental EIS. Rather, in reviewing the Bureau decision not to prepare an EIS ab initio, the court chose to follow what it perceived as the less deferential "reasonableness" standard.

The court found the FONSI to be insufficient in several respects. First, the court, finding no applicable reference in CEQ regulations or guidance, saw "no apparent reason" to believe that incorporating information by reference is appropriate in the context of an EA. The court noted that even if incorporation by reference was permissible in an EA, the requirements set forth in the applicable CEQ regulation (40 CFR § 1502.21) were not adhered to in this instance.

Second, the court disputed economic assumptions made in the EA, as well as the reasonableness of the EA conclusions. The court also faulted the analysis for focusing primarily on California case studies, although the regulations would apply in other western states. Finally, the court found that in light of explicit language in the Reform Reclamation Act requiring the Secretary of the Interior to consider water measures, the EA improperly omitted consideration of an alternative regulatory program that focused on water conservation measures. The United States has appealed the decision to the Ninth Circuit.

NEPA Trends

A total of 94 cases involving a NEPA claim were filed against federal departments and agencies in 1991, up slightly from 1990. In those cases, 14 injunctions were issued. The Department of Transportation was the most frequent defendant with 24 cases filed against it. The Department of the Interior was second with 13 cases, and the Army Corps of Engineers was third with 11 cases.

As in previous years, the most common complaint, the cause of action in 41 cases, was no EIS when one should have been prepared. An inadequate EIS was the second most common complaint, raised in 26 cases. Individuals or citizens groups were the most common plaintiffs (41 cases), followed by environmental groups (37 cases).

In 1992 the number of draft, final, and supplemental EISs filed by federal agencies increased over those filed in 1991, with federal agencies filing a total of 512 EISs.

NEPA Glossary

The National Environmental Policy Act of 1969 (NEPA) requires federal agencies to consider the environmental impacts of major federal actions, weigh alternatives, and seek public participation early in the planning process. The following are terms introduced by NEPA regulations promulgated by CEQ:

- **Environmental Assessment (EA).** This concise public document analyzes the environmental impacts of a proposed federal action and provides sufficient evidence to determine the significance of impacts.

- **Findings of No Significant Impact (FONSI).** This public document briefly presents the reasons why an action will not have a significant effect on the human environment and therefore will not require the preparation of an environmental impact statement.

- **Environmental Impact Statement (EIS).** This is the "detailed statement" required by NEPA when an agency proposes a major federal action that significantly affects the quality of the human environment. An EIS must include an analysis of reasonable alternatives to the proposed action.

- **Record of Decision (ROD).** This public document reflects an agency's final decision, rationale behind that decision, and commitments to monitoring and mitigation.

- **Cumulative Impact.** This impact on the environment results from the incremental impact of the action when added to other past, present, and reasonably foreseeable future actions regardless of what agency (federal or nonfederal) or person undertakes such other actions. Cumulative impacts can result from individually minor but collectively significant actions taking place over a period of time.

ENFORCEMENT

Strong enforcement supports environmental progress in two ways: by correcting specific violations of environmental law and, more generally, by deterring potential violations. Firm and fair enforcement ensures a level playing field for private-sector competitors and spurs polluters to look beyond mere compliance to preventing pollution in the first place.

In 1992, continuing a 3-year trend, civil and criminal penalties for federal environmental violations reached an all-time high. This trend likely will continue, since the Pollution Prosecution Act of 1990 authorized additional civil and criminal investigators for the Environmental Protection Agency (EPA) through 1995. In addition, EPA and the Department of Justice (DOJ) introduced new enforcement techniques and market incentives to gain the greatest environmental benefits as rapidly as possible.

Conditions and Trends

To implement an Enforcement Four-Year Strategic Plan, EPA pursued a risk-based approach to environmental enforcement. The plan has the following objectives:

- Identify violations that involve the most significant environmental and health risks

- Employ, where appropriate, several statutes in an integrated, comprehensive enforcement action

- Use innovative enforcement tools such as environmental auditing to promote pollution prevention

- Choose enforcement actions that maximize the deterrent effect on other potential violators.

Historically most environmental enforcement has been media-specific, focusing, for example, on violations affecting air or water in specific locations. A large part of the enforcement effort will continue to be media-specific, seeking penalties through provisions of the Clean Air Act (CAA), Clean Water Act (CWA), RCRA, or other statutes. However, these media-specific priorities increasingly will be supplemented by targeted multimedia initiatives to deal with violations caused by specific pollutants or industries across entire geographic areas and ecosystems. For example, multimedia cases were filed in 1992 affecting benzene emissions, the import and export of hazardous wastes, pulp and paper manufacturing, metal manufacturing and smelting, organic chemical manufacturing, and cleanups along the U.S.-Mexico border.

Because of the vast number of potential prosecutions, the federal government encouraged greater self-auditing and policing by regulated entities. In 1991 DOJ announced an environmental audit policy that views such voluntary actions as mitigating factors in the exercise of enforcement discretion. Responding to the concern that aggressive enforcement might deter voluntary self-audits and disclosure, the policy describes the factors DOJ considers in bringing criminal prosecutions.

Policies and Programs

The following are highlights of federal environmental enforcement in 1992.

Multimedia Enforcement

EPA has modified procedures for inspecting facilities and for screening to identify cases for multimedia enforcement. To provide the capability to find patterns of noncompliance within or across environmental media, EPA developed a system linking various compliance databases and made it available for use by the states. Major multimedia enforcement efforts in 1992 are described here:

Industrial Enforcement. EPA launched the first coordinated enforcement initiative involving specific industrial sectors. On September 10, 1992, EPA and DOJ filed a series of enforcement cases against industrial facilities in the pulp and paper manufacturing, metal manufacturing and smelting, and industrial organic chemical manufacturing sectors—sectors which EPA identified as having a significant incidence of noncompliance with environmental standards and large amounts of reported releases of toxic substances to the environment.

The 22 civil and administrative enforcement actions filed involved violations of several environmental statutes including the CWA, RCRA, CAA, Toxic Substances Control Act, and Emergency Planning and Community Right-to-Know Act. By concentrating enforcement authorities on noncompliant industries, EPA expects to increase compliance rates with all environmental statutes throughout these industrial sectors and to foster greater use of pollution prevention, pollution reduction, and other innovative solutions to compliance problems.

Benzene Initiative. On August 5, 1992, EPA and DOJ announced the settlement of a $1 million benzene enforcement action against a Chevron petroleum refinery in Philadelphia. The settlement was by far the largest penalty to date obtained for violations of hazardous air pollutant standards for benzene. At the same time, EPA and DOJ entered into a settlement with a Sharon Steel plant in Monesson, Pennsylvania, for a benzene-related civil action.

EPA also instituted seven administrative enforcement actions against companies alleged to be in violation of federal requirements aimed at preventing releases of benzene into the environment. These simultaneous filings under the CAA, RCRA, and

Comprehensive Environmental Response, Compensation, and Liability Act (CERCLA also known as Superfund), signaled an EPA crackdown on violators of environmental standards designed to protect air, water, and land from the hazards of benzene pollution.

Mexican Border. In June 1992 EPA announced the first enforcement actions in the United States resulting from efforts outlined in the Integrated Environmental Plan for the U.S.-Mexico Border Area. A total of 17 federal and state actions against facilities on the border sought $2 million in penalties. These actions involved two criminal indictments and ten civil actions for violations of federal air, toxic substance, community right-to-know, and waste laws.

Bankruptcies. The federal government used a coordinated multifacility, multimedia approach to environ- mental enforcement when companies with violations or liabilities at numerous sites filed for bankruptcy and sought to require the government to adjudicate environmental claims during the course of the bankruptcy proceeding. In 1992 the United States was involved in two complex estimation hearings involving Superfund sites: In re National Gypsum Co., (N.D. Tex.) involved four Superfund sites, and In re CF&I Fabricators of Utah, Inc., (D. Utah). The federal government also entered into comprehensive settlements for In re U.E. Systems, Inc. (N.D. Ind.) and In re Insilco Corp. (W.D. Tex.), which resolved claims at numerous Superfund sites and provided a mechanism for addressing future claims arising from pre-bankruptcy conduct.

Media-Specific Enforcement

The following are highlights of 1992 media-specific enforcement.

• **Air.** In addition to implementing the Clean Air Act Amendments of 1990, EPA settled several major cases and recovered $15 million in penalties. For example, in a case against Bethlehem Steel, EPA addressed unlawful emissions presenting significant risks to human health and the environment in densely populated areas of Pennsylvania. The action significantly reduced benzene emissions and recovered $5 million, a record monetary settlement in stationary-source enforcement.

In addition, on May 20, 1992, EPA announced the coordinated nationwide filing of over 50 Clean Air Act administrative penalty cases. Filed in 26 states and Puerto Rico, the cases enforced various Clean Air Act provisions and regulations, including new requirements for continuous emissions monitoring equipment at petroleum refineries, national emissions standards for asbestos, benzene and uranium mining waste piles, and state standards for smoke density and airborne particle emissions. The initiative helped inform the regulated community of the new administrative penalty authority granted to EPA under the Clean Air Act Amendments. This authority is one of several new enforcement tools provided to EPA by the amendments.

• **Hazardous Waste.** Increased and consistent environmental enforcement has reduced instances of deliberate improper waste disposal. Companies now carefully investigate brokers and waste disposal operations before dealing with them, and disposal companies carefully sample and inspect what they receive. This extra care in the system reflects vigorous enforcement of RCRA requirements and of the Superfund liability scheme.

• **Illegal Operator Initiative.** In 1992 EPA, through a RCRA enforcement program, targeted facilities that failed to provide required notification to the federal government regarding their handling of hazardous wastes. Non-notifiers included violators of the new toxicity characteristic rule. On February 4, 1992, EPA announced the Illegal Operator Initiative, which consists of 27 federal administrative cases and 24 state civil actions. In this effort, the first federal hazardous waste enforcement initiative to include state cases, the governments assessed $20 million in total penalties. EPA also announced federal criminal enforcement actions, including five guilty pleas and six indictments involving hazardous waste violations.

• **Superfund Response Work.** Under CERCLA, EPA has broad enforcement powers to protect human health and the environment from hazardous substances releases. The proportion of Superfund response work conducted through enforcement actions has grown steadily since CERCLA was amended in 1986. In 1992 responsible parties accounted for two-thirds of new Superfund response actions being obtained by EPA, up from 38 percent three years ago. The total amount of response work obtained from responsible parties now exceeds $5 billion, with recoveries for each of the past two years exceeding $1 billion. Each dollar spent for enforcement has produced $8 in response work or cost recovery.

• **CERCLA Settlements.** In fiscal 1992 DOJ entered into a series of settlements—valued at $285 million—in three major Superfund lawsuits that were first filed in the infancy of hazardous waste litigation—(Midco in 1979 and Stringfellow and AVX in 1983). In succeeding years, the cases were heavily litigated. In United States v. Midwest Solvent Recovery, Inc. ("Midco") (N.D. Ind.), the settlement involved implementing hazardous waste cleanups at the Midco I and II sites in Gary, Indiana, reimbursement of the United States for past costs, and payment of a fine for noncompliance with an administrative order. The settlement in United States v. J.B. Stringfellow, Jr., (C.D. Cal.), was valued at $150 million and covered past costs and injunctive relief. In United States v. AVX Corporation, et al. (D. Mass.), the settlement involved cleanup of PCB contamination and restoration of New Bedford Harbor.

• **NPL Sites.** In another Superfund development, a federal district court upheld EPA authority over that of states to set cleanup priorities and choose final remedies at federal facilities on the National Priorities List (NPL) (United States v. State of Colorado, et al. (D. Colo. 1991)). The case is now on appeal to the United States Court of Appeal for the Tenth Circuit. If upheld, this decision will expedite decisionmaking and decrease time-consuming litigation over CERCLA sites at federal facilities.

• **Toxins.** To protect human health and the environment from harmful substances, a number of federal agencies cooperate on enforcement actions. A sampling follows of cooperative efforts by such agencies as EPA, the Occupational Health and Safety Administration (OSHA), the Food and Drug Administration (FDA), the Department of Agriculture, and the U.S Customs Service.

• **Toxic Exports.** EPA, with FDA input, sets permissible tolerances for pesticide residues on food products. FDA is responsible for analyzing food imported into the United States, and EPA is responsible for ensuring that any pesticides exported from the United States meet certain export requirements. In the past, the agencies have found beef, winter fruits, and even coffee to be contaminated with toxic pesticides. Existing pesticide export requirements include bilingual labeling instructions to promote foreign worker protection and proper pesticide applications and foreign purchaser acknowledgments, issued to EPA and the nation of destination and use. These requirements help prevent illegal or unwanted American pesticides from being illegally exported or improperly used. In the past two years, EPA enforcement actions against exporters collected $700,000 in fines.

• **Food Safety.** To set food residue tolerances, FDA and EPA rely on timely and accurate data submitted by pesticide manufacturers. Residue data analyses allow EPA to make scientific judgments regarding the long-term fate or degradability of certain commercial chemicals as used in actual field situations. In 1991, as part of the National Food Basket Survey, EPA discovered that pesticide residue data for more than 25 registered chemicals appeared to be unreliable and might have been deliberately falsified. On August 25, 1992, following an extensive investigation, EPA and DOJ announced the indictment of the president of Craven Laboratories, in Austin, Texas, along with indictments or criminal informations for ten present and former staff members of the lab. Data for more that 200 pesticide residue studies may turn out to be suspect.

• **Worker Health.** Workplace injury from chemicals and chemical exposure remains a serious threat to many Americans. In a formal working relationship, EPA and OSHA focus attention on the potential threats that chemical use, distribution, or disposal create for American workers. The agencies conduct joint inspections, inspector training, and data exchange on target problem areas. EPA is litigating its first case involving a fatality from a toxic chemical. The violation was failure to report the death to EPA as required by the Toxic Substances Control Act.

• **Asbestos in Schools.** Prior to 1976 many schools were constructed with asbestos because of its fire-resistant properties. When the potential carcinogenic properties of airborne asbestos became apparent, EPA established requirements for schools to develop plans to either remove or manage asbestos in a way that protects human health and worker safety. Many schools, lacking the specialized expertise to develop management plans, retained private contractors to develop school management plans. However, EPA learned that some contractors developed plans that were inadequate for purposes of managing asbestos. One company, Hall-Kimbell, prepared management plans for 20 percent of all school districts in the United States, representing more than 25,000 separate school buildings. Litigation resulted, and, to date, 21 cases have been filed, seeking civil penalties of $6.5 million dollars and correction of each management plan.

• **Chemical Health and Safety Data.** Over 100 companies have registered for the EPA Toxic Substances Control Act Compliance Audit Program. Participating companies voluntarily submit late health and safety reports on chemicals and pay stipulated penalties that average $5,000 per late report. Such penalties are lower than the statutory maximums that EPA could impose. The program has increased compliance with the data-reporting requirement of TSCA and has provided valuable health and safety data. To date, EPA has received 5,500 late reports.

• **Water.** In 1992 EPA enforcement under the Clean Water Act focused on environmental areas with cases of major noncompliance. In addition, EPA regional offices developed geographic enforcement initiatives, such as the Grand Calumet River initiative, which impacted the Great Lakes. Major cases involving CWA enforcement included:

United States v. City of Beaumont, Texas (E.D.Tex). The district court awarded a penalty of $400,000 against the city. The penalty represented the savings to the city of noncompliance ($316,000) and a gravity component ($84,000). The court found that the city had failed to complete key pretreatment tasks on time, including sampling and analysis of industrial users, issuing permits requiring industrial self-monitoring, taking enforcement actions, and publishing a list of significant violators in the newspaper.

United States v. Butte Water Co. (D.Mont.) and United States v. Silver Bow Water, Inc. et. al (D.Mont.). In these cases, EPA enforced the Public Water Supply requirements of the Safe Drinking Water Act (SDWA) against larger water systems. On May 15, 1992, the court entered a consent decree, which settled all injunctive relief necessary to ensure that the owners and operators of the drinking water system come into full compliance with SDWA as quickly as possible.

Wildlife and Fisheries

A strong enforcement program in this area continued in 1992.

Black-Market Walrus Ivory. In Alaska, 29 individuals were charged with illegally killing walrus and trafficking in walrus ivory and drugs. An investigation by the U.S. Fish and Wildlife Service found that killings of walrus for their ivory increased after the United States banned the import of African elephant ivory in 1989. A 2-year undercover sting operation revealed a black-market trade in walrus ivory, including the trading of ivory for drugs. To date, 25 of the defendants have been tried and convicted or have pleaded guilty, typically of multiple felony charges.

Exotic Bird Smuggling. Convictions and guilty pleas to multiple felonies resulted from prosecutions of individuals involved in exotic bird smuggling from Mexico, Australia, and New Zealand into the United States. Additional prosecutions are expected to grow out of the 3-year undercover Operation Renegade.

Fishing Laws. In 1992 the U.S. Coast Guard continued air and surface patrols to detect and deter violations of fishing laws and regulations.

Highlights of Environmental Enforcement

In the Exxon *Valdez* case, the federal government collected the criminal penalty and the first payments of the civil settlement—the largest environmental settlement in history. In 1992 environmental enforcement escalated over previous years with the following accomplishments:

• For the fourth straight year, the value of civil, administrative, and judicial enforcement efforts exceeded $1 billion, with the figure for 1992 approaching $2 billion

• Civil judicial enforcement actions returned $41 for every $1 spent; with each DOJ environmental enforcement attorney averaging $12.6 million in civil penalties, natural resource damages, CERCLA (Superfund) cost recovery, and commitments to site cleanup work by defendants

• Civil penalties came to $65.6 million, doubling the previous record of $32.5 million in fiscal 1991

• Recoveries for natural resource damages rose from $45 million in fiscal 1991 to $923 million in fiscal 1992

• A record number of criminal indictments, 174, were handed down; and criminal enforcement set a record of $163 million in fines, restitution, and forfeitures, including $125 million from the Exxon plea

• Of jail time imposed, 91 percent was served

The following achievements in the federal criminal enforcement program had major deterrent effects:

• The Rockwell Corporation agreed to pay the largest Resource Conservation and Recovery Act (RCRA) criminal fine in history, $18.5 million, for RCRA and Clean

Water Act felonies in managing hazardous wastes at the Department of Energy facility at Rocky Flats, Colorado.

• Chevron pled guilty to 65 Clean Water Act violations, paying $8 million in civil and criminal fines ($6.5 million in criminal)—the third largest criminal penalty assessed under any environmental statute.

International Cooperation

Recognizing the increasingly transnational aspects of environmental protection, the federal government in 1992 fostered international cooperation in environmental enforcement. Examples of federal efforts follow.

International Treaties. The United States enforces U.S. laws that implement international environmental treaties. To enforce Annex V of the International Convention for the Prevention of Pollution from Ships (MARPOL), which prohibits discharge of plastics, the U.S. Coast Guard will bring enforcement actions against ships entering U.S. ports for all suspected violations that occur within the U.S. 200-mile exclusive economic zone. In the area of toxic exports, in June 1992 DOJ brought criminal charges against several companies and individuals for illegal export of hazardous waste to Bangladesh and Australia.

Assistance to Developing Nations. The United States assists developing nations in strengthening their own environmental enforcement capabilities.

Bilateral Linkages. Building bilateral linkages with counterparts promotes cooperation in enforcement in such areas as transboundary pollution, illegal traffic in hazardous wastes and toxic substances, and smuggling of endangered species. In 1991 the United States and Mexico established a Cooperative Environmental Enforcement Work Group to plan and carry out information exchanges and coordinated enforcement actions. In addition to training U.S. and Mexican investigative personnel, EPA is establishing a compliance tracking system to support multimedia environmental enforcement in the U.S.-Mexico border area. DOJ has represented the United States and the International Boundary and Water Commission in judicial proceedings pertaining to the U.S.-Mexico border environment.

Global Networks. The nation is developing global environmental enforcement networks. For example, in 1992 the United States participated in international meetings such as the Twelfth Annual Attorneys General Border Conference, held in Phoenix, Arizona, and the Second International Conference on Environmental Enforcement held in Budapest.

FEDERAL FACILITIES MANAGEMENT

In 1992 President Bush signed into law the Federal Facilities Compliance Act, which waives federal sovereign immunity regarding enforcement actions taken by EPA and the states under the Resource Conservation and Recovery Act (RCRA). EPA and the states now have explicit authority to impose fines and other penalties against federal agencies for hazardous waste violations. The law ensures that federal facilities adhere to the same environmental regulations as the private sector.

Conditions and Trends

The current environmental focus of federal facilities management is pollution prevention, control, and cleanup. Efforts range from recycling office copier paper to multi-billion dollar hazardous waste cleanup programs.

Pollution Prevention

The federal government has endorsed recycling and energy efficiency to benefit the environment and provide cost-savings. Federal facilities are using their purchasing power to develop markets for recycled products and to reduce energy consumption in government buildings.

Pollution Control

Federal agencies continued to clean up hazardous waste sites located at their facilities in accordance with the Comprehensive Environmental Response, Compensation, and Liability Act (CERCLA or Superfund). The National Priorities List (NPL), maintained by EPA to identify the nation's most seriously contaminated toxic waste sites, included 116 sites at federal facilities.

The Department of Defense (DOD) and the Department of Energy (DOE) lead in numbers of NPL sites, although other agencies, including the Department of the Interior, Department of Transportation, and General Services Administration, also are cleaning up NPL sites. EPA and the states continued to work with federal agencies to hasten environmental remediation and restoration.

Environmental Planning

In addition to the cleanup of facilities, federal agencies instituted environmental quality goals. To move beyond compliance with environmental regulations, federal agencies took the following actions:

• Collected and monitored baseline environmental information at federal facilities

• Monitored environmental requirements

• Assessed environmental management practices and revised as needed

• Instituted source reduction and waste minimization

• Instilled an environmental ethic in employees through training and education

• Developed and implemented comprehensive environmental audits

Policies and Programs

Two executive orders issued by the President in 1991—one on waste reduction and recycling at federal facilities and the other on reductions in energy use at federal facilities—set the tone for environmental programs in all federal departments and agencies. The General Services Administration (GSA) is the lead agency for waste reduction, recycling, and energy management by federal facilities. In addition to such GSA programs, federal agencies undertook a number of other environmental projects. A sampling of efforts made by federal facilities in 1992 to comply with the executive orders, to prevent pollution, and to clean up existing pollution follows.

General Services Administration

GSA administers an array of environmental programs, among them, waste reduction, gas and diesel restrictions, recycling and the use of recycled products, CFC reductions, and energy efficiency.

Waste Reduction. Federal agencies are reducing paper consumption by making use of advanced technologies, such as electronic data interchange, computer networks, electronic mail systems, and voice mail. An example is the Multi-User File for Interagency News (MUFFIN) that makes information about supply programs available by computer and modem and has the potential to reduce federal paperwork. MUFFIN could eliminate 350,000 hardcopy documents, ranging from requisitions to discrepancy reports, sent from client agencies to GSA each year. Since 1983 GSA alone has saved 1,604 tons of paper through the use of computer technology. In addition, federal agencies are using environmentally sound packaging and 2-sided duplicating machines.

Gas and Diesel Restrictions. In 1992 GSA issued a regulation on energy management of motor vehicles in the federal fleet to assist federal agencies that operate 300 or more vehicles to reduce gasoline and diesel consumption 10 percent below fiscal 1991 levels by fiscal 1995. The regulation also mandates the use of a modified life-cycle costing formula when evaluating bids for passenger vehicles and light trucks up to 8,500 pounds.

Recycling. In 1991 GSA implemented recycling in 345 buildings owned by the U.S. Government nationwide, involving 400,000 federal employees, recovery of 22,000 tons of recyclable materials, and generation of $500,000 through sales contracts. Reduction in handling and dumping fees resulted in savings of $1 million. In 1992 the programs expanded to 728 buildings owned and leased by the U.S. Government with 500,000 federal employees participating. In addition to recycling wastepaper, GSA is composting vegetative materials and recovering items such as copper wire, carpet, ceiling tile, paint, cardboard, telephone books, and glass and plastic food and beverage containers. In 1992 the program recovered 36,000 tons of material, which represents a 63-percent increase over fiscal 1991.

Recycled Products. To stimulate the U.S. market for recyclables, GSA procured 168,000 tons of recycled content paper and paper products in 1992, including paper towels, toilet paper, cardboard boxes, stationery items, and file folders. GSA awarded 271 contracts for 802 recycled-content paper and paper products totaling $169 million—a third of all paper supplied by the agency. Committed to products with high percentages of post-consumer recovered material, GSA stocked scores of products that exceed guidelines issued by EPA under RCRA. For example, in 1992 GSA introduced as a stock item recycled-content copier paper that is not required by the EPA guidelines. The paper contains 50-percent recycled content, including 10-percent post consumer-recovered material.

Other products include retread bus and truck tires, thermal building insulation products made with 9 to 50-percent recovered materials, and recycled wiping rags, toner cartridges used in office equipment, and paint. In fiscal 1991 GSA awarded 70 contracts for retread bus and truck tires valued at $200,000; in fiscal 1992 the numbers increased to 232 contracts valued at $650,000. Remanufactured toner cartridges that reduce disposal and save 50 percent in replacement costs have resulted in savings for federal agencies of $2 million since 1990.

In 1991 GSA highlighted environmentally sound products in two publications: GSA Recycled Products Guide, which lists products available from GSA supply sources and provides a means for agencies to identify and purchase recycled products; and GSA Supply Catalog, which highlights stock items with recycled content. In 1992 GSA expanded the catalog to include 2,100 environmentally sound products, such as lead-free, energy-saving, and water-saving items.

Chlorofluorocarbon Reduction. To meet requirements of the Clean Air Act, GSA is eliminating the use of products containing ozone-depleting chemicals. Federal facilities are reducing the use of chlorofluorocarbons (CFCs) in their operations, and they are servicing vehicles in the federal fleet to capture CFCs. After eliminating aerosol paints that used CFCs as a propellant 15 years ago, GSA now has eliminated plastic foam cups manufactured with CFCs. Halon fire extinguishers also are no longer in the federal supply system.

Energy Efficiency. In 1992 GSA continued to develop and implement projects to reduce energy use in buildings owned, operated, or leased by the U.S. Government. In fiscal 1990 GSA spent $10 million on 90 energy conservation projects; in fiscal 1991 those figures had increased to $30 million for 320 projects. In fiscal 1992 GSA approved an additional $37.5 million for 209 projects that include retrofitting existing lighting systems and installing motion sensors to control lighting, energy efficient motors, improved windows, and energy management and control systems. GSA also is replacing heating, ventilation, and air conditioning systems. Such actions have resulted in a 6-percent reduction in energy consumption in the GSA energy program since 1985.

Department of the Interior

In 1992 the Department of the Interior (DOI) expanded a paper recycling initiative at its Washington, D.C., headquarters to include colored paper, newspaper, cardboard, aluminum cans, and glass containers. Through July 1992 DOI headquarters had collected 466.5 tons of high grade white paper, 126.6 tons of grade 2 colored paper, 96.9 tons of newspaper, 50.2 tons of cardboard, 6.3 tons of aluminum cans, 18.5 tons of glass, and 20.5 tons of scrap metal. The paper collection alone represents estimated savings of 2,443 cubic yards of landfill space, 12,583 trees, and 5.2 million gallons of water.

As an additional benefit, the frequency of trash collection and associated landfill costs have been reduced for an estimated annual savings of $63,000. Under the President's Partnership in Education program, DOI donated $942 in proceeds from the collection of aluminum cans and glass containers to its adopted elementary schools in the area to purchase supplies and materials.

U.S. Postal Service

In 1992 the Postal Service integrated environmental strategies with its business activities and revised specifications for retail products. Use of recycled products is gaining momentum in the postal retail product line and in office products. Although current higher procurement costs of recycled copier paper can be an economic disincentive in large orders, greater market demand for such products should lower procurement costs and increase overall purchases. Examples of recycling include the following:

• **Paper Recycling.** In April 1992 the Postal Service undertook a mixed-paper recycling program in the New York metropolitan area, which will reduce hauling and landfill costs significantly. Wastepaper from postal facilities in Manhattan, Long Island, Queens, and the Bronx will be recycled annually into tissues, napkins, and paper towels.

• **Recyclable Windows on Mailing Envelopes.** Plastic windows on mailing envelopes were a roadblock in recycling mail, until development of glassine, a cellulose material that is re-pulpable and can be pasted onto envelopes with a water-based adhesive.

• **Recycling Solid Waste.** In 1992 the Postal Service recycled an estimated 25,945 tons of solid waste materials, 5,676 batteries, 1,941 tires, and 93,553 gallons of antifreeze, ink, solvents, used oil, and photochemicals.

Department of Defense

The Department of Defense is committed to becoming a leader in environmental compliance. A summary of DOD efforts in environmental restoration and pollution prevention follows. For a more detailed discussion, see the Department of Defense section.

Environmental Restoration. In 1992 the Defense Environmental Restoration Program (DERP), operating with a billion dollar budget, improved management and accelerated characterization and cleanup of 1,800 contaminated military sites, 94 of them on the National Priorities List. DOD analyzed cost-reduction initiatives for a comprehensive restoration strategic plan to be completed by 1995.

Pollution Prevention. The DOD pollution prevention strategy emphasizes systems acquisition, material substitution, process improvement, improved hazardous materials management, and alternative energy sources. By instituting control measures, DOD has reduced hazardous waste disposal by over half since 1987. This reduction was achieved despite the increased production and maintenance related to Operation Desert Shield/Desert Storm.

Department of Energy

In 1992 the Department of Energy implemented programs stressing the compatibility of environmental protection and energy production. In addition, DOE had two major efforts underway to improve management of the U.S. nuclear weapons program, which it has supervised for the past 30 years. The agency also proceeded with plans to manage the nuclear waste generated by commercial nuclear energy plants.

Waste Minimization. In May 1992 DOE introduced a plan to coordinate waste minimization at its facilities. Objectives include creating a staff bias for conserving resources and minimizing wastes, identifying waste minimization options and technologies, and incorporating waste minimization principles in design, development, and production.

DOE also published a multi-year plan for support of industrial waste minimization technologies. The department supports technology R&D cost-shared with industry to reduce waste and save energy. In 1991 DOE and industry representatives signed the first cooperative R&D agreement to reduce wastes associated with industrial soldering operations. DOE and EPA initiated discussions on ways to minimize industrial waste through regulations designed to encourage technology development. In cooperation with the industry-based Center for Waste Reduction Technologies, DOE identified R&D requirements to exploit waste reduction opportunities in industry, with an emphasis on chemical production and use.

Waste Management. In August 1992 DOE issued the fourth annual update of its Environmental Restoration and Waste Management Five-Year Plan. The objectives are to achieve timely compliance with environmental requirements and to complete cleanup of the 1989 inventory of inactive DOE sites and facilities by the year 2019. DOE is preparing a programmatic environmental impact statement (EIS), also scheduled for completion in 1994, to analyze potential environmental impacts of cleanup and compliance at all of its sites. The DOE budget for environmental restoration and waste management has increased from $2.3 billion in fiscal 1990 to $4.8 billion approved for fiscal 1993.

In 1992 DOE began using environmental restoration management contractors to clean up radioactive and hazardous contamination at its facilities. Remediation contractors manage site investigation, cleanup, decontamination, and decommissioning. DOE awarded an environmental restoration management contract in 1992 for the cleanup of its Fernald, Ohio, site.

Internal Audit Programs. The DOE environmental assessment program identifies strengths, weaknesses, corrective actions, and lessons learned. Environmental audits provide the foundation for efficient and cost-effective management planning.

Nuclear Weapons Complex Reconfiguration. To consolidate and modernize existing facilities, DOE issued a Reconfiguration Study in January 1991. The study included plans for constructing and operating a reorganized complex in compliance with federal, state, and local environmental laws. To comply with the National Environmental Policy Act (NEPA), DOE is preparing an EIS on the reconfiguration plans—to be completed in 1994. The EIS will analyze the environmental consequences of alternative configurations of the nuclear weapons complex.

Permanent Nuclear Waste Repository. As directed by the Nuclear Waste Policy Act (as amended in 1987), DOE is investigating the suitability of Yucca Mountain in southern Nevada for a permanent nuclear waste repository. The site is under consideration for the safe disposal of high-level radioactive waste from commercial nuclear powerplants. Establishment of a permanent repository could help resolve the issue of nuclear waste disposal.

Resolution of a 3-year legal dispute allowed DOE to resume site characterization studies with environmental permits from the State of Nevada. These studies, to be completed before 2001, will determine whether the geology and hydrology of Yucca Mountain will allow for permanent disposal of nuclear wastes. Progress continued on surface testing, development of specialized drilling equipment, and design of an underground exploratory studies facility to ascertain the geological characteristics of the site. In this pioneering effort, DOE, the Nuclear Regulatory Commission, and the congressionally designated Nuclear Waste Technical Review Board worked together to assure that planning accords with scientific review and high standards of nuclear safety.

Monitored Retrievable Storage. To facilitate DOE acceptance of commercial spent nuclear fuel by 1998, the department has planned for a monitored retrievable storage (MRS) facility for temporary above-ground storage of nuclear waste. To date, a designated DOE nuclear waste negotiator has received 20 applications from municipalities and Indian tribes for grants to study the hosting of an MRS site. The department expects to select an MRS site in 1993. In addition, federal and state efforts continue toward the siting of low-level radioactive waste disposal facilities, as mandated by the Low-Level Radioactive Waste Policy Amendments of 1985. Numerous interstate compacts for developing facilities for waste generated within the borders of a state are in effect.

Executive Orders on Federal Facilities Management

In 1992 federal departments and agencies initiated programs to comply with two executive orders issued by the President in 1991.

Federal Energy Management. Federal facilities took actions to comply with Executive Order No. 12759 signed by the President in April 1991:

Energy Use Reduction. Federal agencies are taking steps to reduce energy use in federal buildings to 20 percent below 1985 levels by the year 2000

Gas and Diesel Reductions. Federal agencies are reducing gas and diesel use in federal fleets of 300 or more vehicles to achieve a goal of 10 percent below 1991 levels by 1995

Alternative-Fuel Vehicles. Federal agencies are moving to procure the maximum practicable number of alternative-fuel vehicles (AFVs) by the end of 1995

Federal Waste Reduction and Recycling. In October 1991 the President signed Executive Order No. 12780, requiring federal agencies to reduce wastes and increase recycling through the following programs:

Waste Reduction and Recycling. Federal agencies are promoting cost-effective waste reduction and recycling of reusable materials in operations and facilities; and

Recycled Products. Federal agencies are procuring items made from recycled materials in accordance with the Resource Conservation and Recovery Act (RCRA).

Recycling Coordinators. In 1992 the Environmental Protection Agency designated a federal recycling coordinator to assist federal agencies with waste reduction, recycling, and procurement of recycled materials and to report annually to the Office of Management and Budget (OMB) on federal waste reduction and recycling. To date, 50 agencies have named federal recycling coordinators and 36 have submitted status reports on affirmative procurement programs to purchase products made from recycled materials.

Council on Federal Recycling and Procurement Policy. In June 1992 this council, created by the Executive Order, sponsored a 2-day Recycled Products Trade Fair and Showcase. The theme, "Closing the Loop," stressed that until materials recovered from the waste stream are turned into products that someone can use, they have not really been recycled. Over 3,000 vendors demonstrated the diversity and quality of products made with recycled materials and attended workshops with government procurement officers.

Alternative-Fuel Vehicles

The U.S. government is purchasing alternative-fuel vehicles (AFVs) in compliance with the Alternative Motor Fuel Act of 1988, the Clean Air Act Amendments of 1990, and the 1991 Executive Order on Federal Energy Management. AFVs are fueled by methanol, compressed natural gas, ethanol, and electric power. Federal agencies plan to purchase an additional 5,000 AFVs by the end of fiscal 1993, to have more than 20,000 federal AFVs operating in 1995, and to have half of the federal fleet in AFVs by 1998.

The U.S. Government is moving forward with its alternative-fuel vehicle program to meet demands for national energy security and reduced emissions. The federal presence helps stimulate the market and establish the groundwork for state and private sector participation in years to come.

U.S. Postal Service. The Postal Service has the nation's largest AFV fleet.

Compressed Natural Gas. With a CNG delivery fleet of more than 1,200 vehicles, the Postal Service in 1992 added CNG-powered vehicles in Staten Island, New York; Tulsa and Oklahoma City, Oklahoma; and Huntington Beach, Sacramento, and

Concord, California. Current testing indicates that CNG is the best alternative fuel for the purposes of the Postal Service. Its major advantages over gasoline are savings of about 3 cents per mile; less pollution; better safety record than gasoline; and less vehicle maintenance. Primary disadvantages—initial cost and availability of fueling stations—have limited CNG use to areas where public utilities provide refueling equipment.

Electric Power. The Postal Service and the Ford Motor Company are discussing tests for six electric vans in Southern California, which could begin as early as July 1993.

General Services Administration. In 1992 GSA brought 3,287 original-equipment-manufacturer AFVs into the federal fleet. These vehicles use the following fuels:

Methanol Flexible Fuel. GSA purchased 2,500 methanol flexible fuel compact sedans from Chrysler Corporation and 25 methanol flexible fuel vans from Ford Motor Company in 1992. These vehicles, which operate on methanol (M85), unleaded gasoline, or any combination of the two, will be operating in numerous locations nationwide. GSA previously operated 65 methanol flexible fuel sedans.

Compressed Natural Gas. GSA purchased 600 CNG three-quarter ton pickup trucks from General Motors and an additional 25 CNG 8-passenger vans from Chrysler Corporation in 1992. These vehicles will be added to the GSA fleet of 52 CNG 8-passenger vans operating throughout the nation.

Ethanol Flexible Fuel. GSA operates 25 ethanol flexible fuel General Motors sedans in Washington, D.C., and Chicago. These cars use ethanol (E85), unleaded gas, or any combination of the two.

Innovative Technology and Federal Facility Cleanup

The departments of the Interior, Energy, and Defense and the Environmental Protection Agency signed a Memorandum of Understanding with the Western Governors' Association in July 1991 to streamline the cleanup of contaminated federal facilities and sites in the West. Under the 5-year agreement, federal agencies and western states will plan and implement demonstrations of innovative technology at federal nuclear facilities, defense bases, abandoned mine sites, and perhaps eventually private sites throughout the western states.

In October 1992 in Denver, the secretary of Energy announced plans for the start up of pilot demonstration projects in 1993. DOE is requesting funding for three to six innovative cleanup projects through the Strategic Environmental Research and Development Program (SERDP). Such projects could become national models, providing the United States with a world lead in restoration and waste management technologies. Pilot projects have the following objectives:

• Address environmental restoration and waste management problems in western states in an efficient, cost-effective manner;

• Speed the development of new environmental restoration technologies and new waste management technologies or new applications of existing technologies; and

• Ensure a foundation for an adequately sized and appropriately trained work force.

Examples of innovative uses of technology that could be applied in demonstration projects include:

Remote Sensing. Use of thermal and magnetic sensors on helicopters to identify soil and groundwater contamination at a fraction of the cost of a walking survey.

Cone Penetrometers. Replacing old-fashioned drilling with cone penetrometers that can punch into the soil and analyze contaminated soil down to 500 feet or more. The penetrometers can analyze ten times more locations for the same cost, for projected savings of $100 million.

Air Stripping. Using in-situ air stripping to clean contaminated groundwater without pumping it to the surface. This process promises to be five times more effective and to cost 40 percent less than traditional technology.

DEPARTMENT OF DEFENSE

National security depends on the wellbeing of the environment, and federal agencies charged with protecting national security are also responsible for protecting the environment. The linkage between these responsibilities is evident in the environmental program of the Department of Defense (DOD). In 1991, for the first time, the United States recognized environmental issues as a national security concern. The National Security Strategy, stated:

> Global environmental concerns include such diverse but interrelated issues as stratospheric ozone depletion, climate change, food security, water supply, deforestation, biodiversity, and treatment of wastes. A common ingredient in each is that they respect no international boundaries. The stress from these environmental challenges is already contributing to political conflict. Recognizing a shared responsibility for global stewardship is a necessary step for global progress. Our partners will find the United States a ready and active participant in this effort.

DOD is steward of 25 million acres of public land at 600 major installations in the United States and of an additional 2 million acres abroad. Military installations range from a few acres for weather stations and radar sites to a million acres for training installations and bombing ranges. In spite of intensive military use for many years, DOD lands contain a rich diversity of flora and fauna.

While accomplishing its primary mission—defending the national security interests of the United States—DOD seeks to be a leader in federal agency compliance with environmental laws and regulations and stewardship of natural resources on military installations.

Conditions and Trends

In 1992 the Secretary of Defense set the pace for improving environmental quality in all DOD activities by providing that the department should:
Fund environmental compliance, restoration and pollution prevention sufficient to achieve sustainable compliance with federal and state environmental laws and governing standards overseas; and to minimize negative mission impacts and future costs and to provide federal leadership in environmental protection.

Compliance

To conduct environmental programs and to accelerate compliance and cleanups of past contamination, DOD requested and received a supplemental appropriation of $1.1 billion which brought total fiscal 1992 environmental funding to $4.2 billion.

Restoration

In fiscal 1991 DOD invested $1 billion in cleanup efforts through the Defense Environmental Restoration Program (DERP), an increase of $464 million over the previous year. Of the 1,800 military installations in DERP, only 94 have been listed on or proposed for the National Priorities List (NPL) by EPA. In 1991 EPA added one military installation to the NPL, the Pearl Harbor Naval Complex on the island of Oahu, Hawaii. DOD has completed preliminary assessments for DOD sites on the NPL and is moving from the remedial investigation and feasibility study phase into remediation and cleanup. In 1991 the Defense Environmental Restoration Program recorded the following accomplishments:

- Cleanup at a third of the 1,800 DERP sites with no further restoration actions required

- Environmental training for 2,000 military and civilian personnel

- A 63-percent increase in DERP sites with completed studies quantifying the amount and extent of contamination and a 26-percent increase in sites with remedial action completed.

Pollution Prevention. In addition to cleaning up sites contaminated in the past, DOD has adopted a pollution prevention strategy, in which the reduction of hazardous wastes generated by military installations figures prominently. In June 1987 DOD established the goal of reducing hazardous waste disposal by 50 percent before the end of 1992. By 1991 DOD had reduced hazardous waste disposal by 53.9 percent—a year ahead of schedule—and much of the reduction occurred during a time of increased production and maintenance activity related to Desert Shield and Desert Storm. More than 85 percent of the reduction was attributable to improvements in waste minimization efforts at DOD shipyards, maintenance depots, and air logistics centers. Such facilities account for 60 percent of all hazardous waste generated by the department. The remaining 40 percent is generated by such areas as daily installation operations and training facilities.

Policies and Programs

The DOD environmental program has four components: compliance, restoration, pollution prevention, and natural and cultural resource conservation.

Environmental Compliance

A primary element of the DOD compliance strategy is implementation of the National Environmental Policy Act (NEPA). A vehicle for minimizing environmental effects of federal actions, NEPA provides DOD decisionmakers with choices on the manner in which the military mission is to be accomplished. To address actions that may affect the environment, each of the military services has promulgated NEPA regulations.

DOD environmental compliance projects are designed to meet regulatory standards that protect human health and the environment. Projects address the following areas:

- Hazardous waste management
- Underground storage tanks
- Solid waste management
- Air pollution abatement
- Water quality management and safe drinking water
- Requirements based on specific environmental statutes

To achieve compliance in these areas, the department is emphasizing education and training, an environmental ethic for the defense community, and public awareness and participation in DOD environmental activities. Each of the military departments has implemented a comprehensive environmental audit program to highlight problems at facilities before they become violations. Environmental audits enable installations to plan and budget for environmental projects needed to maintain compliance.

Pollution Prevention

Recognizing pollution prevention through source reduction as the key to a cleaner environment, DOD has adopted a pollution prevention strategy. Components of the strategy, accompanied by examples, follow:

Systems Acquisition

The Navy is conducting Logistics Review Group audits of hazardous material use and hazardous material control plans for all new weapon systems and for major system modifications to ensure reduction of hazardous material and hazardous waste. The Air Force has prohibited use of certain hazardous materials, including chlorofluorocarbons, cadmium, and chromium, in the design, manufacture, and operation of the new F-22 fighter plane with no sacrifice of cost, schedule, or performance.

Material Substitution

Kelly Air Force Base, Texas, replaced cyanide stripping baths with non-cyanide strippers, thus saving $390,000 annually on treatment and disposal costs. Scranton

Army Ammunition Plant, Pennsylvania, eliminated the use of chromic acid rinse in preparing steel surfaces for painting.

Process Improvement

By substituting an alternative cleaning process for vapor degreasers, the Naval Aviation Depot at Jacksonville, Florida, eliminated 300,000 pounds of hazardous waste annually and reduced volatile organic compounds emissions by 66 percent. The Naval Air Station at Mirimar, California, designed and built a totally enclosed system for recycling more than 20,000 pounds of solvent, grease, and hydraulic fluid-contaminated rags. The Navy estimates savings of $350,000 a year by using this recycling system.

Improved Material Management

The Naval Air Station at Point Mugu, California, developed a centralized hazardous material inventory control system. In the first year, purchases of hazardous materials decreased 49 percent, and hazardous waste disposal decreased 73 percent.

Alternative Energy Sources

To reduce air pollution from vehicles, DOD and DOE purchased 2,000 alternate fuel vehicles that run on electric batteries, 100-percent compressed natural gas, or 85-percent methanol/15-percent gasoline. For more information, see the Federal Facilities Management section.

The DOD Pollution Prevention Strategy is reducing pollution, improving worker protection, providing more efficient use of natural resources, and saving money. For these reasons, DOD is shifting management emphasis to pollution prevention as a low-cost, environmentally sound alternative to traditional pollution control strategies and contamination cleanup. A sampling of other DOD efforts follows.

Pollution Prevention Awards. DOD installations were recipients of several of the EPA Administrator's Pollution Prevention Awards for 1991:

• **Single-Coat Paint.** The Navy Exploratory Development Program, Warminster, Pennsylvania, won for developing a new single-coat paint that reduces volatile organic compounds and hazardous waste by 67 percent.

• **Planning and Implementation.** Fairchild Air Force Base, Washington, won as a model for comprehensive pollution prevention planning and implementation.

- **R&D Pool.** The Army Depot Systems Command, Chambersburg, Pennsylvania, was a runner-up for a program that pools pollution prevention R&D among several depots.

- **Ozone-Depleting Substances.** In February 1992 President Bush announced that the United States would accelerate the phaseout of the production of ozone-depleting substances by 1995. DOD developed a program to help achieve the President's goal through initiatives such as the following:

- **Army.** To reduce reliance on chlorofluorocarbons (CFCs) and halons, the Army is developing programs to apply current-technology chemicals and processes to weapon systems applications.

- **Navy.** To focus on use reduction through alternative non-CFC technologies, chemical conservation, and inventory management, the Navy is supporting research on substitute materials in two industry/government consortia—Halon Alternatives Research Corporation and the Solvents Working Group.

- **Air Force.** In addition to use reduction, the Air Force is working with industry to find or develop CFC replacements, with emphasis on alternatives to substance CFC-113 and methyl chloroform.

Environmental Restoration

In 1992 DOD accelerated characterizations of remaining contaminated sites and developed cost-reduction initiatives in an effort to produce a comprehensive restoration strategic plan to be completed by 1995. The department conducted 180 remedial actions at military installations on the National Priorities List, which included the following:

- Immediate response actions to protect public health
- Immediate removal of contaminants
- Alternative drinking water supplies
- Installation of permanent remedies such as groundwater-treatment facilities

To date, 40 states and territories have entered into memoranda of agreement with DOD to support environmental restoration at over 400 military installations. Through cooperative agreements, DOD has reimbursed states more than $16 million for services to expedite the review and approval of studies and cleanup decisions.

Natural and Cultural Resources

DOD sponsors the Legacy Resource Management Program to integrate biological, cultural, and geophysical resources with the demands of the military mission. The program emphasizes the stewardship of DOD lands and gives priority to identifying, conserving, and restoring natural and cultural resources. To demonstrate more effective conservation techniques, the program evaluates natural and cultural resources for their significance to such values as biodiversity and historic interpretation. DOD works through partnerships with federal, state, and local agencies and private groups.

In 1992 DOD received 1,000 Legacy proposals from the military services and another 100 proposals from private groups or other agencies. The department added 300 new Legacy demonstration projects for a total of 400 conservation projects throughout the United States and U.S. territories. Examples of Legacy natural resource projects follow.

Management Data

The Army has developed an Integrated Training Area Management Program (ITAM) to help military trainers manage natural resources. ITAM generates data on soil types, vegetative cover, and equipment size and weight useful in management decisions regarding the rest and rotation of training lands. Implemented on 53 Army and Marine Corps installations in the United States and abroad, ITAM assists planners of base realignment and closures determine the capability of land to sustain reuse proposals.

Revitalizing Wetlands

In another Legacy project, DOD is working with the U.S. Fish and Wildlife Service and 17 conservation groups to revitalize wetlands on military installations. For example, Vandenberg Air Force Base, California, and the Nature Conservancy developed an accord to protect 5,000 acres of wetlands, 9,000 acres of undisturbed dunes, relic stands of native bunchgrass, and the largest existing remnant of globally rare Burton Mesa chaparral. The Navy joined with the Defenders of Wildlife to establish a Watchable Wildlife area at the Sprague Neck Bar Ecological Reserve, on land held by Naval Communications in Cutler, Maine. The area, a long cobble beach reaching into Machias Bay, hosts bald eagles and migrating shorebirds and waterfowl.

Cultural Resources

Legacy projects also include restoring historic buildings and sites such as a World War II chapel at Adak, Alaska; public awareness; archeological procedures; ethnohistories and oral histories; rock art; historic preservation technology; and conservation of Cold War artifacts.

DOD Environmental Trends

The Department of Defense is restructuring its environmental program to accelerate development of new technologies and to institute long-range planning. A recent review of environmental management led DOD to institute the following changes:

- Organizational structure, with particular emphasis on unmet needs at the regional, state, and local level

- Education and training across all military services and members of the DOD community

- Improvements in management of the restoration program to provide environmental protection at lower cost

- An environmental research and development (R&D) strategic plan to focus efforts and accelerate payback from DOD environmental technology assets

- Systems to correct deficiencies in environmental information management

Examples of initiatives to improve environmental management include the following.

Research and Development

DOD is working with the private sector and other government agencies on environmental research. In 1992 DOD established the Strategic Research and Development Program (SERDP) to address environmental matters of concern to DOD and the Department of Energy (DOE). The program identifies priorities in environmental research, technology, and information developed by the two departments for defense purposes. SERDP will update the DOD 5-year strategic environmental R&D plan and will provide government agencies and the private sector access to DOD environmental data. It also will identify energy technologies developed for national defense purposes that have environmentally sound, energy efficient applications for other DOD programs, for DOE and other government programs, and for industrial and commercial applications.

Long-Range Planning

In anticipation of future environmental requirements, DOD is undertaking long-range planning. In 1992 the department issued a Report on Environmental Requirements and Priorities, which examines prospects for environmental activities through the 1990s. The Department's first environmental R&D strategic plan will provide input to the environmental restoration strategic plan.

DOD and NEPA

Recent fundamental changes in threats to national security have prompted a major restructuring of the Department of Defense. The easing of international tensions has precipitated base closures in the United States and abroad. In 1992 proposed base closures predominated as the major actions requiring the Department of Defense (DOD) to prepare environmental impact statements (EISs) as called for by the National Environmental Policy Act (NEPA).

Base Closures

DOD worked closely with community reuse committees to develop alternatives for the reuse of closed bases and to ensure that the public was an integral part of the decisionmaking process. Base closures can have the following environmental implications:

• **Consolidation and Restationing.** The military is consolidating more activities on fewer installations with potentially greater pressure on environmental resources. Restationing and consolidation has the potential to affect regional environmental quality.

• **Sale and Reuse of Closed Facilities.** The nation has choices to make regarding the use of closed military installations.

• **Cleanups.** Base closures may affect cleanup priorities for DOD and the federal government.

• **Other NEPA Actions.** Other actions taken by the military services in compliance with NEPA include the following examples:

• **Navy/Marines.** The Navy integrated NEPA training into the official Navy Training Plan and required it for all personnel. A Supplemental EIS for new dredging operations at two naval installations in San Francisco Bay is the result of several years of research and cooperation among the Navy, the Army Corps of Engineers, EPA, and state regulatory agencies. It will determine the most environmentally sound manner to dispose of dredge material while maintaining adequate depth for ship transit. This EIS is the major database in the designation process of a regional disposal site for dredging in San Francisco Bay. The Marine Corps, to ensure inter- disciplinary participation in NEPA decision-making, established environmental impact review boards at headquarters and at each installation. The boards review EISs and forward recommendations to decisionmakers.

• **Army.** The NEPA Compliance Manual for Base Realignment and Closure, prepared by the Army, guides preparers of NEPA analyses. It also helps ensure that military and civilian decisionmakers have adequate environmental information upon which to make decisions about the Army base structure of the twenty-first century. A 1987 environmental assessment (EA) of plans for testing the Strategic Defense Initiative on Kwajalein Atoll revealed existing environmental problems on the island and potential cumulative impacts, if the project proceeded as planned. The Army prepared an EIS to address cumulative environmental effects and detail an extensive mitigation plan to reduce effects of the SDI proposal and mitigate environmental effects of ongoing operations.

• **Army and Air Force.** To provide information for decisionmakers, the Army and Air Force integrated the NEPA process with the Comprehensive Environmental Response, Compensation, and Liability Act (CERCLA or Superfund) process in a unified analysis and documentation. The Air Force integrated its environmental impact analysis process under NEPA and the CERCLA process into a single effort for the BOMARC Missile accident site cleanup. The integration increased public input in developing a method and level of cleanup for the site and enhanced information available to the decisionmaker so that the best alternative could be selected.

• **Air Force.** To involve the public in the NEPA process, the Air Force recently declassified aspects of the Space Nuclear Thermal Propul-sion Program. The Air Force is using the NEPA process as a forum to address public concerns about the development and testing of a new nuclear program. In 1992 the Air Force held public scoping meetings in Alaska and prepared an EIS on the environmental effects of a High Frequency Active Auroral Research Program that would build and operate an Ionosphere Research Instrument to study the basic properties of the Arctic ionosphere.

CHAPTER 2

GLOBAL ENVIRONMENTAL CRISIS

GLOBAL ENVIRONMENTAL CRISIS

SHORT LIST OF SOURCES
FOR GENERAL RESEARCH

Books

Asimov, Isaac; and Pohl, Frederick. *Our angry Earth*. New York: St. Martin's, 1991. 323 p. (TD170 .A75 1991)
> Discusses the nature of the environmental crisis, its urgency, and what individuals and communities are doing to change things. An introduction to environmental damage control and repair.

Caldicott, Helen. *If you love this planet: a plan to heal the Earth*. New York: W. W. Norton, 1992. 231 p. (TD174 .C33)
> A general introduction to environmental problems. Uses a diagnosis and prescription format to delineate the Earth's environmental ills and suggest solutions. Discusses global warming, deforestation, toxic pollution, and other issues, and maintains that managing the overpopulation problem is paramount to solving them.

Ehrlich, Paul R; and Ehrlich, Anne H. *Healing the planet: strategies for resolving the environmental crisis*. Reading, Mass.: Addison-Wesley, 1991. 366 p. (GF75 .E4 1991)
> Covers the primary current environmental issues, including overpopulation, ozone depletion, global warming, land destruction, and water pollution; maintains that too much emphasis has been given to environmental distress rather than the causes. Offers solutions to these problems, as well as a concise summation of the Gaia concept.

Marowski, Daniel G., ed. *Environmental viewpoints: selected essays and excerpts on issues in environmental protection*. Detroit, Mich.: Gale, 1992. 544 p.
> Presents reprinted excerpts from one hundred magazines and journals on a wide range of environmental topics. Each topic has a brief overview, in-depth information from the reprints, and a bibliography.

Porritt, Jonathon. *Save the Earth*. Atlanta, Ga.: Turner Publishing, 1991. 208 p.
  Discusses air, water, and land pollution, and their effects on human health. Includes numerous photographs and charts. Examines the connection between poverty, politics, and pollution. Includes an action packet that makes several suggestions for citizen involvment. Begins with a foreword by the Prince of Wales and an introduction by Robert Redford.

Silver, Cheryl Simon. *One Earth, one future: our changing global environment*. Washington, D.C.: National Academy Press, 1990. 196 p. (GF75 .S55 1990)
  Explains the basic science behind global environmental problems and examines policy implications aimed at solutions. Explores the interactions between the land, water, and atmosphere, and the snowballing impact that human activity is having on that system. Includes discussions of global warming and the implications for agriculture and for sea level rise in coastal areas, ozone depletion in the stratosphere, forest clearing in the tropics and the impacts on species living there, and acid deposition.

General interest periodicals

Allman, William F. "Rediscovering planet Earth." *U.S. News & World Report* 105 (October 31, 1988): 56-61, 63, 65, 67-68.
  Using new techniques to explore past Earth climates and computers to simulate the future, researchers are beginning to see Earth as a complex, interdependent system of oceans, atmosphere, and life.

Bailey, Ronald; and Ames, Bruce N. "Raining in their hearts: nuclear winter, population explosion, non-renewable resources, vanishing species--some of them are genuine problems, others are made up out of whole cloth." (includes related article on cancer and pollution) *National Review* 42,23 (December 3, 1990): 32(5).

Carpenter, Betsy. "Living with our legacy." *U.S. News & World Report* 108 (April 23, 1990): 60-62, 64-65.
  Looks at the progress made in environmental problems since the first Earth Day in 1970, and outlines the key national and global problems to be resolved today that were unimaginable twenty years ago.

Easterbrook, Gregg. "Everything you know about the environment is wrong." *New Republic* 202 (April 30, 1990): 14, 16-27.
  Examines the following issues: acid rain, asbestos, Clean Air Act, climate change, cost-benefit analysis, emissions trading, the economic scorecard, the enviro lineup, enviro politics, the garbage crisis, global warming, nature, ozone depletion, pesticides, reformulated gasoline, species rights, Third World resources and pollution, and zero emissions.

McKibben, Bill. "Reflections: the end of nature." *New Yorker* 65 (September 11, 1989): 47-48, 50, 52, 54-58, 67-76, 78-93, 96-105.

> Concludes that man, by his global pollution of carbon dioxide and methane, has changed the Earth's atmosphere and consequently its weather, and in doing so, has made the idea of nature--"the wild province, the world apart from man, under whose rules he was born and died"--extinct. Furthermore, to live under greenhouse conditions, humans may find it necessary to use biotechnology, lending "to the second end of nature: the imposition of our artificial world in place of the broken natural one."

Environmental and professional journals

Holdgate, Martin W. " '92: the environment of tomorrow." *Environment* 33 (July-August 1991): 14-20, 40-42.

> "What will Earth's environment be like in 40 or 80 years? According to the director general of the World Conservation Union, some environmental trends, such as deforestation, population growth, and loss of biodiversity, are virtually impossible to stop. Strong international cooperation, however, could guide humanity toward a more sustainable relationship with the natural world."

Videos

Can polar bears tread water? Produced by Better World Society. Cinema Guild (1697 Broadway, Suite 802, New York 10019), 1990. 60 min.

> Focuses on the manmade dangers threatening the Earth and discusses the ecological changes that could destroy the planet, including the greenhouse effect and climatic change. Offers suggestions for reversing these trends. Also examines the sociological and political complexities.

Help save the planet. MCA/Universal Home Video (70 Universal City Plaza, Universal City, CA 91608), 1990. 70 min.

> Presents ecological tips, such as recipes for environmentally sound cleaners and suggestions on how to select products with less packaging with comical celebrity performances. Includes phone numbers and addresses of pertinent agencies.

Only one Earth: the sinking ark. Films, Inc. (5547 N. Ravenswood Avenue, Chicago, IL 60640), 1987. 58 min.

> Maintains that the short-term economic benefits of environmental exploitation are far outweighed by the long-term benefits of environmental preservation. Focuses on positive attempts to preserve the environment.

INTRODUCTORY SOURCES

Books (annotated)

Allaby, Michael. *Green facts: the greenhouse effect and other key issues*. London: Hamlyn, 1989. 192 p.
     A general discussion of ecological and technical issues, including global warming, nuclear power, marine mammals, population, hunger, and poverty.

Angell, D. J. R.; Comer, J. D.; and Wilkinson, M. L. N., eds. *Sustaining Earth: response to the environmental threat*. London: Macmillan, 1990. 226 p.
     An outline of global environmental concerns, which was originally part of a series of public lectures delivered at Cambridge University. Addresses such topics as climatic change, deforestation, acid precipitation, and the ozone layer.

Anzovin, Steven. *Preserving the world ecology*. New York: H. W. Wilson, 1990. 236 p. (QH541.145 .P74 1990)
     Examines "world ecological-environmental issues and their political and possible preservational dimensions." Contents: "A world at risk"; "The debate over global warming and climatic change"; "The politics of ecology"; "Preserving the future."

Asimov, Isaac; and Pohl, Frederick. *Our angry Earth*. New York: St. Martin's, 1991. 323 p. (TD170 .A75 1991)
     Discusses the nature of the environmental crisis, its urgency, and what individuals and communities are doing to change things. An introduction to environmental damage control and repair.

Bach, Julie S.; and Hall, Lynn, eds. *Environmental crisis: opposing viewpoints*. St. Paul, Minn.: Greenhaven, 1986. 263 p.: ill. (HC110.E5 E49835 1986)
     Discusses the environmental crisis, corporate responsibility for environmental disasters, effectiveness of pollution regulations, nuclear power, toxic wastes, and acid rain. Partial contents: "The greenhouse effect is potentially disastrous," by Lewis M. Steel; "The greenhouse effect is exaggerated," by H. E. Landsberg; "Corporations must be held responsible for environmental disasters," by Richard Asinoff; "Consumers must pay for a cleaner environment," by Milton and Rose D. Friedman.

Brower, Kenneth. *One Earth*. New York: HarperCollins, 1990.
     A photo review of various environmental problems, from trees colored by acid rain to the tusks of endangered elephants. Includes an annotated list of environmental resources.

Brown, Lester R., et al. *Vital signs 1992: the trends that are shaping our future.* New York: W. W. Norton, 1992. 132 p.

This companion to the *State of the world series* presents a vast array of data designed to complete the picture of the global environment and counterbalance the slant of media coverage on environmental and economic issues. Seeks to identify trends in agriculture, energy, food, the atmosphere, and economic changes, among other areas.

Caldicott, Helen. *If you love this planet: a plan to heal the Earth.* New York: W. W. Norton, 1992. 231 p. (TD174 .C33)

A general introduction to environmental problems. Uses a diagnosis and prescription format to delineate the Earth's environmental ills and suggest solutions. Discusses global warming, deforestation, toxic pollution, and other issues, and maintains that managing the overpopulation problem is paramount to solving them.

Courrier, Kathleen, ed. *Life after '80: environmental choices we can live with.* Andover, Mass.: Brick House Publishing, 1980. 280 p.

Includes articles on deforestation, soil erosion, population growth, toxic chemicals, air and water pollution, and alternative sources of energy.

Diagram Group. *Environment on file.* New York: Facts on File, 1991.

Covers topics such as agriculture and food, air, water, energy, the natural environment, and the human factor; comprised mostly of charts, maps, and other illustrative material. Charts cover such issues as the history of whaling and radioactive contamination pathways. Summarizes each issue and offers specific solutions. Contains a list of suggested readings and an index.

Edberg, Rolf; and Yablokov, Alexei. *Tomorrow will be too late.* Tucson: University of Arizona, 1991. 210 p. (GF80 .E3 1991)

Swedish environmentalist Edberg and Russian ecologist Yablokov exchange views on the dangers of nuclear war, the problems of toxic waste disposal, the need for a global environmental policy, the destruction of rain forests, the role of corporations in the environmental movement, the consequences of pollution, and other social issues related to the environment. These exciting conversations, made possible by glasnost in 1987, have lost some of their edge but are nonetheless sound in their arguments.

Ehrlich, Paul R.; and Ehrlich, Anne H. *Earth.* New York: Franklin Watts, 1987.

Intended for a popular audience. Begins with a basic ecological overview of the globe, then analyzes the current state of the environment and offers dire predictions for the future of the planet. Contains numerous illustrations.

Ehrlich, Paul R; and Ehrlich, Anne H. *Healing the planet: strategies for resolving the environmental crisis.* Reading, Mass.: Addison-Wesley, 1991. 366 p. (GF75 .E4 1991)

Covers the primary current environmental issues, including overpopulation, ozone depletion, global warming, land destruction, and water pollution; maintains that too much emphasis has been given to environmental distress rather than the causes. Offers solutions to these problems, as well as a concise summation of the Gaia concept.

Ellis, Derek. *Environments at risk: case histories of impact assessment.* New York: Springer-Verlag, 1989.

Analyzes specific environmental incidents and their local impacts. Examples range from efforts to clean the River Thames to the accident at Chernobyl. In each case what happened and why it happened are examined, as well as potential lessons to be learned from the experience. The roles of the media and the public are also assessed.

Erickson, Jon. *World out of balance: our polluted planet.* Blue Ridge Summit, Pa.: TAB, 1992. 171 p.

Discusses air, water, and ground pollution, as well as overpopulation and endangered species. Also describes the possible value of using satellite remote sensing to evaluate global environmental problems. Contains a bibliography and an index.

Foster, Ian. *Environmental pollution.* Oxford: Oxford University Press, 1991. 48 p.

This work is divided into three sections: the Pollution Problem, Case Studies, and Exercises. Defines the nature and extent of the problem and examines British examples. Presents activities for students such as data collection.

Friday, Laurie; and Laskey, Ronald, eds. *The fragile environment.* Cambridge, U.K.: Cambridge University Press, 1989. 198 p.

Contains chapters on the history of environmental destruction, the future of forests and relationship between deforestation and global warming, famine, and changing climates and the chemical composition of the atmosphere.

Fumento, Michael. *Science under siege.* New York: Morrow, 1993. 384 p.

Questions the current public fear of much-publicized environmental threats such as Alar, dioxin, food irradiation, electromagnetic fields, and video display terminals. Argues that the media, businesspeople, and even scientists themselves often jump to conclusions about environmental issues before definitive evidence can be found. Maintains that science should no longer be influenced by political policy.

Goldsmith, Edward; and Hildyard, Nicholas, eds. *Earth report: monitoring the battle for our environment.* London: Mitchell Beazley, 1988. 240 p.

Documents global environmental destruction and pollution as well as escalating human problems such as the population explosion, urbanization, poverty, famine, malnutrition and disease. Partial contents: "Man and the natural order," by

Donald Worster; "The politics of food aid," by Lloyd Timberlake; "Nuclear energy after Chernobyl," by Peter Bunyard; "Man and Gaia," by James Lovelock; "Acid rain and forest decline," by Don Hinrichsen; "Water fit to drink?," by Armin Maywald, Uwe Lahle and Barbara Zeschmar-Lahl.

Goldsmith, Edward. *The way: an ecological world-view.* Boston: Shambhala, 1993. 464 p.
Maintains that science and technology are responsible for the current environmental crisis and that the solution lies in formulating a holistic world-view that recognizes the interrelatedness of living things. In arguing for the intrinsic inadequacies of the scientific method, the author presents a perspective that has elements of the Gaian hypothesis, deep ecology, and social ecology.

Gordon, Anita; and Suzuki, David. *It's a matter of survival.* Cambridge, Mass.: Harvard University Press, 1991. 278 p. (GF75 .G66 1991)
Interviews the authors conducted for a program on the Canadian Broadcasting Service in 1989. Analyzes the economic, political, and relgious forces that have influenced environmental problems. Argues for fundamental changes in the policy and lifestyle of consumer nations, proposes giving developing countries more resources, and questions the validity of government's concept of sustainable development.

Gore, Albert, Jr. *Earth in the balance: ecology and the human spirit.* Boston: Houghton Mifflin, 1992. 407 p.: ill. (GF41 .G67 1992)
Employs scientific data to portray a grim future for the planet. Proposes a global Marshall Plan to ensure environmental relief around the world in response to environmental threats.

Hansen, P. E.; and Jørgensen, S. E., eds. *Introduction to environmental management.* New York: Elsevier Science Publishing, 1991. 403 p.: ill. (TD170 .I584 1990)
Offers a broad introduction to environmental management. Examines, using a multidisciplinary approach, a wide variety of issues, such as different kinds of pollution and the effects of law and economics on the planning process. Contains numerous European case studies.

Hirschhorn, Joel S.; and Oldenburg, Kirsten U. *Prosperity without pollution: the prevention strategy for industry and consumers.* New York: Van Nostrund Reinhold, 1991. 386 p.: ill. (TD174.H57 1990)
This work maintains that prevention is the key to solving environmental problems, instead of controlling emissions, working toward sustainable growth, or energy conservation. Emphasis is placed on what individuals can do to see that research and legislation effectively prevent pollution from occurring--including taxing companies that continue to produce pollutants. Offers case studies to illustrate the corresponding concepts.

Kemp, David D. *Global environmental issues: a climatological approach.* New York: Routledge, 1990. 220 p.: ill. (TD177 .K46 1990)

Explores the causal link between environmental problems and geographic events--such as the link between the greenhouse effect and the destruction of the ozone layer, and on a smaller scale, the link between drought, famine, desertification in the Sahel, and Brazilian deforestation. Maintains that long-term international cooperation is necessary to resolve these problems. Also discusses other topics, including nuclear winter, the limits of time-series data, and the uncertainties of atmospheric turbidity. Contains diagrams, an index, and a bibliography.

Kormondy, Edward J., ed. *International handbook of pollution control.* New York: Greenwood, 1989. 466 p. (HC79 .P55I44 1989)

Focusing on international pollution control, this work is divided into major geographic regions and then subdivided into individual countries. The major pollution problems, and the attempts to correct them, in each country are presented. The role of the U.N. in pollution regulation is also addressed. Contains a good bibliography. A good source for specialized information on a particular country.

Marowski, Daniel G., ed. *Environmental viewpoints: selected essays and excerpts on issues in environmental protection.* Detroit, Mich.: Gale, 1992. 544 p.

Presents reprinted excerpts from one hundred magazines and journals on a wide range of environmental topics. Each topic has a brief overview, in-depth information from the reprints, and a bibliography.

Meadows, Donella; et al. *Beyond the limits: confronting global collapse, envisioning a sustainable future.* Post Mills, Vt: Chelsea Green, 1992. 320 p.

Argues that the world has already surpassed many of its physical limits in supporting life, then use computer models to show how global collapse can be avoided, including minimizing the use of nonrenewable resources, preventing overuse of renewable resources, stopping population growth, and curbing consumerism.

Mitchell, George J. *World on fire: saving an endangered Earth.* New York: Scribner, 1991. 247 p. (QC912.3 .M57 1991)

Written by the majority leader of the U.S. Senate, this work points to five major threats to future of the planet: acid rain, the greenhouse effect, the destruction of tropical rain forests, the rift in the stratospheric ozone layer, and overpopulation. Maintains that the only way to save the planet is for the U.S. to take a leadership role on environmental issues. Asserts that both developed and developing countries should be involved in the process of assuring that the planet is healthier in the next century. Faults the U.S. for failing to address environmental issues in the last ten years. Contains a bibliography and an index.

Nisbet, E. G. *Living Earth: a short history of life and its home.* New York: HarperCollins, 1992. 237 p.

 > Traces the history of the Earth from its origins, to the beginnings of its occupation by man, up to the modern world. Each chapter contains a suggested reading list.

Piel, Gerard. *Only one world: our own to make and to keep.* New York: W. H. Freeman, 1992. 367 p.

 > Focuses on the crises of poverty and the environment and maintains that we need greater economic development of the Third World in order to limit population growth, improve the standard of living, and protect the environment. Distributed to the delegates at the Earth Summit in Rio in June of 1992.

Porritt, Jonathon. *Save the Earth.* Atlanta, Ga.: Turner Publishing, 1991. 208 p.

 > Discusses air, water, and land pollution, and their effects on human health. Includes numerous photographs and charts. Examines the connection between poverty, politics, and pollution. Includes an action packet that makes several suggestions for citizen involvment. Begins with a foreword by the Prince of Wales and an introduction by Robert Redford.

Preserving the global environment: the challenge of shared leadership. New York: W. W. Norton, 1991. 362 p.: ill. (HC110.E5 P68 1991)

 > "On April 19, 1990 (through the 22nd), 76 men and women from 18 countries, representing a spectrum of government, business, labor, academia, the media, and the professions, gathered at Arden House, Harriman, New York for the Seventy-seventh American Assembly entitled Preserving the Global Environment: The Challenge of Shared Leadership. For three days the participants discussed how the United States should reorient its policies and relations toward other countries and international institutions to preserve our global environment." This final report is a statement of general agreement issued by the participants.

Ramphal, Shridath S. *Our country, the planet: forging a partnership for survival.* Washington, D.C.: Island Press, 1992. 330 p. (TD195.E25 R36 1992)

 > Asks "if the development of the quarter of the world's people who are now rich has brought all so close to the limits of sustainable living on Earth, how is the development of the three-quarters who are poor to be accommodated?" Discusses global warming, ozone depletion, fossil fuels, deforestation, water pollution, biodiversity, and nuclear waste.

Ray, Dixie Lee. *Trashing the planet: how science can help us deal with acid rain, depletion of the ozone, and nuclear waste (among other things).* Washington, D.C.: Regnery Gateway, 1990. Distributed by National Book Network, Lanham, Md. 206 p. (TD174 .R39 1990)

 > Presents a proscience, protechnology perspective on the environment, maintaining that only science and technology can correct problems created by science and

technology. Moreover, antitechnical solutions are not only temporary, but possibly even disastrous.

Rolling Stone environmental reader. Washington, D.C.: Island Press, 1992. 268 p. (QH541.145 .R64 1992)
> Collects *Rolling Stone*'s environmental essays, including media coverage of environmental issues, the U.S. government's contributions to the environmental crisis, and the implications of rain forest destruction.

Silver, Cheryl Simon. *One Earth, one future: our changing global environment*. Washington, D.C.: National Academy Press, 1990. 196 p. (GF75 .S55 1990)
> Explains the basic science behind global environmental problems and examines policy implications aimed at solutions. Explores the interactions between the land, water, and atmosphere, and the snowballing impact that human activity is having on that system. Includes discussions of global warming and the implications for agriculture and for sea level rise in coastal areas, ozone depletion in the stratosphere, forest clearing in the tropics and the impacts on species living there, and acid deposition.

State of the environment. Paris: Organisation for Economic Co-operation and Development, 1991. 297 p.
> "This third report on the state of the environment reviews the progress achieved in OECD countries in attaining environmental objectives over the past two decades--the lifetime of most environmental policies and institutions. The report also examines the agenda for the 1990s: global atmospheric issues, air, inland waters, the marine enviornment, land, forests, wildlife, solid waste and noise. While focused on the relationships between the state of the environment, economic growth and structural change in OECD countries, the report places its analysis in the context of world ecological and economic interdependence and the need for sustainable development."

State of the world 1992: a Worldwatch Institute report on progress toward a sustainable society. New York, W. W. Norton, 1992. 256 p.
> Includes essays on nuclear waste, mining, sustainable jobs, biodiversity, reproductive health and global environmental governance.

State of the world 1991: a Worldwatch Institute report on progress toward a sustainable society. New York: W. W. Norton, 1991. 254 p.
> Maintains that if the New World Order decides to focus on the environmental crisis, environmental sustainability will become the guiding principle of the global community. Discusses energy, waste reduction, urban transportation, forestry, Eastern European and Soviet environmental reconstruction, military pollution, overconsumption, and the global economy.

State of the world 1990: an Worldwatch Institute report on progress toward a sustainable society. New York, W. W. Norton, 1990. 253 p.

Includes "Slowing global warming," by Christopher Flavin; "The illusion of progress," by Lester R. Brown; "Slowing global warming," by Christopher Flavin; "Saving water for agriculture," by Sandra Postel; "Feeding the world in the nineties," by Lester R. Brown and John E. Young; "Holding back the sea," by Jodi L. Jacobson; "Clearing the air," by Hilary F. French; "Cycling into the future," by Marcia D. Lowe; "Ending poverty," by Alan B. Durning; "Converting to a peaceful economy," by Michael Renner; and "Picturing a sustainable society," by Lester R. Brown, Christopher Flavin, and Sandra Postel.

State of the world 1988: a Worldwatch Institute report on progress toward a sustainable society. New York: W. W. Norton, 1988. 237 p.

Addresses the Earth's vital signs; creating a sustainable energy future; raising energy efficiency; shifting to renewable energy; reforesting the Earth; avoiding a mass extinction of species; controlling toxic chemicals; assessing SDI; planning the global family; and reclaiming the future.

State of the world 1986: a Worldwatch Institute report on progress toward a sustainable society. New York: W. W. Norton, 1986. 263 p.

Contents: "A generation of deficits," by L. Brown; "Assessing ecological decline," by L. Brown and E. Wolf; "Increasing water efficiency," by S. Postel; "Managing rangelands," by E. Wolf; "Moving beyond oil," by C. Flavin; "Reforming the electric power industry," by C. Flavin; "Decommissioning nuclear power plants," by C. Pollock; "Banishing tobacco," by W. Chandler; "Investing in children," by W. Chandler; "Reversing Africa's decline," by L. Brown and E. Wolf; "Redefining national security," by L. Brown.

State of the world 1984: a Worldwatch Institute Report on progress toward a sustainable society. New York: W. W. Norton, 1984. 252 p.

This report, the first of an annual series, "contains news on innovative or particularly successful actions to create a sustainable society; emphasizes global economic connections that policymakers often overlook; reviews national policies and programs, including progress toward specific national goals; and surveys major financial commitments by governments and international development agencies."

Stead, W. Edward; and Stead, Jean Garner. *Management for a small planet: strategic decision making and the environment*. Newbury Park, Calif.: Sage, 1992. 212 p.

Maintains that humankind needs a new economic paradigm that is less consumptive of our limited natural resources and that fosters sustainability in the business sector. Highlights the need to focus on the detrimental ecological results of unlimited economic growth and to balance energy input and waste output.

Wild, Russell, ed. *The Earth Care Annual 1990*. Emmaus, Pa.: Rodale, 1993.
> Compiles environmental journalism published between 1988 and 1990. Covers many issues, including acid rain and wildlife. Concludes with a discussion of what can be done by the individual to help solve the environmental crisis. Contains appendices of environmental organizations and hazardous chemicals.

Wolbarst, Anthony B., ed. *Environment in peril*. Washington, D.C.: Smithsonian Institution, 1991. 232 p.: ill. (GF75 .E56 1991)
> Based on lectures presented to the Environmental Protection Agency, including essays by Paul Ehrlich, Ralph Nader, Barry Commonor, and Carl Sagan. Topics include overpopulation, the world's oceans, and ways to incorporate public thinking into environmental legislation.

World resources 1987: a report by the International Institute for Environment and Development and the World Resources Institute. New York: Basic Books, 1987. 369 p.
> Aims to provide an "objective, current, global assessment of the natural resource base that supports the world economy." Contains issue analysis of hazardous waste management and sustainable development in sub-Saharan Africa. Includes reviews and data tables concerning population and health, human settlements, food and agriculture, forests and rangelands, wildlife and habitat, energy, freshwater, oceans and coasts, atmosphere and climate, global systems and cycles, and policies and institutions.

World resources 1986: a report by the World Resources Institute and the International Institute for Environment and Development. New York: Basic Books, 1986. 349 p.
> Contains concise accounts of the latest global, regional and national resource trends and highlights six emerging issues including multiple pollutants, temporate and tropical forest decline, , soil degradation, the atmosphere as a shared resource, and population growth.

Books (unannotated)

Adler, Bill. *The whole Earth quiz book: how well do you know your planet?* New York: Quill, 1991. 200 p.: ill. (QH541.13 .A28 1991)

Appleman, Milo Don. *Epitaph for planet Earth: how to survive the approaching end of the human species*. New York: F. Fell, 1982. 191 p. (GF41 .A66 1982)

Bear, Firman E. *Earth: the stuff of life*. 2d ed. Norman: University of Oklahoma Press, 1986. 318 p.: ill. (S591 .B33 1986)

Being in the world: an environmental reader for writers. New York: Macmillan, 1993.

Berry, R. J., ed. *Environmental dilemmas: ethics and decisions*. New York: Chapman and Hall, 1993.

Bookchin, Murray. *Defending the Earth: a dialogue between Murray Bookchin and Dave Foreman*. Boston: South End, 1991. 147 p. (JA75.8 .B66 1990)

Campolo, Anthony. *How to rescue the Earth without worshiping nature*. Nashville, Tenn.: T. Nelson, 1992. 211 p. (TD171.7 .C35 1992)

Changing the global environment: perspectives on human involvement. Boston: Academic Press, 1989. 459 p.: plates. (some color) ill. (GF75 .C47 1989)

Commoner, Barry. *Making peace with the planet*. New York: Pantheon, 1990. 292 p. (QH545.A1 C64 1990)

Craig, James R. *Resources of the Earth*. Englewood Cliffs, N.J.: Prentice-Hall, 1988. 395 p.: plates. (some color) ill. (HC21 .C72 1988)

DiSilvestro, Roger L. *Audubon perspectives: fight for survival: a companion to the Audubon television specials*. New York: Wiley, 1990. 284 p. (some color) ill. maps. (QH75 .D57 1990)

DiSilvestro, Roger L. *Audubon perspectives: rebirth of nature: a companion to the Audubon television specials*. New York: Wiley, 1992. 276 p. (some color) ill. color maps. (QH75 .D58 1992)

Durrell, Lee. *State of the ark*. Garden City, N.Y.: Doubleday, 1986. 224 p.: (some color) ill. (QH75 .D87 1986)

Earth and the human future: essays in honor of Harrison Brown. Boulder, Colo.: Westview, 1986. 258 p.: ill. (HC59 .E33 1986)

Earth report 2: monitoring the battle for the environment. Rev. ed. London: Mitchell Beazley, 1990. 176 p.: color ill. (TD170 .E172 1990)

Earth report: the essential guide to global ecological issues. Los Angeles: Price, Stern, Sloan, 1988. 240 p.: (some color) ill. color maps. (TD170 .E17 1988)

Earth's fragile systems: perspectives on global change. Boulder, Colo.: Westview, 1988. 109 p.: ill. (QH541 .E27 1988)

Earth's threatened resources: timely reports to keep journalists, scholars, and the public abreast of developing issues, events, and trends. Washington, D.C.: Congressional Quarterly, 1986. 205 p.: ill. (GF75 .E17 1986)

Ehrenfeld, David. *Beginning again: living in the environment of the new millennium*. New York: Oxford University Press, 1993.

Ending war against the Earth. Hudson, Wis.: G. E. McCuen, 1991. 176 p. (HC79.P55 E52 1991)

Environment at risk: responding to growing dangers. Dubuque, Iowa: Kendall/Hunt, 1989. 44 p.: (some color) ill. (TD180 .E46 1989)

Environment in peril. Washington, D.C.: Smithsonian Institution Press, 1991. 232 p.: ill. (GF75 .E56 1991)

Environment--international. Berlin: E. Schmidt, 1982. 154 p. (HC79.E5 E5728 1982)

Environmental concerns: an inter-disciplinary exercise. New York: Elsevier Science, 1991. 297 p.: ill. (QH540 .E564 1991)

Environmental issues in the 1990s. New York: Wiley, 1992. 349 p.: ill. maps. (GF41 .E53 1992)

Environmental scarcity: the international dimension. Tubingen, Germany: J. C. B. Mohr, 1991. 216 p. (HC79.E5 E59 1991)

Ericson, Harold L. *Handbook for survival, 1990.* Castle Park, Minn.: Futures Foundation, 1989. 158 p. (HC59 .E745 1989)

Essays in the economics of renewable resources. New York: North-Holland, 1982. Distributed by Elsevier Science, New York. 287 p.: ill. (SH334 .E8 1982)

Evernden, Lorne Leslie Neil. *The natural alien: humankind and environment.* Toronto, Canada: University of Toronto Press, 1985. 160 p. (GF21 .E86 1985)

Faber, Malte Michael. *Evolution, time, production, and the environment.* New York: Springer-Verlag, 1990.

Fox, Michael W. *The new Eden: for people, animals and nature.* Santa Fe, N.M.: Lotus, 1989. 77 p.: ill. (GF80 .F65 1989)

Fortner, Diane M. *Environmental studies.* Englewood Cliffs, N.J.: Salem, 1993.

Geller, E. Scott. *Preserving the environment: new strategies for behavior change.* New York: Pergamon, 1982. 338 p.: ill. (TJ163.3 .G44 1982)

Global 2000 revisited: mankind's impact on spaceship Earth. New York: Paragon House, 1991.

Global 2000: the report to the President--entering the twenty-first century. Rev. ed. Cabin John, Md.: Seven Locks Press, 1991.

Global change: geographical approaches. Tucson: University of Arizona Press, 1991. 289 p.: ill. maps. (GF75 .G56 1991)

Global environmental change: understanding the human dimensions. Washington, D.C.: National Academy, 1992. 308 p.: ill. (GF75 .G57 1992)

Global resources: opposing viewpoints. San Diego, Calif.: Greenhaven, 1991. 288 p.: ill. (HD75.6 .G56 1991)

Hardin, G. *Living within limits: ecology, economics, and population taboos.* New York: Oxford, 1993.

Hinnawi, Essam E. *The state of the environment.* Boston: Butterworths, 1987. 182 p.: ill. (TD170.2 .H56 1987)

Hynes, H. Patricia. *Earthright.* Rocklin, Calif.: Prima Publishing, 1990. Distributed by St. Martin's Press, New York. 236 p.: ill. (TD171.7 .H96 1990)

In defence of the Earth, the basic texts on environment: Founex, Stockholm, Cocoyoc. Nairobi: United Nations Environment Programme, 1981. 119 p. (HC79.E5 I512 1981)

In praise of nature. Washington, D.C.: Island Press, 1990. 258 p. (QH541.145 .I5 1990)

International dimensions of the environmental crisis. Boulder, Colo.: Westview, 1982. 298 p.: ill. (GF49 .I57 1982)

Jansma, Pamela E. *Reading about the environment: an introductory guide.* Englewood, Colo.: Libraries Unlimited, 1992.

Kaufman, Donald G. *Biosphere 2000.* New York: HarperCollins, 1993.

Korten, David C. *Getting to the 21st century: voluntary action and the global agenda.* West Hartford, Conn.: Kumarian Press, 1990. 253 p.: ill. (HC60 .K67 1990)

Leinwand, Gerald. *The environment.* New York: Facts on File, 1990. 122 p. (TD180 .L45 1990)

Luoma, Samuel N. *Introduction to environmental issues.* New York: Macmillan, 1984. 548 p.: ill. (GF50 .L86 1984)

Maclean, Kenneth. *World environmental problems.* Edinburgh, Scotland: Holmes McDougall, 1981. 67 p.: (some color) ill. color maps. (TD170.2 .M33 1981)

Managing planet Earth: readings from "Scientific American" magazine. New York: W. H. Freeman, 1990. 146 p.: color ill. color maps. (HC59 .M244 1990)

McKibben, Bill. *The end of nature.* New York: Anchor, 1990. 226 p.

Mending the Earth: a world for our grandchildren. Berkeley, Calif.: North Atlantic Books; San Francisco, Calif.: Environmental Rescue Fund, 1991. 219 p.: ill. (TD170 .M46 1990)

Moore, James W. *The changing environment.* New York: Springer-Verlag, 1986. 239 p.: ill. (TD170 .M65 1986)

Morgan, Sarah. *The endangered Earth: readings for writers.* Boston: Allyn and Bacon, 1992. 480 p. (TD170.3 .M67 1992)

Officer, C.; and Page, J. *Tales of the Earth: global change, natural and human.* New York: Oxford, 1993.

Oremland, R. S., ed. *Biogeochemistry of global change: radiatively active trace gases.* New York: Chapman and Hall, 1993.

Our food, air, and water: how safe are they? New York, N.Y.: Facts on File, 1984. 231 p.: ill. (RA566.3 .O93 1984)

Philip, Prince, consort of Elizabeth II, Queen of Great Britain. *Down to Earth: speeches and writings of His Royal Highness Prince Philip, Duke of Edinburgh, on the relationship of man with his environment.* Lexington, Mass.: Stephen Greene, 1989. Distributed by Viking Penguin. 240 p.: (some color) ill. (GF49 .P46 1989)

Planet Earth: egotists and ecosystems. New York: Rosen, 1991.

Planet under stress: the challenge of global change. New York: Oxford University Press, 1991. 344 p.: ill. (GF41 .P59 1991)

Practical handbook of environmental control. Boca Raton, Fla.: CRC Press, 1989. 537 p.: ill. (TD176.4 .P73 1989)

Prescott-Allen, Robert. *How to save the world: strategy for world conservation.* Totowa, N.J.: Littlefield, Adams, 1981. 144 p.: ill. (S936 .P73 1981)

Rational readings on environmental concerns. New York: Van Nostrand Reinhold, 1992. 841 p.: ill. map. (TD176.7 .R38 1992)

ReVelle, Penelope. *The environment: issues and choices for society.* 3d ed. Boston: Jones and Bartlett, 1988. 749 p.: plates: (some color) ill. (TD174 .R49 1988)

Sale, Kirkpatrick. *Human scale.* 2d ed. New York: Putnam, 1982. 558 p.: ill. (HC106.7 .S24 1982)

Schnaiberg, Allan. *The environment, from surplus to scarcity.* New York: Oxford University Press, 1980. 464 p. (HC79.E5 S29)

Starke, Linda. *Signs of hope: working towards our common future.* New York: Oxford University Press, 1990.

Stead, W. Edward. *Management for a small planet: strategic decision making and the environment.* Newberry Park, Calif.: Sage, 1992. 212 p.: ill. (HD75.6 .S74 1992)

Stone, C. D. *The gnat is older than man: global environment and human agenda.* Princeton, N.J.: Princeton University Press, 1993.

Sustaining Earth: response to the environmental threat. New York: St. Martin's, 1991. 226 p. (HD75.6 .S89 1990)

Taking sides: clashing views on controversial environmental issues. 3d ed. Guilford, Conn.: Dushkin, 1989. 354 p.: ill. (TD170 .T34 1989)

Tiezzi, E. *The horse to Samarra: ecological consequences of unlimited growth.* Baltimore, Md.: Columbia, 1993.

Timberlake, Lloyd. *Only one Earth: living for the future.* New York: Sterling, 1987. 168 p.: (some color) ill. (GF41 .T56 1987)

Udall, Stewart L. *The quiet crisis and the next generation.* Salt Lake City, Utah: Peregrine Smith, 1988. 298 p. (S930 .U3 1988)

Voices in defence of the Earth. Nairobi: United Nations Environment Programme, 1982. 185 p.: (some color) ill. (HC59.72.E5 V65 1982)

World environment 1972-1982: a report. Dublin, Ireland (United Nations Environment Programme): Tycooly International, 1982. 637 p.: ill. (HC79.E5 W668 1982)

World guide to environmental issues and organizations. Harlow, Essex, U.K.: Longman Current Affairs, 1990. Distributed by Gale Research, Detroit, Mich. 386 p.: ill. (TD170 .W68 1990)

World resources 1988-89: a report. New York: Basic Books, 1988.

World watch reader on global environmental issues. New York: W. W. Norton, 1991. 336 p.: ill. (TD174 .W68 1991)

Worster, D. *The wealth of nature: environmental history and the ecological imagination.* New York: Oxford, 1993.

Young, John. *Sustaining the Earth*. Cambridge, Mass.: Harvard University Press, 1990. 225 p. (QH75 .Y68 1990)

Zhivkov, Z. *Can we save the natural environment*. Sofia, Bulgaria: Sofia Press, 1986. 84 p. (TD171.5.B8 Z48 1986)

General interest periodicals (annotated)

"20th Anniversary issue on the environment." *Smithsonian* 21 (April 1990): whole issue (33-190 p.)
> Issued to commemorate Earth Day's anniversary and the magazine's birthday; devoted to the subject of the environment. Partial contents include "The past: knowing the wilderness was to love it," by Wallace Stegner; "The present: the road from Earth Day 1," by Bil Gilbert; "On the Connecticut, a sweet smell of success," by Richard Conniff; "A flight to survival for endangered raptors," by Don Moser; " 'Alternative agriculture' is gaining ground," by Jeanne McDermott; "Rails-to-trails: an exercise in linear logic," by Judy Mills; "Down in the dumps," by Richard Wolkomir; "Trees aren't mere niceties--they're necessities," by Jon Krakauer; "Flushed with pride in Arcata, California," by Doug Stewart; and "No compromise in defense of Mother Earth," by Michael Parfit.

Allman, William F. "Rediscovering planet Earth." *U.S. News & World Report* 105 (October 31, 1988): 56-61, 63, 65, 67-68.
> Using new techniques to explore past Earth climates and computers to simulate the future, researchers are beginning to see Earth as a complex, interdependent system of oceans, atmosphere, and life.

Andreski, Stanislaw. "Why not enough is being done about ecological dangers." *World & I* 4 (April 1989): 297-303.
> Analyzes the need for but resistance to cooperation in global environmental protection and discusses "the asymmetry between costs and benefits: the costs of the remedies are direct and immediate, while the benefits are remote and diffuse."

Carpenter, Betsy. "Living with our legacy." *U.S. News & World Report* 108 (April 23, 1990): 60-62, 64-65.
> Looks at the progress made in environmental problems since the first Earth Day in 1970, and outlines the key national and global problems to be resolved today that were unimaginable twenty years ago.

Commoner, Barry. "The environment." *New Yorker* 63 (June 15, 1987): 46-47, 50-54, 56-71.
> Reviews trends in the state of the environment over the last fifteen years and discusses the environmental movement.

Easterbrook, Gregg. "Everything you know about the environment is wrong." *New Republic* 202 (April 30, 1990): 14, 16-27.

Examines the following issues: acid rain, asbestos, Clean Air Act, climate change, cost-benefit analysis, emissions trading, the economic scorecard, the enviro lineup, enviro politics, the garbage crisis, global warming, nature, ozone depletion, pesticides, reformulated gasoline, species rights, Third World resources and pollution, and zero emissions.

Eisenberg, Evan. "The call of the wild: nature's four lessons for ecologists." *New Republic* 202 (April 30, 1990): 30-38.

"For some ecologists, global warming is a spiritual crisis, for others it is a problem in science and public policy. But there are paths between the Deep Ecologists' apocalyptic pessimism and the Planet Managers' dreams of total control. The key is to let nature be our guide. Four organic principles show the way."

"Environment: Earth Day 20." *Roll Call* 35 (April 19, 1990): 11-29. (*Roll Call* policy briefing no. 13)

Contents: "Seize the moment, change the course of history," by Gaylord Nelson; "The solid waste crisis," by John Porter; "A great success story," by John Dingell; "Indoor air pollution," by John Chafee; "The loss of natural forests," by Wyche Fowler; "A national and worldwide policy," by Pat Quinn; "The threat of global warming," by Tim Wirth; "A long habit of not thinking a thing wrong," by Claudine Schneider; "Mass transit," by John Heinz; "What the United Nations is doing to save the Earth," by Joan Martin-Brown; "The dark Reagan years," by Sid Yates; "This is next year," by Mo Udall; "The American farm and the environment," by Kika de la Garza; and "The greening of the congressional consciousness is an indisputable fact," by Amy Barrett.

Katsuya, Fukuoka. "Making environmental protection pay." *Japan Echo* 17 (Spring 1990): 69-74.

Discusses the environmental problems of global warming, acid rain, automobile pollution and the overly liberal application of pesticides in farming. Suggests policy solutions.

Linden, Eugene. "Is the planet on the back burner?" *Time* 136 (December 24, 1990): 48-51.

"War and recession may be grabbing the headlines, but the relentless trashing of the world's air, land and seas continues apace." Includes a sidebar, "1990 scorecard," which lists environmental winners, losers, and split decisions for the year, and an attached article, "The ecokid corps," by Philip Elmer-Dewitt.

McKibben, Bill. "Reflections: the end of nature." *New Yorker* 65 (September 11, 1989): 47-48, 50, 52, 54-58, 67-76, 78-93, 96-105.

Concludes that man, by his global pollution of carbon dioxide and methane, has changed the Earth's atmosphere and consequently its weather, and in doing so, has made the idea of nature--"the wild province, the world apart from man, under whose rules he was born and died"--extinct. Furthermore, to live under greenhouse conditions, man may find it necessary to use biotechnology, leading "to the second end of nature: the imposition of our artificial world in place of the broken natural one."

Sancton, Thomas A.; Linden, Eugene; and Lemonick, Michael D. "Endangered Earth update." *Time* 134 (December 18, 1989): 60-65, 68, 71.
> *Time* invited 14 environmental experts and policymakers to examine the environmental progress made around the world during the year, and to develop an agenda for the future. Partial contents: "The fight to save the planet"; "Get going, Mr. Bush"; "Scrub that smokestack"; "It's not easy being green"; and "Let Earth have its day."

Sancton, Thomas A. "Planet of the year: with drought, famine and fouled beaches, the Earth warns of environmental disaster." *Time* 133 (January 2, 1989): 26-30, 32-39, 41-42, 44-45, 47-50, 54, 63, 65-66, 68-78.
> Discusses ecological and environmental problems in a global context, including overpopulation, pollution, waste of resources and destruction of natural habitats. Explains the complexities of these interlocking problems and fashions an agenda for environmental action.

"Struggle to save our planet." *Discover* 11 (April 1990): 35-43, 46-49, 51-53, 55-59, 61-62, 64, 66-70, 72, 74, 76-78.
> Contents: "Down in the dumps," by Dan Grossman and Seth Shulman; "The numbers game," by David Berreby; "Ecoglasnost," by Tom Waters; "Playing dice with megadeath," by Jared Diamond; "Green giants," by Doug Stewart; "Invisible garden," by Robert Kunzig; and "The Poison eaters," by Susan Chollar.

Tierney, John. "Betting the planet." *New York Times Magazine* (December 2, 1990): 52-53, 74, 76, 78, 80-81.
> "Ten years ago, an ecologist and an economist with bitterly opposing world views made a $1,000 wager over an old question: was the Earth's growing population running out of natural resources? It was the doomster against the boomster, and this fall one of them had to pay up."

"War on the Earth's environment." (Moscow) *New Times* 46 (November 14-20, 1989): 35-37, 39.
> Contents: "What shall the meek inherit?," by Edward Goldsmith; "A two-pronged plan against debt and destruction: can North meet South to protect the environment?," by Karl Ziegler; and "A catalytic kick: EC gets environmental nudge from Netherlands," by Ronald van de Krol.

Yoichi, Kaya. "Threats to the global ecosystem." *Japan Echo* 17 (Spring 1990): 65-68. Yoichi, a Japanese scientist and member of the United Nations Science and Technology Advisory Committee for Development, expresses his opinions about global warming and ozone depletion.

General interest periodicals (unannotated)

Adler, Jerry. "Survival." (environmentalism) (Special Report: The Age of Anxiety: A Survival Guide to the '90s) *Newsweek* 116,27 (December 31, 1990): 30(2).

Aga Kahn, Sadruddin. "Natural sovereignty." *New Perspectives Quarterly: NPQ* 8,4 (Fall 1991): 32(2)

Alexander, Benjamin H. "Why is the environmental crisis happening? Some scientific causes and effects." *Vital Speeches of the Day* 56,4 (December 1, 1989): 124(5).

Antrobus, Peggy; and Peacocke, Nan. "Who is really speaking in the environment debate?" *UNESCO Courier* (March 1992): 39(4).

Batisse, Michel. "Our small blue planet." (environmental problems) *UNESCO Courier* (November 1990): 46(4)

Bailey, Ronald; and Ames, Bruce N. "Raining in their hearts: nuclear winter, population explosion, non-renewable resources, vanishing species--some of them are genuine problems, others are made up out of whole cloth." (includes related article on cancer and pollution) *National Review* 42,23 (December 3, 1990): 32(5).

Begley, Sharon. "Is it apocalypse now?" (computer simulation of the environmental future) *Newsweek* 119,22 (June 1, 1992): 36(4).

Brown, Lester R.; Flavin, Christopher; and Postel, Sandra . "A planet in jeopardy." (the environmental crisis) *Futurist* 26,3 (May-June 1992): 10(5).

Brown, Lester R.; Flavin, Christopher; and Postel, Sandra. "A world at risk: our endangered environment poses a threat, and renewed hope, for humanity." *Country Journal* 16,4 (May-June 1989): 44(5).

Brown, Lester R.; and Fornos, Werner. "The environmental crisis: a humanist call for action." *Humanist* 51,6 (November-December 1991): 26(7).

Cetron, Marvin; and Davies, Owen. "50 trends shaping the world." *Futurist* 25,5 (September-October 1991): 11(11).

Cox, Murray; et al. "Save the planet." *Omni* 11,12 (September 1989): 34(13).

Dyer, Gwynne. "Doorway to a cleaner world." (global survival) *Reader's Digest* 136,816 (April 1990): 135(4)

"Guide to some of the scariest things on Earth." (global environmental problems) *Scholastic Update* 121,16 (April 21, 1989): 4(2)

Livingston, John A. "The environment: what have we learned?" *Canadian Geographic* 109,6 (December-January 1989): 107(7).

MacLeish, William H. "Where do we go from here? We'll never regain Eden but we may atone for our errors if we find new methods for doing things in the future." (special issue on the environment) *Smithsonian* 21,1 (April 1990): 58(10).

Mohler, Mary; Bliss, Jeff; and Rosen, Margery D. "Will the world be here for our kids?" (a special LHJ roundtable on the environment) (includes related articles on where to get more information and how to make a difference) *Ladies' Home Journal* 106,6 (June 1989): 122(8).

Pollan, Michael; et al. "Only man's presence can save nature." *Harper's Magazine* 280,1679 (April 1990): 37(10).

Reiger, George. "My brother's keeper: it's no longer enough simply to mind the way we ourselves behave." (Conservation) *Field and Stream* 95,2 (July 1990): 20(2).

Reiger, George. "The basket-weavers." (futility of environmental protection efforts) *Field and Stream* 97,3 (July 1992): 19(3).

Schell, Jonathan. "Our fragile Earth: only now are we beginning to fathom the gravity of our environmental dilemma." *Discover* 10,10 (October 1989): 44(5).

Seligman, Daniel. "Cuddling up to Mother Earth." (reaction to television program *Spirit and Nature with Bill Moyers*) *Fortune* 124,3 (July 29, 1991): 177(2).

Singer, S. Fred. "The science behind global environmental scares." *Consumers' Research* 74,10 (October 1991): 17(5).

Tokar, Brian; et al. "Eight visions of our ecological future." *Utne Reader* 36 (November-December 1989): 92(4).

Toolan, David S. "A new song of Earth." (appreciation of the environment) (editorial) *America* 163,19 (December 15, 1990): 467(2)

Walljasper, Jay. "Just my imagination: envisioning the world in 2009." (environmentalism) *Utne Reader* 36 (November-December 1989): 142(2).

Weeden, Robert B. "An exchange of sacred gifts." (thoughts about ecology) *Alternatives* 16,1 (March-April 1989): 40(10).

Environmental and professional journals (annotated)

"1991 global report." *Buzzworm: The Environmental Journal* 3 (January-February 1991): 30-45.

Buzzwork asked 13 leading specialists and scientists to offer a progress report on 13 environmental challenges facing the world. Contents include "A world at risk," by James J. MacKenzie; "Sustainable agriculture," by Terry Gips; "Toxic pollution," by Joel S. Hirschhorn; "Population," by Garrett Hardin; "Deforestation," by Thomas E. Lovejoy; "Desertification," by Michael Glantz; "Ozone depletion," by Richard Elliot Benedick; "Global warming," by Stephen H. Schneider; "Loss of diversity," by Peter H. Raven; "Energy, by James J. MacKenzie; "Indigenous people," by Mary George Hardman; "Ocean pollution," by Justin Lancaster; "Air pollution," by Michael Oppenheimer and Leonie Haimson; and "Groundwater pollution," by Ann Maest and Lois Epstein.

Brown, Lester. "State of the world: an interview with Lester Brown." *Technology Review* 91 (July 1988): 51-58.

In this interview by Sandra Hackman and Marc S. Miller, the head of the Worldwatch Institute assesses global environmental issues.

Bunyard, Peter. "Gaia: the implications for industrialised societies." *Ecologist* 18,6 (1988): 196-206.

"The destruction of the world's forests, the death of the seas, the increase in droughts and floods, the spread of deserts, all point to a major disruption of the Gaian mechanisms that make life on Earth tolerable for humans. The cause is undoubtedly our way of life: but how will Gaia respond?"

Cooper, Mary H. "Setting environmental priorities." Congressional Quarterly's *Editorial Research Reports* 2, 21 (1988): 614-627.

Contents: "The problems"; "Global warming"; "Ozone depletion"; "Acid rain"; "Air pollution"; "Indoor pollution"; "Garbage"; "Nuclear waste"; "The solutions"; "Energy efficiency"; "Worldwide effort"; "More regulation?" Includes pro and con discussion between H. H. Woodson and Ken Bossong over whether nuclear power should be increased to meet energy needs and protect the environment.

"Earth Day and the future of our planet." *National Parks* 64 (March-April 1990): 16-28, 42-43.

Contents: "Global prescription: leading conservationists look to the future and speak their mind"; "Earth Day," by Elizabeth Hedstrom; "Planet at the crossroads: eight issues that will determine Earth's future: species loss, global warming, acid rain, ozone depletion, rain forests, protected areas, marine management, and land abuse," by John Kenney.

Ehrlich, Anne H.; and Ehrlich, Paul R. "Back from the abyss." *Sierra* 72 (March-April 1987): 55-60.
> The collapse of civilization "could come about gradually over the next half century as a consequence of continuing human population growth and behavior patterns that degrade the planet's ecosphere. Or it could occur with lightning swiftness any day" as the result of nuclear war. However "the great hope for civilization lies in this: that people can recognize how the human predicament evolved and what changes need to be made to resolve it. No miracles, no outside intervention, no new inventions are required."

Ehrlich, Anne H.; and Ehrlich, Paul R. "Needed: an endangered humanity act?" *Amicus Journal* 7 (Spring 1986): 12-14.
> "One way or another, the tight connection between our own future and that of our fellow passengers on Spaceship Earth must be made clear to everyone."

"Engineering the global environment." *American Consulting Engineer* 1 (Fall 1990): 32 p.
> Contents: "U.S. engineering firms preparing for EC 1992," by Howard Schirmer, Jr; "East European countries offer attractive markets," by Derish M. Wolff; "Hazardous waste market can be hazardous to CE's," by James H. Kleinfelder; "Engineers to play lead in environmental decade," by Ronald Gobbell and Steve M. Burch; "Mass transit is key to cutting air pollution," by Lisa G. Nungesser; "Global effort needed to combat greenhouse effect," by James E. Hansen.

Fairclough, A. J. "Global environmental and natural resource problems--their economic, political, and security implications." *Washington Quarterly* 14 (Winter 1991): 81-98.
> "Current policies and approaches are unsustainable. Business as ususal is no longer a viable option. Major changes throughout all our societies are needed to set the world on the path to a sustainable future."

"Forum: the global environment." *Congressional Research Service Review* 10 (August 1989): 1-27.
> Discusses global warming, stratospheric ozone depletion, pollution threats to the oceans, tropical deforestation and loss of biological diversity, environmental dilemmas in developing countries and the United States' response, the 'greening' of world politics, and economic implications of environmental threats: a central policy dilemma.

Hileman, Bette. "Industrial ecology route to slow global change proposed." *Chemical and Engineering News* 70 (August 24, 1992): 7-14.

"For two weeks late last month, 51 invited participants from a wide variety of disciplines and 10 countries met in Snowmass Village, Colorado to consider broad questions concerning industrial ecology and global change and to recommend actions and research in this area. The primary aim of the Snowmass meeting was to think about ways of reconfiguring industrial activity to reduce harm to the human and natural environment at the global level. 'Industrial ecology arises from the perception that human economic activity is causing unacceptable changes in basic environmental support systems'. It seeks alternative approaches that will reduce greenhouse gas buildup, dissipative uses of heavy metals, dispersion of long-lived toxic chemicals, and excessive fertilization in agriculture."

Holdgate, Martin W. "'92: the environment of tomorrow." *Environment* 33 (July-August 1991): 14-20, 40-42.
"What will Earth's environment be like in 40 or 80 years? According to the director general of the World Conservation Union, some environmental trends, such as deforestation, population growth, and loss of biodiversity, are virtually impossible to stop. Strong international cooperation, however, could guide humanity toward a more sustainable relationship with the natural world."

"Managing planet Earth." *Scientific American* 261 (September 1989): whole issue (190 p.)
Partial contents include "The changing atmosphere," by Thomas E. Graedel and Paul J. Crutzen; "The changing climate," by Stephen H. Schneider; "The growing human population," by Nathan Keyfitz; "Strategies for energy use," by John H. Gibbons, Peter D. Blair and Holly L. Gwin; "Strategies for manufacturing," by Robert A. Frosch and Nicholas E. Gallopoulos; and "Strategies for sustainable economic development," by Jim MacNeill.

McElroy, Michael. "Challenge of global change." *New Scientist* 119 (July 28, 1988): 34-36.
Calls for the subordination of national prerogatives and parochial interests to protect the rights of all the living elements of the planet.

"New day must dawn." *National Wildlife* 28 (February-March 1990): 5-15, 18-43.
Partial contents include "A new day must dawn," by Thomas A. Lewis; "It's enough to make you sick," by Susan Q. Stranahan; "How safe is your world?: you have a right to know," by Bill Lawren; "Innocent victims of a toxic world," by Peter Steinhart; and "Pollution knows no boundaries," by Sharon Begley.

Orians, Gordon H. "The place of science in environmental problem solving." *Environment* 28 (November 1986): 12-17, 38-41.
"Policy action often cannot wait for adequate scientific information . . . however, it is abundantly clear that there is considerable scope for expanding the role of science and scientists in environmental problem solving."

Parker, Jonathan; and Hope, Chris. "The state of the environment: a survey of reports from around the world." *Environment* 34 (January-February 1992): 18-20, 39-44.
> "All countries participating in the United Nations Conference on Environment and Development this coming June in Rio de Janeiro are supposed to have submitted a report on the environmental conditions, problems, and progress within their borders. The authors have compiled a reasonably comprehensive list of the most current reports worldwide."

Pirages, Dennis. "Toward a theory of global change." *Futures Research Quarterly* 7 (Fall 1991): 57-72.
> Reviews changes in perspectives and policies in the nearly two decades since publication of the initial controversial study that gave a large boost to research on global futures.

Rowlands, Ian. "Security challenges of global environmental change." *Washington Quarterly* 14 (Winter 1991): 99-114.
> The response of the international community to the challenge of global environmental change has thus far been mixed: considerable progress on the ozone layer case but little of tangible value coming out of the global warming negotiations.

Rubin, Debra K. "Making the world a cleaner place." *Engineering News Record* 223 (November 23, 1989): 28-33.
> Describes the growing international concern for environmental protection and eco-consciousness which is creating more markets for services world-wide.

Weiss, Edith Brown. "In fairness to future generations." *Environment* 32 (April 1990): 7-11, 30-31.
> Maintains that each generation has the right to use and benefit from the natural and cultural legacy of its ancestors but also has an obligation to conserve the environment and natural and cultural resources for future generations.

Environmental and professional journals (unannotated)

Carey, John; et al. "Fighting to save a fragile Earth." (environmental protection) *International Wildlife* 20,2 (March-April 1990): 4(18).

"Earth matters." (Editorial) *The Lancet* 339,8805 (May 30, 1992): 1325(2).

"Good planets are hard to find." (excerpts from the book, *State of the World 1989*) *Mother Earth News* 117 (May-June 1989): 126(2).

Holdgate, Martin W. "The environment of tomorrow." (trends in environmental changes) *Environment* 33,6 (July-August 1991): 14(10)

Kenney, John. "Planet at the crossroads: eight issues that will determine Earth's future: species loss, global warming, acid rain, and others." *National Parks* 64,3-4 (March-April 1990): 24(7).

Lamm, Richard D. "The future of the environment." *Journal of the American Academy of Political and Social Science* 522 (July 1992): 57(10).

Monastersky, Richard. "Global change: the scientific challenge." (understanding environmental problems of the next century; includes related article) *Science News* 135,15 (April 15, 1989): 232(4)

Myers, Norman. "What happened to utopia." (Maurice Strong looks at the state of the Earth) (interview) *International Wildlife* 17 (July-August 1987): 36(2).

Ohlendorf-Moffat, Pat. "The air we breathe, the water we drink, the food we eat; how safe are they?" *Chatelaine* 60 (July 1987): 36(3)

"Planet strikes back." (21st Environmental Quality Index) *National Wildlife* 27,2 (February-March 1989): 33(8).

Price, Martin F. "Global change: defining the ill-defined." (environmental change) *Environment* 31,8 (October 1989): 18(6)

Quinney, John. "Out of the ark and into the world." (New Alchemy Institute) *Whole Earth Review* 62 (Spring 1989): 33(3).

Scott, Geoff. "Our planet, our health." (includes related articles) *Current Health* 2 16,4 (December 1989): 4(7)

Waldrop, M. Mitchell. "An inquiry into the state of the Earth." *Science* 226 (October 5, 1984): 33(3)

Law journals

Jones, Owen Donald. "Box H problem: a justification for unilateral international coercion." *Yale Journal of International Law* 15 (Summer 1990): 207-275.
    Discusses environmental problems that combine irreversibility, universality, and a time lag between action and effect. Maintains that since such problems will first be recognized and understood, if at all, in countries with superior scientific resources, measures to achieve action through cooperation alone will prove inherently cumbersome and inadequate. Concludes that in the end, such a circumstance may require unilateral action.

Lamm, Richard D. "The heresy trial of the Reverend Richard Lamm." *Environmental Law* 15,4 (1985): 755-766.

A dramatic plea to an imaginary court hearing in the year 2000, concludes that even in the 1980's it was almost too late to avert the shadow of starvation and chaos which threatened the human race. Presents the concept of "Reality Theology."

Reports

Clark, William C. *Human ecology of global change.* Cambridge, Mass.: Energy and Environmental Policy Center, John F. Kennedy School of Government, Harvard University, 1990. 77 p. (Discussion paper G-90-01)
>Examines the place of human-related studies in the analysis of global environmental change, identifies the principal elements of the human system involved, summarizes the major unresolved questions pertaining to the human dimensions of global change, and highlights a small number of research challenges that merit priority attention.

Clark, William C. *Visions of the 21st century: conventional wisdom and other surprises in the global interactions of population, technology, and environment.* Cambridge, Mass.: Energy and Environmental Policy Center, John F. Kennedy School of Government, Harvard University, 1989. 36 p. (Science, Technology and Public Policy Program. Discussion paper 89-09)
>"This essay is undertaken as part of a larger effort to illuminate the opportunities and constraints that bear on the management of sustainable development in the next century." Sketches "the physical context of the management problem: the ways in which human populations, their technologies, and the environments they transform set the stage on which economic, social, and political changes are played out."

Earth's threatened resources. Congressional Quarterly's *Editorial Research Reports,* 1986. 205 p.
>Presents ten previously published reports which illustrate some of the threats to our planet's resources. Taken together they offer "a troubling picture of what lies ahead--and sometimes atribute to human resourcefulness in meeting challenges."

For Earth Day, nine scientists offer data on the state of the environment. Washington, D.C.: Heritage Foundation, 1990. 21 p.
>Brief articles on environmental issues from a conservative viewpoint. Includes "The politics of global warming," by Patrick J. Michaels.

Simon, Julian Lincoln. *Global 2000 revised.* Washington, D.C.: Heritage Foundation, 1982. 46 p. (HC79.E5 S442 1982)

State of the world environment 1991. Nairobi: United Nations Environment Programme, 1991. 48 p.

Annual report summarizing socio-economic environment, environmental quality, biological diversity, shared water resources, and marine environment.

State of the world environment 1989. Nairobi: United Nations Environment Programme, 1989. 43 p.
Each year the UNEP issues a report on the state of the environment; this issue of the annual publication includes a focus section on greenhouse gases and climate. Partial contents: "Summary"; Part I.: "The state of the environment"; "Environmental quality"; "Development and environment"; Part II. "Focus on: greenhouse gases and climate"; Part III. "Focus on: hazardous waste."

State of the environment 1985. Paris: Organisation for Economic Co-operation and Development, 1985. 271 p.
"The second OECD report on the state of the environment identifies progress as well as remaining and new problems of pollution and of natural resource management. It examines the pressures on the environment from agriculture, energy, industry and transport activities and the responses of the public, of enterprises and of governments to these pressures."

Tolba, Mostafa Kamal. *Earth matters.* Nairobi: United Nations Environment Programme, 1983. 163 p. (TD170.3 .T652 1983)

Train, Russell. *Environmental concerns in the year 2000.* Washington, D.C.: Congressional Research Service, 1989. 19 p. (Congress in the year 2000)
Discusses the U.S. environmental legislative agenda and recommends four overarching environmental policy initiatives. Among the topics examined are global warming, energy efficiency, solid waste, land use, and global environmental problems.

Uniting nations on the Earth: an environmental agenda for the world community: final report of the United Nations Association of the USA and the Sierra Club. New York: UNA-USA, 1991.
"This brochure summarizes the recommendations of Uniting Nations for the Earth, the final report of the 1990 nationwide citizens' study project on international institutions and the global environment organized by the United Nations Association of the United States (UNA-USA) in collaboration with the Sierra Club."

World environment 1972-1982: a report by the United Nations Environment Programme. Dublin, Ireland: Tycooly International Publishing, 1982. 637 p.
Reviews "the changes in the world environment and in human understanding of it during the 1970s." Evaluates the significance of those developments, and provides a foundation for international and national action in the years ahead. Contains a large number of tables but warns that "the data base is of very variable quality."

Federal government reports

Interparliamentary Conference on the Global Environment: final proceedings; report to the Senate of the United States of America. Washington, G.P.O., 1990. 200 p.

Selected major international environmental issues: a briefing book: report prepared for the Committee on Foreign Affairs, U.S. House of Representatives by the Congressional Research Service, Library of Congress. Washington, D.C.: G.P.O., 1991. 52 p.

Conference proceedings

The biosphere, problems and solutions: proceedings of the Miami International Symposium on the Biosphere, 23-24 April 1984, Miami Beach, Florida, U.S.A.. New York: Elsevier, 1984. 712 p.: ill. (TD169 .M53 1984)

Crawford, David L., ed. *Light pollution, radio interference, and space debris: proceedings of the International Astronomical Union Colloquium no. 112, held August 13-16, 1989, in Washington, D. C.* San Francisco: Astronomical Society of the Pacific, 1991. 331 p.: ill. (QB476.5 .I58 1989)
> Based on a conference and colloquium, the papers which comprise this work are not overly technical. Offer a good introduction and a short summary.

Global change: the proceedings of a symposium sponsored by the International Council of Scientific Unions (ICSU), held during its 20th General Assembly in Ottawa, Canada, on September 25, 1984. Miami, Fl.: ICSU Press, 1984. 357 p.: ill. (some color) (QE1 .G716 1984)

Interactions between climate and biosphere: transactions of the C.E.C. symposium in Osnabruck, held on March 21-23, 1983. Lisse: Swets & Zeitlinger, 1984. 392 p.: ill. (QC980 .I55 1984)

International Symposium on Soil Biology and Conservation of the Biosphere (9th: 1985: University of Forestry and Timber Industry) Proceedings of the 9th International Symposium on Soil Biology and Conservation of the Biosphere. Budapest: Akademiai Kiado, 1987. 2 v. ill. (QH84.8 .I57 1985)

Interparliamentary Conference on the Global Environment: final proceedings; report to the Senate of the United States of America. Washington, D.C.: G.P.O., 1990. 200 p. (Document, Senate, 101st Congress, 2nd session, no. 101-31)

Maintenance of the biosphere: proceedings of the Third International Conference on Environmental Future. Edinburgh, Scotland: Edinburgh University Press, 1990. 228 p.: ill. (GF3 .I37 1987b)

Proceedings of the Fourth Symposium on Our Environment, held in Singapore, May 21-23, 1990. Boston: Kluwer Academic, 1991.

Study Week on a Modern Approach to the Protection of the Environment, November 2-7, 1987. New York: Pergamon, 1989. 606 p.: ill. (TD169 .S78 1987)

Vittachi, Anuradha. *Earth Conference one: sharing a vision for our planet.* Boston: New Science Library, 1989. Distributed by Random House, 1989. 146 p. (GF3 .V57 1989)

Bibliographies

Core bibliography on global issues related to environment, resources, population, and sustainable development. Washington, D.C.: Global Tomorrow Coallition, 1991. 29 p.

Jacob, K. C. *Ecological crisis: a select bibliography.* Trivandrum, India: K. C. Jacob, 1989. 44 p. (Z5322.E2 J33 1989)

Jansma, p. E. *Reading about the environment: an introductory guide.* Englewood, Colo.: Libraries Unlimited, 1992.

Link, Terry. "Sources for a small planet." (includes environmental periodicals and sources) *Library Journal* 115,10 (June 1, 1990): 81(8).

Merideth, Robert W. *The environmentalist's bookshelf: a guide to the best books.* New York: G. K. Hall, 1992.

Sinclair, Patti K. *E for environment: an annotated bibliography of children's books with environmental themes.* New Providence, N.J.: R. R. Bowker, 1992. 292 p. (Z1037.9 .S57 1992)
> An annotated bibliography to environmental books (including fiction) aimed at readers between preschool and fourteen. Five hundred books are grouped into five main chapters which are further subdivided. Includes author, subject, and title indices.

Steinhart, Peter. "The longer view." (environmental protection periodicals) *Audubon* 89 (March 1987): 10(3).

Watson, Tom. "Finding the trees in the forest: environmental information sources." *Wilson Library Bulletin* 65,6 (February 1991): 34(6).

Reference books (annotated)

(See also *The Environmental Movement: Reference*) 512

1992 information please environmental almanac. Boston: Houghton Mifflin, 1991. 606 p.
"Aims to provide factual, accurate, practical information about every aspect of our environment, from the food we eat and the homes we live in to the environmental costs of war." Discusses the state of the planet, food, energy, water, waste, forests and wetlands, air pollution, recreation and leisure, green cities, state comparisons and profiles, greenhouse warming, tropical forests, country comparisons, Africa, Antarctica, Asia, Europe, North America, Oceania, and South America.

Allaby, Michael. *Dictionary of the environment.* 3d ed. New York: New York Unviersity Press, 1989. 423 p. (QH540.4 A44 1989)
Lists most of the important geographical, biological, geological, and sociological terms which pertain to the environment. Also includes a listing of major environmental disasters, environmental agencies, and organizations, and titles of legislation. Contains references.

Ashworth, William. *The encyclopedia of environmental studies.* New York: Facts on File, 1991. 470 p.: ill. maps. (TD9 .A84 1991)
Provides explanations for most of the complex terms used when discussing environmental issues. Lists three thousand entries, coming from such fields as seismology, geology, botany, and oceanography. Also includes selected essays from prominent individuals who are from the environmental community, related government agencies, or trade organizations. Contains cross-references and an index.

Conservation directory, 1990: a list of organizations, agencies, and officials concerned with natural resource use and management. Washington, D.C.: National Wildlife Federation, 1990.
Lists, for example, colleges with environmental programs and national and international environmental organizations in the U.S. and Canada.

Darnay, Arsen J., ed. *Statistical record of the environment.* Detroit, Mich.: Gale Research, 1992. 855 p.: ill (TD180 .D37 1991)
Compiled from government and trade organization documents. Provides reference for a wide variety of issues pertaining to the environment. Covers topics such as pollution, the role of the media, politics, government regulation, and public opinion. Includes over eight hundred statistical tables, which cover a diverse range of statistics, from fertilizer use in various Asian countries, to the percentage of Christmas trees recycled each year, to the fifty worst toxic air polluters. Contains an index of important terms.

Dictionary of environmental quotations. Compiled by Barbara K. Rodes and Rice Odell. New York: Simon and Schuster, 1992. 355 p.
Presents over 3,700 quotations related to environmental issues. Arranged into over one hundred alphabetical categories. Within each category the quotations are

arranged chronologically. When available the author and source of the quotation is provided.

Earth journal 1992: environmental almanac and resource directory. Boulder, Colo: Buzzworm, 1992.
> An annual, ready-reference guide to environmentally sound lifestyles. Offers a synopsis of the year in review "in this case the environmental consequences of the war in the Persian Gulf and the California drought" and brief sections on specific environmental issues, including culture, ecofeminism, ecotravel, and biosphere politics.

Environmental data report. 3d edition. London: Blackwell Reference, 1991. 408 p.
> Prepared for the U.N.E.P. by the GEMS Monitoring and Assessment Center in London in cooperation with the World Resources Institute in Washington and the U.K. Department of the Environment in London. Discusses pollution, climate, natural resources, population, health, energy, tourism, wastes, natural disasters, and international cooperation. Most data is given in tabular form, with some pie charts and bar graphs. Includes a few isopleth and choropleth maps. Provides data on national totals, which are localized by site, and climate, which go back one hundred years. Contains appendices of acronyms, contributors, and tables and figures.

The environment index: volume 1. Ann Arbor, Mich.: UMI, 1992.
> Presents environmental articles abstracted from newspapers, popular magazines, and some journals. Aimed at the secondary school libraries and the general public. Contains a subject index.

Environmental directory: products, technologies, services, manufacturers, organizations. Vancouver, Canada: Stewart's Green Line, 1991. 122 p.
> Covers companies that have an environmental product or service or companies making an impressive effort on behalf of the environment. Includes a subject index.

Franck, Irene; and Brownstone, David. *The green encyclopedia.* Englewood Cliffs, N.J.: Prentice Hall, 1992. 512 p.
> A comprehensive guide to environmental issues, ranging from endangered species to environmental disasters. Presents brief biographies of leaders in the environmental movement and descriptions of specific incidents (such as Bhopal) and individual dangers (such as pesticides). There are cross references throughout and lists of environmental organizations and governmental agencies.

Frick, William; and Sullivan, Thomas, F. p., eds. *Environmental regulatory glossary.* 5th ed. Rockville, Md.: Government Institutes, 1990. 449 p. (TD169.3 .E58 1990)
> Provides understandable translations of legal terminology used in pollution regulations to help advanced researchers interpret legal arguments.

Goldsmith, Edward; and Hildyard, Nicholas eds. *The Earth report: the essential guide to global ecological issues.* Los Angeles: Price, Stern, Sloan, 1988. 240 p.: (some color) ill. (TD170 .E17 1988).

>An alphabetical list defining a wide range of terms that present a global perspective of the environmental crisis. Contains useful charts.

Jessup, Deborah Hitchcock, ed. *Guide to state environmental programs.* 2d ed. Washington, D.C.: Bureau of National Affairs, 1990. 700 p. (HC110.E5 J47 1990)

>Presents profiles, addresses, and telephone numbers for agencies involved in environmental regulation in each state.

Jones, Gareth; et al. *Collins reference dictionary of environmental science.* London: Collins, 1989. 473 p.

>Includes most terms associated with the environment. Contains maps and diagrams.

Nierenberg, William A., ed. *Encyclopedia of Earth system science.* 4 vols. San Diego, Calif.: Academic Press, 1991.

>Covers the ocean, the atmosphere, the solid Earth, the biosphere, and near space. Discusses fields such as ecology, agriculture, atmospheric science, geology, seismology, wildlife habitat management, and natural resource exploitation in articles by 257 different specialists. Each article contains its own bibliography and glossary. Includes numerous graphs and charts to illustrate concepts in the text, as well as a cumulative subject index.

OECD environmental data compendium 1991. Washington, D.C.: OECD Publications, 1991.

>Provides data on environmental issues in all OECD countries, including air and water pollution, forests, land, wildlife, and solid waste. Also addresses pressures on the environment caused by energy use, transportation, industry and agriculture.

Rodes, Barbara K.; and Odell, Rice. *A dictionary of environmental quotations.* New York: Simon and Schuster, 1992. 335 p.

>Categorizes 143 environmental topics and presents quotations chronologically from ancient times to the present. Contains useful subject and author indices.

Stevenson, Harold L. *The Facts on File dictionary of environmental science.* New York: Facts on File, 1991. 294 p. (TD9 .S74 1990)

>Includes current terms and acronyms, equations, organizations, legislation, and terms related to the environment. Useful in identifying terms and for beginning basic research.

Wenner, Lettie McSpadden. *U.S. energy and environmental interest groups: institutional profiles.* New York: Greenwood, 1990. 358 p. (HD9502.U52 W45 1990)

A practitioner's guide to influential special-interest congressional lobbyists. Separates organizations into three distinct groups: business corporations or their trade associations, nonprofit public interest groups, and government, professional, and research organizations. The name address, agenda, and tactics of each subject is given, along with a brief description. Suggests further readings.

World environmental directory. 4th ed. Silver Spring, Md.: Business, 1980. 965 p. Provides directory information for government agencies, university and other educational institutions, professional, scientific and public interest organizations dealing with environmental problems and protection in the U.S. and world wide. Includes information on manufacturers of pollution control equipment.

Reference books (unannotated)

Art, Henry, ed. *Dictionary of ecology and environmental science.* New York: Henry Holt, 1993.

Brackley, Peter G. *Energy and environmental terms: a glossary.* Brookfield, Vt.: Gower, 1988. 189 p.: ill. (TJ163.16 .B73 1988)

Dictionary and thesaurus of environment, health and safety. Boca Raton, Fla.: C. K. Smoley, 1992.

Dictionary of environmental protection technology: in four languages, English, German, French, Russian. New York: Elsevier Science, 1988. 527 p. (TD169.3 .D53 1988)

Directory of environmental information sources. 3d ed. Rockville, Md.: Government Institutes, 1990. 299 p. (TD169.5 .D57 1990)

Encyclopedia of environmental information sources. Detroit, Mich.: Gale Research, 1992.

Environmental information sources. Rockville, Md.: Government Institutes, 1986. 218 p. (TD169.5 .E57 1986)

Environmental regulatory glossary. 5th ed. Rockville, Md.: Government Institutes, 1990. 449 p. (TD169.3 .E58 1990)

Franck, Irene M. *The green encyclopedia.* New York: Prentice-Hall, 1992.

Jeryan, Christine, ed. *Environmental encyclopedia.* Detroit, Mich.: Gale, 1993.

King, James J. *The environmental dictionary.* New York: Executive Enterprises, 1989. 663 p. (KF3775.A68 K56 1989)

Lee, C. C. *Environmental engineering dictionary*. Rockville, Md.: Government Institutes, 1989. 629 p. (TD9 .L44 1989)

Paenson, Isaac. *Environment in key words: a multilingual handbook of the environment: English-French-German-Russian*. 2 vols. New York: Pergamon Press, 1990. ill. (QH540.6 .P34 1990)

Porteous, Andrew. *Dictionary of environmental science and technology*. Rev. ed. New York: Wiley, 1992.

Wennrich, Peter. *Anglo-American and German abbreviations in environmental protection*. New York: K. G. Saur, 1980. Distributed by Gale Research. 624 p. (TD170.2 .W46 1980)

Whole Earth ecology: the best of environmental tools and ideas. New York: Harmony Books, 1990. 128 p.: ill. (GF50 .W46 1990)

World directory of environmental organizations: a handbook of national and international organizations and programs--governmental and non-governmental--concerned with protecting the Earth's resources. 3d ed. Claremont, Calif.: California Institute of Public Affairs in cooperation with the Sierra Club and IUCN-The World Conservation Union, 1989. 176 p.

Books for young adults (annotated)

Bach, Julie S.; and Hall, Lynn, eds. *The environmental crisis: opposing viewpoints*. St. Paul, Minn.: Greenhaven Press, 1986. 263 p.: ill. (HC110.E5 .E498 35 1986)
> After a discussion of whether or not an environmental crisis actually exists, this work presents articles from various experts on the crucial issues of toxic wastes, pollution regulation, and acid rain. Grades nine through twelve.

Donnelly, Judy; and Sydelle Kramer. *Space junk: pollution beyond the Earth*. New York: Morrow Junior Books, 1990. 106 p.: ill. (TL1489 .D66 1990)
> Discusses the pollution which is in orbit around the Earth and its potential threats to astronauts.

Middleton, Nick. *Atlas of environmental issues*. New York: Facts on File, 1989. 63 p.: color ill. (QH75 .M486 1989)
> Aimed at younger readers, this international overview offers short articles on a wide range of environmental issues, from soil erosion, to nuclear power, to pollution in Antarctica. Each article contains both global and local maps of affected areas. Contains an explanation of the phenomenon along with explanatory charts.

Middleton, Nick. *Atlas of world issues*. New York: Facts on File, 1989.

Aimed at younger readers, this international overview offers two-page spreads on a wide variety of issues relating to the world, from war and the environment, to unemployment and pollution. Each article contains both global and local maps of affected areas. Contains an explanation of the phenomenon along with explanatory charts. Gives a more general treatment than the author's *Atlas of Environmental Issues*.

Books for young adults (unannotated)

Allison, Rosemary. *Ms Beaver goes west*. Toronto: Women's Press, 1983. 32 p.: ill. (PZ7.A4429 Mq 1983)

Bailey, Donna. *Wasting water*. New York: F. Watts, 1991. 30 p.: color ill. (TD495 .B35 1991)

Barrios, Enrique. *Ami, child of the stars*. Santa Fe, N.M.: Lotus Press, 1989. 113 p.: ill. (PZ7.B2755 Am 1989)

Barron, T. A. *The Ancient One*. New York: Philomel Books, 1992.

Bender, Lionel. *Our planet*. New York: Simon & Schuster Books for Young Readers, 1992. 96 p.: color ill. (TD170.15 .B46 1992)

Berenstain, Stan. *The Berenstain bears don't pollute (anymore)*. New York: Random House, 1991. 1 v.: color ill. (PZ7.B4483 Benb 1991)

Boyle, Doe. *Rick's first adventure*. Norwalk, Conn.: Soundprints, 1992.

Brenford, Dana. *A case of poison*. Mankato, Minn.: Crestwood House, 1988. 62 p.: ill. (PZ7.B7514 Cas 1988)

Burgess, Jeremy. *Endangered Earth*. Vero Beach, Fla.: Rourke, 1988. Distributed by Marshall Cavendish. 45 p.: ill. (some color) (QH75 .B83 1988)

Camp, William G. *Managing our natural resources*. Albany, N.Y.: Delmar Publishers, 1988. 300 p.: ill. (H103.7 .C33 1988)

Cherry, Lynne. *The dragon and the unicorn*. San Diego: Harcourt Brace Jovanovich, 1994.

Cole, Babette. *Supermoo!* New York: G. P. Putnam's Sons, 1992.

Cooper, Clare. *Earthchange*. Minneapolis: Lerner Publications, 1986. 96 p. (PZ7.C78473 Ear 1986)

Dyck, Peter J. *Shalom at last*. Scottdale, Pa.: Herald, 1992. 128 p.: ill. (PZ7.D97 Sh 1992)

Faber, Doris. *Nature and the environment*. New York: Scribner's, 1991. 296 p.: ill. (QH26 .F33 1991)

Geraghty, Paul. *Stop that noise!* New York: Crown Publishers, 1993.

Graham, Ada. *Careers in conservation*. San Francisco: Sierra Club Books, 1980. 166 p.: ill. (QH75 .G678)

Greene, Carol. *Caring for our people*. Hillside, N.J.: Enslow Publishers, 1991. 32 p.: ill. (GF48 .G727 1991)

Hayford, James. *Gridley firing*. Shelburne, Vt.: New England Press, 1987. 152 p.: ill., port. (PZ7.H314889 Gr 1987)

Hebblethwaite, Margaret. *Our two gardens*. Nashville: Oliver-Nelson, 1991. 32 p.: color ill. (TD170.15 .H43 1991)

Hoff, Mary King. *Our endangered planet. Life on land*. Minneapolis, Minn.: Lerner Publications, 1992.

Holland, Barbara. *Caring for planet Earth: the world around us*. Batavia, Ill.: Lion Publishing, 1990. 40 p.: color ill. (QH541.14 .H65 1990)

Kent, Gordon. *All day suckers*. Edina, Minn.: Abdo & Daughters, 1992.

Kerven, Rosalind. *Saving planet Earth*. New York: F. Watts, 1992.

Killingsworth, Monte. *Eli's songs*. New York: Maxwell Macmillan International, 1991. 137 p. (PZ7.K5575 El 1991)

Lambert, David. *Planet Earth 2000*. New York: Facts on File, 1985. 57 p.: ill. (some color) (Q163 .L28 1985)

Lambert, David. *Pollution and conservation*. New York: Bookwright Press, 1986. 32 p.: color ill. (TD176 .L35 1986)

Lambert, Mark. *The future for the environment*. New York: Bookwright Press, 1986. 48 p.: color ill. (TD170.15 .L36 1986)

Landau, Elaine. *Environmental groups: the Earth savers*. Hillside, N.J.: Enslow, 1993.

Lasky, Kathryn. *Home free*. New York: Macmillan, 1985. 245 p. (PZ7.L3274 Ho 1985)

Leinwand, Gerald. *The environment*. New York: Facts on File, 1990. 122 p. (TD180 .L45 1990)

Lipsyte, Robert. *The Chemo Kid*. New York: HarperCollins, 1992. 167 p. (PZ7.L67 Ch 1992)

Luenn, Nancy. *Mother Earth*. New York: Atheneum, 1992. 1 v.: ill. (PZ7.L9766 Mo 1992)

Madan, H. C. *Once in a village--*. New Delhi: National Book Trust, 1988. 24 p.: ill. (PZ7.M2527 On 1988)

Madden, Don. *The Wartville wizard*. New York: Macmillan International, 1993.

Marino, Tony. *Intergalatic grudge match*. Edina, Minn.: Abdo & Daughters, 1992.

Marino, Tony. *Ratchet Hood*. Edina, Minn.: Abdo & Daughters, 1992.

Markham, Adam. *The environment*. Vero Beach, Fla.: Rourke Enterprises, 1988.

Milne, Lorus Johnson. *Dreams of a perfect Earth*. New York: Atheneum, 1982. 120 p.: ill. (QH75 .M5)

Nimmo, Jenny. *Ultramarine*. New York: Dutton Children's Books, 1992. 199 p. (PZ7.N5897 Ul 1992)

O'Neill, Mary Le Duc. *Nature in danger*. Mahwah, N.J.: Troll Associates, 1991.

Pearson, Susan. *The green magician puzzle*. New York: Simon & Schuster, 1991.

Pedersen, Anne. *The kids' environment book: what's awry and why*. Santa Fe, N.M.: John Muir, 1991. Distributed by W. W. Norton. 181 p.: ill. (some color) (TD176 .P43)

Pellowski, Michael. *My father the enemy*. New York, N.Y.: Hollywood Paperbacks, 1992.

Penny, Malcolm. *Pollution and conservation*. Englewood Cliffs, N.J.: Silver Burdett Press, 1989.

Plemons, Marti. *Megan & the owl tree*. Cincinnati, OH: Standard Publishing, 1992. 128 p.: ill. (PZ7.P718 Me 1992)

Rand McNally children's atlas of the environment. Chicago: Rand McNally, 1991. 79 p.: color ill., color maps (G1046.G3 R23 1991)

Ray, Mary Lyn. *Pumpkins*. San Diego: Harcourt Brace Jovanovich, 1992.

Reader, Dennis. *Anthony Anthony's boring day*. New York: Doubleday, 1992.

Santrey, Laurence. *Conservation and pollution*. Mahwah, N.J.: Troll Associates, 1985. 30 p.: color ill. (TD176 .S26 1985)

Sargent, Sarah. *Seeds of change*. New York: Bradbury Press, 1989. 103 p. (PZ7.S2479 Sg 1989)

Schaffer, Ulrich. *Zilya's secret plan to save her threatened valley*. Batavia, Ill.: Lion Publishing, 1991. 32 p.: ill. (PZ7.S3336 Zi 1991)

Sharpe, Susan. *Waterman's boy*. New York: Bradbury Press, 1990. 170 p. (PZ7.S5323 Wat 1990)

Smith, Phillipa. *The cracker crumb rescue*. Chester, Va.: Harbour Duck Specialties, 1992. ill. (some color), color maps (PZ7.S65757 Cr 1992)

St. George, Judith. *Do you see what I see?* New York: Putnam, 1982. 157 p. (PZ7.S142 DO 1982)

Steck-Vaughn Company. *Atlas of the environment*. Austin, Tex.: Raintree Steck-Vaughn, 1993.

Stidworthy, John. *Environmentalist*. New York: Gloucester, 1992. 32 p. (GF48 .S75 1992)

Swift, Carolyn. *Bugsy goes to Cork*. Dublin, Ireland: Poolbeg Press, 1990. 171 p. (PZ7.S976 Bt 1990)

Thompson, Julian F. *Gypsyworld*. New York: H. Holt, 1992.

Turner, Ted. *Captain Planet and the Planeteers*. Atlanta, Ga.: Turner Publishing, 1992

Waddington-Feather, John. *Quill's adventures in Grozzieland*. 2d ed. Santa Fe, N.M.: J. Muir, 1991. Distributed by W. W. Norton. 114 p.: ill. (PZ7.W11375 Qug 1991)

Waddington-Feather, John. *Quill's adventures in the great beyond*. 3d ed. Santa Fe, N.M.: John Muir, 1991. Distributed by W. W. Norton. 85 p.: ill. (PZ7.W11375 Qut 1991)

Waddington-Feather, John. *Quill's adventures in Wasteland*. 2d ed. Santa Fe, N.M.: John Muir, 1991. Distributed by W. W. Norton, 1991. 117 p.: ill. (PZ7.W11375 Quw 1991)

Weitzman, David L. *The mountain man and the president*. Austin, Tex.: Raintree Steck-Vaughn, 1992.

Wilkes, Angela. *My first green book.* New York: Knopf, 1991. 48 p.: color ill. (TD176 .W55 1991)

Videos

Blowpipes and bulldozers. Produced by Bullfrog Films, (Oley, PA 19547). 1989. 60 min.
Focusing on the destruction of the Malaysian Borneo, this work takes a close up look at the Penan, the indigenous, nomadic people of the Borneo. Narrated both by Penan elders and by Swiss artist Bruno Manser, this video aptly portrays the quiet, gentle people whose lives are being destroyed by the corporate greed of logging companies and the wealthy nations who support them.

Can polar bears tread water? Produced by Better World Society. Cinema Guild (1697 Broadway, Suite 802, New York 10019), 1990. 60 min.
Focuses on the manmade dangers threatening the Earth and discusses the ecological changes that could destroy the planet, including the greenhouse effect and climatic change. Offers suggestions for reversing these trends. Also examines the sociological and political complexities.

Earth to kids. Produced by Consumer Reports TV, Consumer Union, and HBO.
Seeks to raise awareness among youngsters. Offers numerous environmentally sound ideas and suggestions. A young host displays products which should be avoided--such as frozen microwave dinners that contain excessive packaging. The host is compared to a bumbling pair of characters who mistakenly chose the wrong products. Makes a case against toothpaste pumps instead of tubes.

Green quiz. Directed by Ray Burley. Filmakers Library (24 E. 40th Street, New York 10016), 1990. 46 min.
Produced by the Canadian Broadcasting Corporation, this film opens with the narrator asking multiple-choice questions about environmental problems. Canadian students respond to the questions, which leads to an explanation of the issues. This platform is enlivened with footage shot worldwide and interviews with celebrities and experts.

Help save the planet. MCA/Universal Home Video (70 Universal City Plaza, Universal City, CA 91608), 1990. 70 min.
Presents ecological tips, such as recipes for environmentally sound household cleaners and suggestions on how to select products with less packaging. Includes phone numbers and addresses of pertinent agencies.

In partnership with the Earth. Produced by Blackwell Productions. Beacon Films (930 Pitner Avenue, Evanston, IL 60202), 1991. 27 min.
Hosted by John Denver and "sponsored by several U.S. corporations." Aims to gently alter public attitudes and behavior regarding the environment. Presents a

wide variety of footage, including commentary from industrialists, commentators, and environmentalists. Concludes that the country must learn to generate less waste and sustain existing resources. Acknowledges the importance of government regulation and industrial advance. Maintains that solving the world's environmental problems relies on the actions and buying decisions of the individual.

Only one Earth: the sinking ark. Films, Inc., (5547 N. Ravenswood Avenue, Chicago, IL 60640), 1987. 58 min.

Maintains that the short-term economic benefits of environmental exploitation are far outweighed by the long-term benefits of environmental preservation. Focuses on positive attempts to preserve the environment.

Paul Ehrlich's Earth watch. Human Relations Media (175 Tompkins Avenue, Pleasantville, NY 10570), 1990. 17 min.

An introduction to the world's ecological problems narrated by a noted environmentalist. Discusses the population explosion in Third World countries, endangered species, and global warming. Blames the U.S. for overconsumption and South America for burning the rain forests.

Pollution: world at risk. National Geographic Educational Services (Dept. 90, Washington, D.C. 20036). 1989. 25 min.

A good, general introduction to pollution, this video examines topics ranging from nuclear waste to toxic landfills and atmospheric pollution.

When the bough breaks. Directed by Lawrence Moore and Robbie Stamp for ITV. Bullfrog Films (P.O. Box 149, Oley, PA 19547), 1990. 52 min.

Focuses on environmental problems affecting the world's children. Presents footage from Poland's toxic areas and Africa's drought ridden Sudan. Topics include poison, debt, relocation, and population. Ages sixteen to adult.

Who's killing Calvert City. Produced by Michael Mierendorf for Frontline. PBS Video (1320 Braddock Place, Alexandria, VA 22314). 1989. 60 min.

Focusing on Calvert City, Kentucky, this video explains the conflict between industry and the environment. Includes interviews with local and national environmental activists and industry spokesmen B. F. Goodrich and GAF, among others. Points out that economically depressed areas will often accept pollution in order to maintain employment. Contains footage of air, water, and soil pollution. Explains the comprehensive joint federal and state pollution audit that is now underway.

Software

Balancing the planet. Chris Crawford Games, Milpitas, Calif.

A computer simulation game in which players attempt to save the Earth by taxing industry and changing policy, such as opting for nuclear power over fossil fuels. Talleys the results to give players some idea about the effectiveness of their decisions.

Enviro/Energyline Abstracts Plus. (software) Bowker A & I Publishing (121 Chanlon Rd. New Providence, N.J. 07974)

Using CD-ROM technology, this software offers access to hundreds of specialized collections. Combines sixteen traditional search entry points, such as author, title, subject, date, and document type, among others. Contains over 200,000 records from 1970 to the present, from *Environment Abstracts*, *Acid Rain Abstracts*, and *Energy Information Abstracts*. Updated with over three thousand new records every three months.

World Resources 1992-1993. World Resources Institute/Oxford University Press.

Contains economic, environmental, natural resource, and population statistics found in the book. Data can be exported to spread sheets, word processing, and desktop publishing programs.

CHAPTER 3

GLOBAL ENVIRONMENTAL POLICY

(Council on Environmental Quality Assessment) 134

continued

GLOBAL ENVIRONMENTAL POLICY

(Council on Environmental Quality Assessment) 134

SHORT LIST OF SOURCES
FOR GENERAL RESEARCH

Books (annotated)

Adamson, David. *Defending the world: the politics and diplomacy of the environment*. New York: St. Martin's, 1990. 240 p. (HC79.E5 A32 1990)
> Focuses on the difficulties of establishing international environmental cooperation. Examines how negotiations have functioned in the past and analyzes current agreements, noting the inherent connection between economic aid and the environment. Discusses issues such as the increased use of coal in India and China, pollution in Eastern Europe, climatic change, and world agriculture.

Cutter, S. L.; Renwick, H. L.; and Renwick, W. H. *Exploitation, conservation, and preservation*. 2d ed. New York: J. Wiley, 1991. 455 p.: ill, maps (HC21 .C96 1991)
> Discusses the social, cultural, economic, and political aspects of the causes and effects of environmental problems. Offers descriptive chapters on rangelands, forests, water resources and quality, air pollution, and energy resources. Each chapter presents related case studies.

Feshbach, Murray; and Friendly, Alfred Jr. *Ecocide in the U.S.S.R: health and nature under siege*. New York: Basic Books/Aurum Press, 1992.
> Discusses the adversarial view the former Soviet Union had of nature. The environmental consequences of this attitude are discussed in the context of the historical, political, and social complexities of Soviet life. Analyzes the link between the environment and public health, employing statistics like a decline in life expectancy and a rise in infant mortality. Maintains that the Chernobyl nuclear accident may lead to some policy changes and greater awareness.

Parkin, Sara, ed. *Green light on Europe*. London: Heretic Books, 1991. 367 p.
> The collapse of the Communist bloc and the official beginning of the European Community provide the backdrop for this economic analysis of how EC goals will lead to environmental disaster, especially in the areas of population increases and consumption.

Pryde, Philip R. *Environmental management in the Soviet Union.* New York: Cambridge
University Press, 1991. 314 p.: ill., maps (HC340.E5 P79 1991)
> Analyzes the environmental problems in the Soviet Union, which have reached
> crisis proportions and threaten the population. Portrays a bureaucracy that has no
> competence to deal with problems such as the shrinking of the Aral Sea, the
> phosphorate moonscapes of Estonia, the Chernoybl accident, and carbon monox-
> ide pollution. The disintegration of the Soviet Union has intensified the problem,
> although in some areas, such as water conservation, the individual states might
> be better equipped to take action. An excellent overview and analysis of the
> subject.

General interest periodicals (annotated)

Brown, Lester R. "Global ecology at the brink." *Challenge* 32 (March-April 1989): 14-22.
> In this interview, Worldwatch Institute President Lester Brown explains why he
> thinks food security will be the main preoccupation of governments in the 1990s.
> He also predicts that climate change, population policy, and energy policy will
> dominate global concerns.

"Environment: striving for global balance." *World & I* 4 (June 1989): 20-53.
> "Beneath the surface agreement that we face a severe environmental problem,
> controversies and conflicts simmer. Developed countries seek to curb pollution,
> while developing countries emphasize growth." Contents: "The coming
> North-South conflict," by Milton R. Copulos; "How to clean up the mess," by
> J. I. Bregman; "Needed: a global response," by James Gustave Speth; "The
> 'Private' approach works" by Kent Jeffreys; and "The siren song of
> environmentalism," by Hugh W. Ellsaesser.

Environmental and professional journals (annotated)

Bolan, Richard S. "Organizing for sustainable growth in Poland." *Journal of the
American Planning Association* 58 (Summer 1992): 301-311.
> "Following the end of Communist rule in 1989, Poland has been facing the
> difficult challenge of responding to the legacy of a devastated environment amid
> an overwhelming economic crisis. This article describes how Poland has been
> responding to this challenge, with particular focus on the country's policy of
> sustainable development within the framework of transition to a democratic
> market economy. The article discusses achievements and difficulties in seeking
> both effective environmental protection and economic restructuring since 1989 and
> examines the country's future needs and problems."

Finger, Matthias. "The military, the nation state and the environment." *Ecologist* 21
(September-October 1991): 220-225.

"The world's armed forces--and the industry upon which they depend--are a major cause of environmental degradation across the globe. Yet, the environmental regulations and agreements now being formulated by nation states (or groups of nation states) rarely apply to the military. On the contrary, with the ecological crisis now confronting us increasingly being defined as a 'threat to national security', the military is seen by many as part of the solution to the crisis rather than one of its major causes."

French, Hilary F. "Eastern Europe's clean break with the past." *Worldwatch* 4 (March-April 1991): 21-27.

"Under the assault of air pollution and acid deposition, Eastern Europe's medieval cities are blackened and crumbling, whole hillsides are deforested, and crop yields are falling. Rivers serve as open sewers, and clean drinking water is in short supply. Life expectancies in the dirtiest parts of the region are as much as five years shorter, and rates of cancer, reproductive problems, and other ailments far higher than in relatively clean areas."

Kabala, Stanley J. "EC helps Czechoslovakia pay 'debt to the environment.' " *Radio Free Europe/Radio Liberty Research Report* 1 (May 15, 1992): 54-58.

"In 1991 the EC's Polish and Hungarian Assistance for the Reconstruction of Europe program made 25.5 million ecu ($34 million) available for environmental protection projects in Czechoslovakia, which range from assessing the country's hazardous waste problem and its air quality control to upgrading safety provisions at nuclear power plants and evaluating the environmental impact of the controversial Gabcikovo-Nagymaros dam project."

Karliner, Joshua. "Central America's other war." *World Policy Journal* 6 (Fall 1989): 787-810.

Discusses the environmental crisis in Central America (Nicaragua, El Salvador, and Costa Rica). "The region's governments feel pressured to continue accelerated exploitation and export of their natural resources to satisfy local oligarchs and cope with huge external debts. U.S. militarization has worsened the crisis."

Luiki, Paul S.; and Stephenson, Dale E. "Environmental laws are stricter in 'green'-influenced Europe." *National Law Journal* 14 (September 30, 1991): 45-47 (6 p.)

"Europe has emerged as a leading force among industrial democracies in promulgating stringent environmental requirements. Growing environmental awareness among the European population has caused a 'green' philosophy to emerge across the political spectrum in many countries, acting as a catalyst for change as parties compete to have the 'greenest' platforms. Environmental law in Europe is dynamic and complex, consisting not only of regulation applicable in individual countries but also of those imposed by a supranational institution."

"Managing the global commons." *Evaluation Review* 15 (February 1991)

Special issue on the global environment. Partial contents include: "The role of international law: formulating international legal instruments and creating international institutions," by Paul C. Szasz; "International agreements and cooperation in environmental conservation and resource management," by Peter S. Thacher; "Bargaining among nations: culture, history, and perceptions in regime formation," by Ronnie D. Lipschutz; "A cultural perspective on the structure and implementation of global environmental agreements," by Steve Rayner; "Cross-national differences in policy implementation," by Sheila Jasanoff; "Global thinking, local acting: movements to save the planet," by Luther P. Gerlach; and "Developmental and geographical equity in global environmental change: a framework for analysis," by Roger E. Kasperson and Kirstin M. Dow.

Robertson, David. "The global environment: are international treaties a distraction?" *World Economy* 13 (March 1990): 111-127.
> Calling for negotiation of international treaties is an easy way for governments to divert attention from their own failure to deal with domestic pollution and environmental damage.

Rowlands, Ian. "Security challenges of global environmental change." *Washington Quarterly* 14 (Winter 1991): 99-114.
> The response of the international community to the challenge of global environmental change has thus far been mixed: considerable progress on the ozone layer case but little of tangible value coming out of the global warming negotiations.

Speth, James Gustave. "Toward a North-South compact for the environment." *Environment* 32 (June 1990): 16-20, 40-43.
> Discusses the importance of nurturing cooperation and agreements between industrialized and developing countries to meet the challenges of the environment and economic progress.

Tarnoff, Curt. "Eastern Europe and the environment." *Congressional Research Service Review* 11 (March-April 1990): 29-31.
> Describes the relationship between the democratization process in Eastern Europe and the political concern over the deteriorating environment. Emphasizes the opportunities for western nations to help Eastern European countries and themselves through expanded trade relations and improved environmental security.

INTRODUCTORY SOURCES

Books (annotated)

Adamson, David. *Defending the world: the politics and diplomacy of the environment.* New

York: St. Martin's, 1990. 240 p. (HC79.E5 A32 1990)

Focuses on the difficulties of establishing international environmental cooperation. Examines how negotiations have functioned in the past and analyzes current agreements, noting the inherent connection between economic aid and the environment. Discusses issues such as the increased use of coal in India and China, pollution in Eastern Europe, climatic change, and world agriculture.

Cutter, S. L.; Renwick, H. L.; and Renwick, W. H. *Exploitation, conservation, and preservation.* 2d ed. New York: J. Wiley, 1991. 455 p.: ill, maps (HC21 .C96 1991)

Discusses the social, cultural, economic, and political aspects of the causes and effects of environmental problems. Offers descriptive chapters on rangelands, forests, water resources and quality, air pollution, and energy resources. Each chapter presents related case studies.

Dorney, Robert S.; and Dorney, Lindsay C., eds. *The professional practice of environmental management.* New York: Springer-Verlag, 1989. 228 p.: ill. (TD159 .D67 1989)

Maintains that a wide range of issues must be considered before decisions on land management should be made, and this includes studying the full range of environmental alternatives. Moreover, such decisions are best left to the professional environmental manager. The philosophical, technical, and conceptual aspects of environmental decision making are presented, as well as suggestions for organizing an actual practice.

Johnston, R. J. *Environmental problems: nature, economy, and state.* London: Belhaven, 1989. 211 p.

Addresses the debate stemming from current environmental problems, including the various political, economic, and environmental issues. Discusses the physical systems, economic systems, and modes of production. Maintains that current political structure is the only setting for change.

Science responds to environmental threats. Paris: Organisation for Economic Co-operation and Development, 1992. Distributed by OECD Publications and Information Centre, Washington, D.C. 455 p.

"Environmental threats have been increasingly taken into account in scientific and technological policy formulation. However, Research-Development (R&D) efforts in that area have been somewhat uneven among countries and there have been markedly different institutional developments. One central issue concerns the place that environmental research should have. Should it be isolated in the general organization of R&D or should countries develop a more integrated approach? A proper use of international co-operation has also become essential, given the global dimension of environmental threats."

Tisdell, Clement A. *Natural resources, growth, and development: economics, ecology, and resource-scarcity.* New York: Praeger, 1990. 186 p.: ill. (HD75.6 .T565 1990)

A good introduction to the relationship between natural resources, development, and ecology that covers a broad range of topics, including resource preservation, sustainable development, foreign aid to small Pacific economies, biological pest control, and rural-urban migration.

Trudgill, Stephen. *Barriers to a better environment: what stops us solving environmental problems?* London: Belhaven, 1990 151 p.: ill. (TD170 .T78 1990)
Maintains that the reasons why modern society cannot solve environmental problems can be divided into six types of barriers: economic, political, social, agreement, technology, and knowledge. Suggests that environmental solutions require a holistic approach. Contains numerous charts.

Westing, Arthur H., ed. *Environmental hazards of war: releasing dangerous forces in an industrialized world.* Newberry Park, Calif.: SAGE, 1990. 96 p. (TD195 .A75E6 1990)
Focusing on so-called "collateral damage," the military's term for inadvertent destruction of nonmilitary targets, this work raises concern about the potential environmental consequences of such damage occurring to dams, and chemical or nuclear plants, all of which are often located near civilian populations. Examines ways in which international law can mitigate this collateral damage. Contains bibliographies, appendices of dams and nuclear/chemcial plants, and an index.

Books (unannotated)

Baker, Ann. *Counting on a small planet: activities for environmental mathematics.* Portsmouth, N.H.: Heinemann, 1991.

Baldwin, John H. *Environmental planning and management.* Boulder, Colo.: Westview Press, 1985. 336 p.: ill., map (HC110.E5 B36 1985)

Barbour, Ian G. *Technology, environment, and human values.* New York: Praeger, 1980. 331 p. (HC110.E5 B37)

Baumol, William J. *The theory of environmental policy.* 2d ed. New York: Cambridge University Press, 1988. 299 p.: ill. (HC79.E5 B375 1988)

Brookins, Douglas G. *Earth resources, energy, and the environment.* Columbus, Ohio: Merrill, 1981. 190 p.: ill. (TN23 .B77)

Brown, Richard D. *National environmental policies and research programs: a comparison among selected countries.* Lancaster, Pa.: Technomic, 1983. 160 p. (HC79.E5 B76 1983)

Buchholz, Rogene A. *Managing environmental issues: a casebook.* Englewood Cliffs, N.J.: Prentice Hall, 1992. 286 p.: ill., 1 map (HC79.E5 B82 1992)

Buchholz, Rogene A. *Principles of environmental management.* Englewood Cliffs, N.J.: Prentice Hall, 1993.

Caldwell, Lynton Keith. *Between two worlds: science, the environmental movement, and policy choice.* New York: Cambridge University Press, 1990. 224 p. (HC79.E5 C33 1990)

Caldwell, Lynton Keith. *International environmental policy: emergence and dimensions.* Durham, N.C.: Duke University Press, 1984. 367 p. (K3585.4 .C34 1984)

Choucri, N., ed. *Global accord: environmental challenges and international responses.* Cambridge, Mass.: MIT Press, 1993.

Cleveland, Harlan. *The global commons: policy for the planet.* Lanham, Md. (Aspen Institute, Queenstown, Md.): University Press of America, 1990. 118 p. (HC79.E5 C59 1990)

Commons without tragedy. Savage, Md.: Barnes and Noble, 1991.

Cotgrove, Stephen F. *Catastrophe or cornucopia: the environment, politics, and the future.* New York: Wiley, 1982. 154 p. (HC79.E5 C668 1982)

Crossroads: environmental priorities for the future. Washington, D.C.: Island Press, 1988. 339 p. (QH75 .C74 1988)

Culture and conservation: the human dimension in environmental planning. New York: Croom Helm, 1985. 308 p. (HC79.E5 C85 1985)

Distributional conflicts in environmental-resource policy. New York: St. Martin's, 1986. 455 p.: ill. (HC79.E5 D57 1986)

Dryzek, John S. *Rational ecology: environment and political economy.* New York: B. Blackwell, 1987. 270 p.: ill. (HC79.E5 D79 1987)

Eagles, Paul F. J. *The planning and management of environmentally sensitive areas.* New York: Longman, 1984. 160 p.: ill., maps (QH75 .E23 1984)

Eckholm, Erik P. *Down to Earth: environment and human needs.* New York: Norton, 1982. 238 p. (GF41 .E24 1982)

Ecosystem health: new goals for environmental management. Washington, D.C.: Island Press, 1992.

Environment and the global arena: actors, values, policies, and futures. Durham, N.C.: Duke University Press, 1985. 188 p.: ill. (HC79.E5 E743 1985)

Environmental policies: an international review. Dover, N.H.: Croom Helm, 1986. 315 p.: ill. (HC79.E5 E57846 1986)

Environmental problems and solutions: greenhouse effect, acid rain, pollution. New York: Hemisphere Publishing, 1990. 525 p.: ill. (QH545.A1 E576 1990)

Environmental protection: standards, compliance, and costs. Chichester, England (Water Research Centre): E. Horwood, 1984. Distributed by Halsted Press, New York. 329 p.: ill. (TD257 .E58 1984)

Environmental protection: the international dimension. Totowa, N.J.: Allanheld, Osmun, 1983. 340 p. (HC79.E5 E5794 1983)

Environmental technology, assessment, and policy. Chichester, England: E. Horwood, 1989. Distributed by Halsted Press, New York. 275 p.: ill. (TD5 .E64 1989)

Faludi, Andreas. *A decision-centred view of environmental planning.* New York: Pergamon, 1987. 240 p. (HC79.E5 F28 1987)

Flynn, Eileen P. *Cradled in human hands: a textbook on environmental responsibility.* Kansas City, Mo.: Sheed and Ward, 1991. 155 p. (GF80 .F57 1991)

Frankel, Boris. *The post-industrial utopians.* Madison, Wis.: University of Wisconsin Press, 1987. 303 p. (HX73 .F73 1987)

Guariso, G. *Environmental decision support systems.* Chichester, England: E. Horwood, 1989. Distributed by Halsted Press, New York. 240 p.: ill. (TD170.2 .G82 1989)

Hahn, Robert William. *A primer on environmental policy design.* New York: Harwood Academic Publishers, 1989. 135 p.: ill. (HC79.E5 H316 1989)

Head, Ivan L. *On a hinge of history: the mutual vulnerability of South and North.* Toronto: University of Toronto Press in association with the International Development Research Centre, 1991. 244 p. (HC59.7 .H395 1991)

Hetzel, Nancy K. *Environmental cooperation among industrialized countries: the role of regional organizations.* Washington, D.C.: University Press of America, 1980. 381 p. (HC79.E5 H48)

Incentives for environmental protection. Cambridge, Mass.: MIT Press, 1983. 355 p. (HC79.E5 I513 1983)

Institutions and geographical patterns. London: Croom Helm, 1982. 331 p.

International environmental negotiation. Newbury Park, Calif.: Sage Publications, 1993.

International politics of the environment, actors, interests, and institutions. New York: Oxford University Press, 1992.

International public policy sourcebook. New York: Greenwood Press, 1989. 2 v. (H97 .I58 1989)

International responsibility for environmental harm. Boston: Graham and Trotman, 1991. Distributed by Kluwer Academic Publishers Group, Norwell, Mass. 499 p.: ill. (K955 .I57 1991)

Kozlowski, Jerzy M. *Planning with the environment: introduction to the threshold approach.* New York: University of Queensland Press, 1986. 82 p.: ill., maps (HT166 .K688 1986)

Kozlowski, Jerzy M. *Threshold approach in urban, regional, and environmental planning: theory and practice.* New York: University of Queensland Press, 1986. 262 p.: ill. (some color) (HT166 .K689 1986b)

Maintaining a satisfactory environment: an agenda for international environmental policy. Boulder, Colo.: Westview Press, 1990. 81 p.: ill. (HC79.E5 M3344 1990)

Managing Leviathan: environmental politics and the administrative state. Lewiston, N.Y.: Broadview Press, 1990. 310 p. (HC120.E5 M354 1990)

Perspectives on environmental conflict and international relations. New York: Pinter Publishers, 1992. 162 p.: ill., maps (HD75.6 .P48 1992)

Pieces of the global puzzle: international approaches to environmental concerns. Golden, Colo.: Fulcrum, 1986. 204 p.: ill. (TD174 .P53 1986)

Planning and ecology. New York: Chapman and Hall, 1984. 464 p.: ill. (TD194.6 .P63 1984)

Porter, Gareth. *Global environmental politics.* Boulder, Colo.: Westview Press, 1991. 208 p.: ill. (HC79.E5 P669 1991)

Reese, Craig E. *Deregulation and environmental quality: the use of tax policy to control pollution in North America and Western Europe.* Westport, Conn.: Quorum Books, 1983. 495 p. (HC79.E5 R433 1983)

Rural environmental planning for sustainable communities. Washington, D.C.: Island Press, 1991. 254 p.: ill. (HT392 .R873 1991)

Ryding, Sven-Olof. *Environmental management handbook.* Boca Raton, Fla.: CRC Press, 1992.

Sadler, B., ed. *Environmental protection and resource development: convergence for today.* Calgary, Alberta, Canada: University of Calgary Press, 1985. 202 p.

Smil, Vaclav. *Energy, food, environment: realities, myths, options.* New York: Oxford University Press, 1987. 361 p.: ill. (TJ163.2 .S62 1987)

Sustainable development and Finland. Helsinki: Finnish Government Printing Centre, 1991. 98 p. (HC340.2.Z65 S84 1991)

Taylor, Ann. *Choosing our future: a practical politics of the environment.* New York: Routledge, 1992. 235 p. (HD75.6 .T39 1992)

Vogel, David. *National styles of regulation: environmental policy in Great Britain and the United States.* Ithaca, N.Y.: Cornell University Press, 1986. 325 p. (HC260.E5 V64 1986)

Watt, Kenneth E. F. *Understanding the environment.* Boston, Mass.: Allyn and Bacon, 1982. 431 p.: ill. (HC79.E5 W38 1982)

Weale, Albert. *The new politics of pollution.* Manchester, England: Manchester University Press, 1992. Distributed by St. Martin's Press, New York. 227 p. (HC240.9.P55 W43 1992)

Young, Oran R. *International cooperation: building regimes for natural resources and the environment.* Ithaca, N.Y.: Cornell University Press, 1989. 248 p. (HC59 .Y685 1989)

General interest periodicals (annotated)

Averous, Christian. "An uneven track record on the environment." *OECD Observer* 168 (February-March 1991): 8-13.
> Discusses the OECD's third report on the state of the environment. The trends observed and analyzed "over the last twenty years clearly demonstrate that environmental policies have to be strengthened, that structural adjustment in member-country economies has to be exploited more fully, and that effective international action is required. They further show the importance of defining a new generation of strategies designed to protect the environment and secure sustainable development in the OECD countries and throughout the world."

Brown, Lester R. "Global ecology at the brink." *Challenge* 32 (March-April 1989): 14-22.
> In this interview, Worldwatch Institute President Lester Brown explains why he thinks food security will be the main preoccupation of governments in the 1990s. He also predicts that climate change, population policy, and energy policy will dominate global concerns.

"Cover story: the environment." *World Press Review* 39 (June 1992): 9-14.
> Contains "Saving the Earth: economic development vs. a clean world," by Horst Bieber; "Delegates' view," by Anna Muggiati; "The West sets a bad example: keep your pollution! says one expert," by Maneka Gandhi; "Stop dumping on the

South: the lure of loose laws," by Pravin Kumar; "Africa should care," by Bertrand Schneider; "Poverty dooms the planet: now is the time to act," by Paulo Nogueira Neto; and "Sharing the burden," by Gro Harlem Brundtland.

"Environment: striving for global balance." *World & I* 4 (June 1989): 20-53.
"Beneath the surface agreement that we face a severe environmental problem, controversies and conflicts simmer. Developed countries seek to curb pollution, while developing countries emphasize growth." Contents: "The coming North-South conflict," by Milton R. Copulos; "How to clean up the mess," by J. I. Bregman; "Needed: a global response," by James G. Speth; "The 'Private' approach works" by Kent Jeffreys; and "The siren song of environmentalism," by Hugh W. Ellsaesser.

"Protecting the environment." *Politics Today* 17 (October 17, 1986): 314-332
Presents British Conservative perspectives on economic development and the environment, wildlife and countryside, pollution control, urban environmental problems, and historic buildings.

General interest periodicals (unannotated)

Baldwin, J. "World game." *Whole Earth Review* 68 (Fall 1990): 30(2)

Sadik, Nafis. "Poverty, population, pollution." *UNESCO Courier* (January 1992)· 18(4)

Tetzeli, Rick. "Ecology battle: North vs. South." (conference of International Chamber of Commerce) *Fortune* 125,5 (March 9 1992): 12(2)

"Too much, too fast." (over-population, endangered species, overuse of land) (Cover Story) *Newsweek* 119,22 (June 1, 1992): 34(2)

Wallis, Victor. "Socialism, ecology, and democracy: toward a strategy of conversion." *Monthly Review* 44,2 (June 1992): 1(22)

"Whose world is it, anyway?" (the effect of environmental issues on international relations; survey of the global environment) *Economist* 323,7761 (May 30 1992): S5(3)

Environmental and professional journals (annotated)

Brown, Lester R.; Flavin, Christopher; and Postel, Sandra. "No time to waste: a global environmental agenda for the Bush administration." *World Watch* 2 (January-February 1989): 10-19.
Suggests the Bush administration pursue a climate-sensitive energy policy, adopt forest conservation and reforestation policies, reverse the global decline in food security, and slow population growth.

Bush, George; and Gorbachev, Mikhail S. "Two world leaders on global environmental policy." *Environment* 32 (April 1990): 12-15, 32-35.

> Contains remarks by George Bush to the Intergovernmental Panel on Climate Change (IPCC) on February 5, 1990, in Washington, D.C., and remarks by Mikhail Gorbachev to the Global Forum on Environment and Development for Survival on January 19, in Moscow.

Chernushenko, David. "Foreign policy and the environment--managing conflict." *Cambridge Review of International Affairs* 4 (Autumn 1990): 28-35.

> "The environmental problems facing the world are well known: drought, desertification, deforestation, flooding and 'global warming.' This article suggests that productive linkages can and should be made between environmental issues and foreign policy-making processes."

Dohlman, Ebba. The trade effects of environmental regulation. *OECD observer*, February-March 1990: 28-32.

> "Public and political attitudes to the environment have changed dramatically, often giving priority to the environment over economic growth. The 'greens' have emerged as a powerful political force. And consumers are increasingly aware of the impact of their choices on the world they live in. These changes present governments with a significant new source of pressure to alter their policies in ways which might have implications for trade and trade policy."

"Environmental challenge." *GAO Journal* 6 (Summer 1989): 15-37.

> These three articles discuss the progress of the United States and the international community in addressing global environmental issues. Contents: "Government and the environment: an interview with Lee Thomas" (participants also included Charles A. Bowsher, J. Dexter Peach, Richard L. Hembra, and Peter F. Guerrero); "Turning point for the Earth: the deterioration of the atmosphere reflects the globalization of our environmental woes," by James Gustave Speth; and "Diplomacy and the ozone crisis: at Montreal, a new mode of international cooperation emerged that may be crucial to the health of the planet," by Richard Elliot Benedick.

"Environmental crisis." *Harvard International Review* 12 (Summer 1990): 4-6, 8-14, 16-27, 56, 61.

> Contents: "Heeding nature's tug: the global environmental agenda," by Mostafa K. Tolba; "Coping with the uncertainties of the Greenhouse effect," by Jessica T. Mathews; "Democracy and the environment in Eastern Europe and the Soviet Union," by Barbara Jancar; "The new context of multilateral environmental negotiations," by Kenneth W. Piddington; "Challenging the international order: the 'Greening' of European politics," by Konrad von Moltke; "Seizing the international environmental initiative," by Albert Gore, Jr.

"Environmental diplomacy." *Gist*, January 1990: 1-2.

"Among the important environmental issues requiring attention are ozone depletion; the potential for global warming and climate change; air pollution and acid rain; international transport and disposal of hazardous wastes; tropical deforestation and sustaining biological diversity; and protection of endangered species. The U.S. is among the countries in the forefront of global environmental protection. President Bush has given high priority to environmental issues, placing the U.S. in a position of international leadership to develop a balanced approach. Within the U.S. Government, the Department of State's Bureau of Oceans and International Environmental and Scientific Affairs is the focal point for dealing with all international environmental issues."

Holdgate, Martin W. "Parliaments and international endeavours to protect the environment." *Inter-parliamentary* 1 (January-March 1989): 19-27.
Analyzes how parliaments should respond to the Earth's environmental crisis, suggesting that their response lies in shaping and managing the sociosphere--"the system of socio-political, economic, and cultural institutions by which human communities are organized."

Kelley, Donald R. "East-West environmental cooperation." *Environment* 22 (November 1980): 29-37.
Surveys the United States-U.S.S.R. bilateral environmental agreements and multilateral cooperation on environmental issues between the capitalist and socialist blocs. Contends that East-West environmental cooperation has produced positive although limited results and that this status will persist until environmental concerns develop stronger political constituencies in both the East and the West.

Livingston, M. L. "Transboundary environmental degradation: market failure, power, and instrumental justice." *Journal of Economic Issues* 23 (March 1989): 79-91.
"Explores the problem of transboundary environmental degradation from an institutionalist perspective. The objective is to (1) demonstrate the inadequacies in a market failure approach to international environmental degradation (2) provide a description of negotiation as it relates to existing power structures, and (3) clarify how the concept of instrumental justice can contribute to understanding the problem and identify prospects for its mitigation."

Malone, Thomas F.; and Corell, Robert. "Mission to planet Earth revisited." *Environment* 31 (April 1989): 6-11, 31-35.
Reports on work of the International Geosphere-Biosphere Programme sponsored by the International Council of Scientific Unions and discusses the institutional, scientific, political, and financial challenges that lie ahead for global research activities.

"Managing the global commons." *Evaluation Review* 15 (February 1991)
Special issue on the global environment. Partial contents include: "The role of

international law: formulating international legal instruments and creating international institutions," by Paul C. Szasz; "International agreements and cooperation in environmental conservation and resource management," by Peter S. Thacher; "Bargaining among nations: culture, history, and perceptions in regime formation," by Ronnie D. Lipschutz; "A cultural perspective on the structure and implementation of global environmental agreements," by Steve Rayner; "Cross-national differences in policy implementation," by Sheila Jasanoff; "Global thinking, local acting: movements to save the planet," by Luther P. Gerlach; and "Developmental and geographical equity in global environmental change: a framework for analysis," by Roger E. Kasperson and Kirstin M. Dow.

Odum, Eugene P. "Input management of production systems." *Science* 243 (January 1989): 177-182.
"Nonpoint sources of pollution, which are largely responsible for stressing regional and global life-supporting atmosphere, soil, and water, can only be reduced (and ultimately controlled) by input management that involves increasing the efficiency of production systems and reducing the inputs of environmentally damaging materials."

Perry, John S. "Managing the world environment." *Environment* 28 (January-February 1986): 10-15, 37-40.
Discusses the roles and effectiveness of governmental and nongovernmental international environmental institutions. Comments that an effective institution must offer adequate information on a worthwhile problem, must be likely to influence policy, and must be widely deemed legitimate.

Robertson, David. "The global environment: are international treaties a distraction?" *World Economy* 13 (March 1990): 111-127.
Calling for negotiation of international treaties is an easy way for governments to divert attention from their own failure to deal with domestic pollution and environmental damage.

Sand, Peter H. "Innovations in international environmental governance." *Environment* 32 (November 1990): 16-20, 40-44.
Discusses what can be expected from global institutions for environmental management, and the need for "innovative approaches in three areas: transnational environmental standard setting, transnational environmental licensing, and transnational environmental auditing."

"Special issue: global environmental change and international relations." *Millennium: Journal of International Studies* 19 (Winter 1990): whole issue (337-476 p.)
Content include discussions on global environmental change, international environmental protection, and environmental degradation and national security.

Speth, James Gustave. "Toward a North-South compact for the environment."

Environment 32 (June 1990): 16-20, 40-43.
> Discusses the importance of nurturing cooperation and agreements between industrialized and developing countries to meet the challenges of the environment and economic progress.

Sterner, Thomas. "An international tax on pollution and natural resource depletion." *Energy Policy* 18 (April 1990): 300-302.
> "To come to grips with global problems we need global institutions and policy instruments. This article proposes an international tax based on pollution and natural resource depletion. Such a tax would not only finance urgent environmental work but would also be a valuable instrument of environmental policy."

Stoel, Thomas B., Jr. "Global tomorrow." *Amicus Journal* 10 (Fall 1988): 21-27.
> Suggests actions the Bush administration could have taken to solve ozone depletion; global warming and, in the Third World, the interrelated problems of population growth, desertification, deforestation, and loss of wildlife species.

Teclaff, Ludwik A.; and Teclaff, Eileen. "International control of cross-media pollution--an ecosystem approach." *Natural Resources Journal* 27 (Winter 1987): 21-53.
> Cross-media pollution, or the transfer of pollutants from one environmental medium to another, is a growing international problem that long escaped attention because treaties were devised to protect only one element of the environment.

Thacher, Peter. "New challenges for international environmental institutions." *Journal of the World Resources Institute* 1985: 35-44.
> "The record of two decades of discussion, policy changes, and action reveals both expensive mistakes and considerable progress. This mixed record suggests a new agenda for global organizations, one built partly on new initiatives and partly on continued commitment to the programs and approaches that have worked."

Wood, William B.; Mofson, Phyllis; and Demko, George J. "Ecopolitics in the global greenhouse." *Environment* 31 (September 1989): 12-17, 32-34.
> Discusses ecopolitics at each government level--from local to global--and the key actors involved in environmental politics.

Environmental and professional journals (unannotated)

McCloskey, J. Michael; and Pope, Carl. "Together in time?" (future of environmental protection) *Sierra* 77,3 (May-June 1992): 96(6)

Law journals

Jones, Owen Donald. "Box H problem: a justification for unilateral international coercion." *Yale Journal of International Law* 15 (Summer 1990): 207-275.

> Discusses environmental problems that combine irreversibility, universality, and a time-lag between action and effect. Since such problems will first be recognized and understood, if at all, only in countries with superior scientific resources, measures to achieve action through cooperation alone will prove inherently cumbersome and inadequate. In the end, such a circumstance may require unilateral action.

Sand, Peter H. "Lessons learned in global environmental governance." *Boston College Environmental Affairs Law Review* 18,2 (1991): 213-277.

> Examines past efforts at international environmental management or governance and identifies progressive ideas which may be beneficial to future policy making.

Reports

Clarke, Robin. *Stockholm plus ten: promises, promises?: the decade since the 1972 U.N. Environment Conference.* London: International Institute for Environment and Development, 1982. Distributed by Earthscan, London. 75 p. (HC79.E5 C57 1982)

Compact for a new world: an open letter to the heads of state and government and legislators of the Americas from the members of the New World Dialogue on Environment and Development in the Western Hemisphere. Washington, D.C.: World Resources Institute, 1991. 26 p.

> "In 1990, with colleagues from Latin America and the Caribbean, Canada, and the United States, the World Resources Institute organized a New World Dialogue on Environment and Development in the Western Hemisphere." As a first step, the Dialogue has drawn up a North-South compact calling for specific international commitments by the hemisphere's governments aimed at securing economic development that is environmentally sustainable and socially equitable. The compact proposes eight sustainable development initiatives for the America's on the following issues: forestry; energy; pollution prevention; anti-poverty; population; science and technology; trade and investment; and financial (funding). The authors urge formalization of the initiatives at the 1992 Earth Summit.

Cooper, Richard N. *U.S. policy toward the global environment.* Cambridge, Mass.: Harvard Institute of Economic Research, Harvard University, 1991. 31 p. (Discussion paper no. 1552)

> Suggests why the U.S.'s relatively cautious attitude toward strong actions in the early 1990s regarding global climate change and toward setting emission targets for carbon dioxide in the near future is entirely warranted. However, "there are important steps which can and should be taken, but to gather more information germane to intelligent decisions on the complicated questions at issue, and to attenuate the rate at which greenhouse gases are emitted into the atmosphere, thus providing additional time at relatively low cost."

Everett, Sidney J. *New approaches to environmental policies, laws, and regulations.* Menlo Park, Calif. (333 Ravenswood Ave., Menlo Park 94025): SRI International, Business Intelligence Program, 1986. 18 p. (KF3775.Z9 E94 1986)

French, Hilary F. *After the Earth Summit: the future of environmental governance.* Washington, D.C.: Worldwatch Institute, 1992. 62 p. (HC79.E5 F72 1992)

Global change in the geosphere-biosphere: initial priorities for an IGBP. Washington, D.C.: National Academy Press, 1986. 91 p.
  Describes the organization and goals of the International Geosphere-Biosphere Program (IGBP), a proposed international program uniting Earth and biological scientists in an effort to understand "the workings of Earth and the living organisms on it as a coupled system."

Human sources of global change: a report on priority research initiatives for 1990-1995. Cambridge, Mass.: Energy and Environmental Policy Center, John F. Kennedy School of Government, Harvard University, 1990. 87 p. (Discussion paper G-90-08)
  "Global Environmental Policy Project." Over the year since the release of its 1988 report (the "green book"), the U.S. Committee on Global Change (CGC) has sought to formulate a research plan for achieving a better understanding of the human sources of global change by the end of the decade. This report outlines the recommended program, focusing "on two principal human sources of global change: industrial metabolism and land transformation. For each source, the program recommends research on integrative models, process studies, and data base development. In addition, a small number of synthetic studies are proposed."

Kopp, Raymond J.; Portney, Paul R.; and DeWitt, Diane E. *International comparisons of environmental regulation.* Washington, D.C.: Resources for the Future, 1990. 50 p. (Discussion paper QE90-22)
  Examines "whether environmental regulations put the United States at a competitive trade advantage or disadvantage with respect to other industrialized countries. The analysis focuses exclusively on policies the U.S. and OECD countries have adopted to control air and water pollution and hazardous wastes."

Myers, Norman. *Not far afield: U.S. interests and the global environment.* Washington, D.C.: World Resources Institute, 1987. 73 p. : ill. (HC59.72.E5 M93 1987)
  Considers the relationship of critical issues in world environment, natural resources, and population to U.S. economic, ecological, and national security interests.

Perspective on transboundary environmental issues. British Columbia: Round Table, 1990. 25 p. (HC117.B8 P47 1990)

Petesch, Patti L. *North-South environmental strategies, costs, and bargains.* Washington, D.C.: Overseas Development Council, 1992. 112 p.: ill. (HC59.72.E5 P48 1992)

Public and leadership attitudes to the environment in four continents: a report of a survey in 16 countries: conducted for the United Nations Environmental Programme: fieldwork, February 1988 to June 1989. New York (630 5th Ave., New York 10111): Louis Harris and Associates, 1989. 232 p. (HC79.E5 P79 1989)

Rompczyk, Elmar. *International environmental policy as a challenge to the politics of development.* Bonn, Germany: Research Institute, Friedrich-Ebert-Stiftung, 1987. 39 p. (HC79.E5 R632 1987)

Seen from the South. Provo, Utah: David M. Kennedy Center for International Studies, Brigham Young University, 1989. 226 p. (HF1480.15.U5 S44 1989)

Uniting nations on the Earth: an environmental agenda for the world community: final report of the United Nations Association of the USA and the Sierra Club. New York: United Nations Association of the United States of America, 1991. 13 p.
> Summarizes recommendations of Uniting Nations for the Earth and suggests topics for the 1992 United Nations Conference on Environment and Development.

Federal government reports

Basket II--implementation of the final act of the Conference on Security and Cooperation in Europe: findings eleven years after Helsinki: report submitted to the Congress of the United States. Washington, D.C.: G.P.O., 1987. 95 p. (HF1411 .B29 1987)

International agreements to protect the environment and wildlife. Washington, D.C.: U.S. International Trade Commission, 1991. 217 p. (USITC publication 2351)
> Categorizes 170 multilateral and bilateral agreements of significance to U.S. interests into 8 groups: marine fishing and whaling; land animals (includes birds) and plant species; marine pollution; pollution of air, land, and inland waters; boundary waters between the U.S. and Mexico and Canada; archaeological, cultural, historical, or natural heritage; maritime and coastal matters; nuclear pollution. "Summary information on all these agreements (when available) includes objectives and obligations, dates signed, literature citations, enforcement and dispute-settlement provisions, information-exchange provisions, current issues, and a listing of parties."

International environment: international agreements are not well monitored: report to congressional requesters. January 27, 1992. Washington, D.C.: General Accounting Office, 1992. 60 p.
> Reviews compliance of countries on the following international agreements: "the Montreal Protocol (ozone depletion), the Nitrogen Oxides (NOx) Protocol (acid rain, air pollution), the Basel Convention (hazardous waste disposal), the London Dumping Convention (marine pollution), the International Convention for the Prevention of Pollution from Ships (MARPOL), the Convention on International

Trade in Endangered Species (CITES), the International Whaling Convention, and the International Tropical Timber Agreement (deforestation)."

Selected major international environmental issues: a briefing book: report. Washington, D.C.: G.P.O., 1991. 52 p. (HC79.E5 S39 1991)

Conference proceedings

2nd International Conference on Environment Protection: proceedings, S. Angelo, Ischia, Italy, 5-7 October 1988. Naples: CUEN, 1988. 2 v.: ill. (TD5 .I49 1988)

Environmental information and communication systems: ECOINFORMA 1: reviewed proceedings of the First International Conference and Exhibition on Environmental Information, Communication and Technology Transfer, Bayreuth, Germany, 16-19 May, 1989. Philadelphia, Pa.: Gordon and Breach Science Publishers, 1991. 292 p.: ill., maps (TD193 .I565 1989)

Interparliamentary Conference on the Global Environment: final proceedings, April 29-May 2, 1990: report to the Senate of the United States of America. Washington, D.C.: The Conference, 1990. 1 v.: ill. (some color), maps (some color) (HC79.E5 I625 1990)

UNITED NATIONS PROGRAMS

Books (annotated)

Taylor, Paul; and Groom, A. J. R., eds. *Global issues in the United Nations' framework.* New York: Macmillan, 1989.
> Examines the ability of the United Nations to handle global problems. Presents the U.N.'s strengths and weaknesses, maintaining that it does not have the capacity to respond on the appropriate scale. Points to the nation-state system as a fundamental barrier.

Books (unannotated)

Croner, Stan. *An introduction to the World conservation strategy.* Gland, Switzerland: IUCN; Nairobi: United Nations Environment Programme, 1984. 28 p.: ill. (S944.5.I57 C76 1984)

Environment Programme: medium-term plan, 1982-1983: report of the Executive Director. Nairobi: United Nations Environment Programme, 1981. 214 p. (HC79.E5 U46 1981)

Global environmental issues: United Nations Environment Programme. Dublin, Ireland (United Nations Environment Programme): Tycooly International, 1982. 236 p.: (some color) ill. (TD170.2 .G56 1982)

Gosovic, Branislav. *The quest for world environmental cooperation: the case of the U.N. Global Environment Monitoring System.* New York: Routledge, 1992. 284 p. (TD193.G66)

Readings in environmental management. Bangkok: United Nations Asian and Pacific Development Institute, 1980. 416 p.: ill. (HD75.6 .R42 1980)

Environmental and professional journals (annotated)

Dyer, M. I.; and Holland, M. M. "UNESCO's Man and the Biosphere Program." *BioScience* 38 (October 1988): 635-641.
> Discusses the structure, budget, and future of the Man and the Biosphere Program. Among observations "MAB, rather than focusing on what is being done in the basic ecological sciences, has had a tendency to ignore new and emerging scientific directions and respond instead to the political process."

Golley, Frank B. "Ten years of MAB: establishing the balance sheet." *Nature and Resources* 17 (April-June 1981): 3-7.
> On the tenth anniversary of the founding of the Man and the Biosphere Programme the author describes two examples of successful projects: the San Carlos humid tropics research operation in Venezuela and the Tai Forest Project in the Ivory Coast.

Gwynne, Michael D. "The Global Environment Monitoring System (GEMS) of UNEP." *Environmental Conservation* 9 (Spring 1982): 35-41.
> Describes the five major monitoring programs supported by the United Nations Environment Programme: climate related monitoring, monitoring of long-range transport of pollutants, health-related monitoring, ocean monitoring, and terrestrial renewable-resources monitoring.

Strong, Maurice. "The United Nations Environment Programme and the future of the global environment." *World Outlook* 9 (Summer 1989): 96-108.
> In this interview, Strong, president of the World Federation of United Nations Association, the Better World Society, and World Industry Forum, discusses U.N. environmental programs.

Law journals

"Enhancing UNEP's role: US proposal." *Environmental Policy and Law* 21 (May 1991): 47-48

Discusses the United States' proposals to the United Nations, as a part of a multi-national effort, concerning "the structure and responsiveness of the United Nations to deal with major environmental issues."

Gray, Mark Allan. "The United Nations Environment Programme: an assessment." *Environmental Law* 20 (1990): 291-319.

Assesses "the United Nations Environment Programme (UNEP) in terms of its history, structure, and achievements. Explores the institutional and political reasons for UNEP's failure to meet all expectations and reviews such successes as the Regional Seas Programme and identifies fundamental philosophical flaws in the 1972 Stockholm Declaration . . . and suggests how and why UNEP must be improved."

Tinker, Catherine. "Environmental planet management by the United Nations: an idea whose time has not yet come?" *New York University Journal of International Law and Politics* 22 (Summer 1990): 793-830.

"Argues that while the goal of expanding the United Nations' role in environmental management is important, the mere creation of new bureaucratic structures would be ineffective in addressing the difficult environmental problems that plague the planet. The primary task of the United Nations must be information-gathering and monitoring dedicated to planet management and the development of international environmental law. Accomplishing these tasks is a matter of strengthening an existing program, the United Nations Environment Programme, and using the preparatory process for the Brazil Conference to prepare drafts on legal obligations, duties, and rights."

Reports

ACCIS guide to United Nations information sources on the environment. New York: United Nations, 1988. 141 p. (TD169.5 .U55 1988)

Framework for the development of environment statistics. New York: United Nations, 1984. 28 p. (HC79.E5 F695 1984)

Index to the decisions and resolutions of the Governing Council of the United Nations Environment Programme: the first through the thirteenth session and the session of a special character, 1982. Nairobi: United Nations Environment Programme, 1986. 146 p. (Z5863.P7 I5 1986)

Integrated physical, socio-economic, and environmental planning. Dublin, Ireland (United Nations Environment Programme): Tycooly, 1982. 199 p.: ill. (HC79.E5 I5175 1982)

Lean, Geoffrey. *Environment, a dialogue among nations*. Nairobi: United Nations Environment Programme, 1985. 28 p.: ill. (GF75 .L43 1985)

Lean, Mary. *United Nations Environment Programme profile.* Nairobi: Information and Public Affairs Branch, United Nations Environment Programme, 1990. 48 p.: color ill. maps. (TD170.2 .L43)

Managing the environment: an analytical examination of problems and procedures. Nairobi: United Nations Environment Programme, 1983. 99 p.: ill. (HC79.E5 M349 1983)

Overview, environmental management. Nairobi: United Nations Environment Programme, 1981. 89 p. (HC79.E5 O9 1981)

Principal governmental bodies dealing with the environment. Nairobi: United Nations Environment Programme, 1983. (HC79.E5 P693 1983

Federal government reports

United Nations: U.S. participation in the environment program: report to congressional requesters. June 21, 1989. Washington, D.C.: General Accounting Office, 1989. 32 p. (GAO/NSIAD-89-142, B-232768)
   Reviews the U.N. Environment Program, examining its accounting practices, financial reserves, and trust funds; the efficiency and effectiveness of its headquarters and field programs; and the level of U.S. influence in the Program.

Bibliographies

Environmental bibliography: publications issued by UNEP or under its auspices, 1973-1980. Nairobi: United Nations Environment Programme, 1981. 67 p. (Z5322.E2 E583)

1992 EARTH SUMMIT

Books (annotated)

Lerner, Steve. *Earth Summit: conversations with architects of an ecologically sustainable future.* Bolinas, Calif.: Common Knowledge Press, 1991. 263 p.
   "Records conversations that Steve Lerner, director of the Commonweal Research Institute, held with nineteen leading environmentalists and social activists from around the world." Focuses on the June 1992 Earth Summit.

Books (unannotated)

Gardner, Richard N. *Negotiating survival: four priorities after Rio.* New York: Council on Foreign Relations Press, 1992.

General interest periodicals (annotated)

Babbit, Bruce. "Earth summit." *World Monitor* 5 (January 1992): 26-31.
> Discusses the possibility that " most of the world's leaders are likely to attend the Earth Summit but not because they have much idea of what to say or do. They simply do not want to risk negative publicity, especially in countries where it is an election year."

"Growth vs. environment: in Rio next month, a push for sustainable development." *Business Week* 3265 (May 11, 1992): 66-75.
> Highlights the problems to be discussed at the upcoming United Nations Conference on the Environment and Development and proposes some solutions. Includes two sidebars: "The road to Rio: plenty of good intentions, but. . . ," by Emily T. Smith, John Carey, and Peter Hong; and "The next trick for business: taking a cue from nature," by Emily T. Smith, David Woodruff, and Fleur Templeton.

Nichols, Mark. "The world prepares for the 'Earth Summit.' " *World Press Review* 39 (March 1992): 22-24.
> Discusses topics and expectations of UNCED. Includes related article, "A controversy over funding," by Bharat Bhushan, which discusses the one-year-old, World-Bank-sponsored Global Environment Facility (GEF)

Scheer, Hermann. "Earth Summit in Rio: will it do more harm than good?" *Nation* 254 (April 20, 1992): 522-524.
> Maintains that UNCED's goals are not inclusive enough to effectively limit future environmental damage.

Weidenbaum, Murray. "Leviathan in Rio." *National Review* 44 (April 27, 1992): 44-45, 56.
> Examines the possible outcomes of UNCED, maintaining that after the "Earth Summit is all over, the U.N. agencies will have achieved a substantial accretion of power and will start planning the next round of such endeavors."

General interest periodicals (unannotated)

Batker, David. "The rift between North and South over development and finances. *America* 166,18 (May 23, 1992): 458(5).

Brown, Lester. "Launching the environmental revolution." *UNESCO Courier* (April 1992): 44(3)

Budiansky, Stephen. "Giving green a bad name." *U.S. News & World Report* 112,24 (June 22 1992): 16(2)

"First Earth Summit." (includes related article; Environment and Development: A Global Commitment) *UNESCO Courier* (November 1991): 39(2)

"Green legacy." *Economist* 323,7763 (June 13, 1992): 39(2).

Holmes, Nigel; Dorfman, Andrea; and Wells, Deborah. "The world's next trouble spots." *Time* 139,22 (June 1, 1992): 64(2).

McGurn, William. "Blame it on Rio." *National Review* 44,11 (June 8, 1992): 23(2)

Environmental and professional journals (annotated)

"1992 Earth Summit in Brazil." *One country* 3 (April-June 1991): 1, 10- 14.
   Three separate articles from the newsletter of the Bahai International Community discuss preparations for the United Nations Conference on Environment and Development to be held in Brazil in June 1992. Discusses the Earth Summit stimulation of NGO activities worldwide; presents excerpts from the Bahai statement on the proposed Earth Charter; and examines how Bahai communities are actively preparing for UNCED worldwide.

Bilger, Burkhard. "Earth Summit." *Earthwatch* 11 (November 1991): 12-13.
   Considers whether the United Nations Conference on Environment and Development, or the Earth Summit, will provide solutions to environmental problems. The Earth Summit "is being billed as the last, best chance for the world's governments to chart a course toward sustainable development."

Brady, Gordon L. "Greenthreats?: the 1992 U.N.Conference on Environment and Development." *International Freedom Review* 4 (Spring 1991): 37-58.
   Predicts that 1992 U.N.Conference on Environment and Development will be driven largely by climate change issues and that action on global warming will be constrained by the less developed countries' demand for the transfer of financial resources and clean technology in order to obtain environmental improvements.

Brown, Janet Welsh. "Road map to Rio: an agenda for ECO '92." *Hemisphere* 4 (Winter-Spring 1992): 2-4.
   Evaluates the Compact for a New World, written by a group of 30 citizens from various countries, which "proposes a set of linked initiatives that are in the mutual interests of both the nations of this hemisphere and the globe." Discusses the Compact and the group responsible for writing it.

Collett, Stephen. "PrepCom 3: preparing for UNCED." *Environment* 34 (January-February 1992): 3-5, 45.
   PrepCom 3, the preparatory committee for the United Nations Conference on

Environment and Development, "came to an agreement on the form and general content of the two central products of the conference: an 'Earth Charter' of principles on general rights and obligations for sustainable development and 'Agenda 21', which lists programs of action integrating all the issues mandated to the conference--both environmental and developmental--as well as financing and other resource transfer mechanisms."

Dobb, Edwin. "Summit on the future." *Audubon* 94 (May-June 1992): 84-86, 88, 90, 92-97.

Discusses the opening of the United Nations Conference on Environment and Development (UNCED) and contains four separate articles written by some major UNCED voices: Ambassador Robert J. Ryan, Jr., the United States coordinator for UNCED; Senator Al Gore, an outspoken environmentalist; Herman E. Daly, a forward-looking economist at the World Bank; and Anil Agarwal, an Indian dissenter who decries what he calls green imperialism.

Feeney, Andy. "Backpeddling to Brazil." *Environmental Action* 23 (September-October 1991): 16-18.

"As the 1992 U.N. Earth Summit in Brazil approaches, President Bush appears to be ignoring key opportunities for global environmental leadership."

French, Hilary F. "From discord to accord." *World Watch* 5 (May-June 1992): 26, 28-32.

"Without an overhaul of the U.N. system to strengthen policing and penalties for treaty breakers, the Earth Summit may not be as far-reaching as hoped."

Graham, Frank, Jr. "Audubon prepares for U.N. conference." *Audubon* 93 (July-August 1991): 124, 126-127.

Discusses Fran Spivy-Weber's (director of National Audubon Society's International Program) appointment to represent U.S. private environmental organizations on the Preparational Committee for the United Nations Conference on Environment and Development, and her current activities relating to UNCED.

Hinchberger, Bill. "The rocky road to Rio." *Development Forum* 19 (November 1991-February 1992): 13(3)

Discusses preparations in Brazil for the United Nations' Earth Summit.

Lohmann, Larry. "Whose common future?" *Ecologist* 20 (May-June 1990): 82-84.

Discusses the political management of the environmental crisis, especially as it pertains to the 1992 United Nations Conference on Environment and Development agenda. Asserts that NGOs (nongovernmental organizations) will play an important, if not critical, role in addressing environmental problems.

MacDonald, Gordon J. "Brazil 1992: who needs this meeting?" *Issues in Science and Technology* 7 (Summer 1991): 41-44.

Concludes that there is no support at the highest levels of the U.S. government in preparing UNCED and that "United States participation should be broader, comprising high-level representatives of agencies with economic, environmental, and developmental responsibilities."

Mayur, Rashmi. "Earth Summit: broadening the NGO role." *Development Forum* 19 (July-August 1991): 19(2)

"Preparations for the United Nations Conference on Environment and Development (UNCED) are under way everywhere . . . perhaps most significantly, non-governmental organizations (NGOs) are participating actively in the official process, whose goal is nothing less than reaching a consensus among governments on long-term policies that will protect the global environment and promote sustainable development."

McCoy, Michael. "High expectations raise pressure on Earth Summit." *Development Forum* 19 (November 1991-February 1992): 14(4)

"Organizers are worried that the governments of the South and the North are still far apart on crucial issues" for the United Nations Conference on Environment and Development. However, "the negotiating impasses can be best broken if the North starts talking economics and the South responds more on ecology." Also includes a sidebar, "NGOs prepare for final PrepCom."

McCoy, Michael. "NGOs pull together logistics for Rio." *Development Forum* 19 (July-August 1991): 20 (2 p.)

Highlights information and an organizational map of the (non-governmental organizations) involved in UNCED.

McCoy, Michael. "PrepCom 3: setting the priorities." *Development Forum* 19 (July-August 1991): 10(3)

"The third PrepCom, meeting in Geneva from 12 August to 4 September 1991, promises to be the most difficult phase in the preparations of UNCED. In addition to agreeing on a set of priorities in approaching the issues on the agenda of the Earth Summit, participants must achieve a rough consensus on the amount of financial resources that will be necessary to carry out the ambitious goals. There is now concern that what can be realistically accomplished by the conference in 1992 has been oversold."

McCoy, Michael. "Now comes the hard part." *Development Forum* 19 (May-June 1991): 9(3)

"After three weeks of deliberations in Geneva, ending in early April 1991, the second session of the Preparatory Committee (PrepCom) of government delegates moved forward in its planning for the U.N. Conference on Environment and Development. Several of the procedural steps necessary to define the framework in which the PrepCom can begin the negotiations on the programme and agenda for UNCED were finally agreed upon by assembled government delegates."

McCoy, Michael. "PrepCom 4: New York dress rehearsal for Rio." *Development Forum* 20 (March-April 1992): 9(3)

"Final negotiations for the United Nations Conference on Environment and Development (UNCED) are taking place from 2 March to 3 April 1992 at United Nations headquarters. For negotiators, this five-week window of opportunity will determine the success of the Earth Summit in Rio in June. As talks begin, however, prospects for progress in the fourth and final meeting of the Preparatory Committee here are dim."

McCoy, Michael. "Whither the new world?" *Buzzworm: The Environmental Journal* 4 (January-February 1992): 26-31.

"Spanning the economic gulf between northern industrial countries and the developing nations of the south is the focus of preparations for the environmental Earth Summit to be held in Rio de Janeiro, Brazil, in June." Includes a sidebar, "Earth Summit."

Metzger, Jennifer. "Progress, of sorts, in U.S. preparation for UNCED." *Interdependent* 17 (April-May 1991): 5(2)

In preparation for UNCED, "the President's Council on Environmental Quality (CEQ) is charged with preparing the national report while the State Department is charged with overseeing the preparation of position papers to take to the Preparatory Committee meetings and, ultimately, to the 1992 conference. Each has asked countless departments, agencies, and interagency groups to contribute to the process."

Metzger, Jennifer. "Rio '92: new players, old problems." *Interdependent* 17,1 (1991): 5(2)

Highlights of two major resolutions passed by the 45th General Assembly relating to the United Nation's Conference on Environment and Development (UNCED). Lists the six major themes approved as the focus of conference deliberations and explains underlying North-South tensions and problems between developed and developing countries.

Metzger, Jennifer. "The road to Brazil 1992." *Interdependent* 16,2 (1990): 5(2)

Reports "on the work of the United Nations system in preparation for this historic conference (UNCED)."

Metzger, Jennifer. "The U.S.: Brazil or bust." *Interdependent* 16,3 (1990): 5(2)

"Thirty-four environmental ministers of the industrialized West met in Bergen, Norway, between May 8 and 16 to translate into action the General Assembly-endorsed recommendations of the United Nation's Commission on Environment and Development. But attempts to go beyond rhetoric were blocked by the United States . . . and since Bergen, environmental and other groups have been highly critical of what they perceive as continued U.S. foot-dragging on the environmental front."

Metzger, Jennifer. "UNCED: the people speak out." *Interdependent* 17 (August-September 1991): 5(3)

> Discusses public interest and involvement in the UNCED policy debate, particularly referring to public citizen testimonies before United States government and United Nations officials in St. Louis. "Whether they were farmers, corporate executives, school children, lawyers, biologists, or community leaders, those who testified had a common message to deliver: The U.S. government--and its citizens--must do more to protect the environment and to promote sustainable development, both at home and abroad."

Metzger, Jennifer. "Words speak louder than action at UNCED meeting." *Interdependent* 17 (June-July 1991): 5(2)

> Delegates at the second PrepCom for the 1992 United Nations Conference on Environment and Development "agreed that at the next session they will have to spend less time discussing environmental problems and the obstacles to sustainable development, and more time negotiating concrete proposals for action."

Nixon, Will. "Earth Summit--can talking heads save an ailing planet?" *E: The Environmental Magazine* 3 (May-June 1992): 36-42, 63.

> Evaluates the Earth Summit's possibilities for success, and reports on the goals of UNCED, including separate sections on climate, biodiversity, forests, and population. Includes a sidebar, "A word from the organizer," by Anne W. Semmes.

Parker, Jonathan; and Hope, Chris. "The state of the environment: a survey of reports from around the world." *Environment* 34 (January-February 1992): 18-20, 39-44.

> A compilation of current " 'state-of the-environment' " reports published by many countries attending the United Nations Conference on Environment and Development in Rio de Janeiro.

Pearce, Fred. "North-South rift bars path to Summit." *New Scientist* 132 (November 23, 1991): 20-21.

> Discusses the problem of the conflict between poor and rich nations at U.N. Earth Summit (United Nations Conference on Environment and Development).

Rauber, Paul. "The rocky road to Rio." *Sierra* 77 (March-April 1992): 30, 32-33.

> "The 'Earth Summit' in Brazil pits haves against have-nots: we want them to clean up, they want us to pay up."

"Report from the Preparatory Committee for the United Nations Conference on Environment and Development." *Environment* 33 (January-February 1991): 16-20, 39-41, 43-44.

> "This article contains the substantive issues on the agendas of the working groups for the (UNCED) conference and the decisions made on those and other issues by the conference preparatory committee."

Smith, John Thomas, II. "The 'Earth Summit'--global bargains better left 'UNCED'." *International Economic Insights* 3 (March-April 1992): 34-36.
> Asserts that the United Nations Conference on Law of the Sea, 1974-1981, experience may instruct the Earth Summit participants. "In particular, it teaches that UNCED will end in disappointment if developing countries insist that the negotiation achieve transfer of wealth, technology, and political authority from North and South."

"Special Earth summit section." *Amicus Journal* 14 (Winter 1992): 15-29.
> Contents "The rocky road to Rio," by Don Hinrichsen; "North meets South," by Steve Lerner; "Voices and visions"; "Beyond the limits," by Donella Meadows.

Strong, Maurice F. "Earth Summit: on the road to Rio 1992." *Development Forum* 19 (March-April 1991): 1(4)
> Discusses the issues to be addressed at UNCED, including: "climate change, transboundary pollution, waste management, the protection of land resources, the conservation of biological diversity, the management of the oceans and coastal areas, and the quality and supply of freshwater resources."

Strong, Maurice F. "Preparing for the U.N. Conference on Environment and Development." *Environment* 33 (June 1991): 5, 39-41.
> The secretary-general of the United Nations Conference on Environment and Development evaluates present progress and future challenges.

Suter, Keith D. "The 1992 U.N. Conference on Environment and Development." *Medicine and War* 7 (April-June 1991): 146-159.
> Examines the precedent of UNESCO, the 1972 U.N. Conference on the Human Environment, and "explains why conferences are so important for devising new strategies for international cooperation."

Tinker, Catherine. "UNCED: sustaining the momentum." *Interdependent* 17 (November-December 1991): 5(2)
> Discusses issues raised at the PrepCom 3 meeting of the committee preparing for the June 1992 U.N. Conference on Environment and Development, especially the United States' emphasis that "no binding principles of international law should be adopted at UNCED and no mandatory contributions demanded of the industrialized nations."

Environmental and professional journals (unannotated)

Gray, Gerald. "Outlook for the Earth Summit." (includes related article; World Forests) *American Forests* 98,5-6 (May-June 1992): 49(3).

Holloway, Marguerite. "Still negotiating." (United Nations Conference on Environment and Development) *Scientific American* 266,6 (June 1992): 17(2).

Lerner, Steve; Pearce, Fred; and Dobb, Edwin. "Summit on the future: will 140 nations agree on anything?" *Audubon* 94,3 (May-June 1992): 84(11).

Sampson, Neil. "Challenge at Rio: facing hard facts." *American Forests* 98,5-6 (May-June 1992): 6(2).

Law journals

"Preparatory committee for the U.N.Conference on Environment and Development." *Environmental Policy and Law* 20 (September-October 1990): 127-133.
> Summarizes the first substantive session of the Prepcom for the 1992 United Nations Conference on Environment and Development.

"UNCED: decisions of the Preparatory Committee." *Environmental Policy and Law* 20 (September-October 1990): 161-165.
> Reports on the decisions made by the Preparatory Committee for the United Nations Conference on Environment and Development during its first substantive session in Nairobi in August 1990.

Reports

Canada's national report: United Nations Conference on Environment and Development, Brazil, June 1992. Ottawa: Minister of Supply and Services Canada, 1991. Distributed by Enquiries Centre, Environment Canada. 149 p.: ill. (HC113.5 .C277 1991)

Environment and international trade: report of the Secretary-General of UNCTAD submitted to the Secretary-General of the Conference pursuant to General Assembly resolution 45/210. New York: United Nations, 1991. 24 p. (United Nations. Document A/CONF.151/-PC/48)
> An analytical study (requested by General Assembly resolution 45/210) examines the relation between environmental issues and international trade; it was submitted to the Preparatory Committee for the United Nations Conference on Environment and Development (third session).

Gacek, Christopher M.; and Malone, James L. *Guidelines for the U.N. environmental conference.* Washington, D.C.: Heritage Foundation, 1992. 13 p.
> Conservative think-tank agenda for the March 1992 global environmental summit in Rio de Janeiro. Advises the U.S. government to limit discussions of global warming; resist a detailed plan for reducing specific quantities of greenhouse gases by a set date; avoid issues dealt with by other international bodies; promote

an understanding of biotechnology that realistically assesses its risks and benefits; protect private intellectual property rights; and oppose proposals to spend more money on environmental problems in developing nations.

In our hands--Earth Summit '92--a reference booklet about the United Nations Conference on Environment and Development. Geneva: United Nations Conference on Environment and Development Secretariat, 1991. 36 p.

Contents: "An overview of the Earth Summit"; "Questions and answers about the Earth Summit"; "Senior officials at UNCED"; "Stockholm to Rio: a journey down a generation," by Maurice F. Strong. Inside front cover lists contacts for information about the Earth Summit and the inside back cover highlights media material about the conference.

Mazur, Laurie Ann. *UNCED briefing packet.* New York: Environmental Grantmakers Association, 1991. 77 p.

Contains information on the UNCED, including foundation strategies, background, calendar, taking action on UNCED, and issue briefs.

Notes for speakers: environment and development. New York: Department of Public Information, United Nations, 1991. 79 p.

Written for the Earth Summit, which was held in Rio de Janeiro in 1992. Maintains that "poverty is dangerous to your planet's health and so is greed." Discusses the relationship between environment, economics, and ecology, as well as treating other topics such as human resource development, science and technology, international environmental law, and the role of private organizations.

Report of the Aspen Institute working group on international environment and development policy. Aspen, Colo.: Aspen Institute for Humanistic Studies, 1991. 34 p.

In July, 1991, "the Aspen Institute convened a group of international leaders to address critical issues related to the environment and economic development in advance of the 1992 U.N. Conference on Environment and Development that will take place in Rio de Janeiro in June 1992. The focus of the meetings was the institutional, technological, financial and legal issues that UNCED will address." Contains the results of the meetings, addressed to all governments and non-governmental organizations involved in the preparation for UNCED.

Sweden national report to UNCED 1992: United Nations Conference on Environment and Development. Stockholm, Sweden: Ministry of the Environment, 1991. 95 p.: color ill., color maps (HC380.E5 S96 1991)

United Nations Conference on Environment and Development: United States of America national report. New York: United Nations, 1992. 423 p.: ill. maps.

United Nations Conference on Environment and Development. Stockholm: Ministry of the Environment, 1991. 48 p. : color ill. color maps. (HC380.E5)

Valentine, Mark. *An introductory guide to the Earth Summit, June 1-12, Rio de Janeiro, Brazil.* San Francisco, Calif.: (300 Broadway, Suite 39, San Francisco 94133) U.S. Citizens Network on the United Nations Conference on Environment and Development, 1991. 44 p. (HD75.6 .V35 1991)

> Provides adequate information on the Earth Summit for general readers, including history, structure, politics, and economics.

Weidenbaum, Murray. *Earth Summit: U.N.spectacle with a cast of thousands.* St. Louis, Mo.: Center for the Study of American Business, Washington University, 1992. 18 p. (Contemporary issues series 50)

> Evaluates the United Nation's preparatory materials for the Earth Summit, which are both "simple-minded propaganda" and "resting on a shaky foundation." Appendix includes a copy of the Earth Charter.

Federal government reports

Bush, George. "America's commitment to the global environment." *U.S. Department of State Dispatch* (June 1, 1992): 421-423.

> Summarizes the environmental accomplishments of his administration and outlines his four-point plan of cooperation for improving the global environment, which he will present at the Earth Summit. Includes sidebar, "Convention on Biological Diversity."

INTERNATIONAL ENVIRONMENTAL LAW

Books (annotated)

Pontecorvo, Giulio. *The new order of the oceans: the advent of a managed environment.* New York: Columbia University Press, 1986.

> Explains the legal, scientific, and economic ramifications of the Law of the Sea, signed by one hundred nations (but not the U.S.) in 1982. The introduction by Elliot Richardson addresses the need for the new law because of new technologies, diminishing sea resources, and new ideas about freedom of navigation.

Weiss, Edith Brown. *In fairness to future generations: international law, common patrimony, and intergenerational equity.* Dobbs Ferry, N.Y.: Transnational Publishers, 1988. 385 p. (K3585.4 .W45 1989)

> Examines how the international legal order has been preoccupied with current environmental threats while in reality the environmental challenges we face will have results that stretch far into the future. Ozone depletion and global warming will have effects for generations, and mass species extinction will leave a

depauperate biosphere for possibly a million years. Examines such topics as nuclear waste, biological resources, water, soil, and climate.

Books (unannotated)

Basic documents of international environmental law. Boston: Graham and Trotman, 1992.

Environmental protection and international law. Boston: Graham and Trotman: M. Nijhoff, 1991. 244 p.: ill. (K3584.6 1990)

Innovation in environmental policy: economic and legal aspects of recent developments in environmental enforcement and liability. Aldershot, Hants., England: E. Elgar, 1992. Distributed by Ashgate Publishing, Brookfield, Vt. 269 p. (HC110.E5 I52 1992)

International environmental law and regulation. Salem, N.H.: Butterworth Legal, 1991. (K3585.4 .I573 1991)

International environmental law: primary materials. Boston: Kluwer Law and Taxation, 1991. 571 p. (K3583 .I57 1991)

Kiss, Alexandre Charles. *International environmental law.* New York: Transnational Publishers, 1991.

Understanding US and European environmental law: a practitioner's guide. Boston: Graham and Trotman, 1989. 176 p. (K3584.6 1987)

Weiss, Edith Brown. *International environmental law: basic instruments and references.* Dobbs Ferry, N.Y.: Transnational Publishers, 1992. 749 p. (K3583 .W45 1991)

Environmental and professional journals (annotated)

Bear, Dinah; and Elkind, Jonathan. "Environmental law: Soviet-United States cooperation." *Environment* 32 (April 1990): 5, 41-43.
   Discusses the development of environmental law in the Soviet Union and its stimulation by United States-Soviet cooperation to protect the environment.

Di Leva, Charles E. "Trends in international environmental law: a field with increasing influence." *Environmental Law Reporter* 21 (February 1991): 10076-10084.
   Examines the difficulties of establishing an effective international environmental system by analyzing several programs, treaties, and institutions, in light of the United Nations Conference on Environment and Development (UNCED)

McLean, Ronald A. N.; and Savoie, Marie-Claude. "A complete environmental management and protection system." *Environmental Claims Journal* 2 (Autumn 1989): 87-98.

    Discusses the provisions of the Canadian Environmental Enforcement Statute Law Amendment Act of 1986 which defines personal liability of directors and officers of corporations whose companies are liable for pollution damages.

Law journals

Cameron, James; and Abouchar, Juli. "The precautionary principle: a fundamental principle of law and policy for the protection of the global environment." *Boston College International and Comparative Law Review* 14 (Winter 1991): 1-27.

    Surveys the development of the precautionary principle, which ensures that a substance or activity posing a threat to the environment is prevented from adversely affecting the environment, as an emerging principle of law, and its potential use as "a comprehensive guide for environmental protection policy."

"Developments in the law: international environmental law." *Harvard Law Review* 104 (May 1991): 1484-1639.

    "Focusing on the components of the emerging international environmental regime, this questions whether and to what degree international law and institutions can overcome two fundamental obstacles to global environmental protection: states' reluctance to cede sovereignty and conflicting state interests." Also suggests legal and political responses that address some of the limitations on global environmental cooperation inherent in a world of sovereign states.

Gaines, Sanford E. "International principles for transnational environmental liability: can developments in municipal law help break the impasse?" *Harvard International Law Journal* 30 (Spring 1989): 311-349.

    Looks at the reasons why establishing international liability principles for environmental damage have reached an impasse. Also reviews "recent municipal developments in four areas of environmental liability and considers their relevance to the transnational context."

Hahn, Robert W.; and Richards, Kenneth R. "The internationalization of environmental regulation." *Harvard International Law Journal* 30 (Spring 1989): 421-446.

    Attempts to establish a paradigm "explaining the formation and shape of individual environmental agreements. Demonstrates the strengths and limitations of international responses to global environmental issues . . . and clarifies current misunderstandings about the forces driving the form and scope of multilateral international agreements, particularly in the environmental arena."

Mushkat, Roda. "International environmental law in the Asia-Pacific region: recent developments." *California Western International Law Journal* 20,1 (1989-1990): 21-40.

    Examines regional initiatives and cooperation in protecting the environment, including regional response to developments in international environmental law on such issues as sustainable development, protection of the ozone layer, transboundary movement of hazardous wastes, marine pollution, and state responsibility for environmental damage.

Conference proceedings

Ross, M.; and Saunders, J. O., ed. *Growing demands on a shrinking heritage: managing resource-use conflicts, essays from the fifth institute conference on natural resources law, 9-11 May 1991.* Calgary, Alberta, Canada: University of Calgary Press, 1992. 431 p.

ENVIRONMENTAL SECURITY

Books (unannotated)

Prins, Gwyn. *Top guns and toxic whales: the environment and global security.* Post Mills, Vt: Chelsea Green, 1991. 165 p.: (some color) ill. (TD170 .P73 1991)

General interest periodicals (annotated)

Dalby, Simon. "Security, modernity, ecology: the dilemmas of post-Cold War security discourse." *Alternatives* 17 (Winter 1992): 95-134.
> Uses insights drawn from contemporary critical theorizing in international relations to explore "the political implications and the limitations of the traditional discourse of security. To come to grips with security discourse, it is necessary to examine the arguments for extending the term to include themes of common security and matters of resources and ecology. Unless considerable care is taken the unquestioned political assumptions of earlier formulations of national security may well end up attached to the new meanings of the term with unintended implications for those who wish to reformulate the term and use it in new ways and in new circumstances."

Satchell, Michael. "The whole Earth agenda." *U.S. News & World Report* 107 (December 25, 1989/January 1, 1990): 50-52.
> Discusses the possibility of broadening the definition of national security to include environmental concerns.

Environmental and professional journals (annotated)

Cohen, Eliot A. "The future of force: and American strategy." *National Interest* 21 (Fall

1990): 3-15.

Disagrees with commentators who suggest that the study of national security should shift away from military matters and toward considering transnational pollution, economic competitiveness, and epidemiology as aspects of national security. Concludes that "in the final analysis, the problems of force will not disappear, and attempts to redefine national security expertise and institutions away from the study of real or potential war will only complicate the urgent task of renovating the study of strategy in the United States."

"Environmental security." *Bulletin of Peace Proposals* 20 (June 1989): 115-142.

Contents: "International organization for environmental security," by Nico Schrijver; "Security and the environment: a preliminary exploration," by Johan Jorgen Holst; "The environmental component of comprehensive security," by Arthur H. Westing; and "North-South trade, resource degradation, and economic security," by Johannes B. Opschoor.

Finger, Matthias. "The military, the nation state, and the environment." *Ecologist* 21 (September-October 1991): 220-225.

"The world's armed forces--and the industry upon which they depend--are a major cause of environmental degradation across the globe. Yet, the environmental regulations and agreements now being formulated by nation states (or groups of nation states) rarely apply to the military. On the contrary, with the ecological crisis now confronting us increasingly being defined as a 'threat to national security', the military is seen by many as part of the solution to the crisis rather than one of its major causes."

Green, Fitzhugh. "The environmental evolution of NATO, 1949-1986." *Atlantic Community Quarterly* 24 (Winter 1986-87): 332-337.

"It has been recognized since the birth of the NATO alliance in 1949 that there is more to defense than military preparedness. In the postwar period, therefore, national security has come increasingly to be defined in political, economic, and social terms as well."

Imber, Mark F. "Environmental security: a task for the U.N. system." *Review of International Studies* 17 (April 1991): 201-212.

Maintains that "there is clearly an environmental dimension to security. Whether this is defined as some new genus of security, or whether the environmental dimension is simply absorbed into an enlarged definition of security is subject to debate. Several potential environmental crises pose threats to conventional, military-territorial concepts of security." Examines the notion that the United Nations has a diplomatic role in environmental security.

Mathews, Jessica Tuchman. "Redefining security." *Foreign Affairs* 68 (Spring 1989): 162-177.

"Global developments now suggest the need for another analogous, broadening definition of national security to include resource, environmental and demographic issues. The assumptions and institutions that have governed international relations in the postwar era are a poor fit with these new realities."

Myers, Norman. "Environment and security." *Foreign Policy* 74 (Spring 1989): 23-41. Discusses how environmental issues relate to international relations and U.S. foreign policy. Notes that Third World nations whose progress is slowed by resource deterioration are more likely to become impoverished and destabilized. As examples of the links between environment and policy, describes how U.S. interests are caught up in deforestation in the Philippines, water deficits in the Middle East, land degradation in El Salvador, and population growth in Mexico.

Porter, Gareth. "Post-cold war global environment and security." *Fletcher Forum of World Affairs* 14 (Summer 1990): 332-344.
The collapse of the Soviet Union accelerated the erosion of the Cold War security system. Downgrading traditional security threats has been accompanied by a second fundamental trend: the emergence of global environmental issues. "The convergence of these two trends is creating a new phenomenon in world politics: the redefinition of national and international security primarily in terms of environmental threats rather than in terms of political-military threats from national or ideological rivals."

Timoshenko, Alexander S. "Ecological security: the international legal aspect." *Renewable Resources Journal* 7 (Winter 1989): 9-12.
Considers the evolving concept of ecological security and its link with the survivability of mankind, and the responsibilities of the international community for mutual cooperation in environmental issues.

Environmental and professional journals (unannotated)

Clark, William C. "National security and the environment." *Environment* 29 (June 1987): COV(2).

Junkin, Elizabeth Darby. "Environment as national security." *Buzzworm: The Environmental Journal* 1,4 (Summer 1989): 38(4)

Myers, Norman. "Linking environment and security." *Bulletin of the Atomic Scientists* 43 (June 1987): 46(2).

Reports

Renner, Michael. *National security: the economic and environmental dimensions.* Washington, D.C.: Worldwatch Institute, 1989. 78 p. (Worldwatch paper 89)

" 'Environmental security' offers a more fruitful basis for cooperation and security among nations than military security. Whereas military security offers at best the continuation of an uneasy status quo and at worst the prospect of annihilation, environmental security seeks to protect or to restore."

GLOBAL RESOURCE MANAGEMENT

(See also *U.S. Environmental Policy: Management*) 402

Books (annotated)

Anderson, S.; and Ostreng, W., eds. *International resource management*. London: Belhaven, 1989.
> Overviews international efforts to manage shared resources, such as Antarctica, the ozone layer, the great whales, and the atmosphere. Considers how the needs of individual nations are weighed against the needs of the international community. Upholds the role of the scientist as government policy advisor.

Owens, Susan; and Owens, Peter L. *Environment, resources, and conservation*. New York: Cambridge University Press, 1991. 112 p.: ill. maps. (TD174 .O94 1990)
> Selects specific issues (acid rain, coal depletion, and wetlands management) to demonstrate the relationship between the use and distribution of resources and the attendant underlying problems. Includes maps, graphs, photos, models, endnotes, and references.

Rees, Judith Anne. *Natural resources: allocation, economics, and policy*. 2d ed. New York: Routledge, 1990. 499 p.: ill. (HC79.E5 R432 1990)
> Maintains that the principal problems of natural resources are not those of scarcity but resource misallocations due to market failure or conflicts. Applies political systems and economic policy to the world mining industry, theoretical perspectives of renewable resources, and environmental management options. Concludes that the real issues are shrouded by "false" issues: resource scarcity, national security, and ecocatastrophe. Maintains that Third World poverty is the central problem that is causing environmental degradation.

Repetto, Robert. *World enough and time: successful strategies for resource management*. New Haven, Conn.: Yale University Press, 1986. 147 p. (HC59 .R43 1986)
> Presents the essential message and findings of the Global Possible Conference held in May, 1984, sponsored by the World Resources Institute.

World Resources, 1990-1991. Oxford: Oxford University Press, 1990.
> This work, the fourth in the series by the World Resources Institute, discusses

environmental issues and resource use around the world. Addresses issues such as wildlife, urbanization, and the atmosphere. Contains many illustrations, charts, and statistical tables.

Books (unannotated)

Conflicts and cooperation in managing environmental resources. New York: Springer-Verlag, 1992. 338 p.: ill. (HC79.P55 C655 1992)

Dasgupta, Partha. *The control of resources.* Cambridge, Mass.: Harvard University Press, 1982. 223 p.: ill. (HC79.E5 D33 1982)

Hinckley, Alden D. *Renewable resources in our future.* New York: Pergamon, 1980. 121 p.: ill. (S934.G7 H56 1980)

International resource management: the role of science and politics. New York: Belhaven, 1989. 301 p. (HC79.E5 I597 1989)

Lang, R., ed. *Integrated approaches to resource planning and management.* Calgary, Alberta, Canada: University of Calgary Press, 1986. 302 p.

Renewable natural resources: economic incentives for improved management. Paris: Organisation for Economic Co-operation and Development, 1989. Distributed by OECD Publications and Information Centre, Washington, D.C. 157 p. (HC79.E5 R452 1989)

World systems of traditional resource management. New York: Wiley, 1980. 290 p.: ill. (HC55 .W68 1980)

Conference proceedings

Bankes, N.; and Saunders, J. O., eds. *Public disposition of natural resources: essays from the first Banff conference on natural resources law, 12-15 April, 1983.* Calgary, Alberta, Canada: University of Calgary Press, 1985. 366 p.

Role of environmental impact assessment in the decision making process: proceedings of an international workshop held in Heidelberg, Federal Republic of Germany, August 1987. Berlin: E. Schmidt, 1989. 336 p.: ill., maps (TD171.8 .R65 1989)

INTERNATIONAL TRADE

(See also *U.S. Environmental Policy and Politics: NAFTA*)   411

Books (unannotated)

Greening of world trade issues. Ann Arbor: University of Michigan Press, 1992. 276 p.: ill. (HF1379 .G74 1992)

General interest periodicals (annotated)

Shrybman, Steven. "International trade and the environment: an environmental assessment of present GATT negotiations." *Alternatives* 17 (July-August 1990): 20, 22-29.

     Evaluates the effects of the renogotiation of General Agreement on Tariffs and Trade (GATT)--primarily the possibility of free trade--and the possibility that the agreement is likely to undermine environmental initiatives.

General interest periodicals (unannotated)

Babbitt, Bruce. "Free trade and environmental isolationism." *New Perspectives Quarterly* 9,3 (Summer 1992): 35(3)

"Freedom to be dirtier than the rest." (different environmental concerns of countries affect international trade) *Economist* 323,7761 (May 30, 1992): S7(4).

"Free trade's green hurdle: countries with tough green standards often want to foist them on their trading partners, usually, that is wrong." *Economist* 319,7711 (June 15, 1991): 61(2)

"GATTery vs. greenery." (use of trade sanctions for environmental reasons threatens the General Agreement on Tariffs and Trade) *Economist* 323,7761 (May 30 1992): S12(4)

Juffer, Jane. "Dump at the border: U.S. firms make a Mexican wasteland." (includes related articles) *Progressive* 52,10 (October 1988): 24(6)

"Should trade go green? How to stop protection for the environment becoming protectionism in trade." *Economist* 318,7691 (January 26 1991): 13(2)

Environmental and professional journals (annotated)

Bown, William. "Trade deals a blow to the environment." *New Scientist* 128 (November 10, 1990): 20-21.

     "International rules governing trade aim to stamp out protectionism, but they may destroy Third World attempts to protect the environment and achieve sustainable development."

Charnovitz, Steve. "Exploring the environmental exceptions in GATT articles." *Journal of World Trade* 25 (October 1991): 37-55.

Reviews the history of GATT Article to see what might have been the intention of those who wrote it, and analyzes some of the issues that arise in the adjudication of the Article.

"Free trade vs. the green wave." *Earth Island Journal* 7 (Winter 1992): 30-35.

Contents include: "Biodiversity: industry's green oil" (reveals hidden agenda of biotechnology), by Vandana Shiva; "A royalty for every potato" (explains how GATT will rob Third World farmers), by Anil Agarwal and Sunita Narain; and "We're on a merry-go-round to hell" (pleas to control science, chemicals, and GATT), by James Goldsmith.

Kulessa, Margareta E. "Free trade and protection of the environment: Is the GATT in need of reform?" *Intereconomics* 27 (July-August 1992): 165-173.

"The GATT negotiations under the Uruguay Round have almost run their course. However, consultations are certain to continue, as critics regard the GATT rules on environmental protection as inadequate. What aspects need to be reformed, and how might the initial reform measures look?"

Petersmann, Ernst-Ulrich. "Trade policy, environmental policy, and the GATT: why trade rules and environmental rules should be mutually consistent." *Aussenwirtschaft* 46 (1991): 197-221.

Discusses the conflict between trade rules and environmental rules, the economic ranking of environmental policy instruments, the political ranking of environmental policy instruments, the legal ranking of environmental policy instruments in GATT law, GATT dispute settlement proceedings over environmental measures, and GATT as a forum for the negotiation of new environmental rules.

Runge, C. Ford; and Nolan, Richard M. "Trade in disservices: environmental regulation and agricultural trade." *Food Policy* 15 (February 1990): 3-7.

Examines how "environmental and health regulations are increasingly being used as international trade barriers."

Van Bergeijk, Peter A. G. "International trade and the environmental challenge." *Journal of World Trade* 25 (December 1991): 105-115.

Examines the connection between trade, economy and environmental degradation. Concludes that the taxes, trade bans, and duties suggested for environmental improvement "are the same instruments that are to be eliminated in the framework of international trade negotiations."

Environmental and professional journals (unannotated)

Truax, Hawley, et al. "Coming to terms with trade." (General Agreement on Tariffs and Trade and other treaties that relax import standards on pesticides and other environmental hazards; terminology used in government explained) *Environmental Action Magazine* 24,2 (Summer 1992): 31(6)

Federal government reports

Trade and environment: conflicts and opportunities: background paper. Washington, D.C. (Office of Technology Assessment): G.P.O., 1992. 109 p. (OTA-BP-ITE-94, May 1992)
Focuses "primarily on issues pertinent to the General Agreement on Tariffs and Trade (GATT), although it gives some treatment to questions related to the proposed North American Free Trade Agreement. The study reviews several issues, including the relation between international environmental agreements and GATT, interactions between trade and national environmental laws, the effects of trade on environmental quality, the impacts of environmental regulations on trade and competitiveness, differences in perspectives between developed and developing countries, and how trade and environment policies might be coordinated."

LENDING ORGANIZATIONS

Books (annotated)

Le Prestre, Philippe. *The World Bank and the environmental challenge.* Toronto: Associated Presses, 1989. 263 p.
Takes a critical look at the environmental policies of the World Bank and shows the disparity between rapid economic development and environmental preservation. Environmental concerns are necessarily in conflict with the World Bank because it uses a cost-benefit decision making process, has the stated goal of economic development, relies on the financial community for funding, lacks environmental expertise, and permits clients to avoid adding costs for environmental preservation. The World Bank has, however, in recent years introduced environmental concerns into its policy process.

Mikesell, Raymond F.; and Williams, Lawrence F. *International banks and the environment.* San Francisco, Calif.: Sierra Club, 1992. 302 p.
Chastises the World Bank and the multilateral development banks for funding projects that have led to environmental degradation, but acknowledges that the banks have become more aware of their influence over development that harms the environment as they attempt to balance economic growth with sustainability.

Searle, Graham. *Major World Bank projects: their impact on people, society and the environment.* Camelford, England: Wadebridge Ecological Centre, 1987. 190 p.

Contains case studies on three World Bank projects: the Narmada River Development, the Polonoroeste Projects, and Indonesian Transmigration. Examines the impacts of the Bank's lending policies on developing countries.

General interest periodicals (unannotated)

Levine, Art. "Bankrolling debacles?" *U.S. News & World Report* 107,12 (September 25, 1989): 43(4).

Environmental and professional journals (annotated)

Rich, Bruce M. "The 'greening' of the development banks: rhetoric and reality." *Ecologist* 19 (March-April 1989): 44-52.
   "Despite the beginnings of major institutional reform in the World Bank, progress in implementing environmental reform in the multilateral development banks has been inadequate. Examples from Brazil and India show that the massive social and environmental disruption which results from the mega-projects which these banks are continuing to fund does not go away after bureaucratic reshuffles or 'concerned' speeches by top-level bank staff."

Environmental and professional journals (unannotated)

Rauber, Paul. "World bankruptcy." (World Bank and environment) *Sierra* 77,4 (July-August 1992): 34(4).

Law journals

Guyett, Stephanie C. "Environment and lending: lessons of the World Bank, hope for the European Bank for Reconstruction and Development." *New York University Journal of International Law and Politics* 24 (Winter 1992): 889-919.
   Asserts that World Bank "programs have played a substantial role in creating pollution problems in developing countries." By contrast, the European Bank for Reconstruction and Development focuses on environmental issues when providing assistance to Eastern Europe.

Joannides, Darryl. "Restructuring the World Bank: the environmental light shines on the funding of development projects." *Georgetown International Environmental Law Review* 2 (Fall 1989): 161-184.
   "Addresses the World Bank's reorganization and increased emphasis on environmental issues; the impact of World Bank development projects on the environment; the Bank's projects specifically aimed at achieving sustainable

development; and the economic realities of developing countries being forced to comply with environmental protection measures. Advocates that the World Bank aid in the development of global standards to achieve sustainable development."

Reports

Conable, Barber B. *The Conable years at the World Bank: major policy addresses of Barber B. Conable, 1986-91.* Washington, D.C.: World Bank, 1991. 177 p. (HG3881.5.W57 C66 1991)

Environmental assessment sourcebook. 3 vols. Washington, D.C.: World Bank (International Bank for Reconstruction and Development), 1991. ill. (TD195.E25 I58 1991)

Striking a balance: the environmental challenge of development. Washington, D.C.: World Bank, 1989. 52 p.
> Describes how the World Bank is seeking ways to strike a balance that will conserve resources while promoting economic development around the world. In a major policy redirection, the World Bank is promoting sustainable development in developing countries.

Warford, Jeremy J.; and Partow, Zeinab. *World Bank support for the environment: a progress report.* Washington, D.C.: World Bank, 1989. 41 p. (Development Committee, no. 22)
> Partial contents include: "The integration of environmental concerns into World Bank operations;" "Environmental issues papers;" "Environmental action plans;" "Economic policy and environmental management;" and "The global commons and sustainability."

DEBT-FOR-NATURE SWAPS

(See also *Rain Forests: Debt-for-Nature Swaps*) 2986

General interest periodicals (annotated)

Swire, Peter P. "Tropical chic." *New Republic* 200 (January 30, 1989): 18, 20-21.
> Lauds debt-for-nature swaps but considers them tiny compared with overall debt problems. Suggests aggressive debt reduction plans that would directly ease pressure on developing countries to exploit their resources.

Woody, Todd. "Conservationists push debt-for-nature swaps: even bank joins movement." *Legal Times* 14 (October 7, 1991): 2, 16, 18(5)
> "Two proposed deals could dramatically widen the scope of such swaps and

bolster efforts to use the same debt that threatens rain forests and wildlife to fund their survival." In the first, "the Bank America Corp. has agreed to forgive up to $6 million in Latin American loans on the condition that governments use the money to support environmental programs." The second involves the junk-bond financed takeover of the Pacific Lumber Co. that made Michael Milken the scourge of the Sequoias. In it, the Trust for Public Land would exchange $214 million in junk bonds for California's Headwaters Forest.

Environmental and professional journals (annotated)

Borrelli, Peter. "Debt or equity?" *Amicus Journal* 10 (Fall 1988): 42-49.
Describes debt-for-nature swaps in Bolivia, Costa Rica, and Ecuador, comparing this transaction with the debt-for-equity swap. Examines whether such swaps are ethical and in the economic best interest of debtor nations. Contrasts the destructive development promoted by industrial lenders versus sustainable development, which emphasizes ecosystem conservation.

"Debt-for-nature swaps: a new conservation tool." *World Wildlife Fund Letter* 1 (1988): 1-9.
"This issue reviews debt-for-nature swapping to date and examines its potential as a link between conservation and development."

Hultkrans, Andrew N. "Greenbacks for greenery." *Sierra* 73 (November-December 1988): 43-44, 46-47.
Explains and describes debt-for-nature swaps in Bolivia (1987), Ecuador (1987), and Costa Rica (1987 and 1988). The swaps involve the forgiveness of a developing country's debt when the debt is purchased by a conservation group. In exchange for its debt, the developing country makes a commitment to conservation.

Moran, Katy. "Debt-for-nature swaps: U.S. policy issues and options." *Renewable Resources Journal* 9 (Spring 1991): 19-24.
"Debt-for-nature swaps offer the critical time to secure environmental protection of the larger ecosystem for the present and future global community. Swaps also protect and promote the sustainable use of natural resources that is necessary for the economic develoment of the Third World."

Law journals

Alagiri, Priya. "Give us sovereignty or give us debt: debtor countries' perspective on debt-for-nature swaps." *American University Law Review* 41 (Winter 1992): 485-515.
Illustrates how developing countries "perceive debt-for-nature swaps as encroachments upon their sovereignty."

Barrans, David. "Promoting international environmental protection through foreign debt exchange transactions." *Cornell International Law Journal* 24 (Winter 1991): 65-95. Examines "the operation of debt-for-nature swaps and related transactions and their potential effectiveness in achieving the dual ends of easing the debt burden and protecting the environment . . . concluding that existing forms of environmentally-oriented debt exchanges may offer limited benefits."

Reports

Debt-for-nature exchange:. Washington, D.C.: Conservation International, 1989. 43 p. (S934.D44 D43 1989)

Occhiolini, Michael. *Debt-for-nature swaps.* Washington, D.C. (1818 H Street NW, Washington 20433): Debt and International Finance Division, International Economics Department, World Bank, 1990. 34 p. (HJ8086.D44 O33 1990)

CANADA

(See also *U.S. Environmental Policy and Politics: NAFTA*) 411

Books (unannotated)

Barney, Gerald O. *Global 2000: implications for Canada.* New York: Pergamon, 1981. 171 p. (HC120.E5 B37 1981)

Canadian resource policies: problems and prospects. New York: Methuen, 1981. 294 p.: ill. (HC120.E5 C37 1981)

Environmental rights in Canada. Toronto: Butterworths, 1981. 447 p. (KE5110 .E58 1981)

Environmentally significant areas of the Yukon Territory. Ottawa: Canadian Arctic Resources Committee, 1980. 134 p.: ill. maps. (QH77.C2 E58)

Howard, Ross. *Poisons in public: case studies of environmental pollution in Canada.* Toronto: J. Lorimer, 1980. 173 p. (QH545.A1 H68)

Jacobs, Peter. *Environmental strategy and action: the challenge of the world conservation strategy with reference to environmental planning and human settlements in Canada.* Vancouver, Canada: University of British Columbia Press, 1981. 99 p.: plates. ill. (GF75 .J33 1981)

Resources and the environment: policy perspectives for Canada. Toronto: McClelland and Stewart, 1980. 346 p.: ill. (HC120.E5 R47)

Status of the industry and measures for pollution control. Ottawa: Beauregard, 1987. 116 p.: ill. (TD899.M45 C36 1987)

General interest periodicals (annotated)

"Defining a sustainable society: values, principles and definitions." *Alternatives* 17 (July-August 1990): 36, 38-46.

> Provides "a working definition of sustainability that can be used to describe a sustainable Canadian society. This description will be used to generate and test one or more scenarios of Canadian society over the period from 1981 to 2031."

Nelson, J. G. "Experience with national conservation strategies: lessons for Canada." *Alternatives* 15 (December 1987-January 1988): 42-49.

> Notes that national conservation strategies are the principle means of implementing the 1980 World Conservation Strategy proposed by the International Union for Conservation of Nature and Natural Resources. Analyzes the initiation, preparation process, philosophy and purposes, comprehensiveness, cross-sectoral approach, multi-disciplinarity, management plans, and implementation of the national conservation strategies of Australia, Belize, Canada, Czechosovakia, India, Senegal, United Kingdom, Vietnam, and Zambia.

General interest periodicals (unannotated)

Dunster, Julian A. "Forest conservation strategies in Canada." *Alternatives* 16,4 (March-April 1990): 44(8).

Pell, David; and Wismer, Susan. "The role and limitations of community-based economic development in Canada's north." *Alternatives* 14 (February 1987): 31(4).

Environmental and professional journals (unannotated)

Dunbrack, Janet. "A gift to last." (donations to the Nature Conservancy of Canada) *Nature Canada* 15 (Spring 1986): 27(8).

Reports

Arctic environmental strategy: an action plan. Ottawa: Minister of Indian Affairs and Northern Development, 1991. 20 p.: color ill. map. (HC117.N48 A84 1991)

Canada's environment: an overview. Ottawa: Environment Canada, 1986. 20 p.: (some color) ill. maps. (HC120.E5 C345 1986)

Environment 2001: strategic directions for British Columbia. Victoria, B.C.: Ministry of Environment, 1991. 71 p. (HC117.B8 B642 1991)

Environmental peacekeepers: science, technology, and sustainable development in Canada: a statement. Ottawa: Science Council of Canada, 1988. 24 p. (HC120.E5 E587 1988)

Integration of environmental considerations into government policy. Hull, Canada: Canadian Environmental Assessment Research Council, 1990. 47 p. (HC120.E5 I58 1990)

Lead in the Canadian environment: science and regulation: final report. Ottawa: Royal Society of Canada, 1986. 374 p.: ill. (TD196.L4 C65 1986)

Maly, Stephen. *The greening of Canadian economic and trade policies.* Indianapolis, Ind.: Universities Field Staff International, 1990. 12 p. (UFSI field staff reports, North America 1990-91/no. 16)
> Discusses Canada's attempt to incorporate factors of environmental protection into Canadian economic and trade policies. Examines green activism in Canada.

Manitoba government response to the federal green plan. Winnipeg, Manitoba, Canada: Sustainable Development Coordination Unit, 1990. 28 p. (HC117.M3 M29 1990)

Pollard, Douglas Frederick William. *World conservation strategy, Canada: a report on achievements in conservation.* Ottawa: Conservation and Protection, Environment Canada, 1986. 61 p.: ill. (QH77.C2 P65 1986)

Protected area vision for Canada. Ottawa: Canadian Environmental Advisory Council, 1991. 88 p.: (some color) ill. (F1011 .P76 1991)

Report on Canada's progress towards a national set of environmental indicators. Ottawa: Environment Canada, 1991. 98 p.: color ill. (HC120.E5 R45 1991)

Survival in a threatened world: submission by the People of Canada to the World Commission on Environment and Development, May, 1986. Ottawa: Environment Canada, 1986. 37 p. (HC113.5 .S95 1986)

Swinnerton, Guy S. *People, Parks, and preservation: sustaining opportunities.* Edmonton, Canada: Environment Council of Alberta, 1991. 89 p.: ill. (SB484.C2 S88 1991)

Conference proceedings

The National Consultation Workshop on Federal Environmental Assessment Reform: report of proceedings. Canada: Federal Environmental Assessment Review Office, 1988. 34 p.

Youth declaration on the Alberta environment to the year 2000: papers and concensus from Youth and the Environment Conference, June 1, 2, 1985, Edmonton, Alberta. Edmonton, Canada: Alberta Environment, 1985. 70 p. (HC120.E5 Y68 1985)

LATIN AMERICA

(Related topics in *Environmental Philosophy: Sustainable Development*) 569
(Related topics in *Rain Forests*) 2929

Books (annotated)

Goodman, David; and Redclift, Michael, eds. *Environment and development in Latin America: the politics of sustainability.* Manchester, U.K.: Manchester University Press, 1991. 238 p.
    Examines sustainable development and how it relates to global inequality. Discusses the Amazon, forest management, hunger, and nuclear power.

Books (unannotated)

Belize: country environmental profile, a field study. Belize City, Belize: R. Nicolait and Associates, 1984. 151 p.: ill. maps. (HC142 .B447 1984)

Bordering on trouble: resources and politics in Latin America. Bethesda, Md.: Adler and Adler, 1986. 448 p.: ill. (HC125 .B594 1986)

Bunker, Stephen G. *Underdeveloping the Amazon: extraction, unequal exchange, and the failure of the modern state.* Urbana: University of Illinois Press, 1985. 279 p.: ill. (HC188.A5 B86 1985)

Burbridge, Peter R. *Environmental guidelines for resettlement projects in the humid tropics.* Rome: Food and Agriculture Organization of the United Nations, 1988. 67 p.: ill. (HC695.Z9 E53 1988)

Ecological development in the humid tropics: guidelines for planners. Morrilton, Ark.: Winrock International Institute for Agricultural Development, 1988.

Environment and diplomacy in the Americas. Boulder, Colo.: L. Rienner, 1992. 149 p.: ill. (HC130.E5 E59 1992)

Environmental profile of Paraguay.. Washington, D.C.: International Institute for Environment and Development, 1985. 162 p.: (some color) ill. (TD171.5.P3 E58 1985)

Johnson, Timothy H. *Biodiversity and conservation in the Caribbean: profiles of selected islands.* Cambridge, U.K.: International Council for Bird Preservation, 1988. 144 p. (QH77.C36 J64 1988)

Mexican topics: questions involving Mexico that have aroused international public interest. Mexico, D. F.: Presidencia de la Republica, Direccion General de Comunicacion Social, 1986. (some color) ill. (HC135 .M52554 1986)

Poverty, natural resources, and public policy in Central America. New Brunswick, N.J.: Transaction Publishers, 1992.

Toward a green Central America: integrating conservation and development. West Hartford, Conn.: Kumarian, 1992.

Visions of nature. Sao Bernardo do Campo, SP, Brazil: Mercedes Benz do Brasil, 1989. 143 p.: color ill. color map. (QH77.B7 V57 1989)

Wallace, David Rains. *The Quetzal and the Macaw: the story of Costa Rica's national parks.* San Francisco, Calif.: Sierra Club, 1992. 222 p.: maps. (SB484.C8 W35 1992)

Weinberg, Bill. *War on the land: ecology and politics in Central America.* Atlantic Highlands, N.J.: Zed, 1991. 203 p.: ill. (HC141.Z9 E59 1991)

General interest periodicals (annotated)

Reilly, William K. "The new context for conservation in Latin America: dealing with the economic challenge." *Vital Speeches of the Day* 53 (October 15, 1986): 26-29.
   The president of the World Wildlife Fund addresses the 25th Conference, Washington, D.C., September 17, 1986. He discusses the link between conservation and development, and the implications of the emergency constituency for conservation, in the countries of the region.

General interest periodicals (unannotated)

Aridjis, Homero. "The death of a masterpiece." (Mexico City pollution) *New Perspectives Quarterly: NPQ* 6,1 (Spring 1989): 40(4)

Quammen, David. "Brazil's jungle blackboard: a test for conservation deep in Amazonas." *Harper's Magazine* 276,1654 (March 1988): 65(6).

Rabben, Linda. "Amazon gold rush: Brazil's military stakes its claim." (gold, other industries and the environment) *Nation* 250,10 (March 12, 1990): 341(2).

Sachs, Ignacy. "What future for Amazonia?" *UNESCO Courier* (November 1991): 32(4).

Weisman, Alan. "Dangerous days in the Macarena." (Colombia nature reserve) *The New York Times Magazine* 138 (April 23, 1989): 40.

Environmental and professional journals (annotated)

Abate, Tom. "Environmental rapid-assessment programs have appeal and critics." *BioScience* 42 (July-August 1992): 486-489.
> "The most visible manifestation of this quick science has been the Rapid Assessment Program (RAP) from Conservation International in Washington, D.C. In four- to six-week surveys of Latin American areas whose ecology appears threatened, Conservation International's four-person RAP team works to identify habitats in need of immediate protection. Critics say the RAP team has not published results of its quick surveys in peer-reviewed journals. They also fear RAP will undermine support for less glamorous work."

Allen, William H. "Biocultural restoration of a tropical forest." *BioScience* 38 (March 1988): 156-161.
> "Architects of Costa Rica's emerging Guanacaste National Park plan to make it an integral part of local culture."

"Backyard Belize." *Sanctuary* 27 (September 1988): whole issue (18 p.)
> Discusses the forests and wildlife of Belize and efforts to protect them.

Buchanan, Al. "Costa Rica's wild west." *Sierra* 70 (July-August 1985): 32-35.
> "The country's much-admired park system faces the most serious threat of its 16-year existence as gold miners wreak havoc in Corcovado National Park."

Bunyard, Peter. "Guardians of the Amazon." *New Scientist* 124 (December 16, 1989): 38-41.
> Describes Colombia's policy of turning over huge areas of Amazon rain forests to the local Indians and other large tracts for the creation of national parks.

"Environmental protection in Costa Rica." *Nature Conservancy News* 34 (January-February 1984): 4-25.
> Contents: "A model for conservation," by S. Beebe; "An interview with Alvaro Ugaldo," by R. Cahn; "Costa Rica's endangered felines," by C. Vaughan; "Costa Rican parks: a researcher's view," by D. Jantzen. Discusses the national park system in Costa Rica and international environment agencies' contribution to conservation in the country.

Fernandez, Lisa. "Private conservation groups on the rise in Latin America and the Caribbean." *World Wildlife Fund Letter* 1 (1989): 1-8.
> Describes the activities of well over 200 private nonprofit conservation organizations (NCOs) that have emerged in Latin America and the Caribbean. Explores the

challenges they face, such as limited traditions of philanthropy and discouraging tax and labor laws.

Gaupp, Peter. "Ecology and development in the tropics." *Swiss Review of World Affairs* 42 (September 1992): 14-19.
Examines Costa Rica, which "provides graphic examples of false handling of natural resources. At the same time, in recent years Costa Rica has become a veritable laboratory for projects aimed at environmental protection, reforestation, and sustainable use of tropical forests."

Jones, Lisa. "Costa Rica: a vested promise in paradise." *Buzzworm: The Environmental Journal* 2 (May-June 1990): 31-35, 37-39.
"Costa Rica has been noted worldwide for its conservation efforts. Twelve percent of its national territory is set aside in a system of national parks and biological reserves prized by scientists and tourists alike. Its teeming biological diversity, stable society and governmental commitment to preserving the environment has a magnetic effect on foreign expertise and funding." But, "as in most tropical countries, Costa Rica's forests are shrinking in the face of exploding population and destructive land use."

Karliner, Joshua. "Central America's other war." *World Policy Journal* 6 (Fall 1989): 787-810.
Discusses the environmental crisis in Central America (Nicaragua, El Salvador, and Costa Rica). "The region's governments feel pressured to continue accelerated exploitation and export of their natural resources to satisfy local oligarchs and cope with huge external debts. U.S. militarization has worsened the crisis."

Karliner, Joshua. "Make parks, not war." *Amicus Journal* 9 (Fall 1987): 8-13.
"In the midst of armed conflict, Nicaragua and its neighbors discover much in common in their rich but threatened environments."

Pang, Eul-Soo; and Jarnagin, Laura. "Brazil's catatonic lambada." *Current History* 90 (February 1991): 73-75, 85-87.
Brazil must move away from its political and economic catatonic dance if Collor's plan is to work.

Rice, Robert A. "A casualty of war: the Nicaraguan environment." *Technology Review* 92 (May-June 1989): 63-71.
Reports on the Sandinista government's "ambitious and innovative projects in pesticide regulation, alternative energy, and rain-forest and wildlife conservation. Nicaraguan environmental policy could serve as a model for Third World countries, but only if the Contra war and the U.S. economic embargo end."

Simons, Paul. "Belize at the crossroads." *New Scientist* 120 (October 29, 1988): 61-65.
Discusses conflicting pressures in the Central American democracy of Belize. The

nation has substantial areas of intact tropical ecosystems and a strong conservation movement, but illegal immigrants are threatening the forests, and economic development, including tourism, is often attempted with little thought for environmental impacts.

Tangley, Laura. "Cataloging Costa Rica's diversity." *BioScience* 40 (October 1990): 633-636.
> "To make tropical biodiversity useful to society--and thus to save it--the first step is finding out what there is to lose."

Whelan, Tensie. "A tree falls in Central America." *Amicus journal* 10 (Fall 1988): 28-38. Describes the growth of environmentalism in the countries of Central America. "Central Americans have taken the goal of sustainable development (the notion that conservation is integral to economic development) and made it theirs." Countries discussed include Costa Rica, Panama, Nicaragua, El Salvador, Honduras, Belize, and Guatemala.

Environmental and professional journals (unannotated)

Ereira, Alan. "Words of warning: a message from the elder brothers." (Kogi tribe of Colombia) *Buzzworm: The Environmental Journal* 4,2 (March-April 1992): 40(6)

Greanville, David P. "El Salvador--air war makes casualty of the country's environment." *Animals' Agenda* 10,3 (April 1990): 31(2).

Mares, Michael A. "Conservation in South America: problems, consequences, and solutions." *Science* 233 (August 15, 1986): 734(6).

Tangley, Laura. "A new era for biosphere reserves." (Sian Ka'an Biosphere Reserve, Quintana Roo, Mexico) *BioScience* 38,3 (March 1988): 148(8).

Watkins, T. H. "The tropical equation." *Wilderness* 53,187 (Winter 1989): 18(2).

Reports

Managing protected areas in the tropics. Gland, Switzerland: International Union for Conservation of Nature and Natural Resources, 1986. 295 p.: ill. (QH75 .M354 1986)

Minimum conflict: guidelines for planning the use of American humid tropic environments. Washington, D.C.: Executive Secretariat for Economic and Social Affairs, Department of Regional Development, 1987. 198 p.: ill. (HC230.E5 M56 1987)

Teitel, Simon, ed. *Towards a new development strategy for Latin America: path ways from Hirschman's thought.* Washington, D.C.: Inter-American Development Bank, 1992. 403 p.

Uses the ideas of Albert Hirschman as a foundation for examining the future of Latin American development. Highlights his concepts of "linkages" and "uneven development." There are also chapters emphasizing trade problems, deregulation, and problems in specific countries.

Conference proceedings

First Interparliamentary Conference on the Environment in Latin America and the Caribbean: Mexico City, March 23 to 25, 1987. Mexico City: Senado de la Republica, 1987. 42 p.

Bibliographies

Environmental management in the tropics: an annoted bibliography 1985-1989. Amsterdam: Royal Tropical Institute, 1990. 236 p. (Z7165.T76 E58 1990)

AFRICA
■■■■■■■

Books (annotated)

Adams, W. M. *Green development: environment and sustainability in the Third World.* New York: Routledge, Chapman, and Hall, 1990. 255 p.
Explains the concepts of sustainable development and nature preservation, especially in African countries with colonial administrations. Argues that environmental problems are primarily political and that sustainable development can be achieved only through political economics and not by environmental science. Discusses desertification in Sahelian Africa, tropical wetland degradation, environmental impact assessment, the ineptness of bureaucracies and greed of developers, and the impact of environmentalists, which is not always productive.

African development sourcebook. New York: UNESCO, 1991. Distributed by UNIPUB, Lanham, Md. 157 p.
This professional sourcebook aims to identify the organizations, both private and governmental, that are working toward international development in Africa. Presents addresses, names of who to contact, purpose, and budget, among other information, for each organization. Also contains two articles on African development and numerous indices. There is a country listing of development organizations and bibliography.

Fratkin, Elliot. *Surviving drought and development: Ariaal pastoralists of northern Kenya.* Boulder, Colo.: Westview, 1992. 152 p.

Maintains that pastoralism in the Sahel region of Africa is not a doomed way of life and that Western efforts at development should not seek to convert those whose lives subsist by it. Uses the Ariaal as a case study to demonstrate the ability of pastoralism to cope with changing conditions. Argues that development has been one of the principal threats to these people. Well documented and readable.

Homewood, K. M.; and Rodgers, W. A. *Maasailand ecology: pastoralist development and wildlife conservation in Ngorongoro, Tanzania.* New York: Cambridge University Press, 1992. 298 p.
Focuses on the Maasai pastoralists of East Africa. Discusses their history, prehistory, and ethnography. Analyzes the ecology of the Tanzania Ngorongoro Conservation Area (NCA) and discusses the politics of resource management in colonial and postcolonial times. Maintains that the Maasai should be involved in management of the NCA.

Leach, Gerald; and Mearns, Robert. *Beyond the woodfuel crisis: people, land, and trees in Africa.* London: Earthscan with IIED, 1989.
Focuses on the problems of woodfuel supply in rural areas and the different kinds of problems associated with the firewood trade in urban areas. Includes case studies.

Lewis, L. A.; and Berry, L. *African Environments and Resources.* Boston, Mass: Unwin Hyman, 1988.
Details the existing resources of Africa, analyzes them with regard to human use, and describes the environmental consequences of alternative development paths, focusing on the interdependence of people and the environment. Emphasizes the importance of soil and water resources and concludes that for the rest of this century Africans will derive their living from the productivity of soil, water, and vegetation resources.

Milner, Chris; and Rayner, A. J., eds. *Policy adjustment in Africa.* New York: St. Martin's, 1992. 249 p.
Presents papers from a 1989-1990 seminar series at the Center for Research on Economic Development and International Trade. Focuses on the nature and impact of structural adjustment programs in Africa. Takes a neoclassical economic perspective that is mostly supportive of World Bank and IMF policies.

Porter, Doug; Allen, Bryant; and Thompson, Gaye. *Development in practice: paved with good intentions.* London: Routledge, 1991. 247 p.
Using the Magarini Settlement Project in Kenya's Coast Project as a case study, examines the environmental and human consequences of internationally funded development. Chronicles thirteen years of good intentions and frustrations.

Rosenblum, Mort; and Williamson, Doug. *Squandering Eden: Africa at the edge.* London:

Bodley Head, 1988. 326 p.
An American journalist and scientist detail the rampant environmental degradation of Africa. Reveals contemporary knowledge about Africa and offers ideas for
solving some of the problems.

Timberlake, Lloyd. *Africa in crisis: the causes, the cures of environmental bankruptcy.*
Washington, D.C.: International Institute for Environment and Development, 1985.
233 p.
"Year after year, African peasants have been forced, in their efforts to survive, to
take more from their forests, soils, and rivers than these natural resources can
provide. Such withdrawals are bankrupting Africa's environment, steadily
undermining virtually every nation's ability to feed itself. Have African governments and foreign aid agencies been adopting the wrong policies?"

Twenty-first-century Africa: towards a new vision of self-sustainable development. Los
Angeles, Calif.: Africa World/African Studies Association, 1992. 330 p.
Report of a task force aiming to help U.S. researchers contribute to sustainable
African development. Presents a new vision of self-sustainable development in
Africa; a new economic strategy; regional cooperation; education and develop-
ment; health; ecological problems; and relevant legislation, among other topics.

*World resources 1987: a report by the International Institute for Environment and
Development and the World Resources Institute.* New York: Basic Books, 1987. 369 p.
Contains issue analysis of hazardous waste management and sustainable
development in sub-Saharan Africa as well as reviews and data tables concerning
population and health, human settlements, food and agriculture, forests and
rangelands, wildlife and habitat, energy, freshwater, oceans and coasts, atmosphere and climate, global systems and cycles, and policies and institutions.

Books (unannotated)

Akuffo, S. B. *Pollution control in a developing economy: a study of the situation in Ghana.*
Accra: Ghana Universities Press, 1989. 117 p.: ill. (TD188.5.G4 A38 1989)

Alemneh Dejene. *Environment, famine, and politics in Ethiopia: a view from the village.*
Boulder, Colo.: L. Rienner, 1990. 150 p.: ill. (HD979 .A65 1990)

Bartelmus, Peter. *Economic development and the human environment: a study of impacts
and repercussions with particular reference to Kenya.* Munich: Weltforum Verlag, 1980.
184 p.: ill. (HD75.6 .B37)

Bennett, John William. *Political ecology and development projects affecting pastoralist
peoples in East Africa.* Madison, Wis.: Land Tenure Center, University of Wisconsin--
Madison, 1984. 150 p.: ill. (HD107 .W52 no. 80)

Beyond structural adjustment in Africa: the political economy of sustainable and democratic development. New York: Praeger, 1992. 190 p.: map. (HC800 .B5 1992)

Blackwell, Jonathan M. *Environment and development in Africa: selected case studies.* Washington, D.C.: World Bank, 1991. 127 p.: ill. maps. (HC800 .B54 1990)

Clarke, James Frederick. *Back to Earth: South Africa's environmental challenges.* Halfway House, South Africa: Southern Book Publisher, 1991. 332 p.: ill. (GF758 .C54 1991)

Conservation in Africa: people, policies, and practice. New York: Cambridge University Press, 1987. 355 p.: ill. maps. (QH77.A4 C66 1987)

Ecology and politics: environmental stress and security in Africa. Uppsala, Sweden: Scandinavian Institute of African Studies, 1989. 255 p.: ill. (HC800.Z9 E539 1989)

Environmental concerns in South Africa: technical and legal perspectives. Cape Town: Juta, 1983. 587 p.: ill. (TD171.5.S6 E58 1983)

Environmental issues and management in Nigerian development. Ibadan, Nigeria: Evans Brothers (Nigeria Publishers), 1988. 410 p.: ill. (HC1055.Z9 E54 1988)

Environmental problems in Sudan: a reader. 2 vols. The Hague: Institute of Social Studies, 1990. 819 p. : ill. maps. (GF728 .E58 1990)

Gaining ground: institutional innovations in land-use management in Kenya. Rev. ed. Nairobi: Acts Press, African Centre for Technology Studies, 1991. 228 p.: ill. maps. (HD983.Z63 G35 1991)

Geography of Nigerian development. 2d ed. Ibadan, Nigeria: Heinemann, 1983. 456 p.: ill. (HC1055 .G46 1983)

Gritzner, Jeffrey A. *The West African Sahel: human agency and environmental change.* Chicago: University of Chicago, 1988. 170 p.: ill. maps. (GF740 .G75 1988)

Huntley, Brian J. *South African environments into the 21st century.* Cape Town: Human and Rousseau: Tafelberg, 1989. 127 p.: (some color) ill. (HC905.Z9 E54 1989)

Jolly, Alison. *A world like our own: man and nature in Madagascar.* New Haven: Yale University Press, 1980. 272 p.: (some color) plates. (QH77.M28 J64)

Lewis, Laurence A. *African environments and resources.* Boston: Unwin Hyman, 1988. 404 p.: ill. maps. (HC800 .L48 1988)

Lovatt Smith, David. *Amboseli, nothing short of a miracle.* Nairobi: East African, 1986. 96 p.: (some color) ill. color maps. (SB484.K4 L68 1986)

MacKinnon, John Ramsay. *Review of the protected areas system in the Afrotropical realm.* Gland, Switzerland: International Union for Conservation of Nature and Natural Resources, 1986. 259 p.: ill. color maps. (QH77.A43 M33 1986)

Nigeria's threatened environment: a national profile. Ibadan, Nigeria: Nigerian Environmental Study Action Team, 1991. 288 p.: ill. (GF746.4 .N54 1991)

Rotating the cube. Durban: Department of Geographical and Environmental Sciences and Indicator Project, University of Natal, 1990. 117 p.:ill (TD171.5.S6 R68 1990)

Timberlake, Lloyd. *Africa in crisis: the causes, the cures of environmental bankruptcy.* London: Earthscan, 1988. 203 p.

Zimbabwe's environmental dilemma: balancing resource inequities. Harare: ZERO, 1991. 165 p.: maps. (HC910.Z65 Z56 1991)

General interest periodicals (annotated)

Bailey, Robert C. "The Efe: archers of the African rain forest." *National Geographic* 20 (November 1989): 664-686.
> Describes the life and culture of the Efe, a seminomadic group known to outsiders as pygmies, who carry on a tradition of hunting and gathering, little affected by the outside world; part of the Ituri Project.

General interest periodicals (unannotated)

Armstrong, Sue. "Pride and prejudice." (South Africa's national parks; includes related information on travel opportunities) *World Magazine* 48 (April 1991): 50(7).

Jeanneret, Charles. "Rwanda: land of a thousand hills." (Environment and Development: A Global Commitment) *UNESCO Courier* (November 1991): 19(4)

Environmental and professional journals (annotated)

El-Ashry, Mohamed T.; and Ram, Bonnie J. "Sustaining Africa's natural resources." *Journal of Soil and Water Conservation* 42 (July-August 1987): 224-227.
> "The term 'integrated resource management' must now take on a dual meaning, referring to both integrated management of the land, water, forest, and energy sectors and the integration of resource management into economic development."

"Experts prescribe many cures for Africa's illness." *Conservation Foundation Letter* (November-December 1985): 1-7.

"Obstacles to economic progress include population growth, constraints on agricultural development, wasted human resources, domestic politics, and external policies. Yet no single problem is beyond the capacity of humanity to remedy."

Haas, Peter, M. "Towards management of environmental problems in Egypt." *Environmental Conservation* 17 (Spring 1990): 45-50.
    "As a poor, developing country, Egypt faces a wide variety of pollution problems that are associated with poverty." Reviews the most severe environmental problems currently facing Egypt.

Mann, R. D. "Time running out: the urgent need for tree-planting in Africa." *Ecologist* 20 (March-April 1990): 48-53.
    "The Sahel is now in its third decade of drought. African governments must make tree-planting a national priority, to recognize the wisdom of traditional farming and to make full use of the extensive ecological knowledge of the African farmer."

Meldrum, Andrew. "The trouble with paradise." *Africa Report* 36 (September-October 1991): 18- 20.
    "In the midst of a protracted civil war, environmentalists are trying to save an Inhaca island off Maputo harbor, now crowded with war refugees who threaten the fragile ecology. Why bother about one little island? Establishing an equilibrium with nature can set an example for the rest of the country."

White, Rodney R. "Environmental management and national sovereignty: some issues from Senegal." *International Journal* 45 (Winter 1989-90): 106-137.
    Looks at environmental impacts on Third World countries, the influence of the industrialized world on Third World decisions to reduce negative impacts on the global environment, and Third World ability to protect itself from its own impacts on the environment. Uses Senegal as a case study.

Environmental and professional journals (unannotated)

Brock, Robert G. "Saving paradise." (Madagascar) *Technology Review* 90 (May-June 1987): 13(3).

Eddy, William. "Rhythms of survival." (Kenya's nature conservation efforts) *National Parks* 61 (September-October 1987): 21(3).

Reports

Ahmad, Yusuf J. *The state of the environment in Africa.* Nairobi (NGO's Environment Network: Environment Liaison Centre (ELC), 1985. 36 p.: ill. (HC800.Z9 E53 1985)

Atta el Moula, Mutasim el Amin. *On the problem of resource management in the Sudan.* Khartoum: Institute of Environmental Studies, University of Khartoum, 1985. 131 leaves: ill. maps. (HC835 .A85 1985)

Brown, Lester R. *Reversing Africa's decline.* Washington, D.C.: Worldwatch Institute, 1985. 81 p.: ill. (HC800.Z9 E536 1985)

Country strategy for strengthening environmental considerations in Danish development assistance to Tanzania. Copenhagen, Denmark: Department of International Development Cooperation, 1989. 70 p.: map. (GF729 .D36 1989)

Country strategy for strengthening environmental considerations in Danish development assistance to Kenya. Copenhagen, Denmark: Department of International Development Cooperation, 1989 116 p.: map. (HC865.Z9 E536 1989)

Durning, Alan B. *Apartheid's environmental toll.* Washington, D.C.: Worldwatch Institute, 1990. 50 p.: ill. maps. (HC905.Z9 E53 1990)

Environmental management problems in resource utilization and survey of resources in the west and central African region. Geneva, Switzerland: United Nations Environment Programme, 1984. 83 p.: ill. (HC1000.Z9 E53 1984)

Gritzner, Jeffrey Allman. *The West African Sahel: human and environmental change.* Chicago: University of Chicago Geography Research Paper 226. 1988. 176 p.
   Topics include the use of brush fires as a tool of warfare; effects of trade on vegetation; settlement location; dune destabilization; introduction of firearms which depleted bush animals; irrigation agriculture; proliferation of cattle; the growth of transportation that encouraged resource exploitation and environmental degradation; and ideas for rehabilitating the desert within the social, environmental, and cultural constraints of the region.

Kuron, John Lado. *The Juba Environment Project report I.* Juba, Sudan: Environment Project, 1986. 39 p.: ill. maps. (TD171.5.S732 J835 1986)

Maputaland: conservation and removals. Pietermaritzburg, South Africa: Association for Rural Advancement, 1990. 60 p.: ill. (S934.S6 M37 1990)

McNamara, Robert S. *The challenges for sub-Saharan Africa.* Washington, D.C.: n.p., 1985. 49 p. (HC800 .M38 1985)

Nature of Zimbabwe: a guide to conservation and development. Gland, Switzerland: Field Operations Division, International Union for Conservation of Nature and Natural Resources, 1988. 87 p.: color ill. (HC910.Z65 J66 1988)

Norwegian aid and the environment in Mozambique: the issues. Bergen, Norway: DERAP,

Development Research and Action Programme, Chr. Michelsen Institute, Department of Social Science and Development, 1990. 82 p.: ill. (HC890.Z9 E56 1990)

Scheepers, J. C. *Conservation of major biotic communities in central and southern Africa.* South Africa: SARCCUS, 1990. 21 p.: color ill. (QH195.C37 S34 1990)

Conference proceedings

Development and the environment: proceedings of a national conference. Ibadan: Nigerian Institute of Social and Economic Research, 1986. 442 p.: ill. (HC1055.Z9 E53 1986)

Environmental policies for sustainable growth in Africa: *Montclair State College, Upper Montclair, New Jersey, May 6, 1991.* Upper Montclair, N.J.: Center for Economic Research on Africa, 1991. 101 p.: ill. '(HC800.Z9 E54 1991)

Food, forestry, and environment, the challenge to rural poverty in Africa: report on the regional observance of the Fifth World Food Day and the fortieth anniversary of FAO, Buea, Cameroon, October 14-16, 1985. Accra, Ghana: United Nations Food and Agriculture Organisation; Douala, Cameroon: Pan African Institute for Development, 1986. 86 p.: ill. (HD9017.A2 F655 1986)

National conference on development, 1990, Imo State University Okigwe, Nigeria: issues in national development. Enugu, Nigeria: ABIC, 1990. 518 p. (HC1055 .N37 1990)

South African Development Coordination Conference: SADCC policies on food and agriculture, and natural resources, and the environment. Gaborone: Southern African Centre for Cooperation in Agricultural Research, 1989. 81 p. (HD2130.Z8 S65 1989)

Bibliographies

Seeley, J. A. *Conservation in Sub-Saharan Africa: an introductory bibliography for the social sciences.* Cambridge: African Studies Centre, 1985. 207 p.: ill. (Z5863.P6 S43 1985)

EUROPE

(See also *Environmental Movement: Europe*) 471

Books (annotated)

Enyedi, Gyorgy, et al.; eds. *Environmental policies in East and West.* London (European Coordination Centre for Research and Documentation in Social Sciences): Taylor

Graham, 1987. 401 p.: ill. (HC240.9.E5 E6 1987)
   Maintains that individual nations must learn the environmental challenges and
   subsequent policies that exist in neighboring countries in order to forge international cooperation. Examines case studies from both the West and the East, as
   well as analyzing the policies of the European Economic Community. Concludes
   that Europeans need a new attitude that allows them to balance nature with their
   economic needs.

Goldsmith, Edward; and Hildyard, Nicholas, eds. *Green Britain or industrial wasteland?*
   Cambridge, England: Polity Press, 1986. 374 p.
   Contents include: "Britain's farm policy: a history of wanton oblivion," by Robert
   Waller; "The destruction of the countryside," by Chris Rose; "Soil erosion in
   Britain," by R. P. C. Morgan; "The environment in forest policy," by Colin Price,
   et al.; "Acid rain and British pollution control policy," by Nigel Dudley; "Pesti-
   cide controls: a history of perfidy," by Chris Rose; "Britain's dirty beaches," by
   Fred Pearce; "Down in the dumps: Britain and hazardous wastes," by Nicholas
   Hildyard; "The hazards of high-voltage power lines," by Hilary Bacon; "Dumping
   nuclear waste at sea," by Jim Slater; and "Pesticide exports: Britain's record," by
   Chris Rose.

Parkin, Sara, ed. *Green light on Europe.* London: Heretic Books, 1991. 367 p.
   The collapse of the Communist bloc and the official beginning of the European
   Community provide the backdrop for this economic analysis of how EC goals will
   lead to environmental disaster, especially in the areas of population increases and
   consumption.

Books (unannotated)

Calder, Jenni. *Scotland in trust.* Washington, D.C.: Preservation Press, 1990. 175 p.:
color ill. (DA873 .C35 1990)

*Conservation and development programme for the U.K.: a response to the world conservation
strategy.* London: Kogan Page, 1983. 496 p.: ill. (S934.G7 C66 1983)

Crawford, Iain. *Held in trust: the National Trust for Scotland.* Edinburgh, Scotland:
Mainstream Publishers, 1986. 183 p.: ill. (DA873 .C68 1986)

Crisis of London. New York: Routledge, 1992. 213 p.: ill. (HT133 .C72 1992)

Crofts, Tony. *The return of the wild: the British countryside and the world-wide rural crisis.*
Waterford, Ireland: Friendly Press, 1987. 162 p., 8 p. of plates: ill. (HD596 .C76 1987)

Eckerberg, Katarina. *Environmental protection in Swedish forestry.* Brookfield, Vt.:
Avebury, 1990. 179 p.: ill., maps (SD387.E58 E28 1990)

Environment and economic development in the regions of the European Community. Brookfield, Vt.: Avebury, 1988. 154 p. (HC240.9.E5 E57 1988)

Environmental management: British and Hungarian case studies. Budapest, Hungary: Akademiai Kiado, 1984. 263 p.: ill. (HC260.E5 E58 1984)

Environmental protection: tasks and results: information from the German Democratic Republic. Berlin: Panorama DDR, 1986. 70 p., 16 p. of plates: ill. (some color) (TD171.5.G35 E4 1986)

Evans, David. *A history of nature conservation in Britain.* New York: Routledge, 1992. 274 p.: ill. (QH77.G7 E9 1992)

Freethy, Ron. *The making of the British countryside.* North Pomfret, Vt.: David and Charles, 1981. 256 p.: ill. (QH137 .F7)

Futures for the Mediterranean Basin: the Blue Plan. New York: Oxford University Press, 1989. 279 p.: ill. maps. (TD186 .F88 1989)

Gaze, John. *Figures in a landscape: a history of the National Trust.* London: Barrie and Jenkins and the National Trust, 1988. 336 p., 16 p. of plates: ill. (DA655 .G39 1988)

Green light on Europe. London: Heretic; East Haven, Conn.: Distributed in North America by Inbook, 1991. 367 p.: ill.

Johnson, Brian D. G. *The conservation and development programme for the UK: a response to the world conservation strategy: an overview, resourceful Britain.* London: Kogan Page, 1983. 104 p. (S934.G7 J64 1983)

Johnson, Stanley. *The environmental policy of the European communities.* Boston: Graham and Trotman, 1989. 349 p. (HC240.9.E5 J626 1989)

Mabey, Richard. *The common ground: a place for nature in Britain's future?* London: Hutchinson in association with the Nature Conservancy Council, 1980. 280 p., 18 leaves of plates: ill. (some color) (QH77.G7 M3)

MacEwen, Ann. *Greenprints for the countryside?: the story of Britain's national parks.* Boston: Allen and Unwin, 1987. 248 p., 16 p. of plates: ill., maps (SB484.G7 M22 1987)

Mallinckrodt, Anita M. *The environmental dialogue in the GDR: literature, church, party, and interest groups in their socio-political context: a research concept and case study.* Lanham, Md.: University Press of America, 1987. 198 p.: ill. (JA76 .M335 1987)

Marcinkiewicz, Jan. *Pollution in the heart of Europe.* London: Polish Society of Arts and Sciences Abroad, 1987. 55 p.: ill. (some color) (TD186.5.P7 M37 1987)

Nature of West Wales: the wildlife and ecology of the County of Dyfed. Buckingham, England: Barracuda, 1986. 164 p., 8 p. of plates: ill. (some color) maps. (QH144.N37)

New politics in Western Europe: the rise and success of green parties and alternative lists. Boulder, Colo.: Westview Press, 1989. 230 p. (JN94.A979 N49 1989)

Participation and litigation rights of environmental associations in Europe: current legal situation and practical experience. New York: P. Lang, 1991. 196 p. (KJC6242 .P37 1991)

Robinson, Mike. *The greening of British party politics.* Manchester, England: Manchester University Press, 1992. Distributed by St. Martin's Press, New York. 246 p.: ill. (HC260.E5 R63 1992)

Siegmann, Heinrich. *The conflicts between labor and environmentalism in the Federal Republic of Germany and the United States.* New York: St. Martin's Press, 1985. 201 p. (HD5178 .S54 1985)

Silent countdown: essays in European environmental history. New York: Springer-Verlag, 1990. 265 p.: ill., maps (GF540 .S55 1990)

Society and the environment: a Swedish research perspective. Boston: Kluwer Academic Publishers, 1992. 321 p.: ill. (HC380.E5 S62 1992)

Strategy for industrial and urban development: plan of action for integration of environmental considerations into Danish development assistance. Copenhagen: DANIDA, 1989. 70 p. (HT395.D44 S77 1989)

Ten years of community environment policy. Paris: Commission of the European Communities, 1984. 1 v. : ill. (HC240.9.E5 C65 1984)

Zucchetto, J. *Resources and society: a systems ecology study of the island of Gotland, Sweden.* New York: Springer-Verlag, 1985. 246 p.: ill. (HC373.5 .Z83 1985)

General interest periodicals (annotated)

"EC at the crossroads: strategic choices in EC policy." *Futures* 23 (September 1991): whole issue (679-784 p.)
> Contents include: "Institutional choices: the rise and fall of subsidiarity," by Guenther F. Schaefer; "EC environmental policy: coping with interdependency," by David Wright; "Social Europe in the 1990s: beyond an adjunct to achieving a common market?" by Patrick Kenis; "The integration dilemma: enlarging and/or deepening the Community," by Francois de la Serre; "The EC in the world context: civilian power or superpower?" by Finn Laursen; and "Upgrading EC strategic choice capacities," by Yehezkel Dror.

General interest periodicals (unannotated)

"Come home, John Muir." (national parks in Scotland) *Economist* 310,7588 (February 4, 1989): 51(2)

Jackman, Brian. "Penwith in peril." (West Penwith, England) *World Magazine* 55 (December 1991): 66(7)

"Murky waters: pollution control." (Britain) *Economist* 320,7724 (September 14 1991): 69(2)

Environmental and professional journals (annotated)

Comolet, Arnaud. "How OECD countries respond to state-of-the-environment reports." *International environmental affairs* 4 (Winter 1992): 3-17.
  "The fact that the role of these documents is often badly defined is the most striking fact of the analysis of the state-of-the-environment reports in the OECD countries. In most cases the documents are very administrative in their orientation and scarcely do more than praise existing policies."

Liberatore, Angela; and Lewanski, Rudolf. "The evolution of Italian environmental policy." *Environment* 32 (June 1990): 10-15, 35-40.
  "Although the Italian environmental movement was slow to take off, the adoption of more environmentally sound policies has been gradually accelerating. Now, most of the relevant policy instruments appear to be in place, but serious problems in their implementation remain to be solved."

Linnerooth, Joanne. "The Danube River Basin: negotiating settlements to transboundary environmental issues." *Natural Resources Journal* 30 (Summer 1990): 629-660.
  Describes "the political and institutional hurdles involved in negotiating and reaching agreements for improving the quality of Danube water" and possible forms for the ensuing cooperation. Discusses "a potential role for the analyst in aiding the bilateral and multilateral negotiation process. . ."

McCarthy, James. "Sites of special scientific interest: the British experience." *Natural Areas Journal* 11 (April 1991): 108-113.
  "The system of protection for Sites of Special Scientific Interest (SSSI) in Great Britain, and specifically Scotland, is described, with particular reference to the new provisions of the 1981 Wildlife and Countryside Act. The nationwide systems have been important in stemming losses of valuable natural habitats and geological localities."

O'Riordan, Timothy. "The politics of environmental regulation in Great Britain." *Environment* 30 (October 1988): 5-9, 39-44.

Looks at how Britain's environmental policies are formed, how public opinion is changing, and what the main political parties are advocating. Also discusses the growing significance of the EC to Britain's environmental regulation and examines the environmental implications of Margaret Thatcher's "enterprise culture."

O'Sullivan, Dermot A. "Environmental concerns gain prominence in Europe." *Chemical and Engineering News* 67 (March 27, 1989): 7-15.
"With a flurry of diverse issues ranging from dumping of industrial wastes in landfills to the buildup of noxious gases in urban air, environmental problems have become a major factor in policy and trade decisions."

Schaffer, Daniel. "Exploring environmental frontiers: integrating solutions East and West." *Survey of Business* 26 (Spring 1991): 5-27.
Describes the pollution problems that confront these nations and considers the obstacles that must be overcome in finding solutions. The value of U.S. environmental policy and assistance is also discussed.

Environmental and professional journals (unannotated)

Liberatore, Angela. "EC environmental policy and the Mediterranean region: an international colloquium." *Environment* 32,1 (January-February 1990): 44(2).

Von Moltke, Konrad. "Three reports on German environmental policy." (Protecting the Earth's Atmosphere: An International Challenge, Protecting the Tropical Forests, Protecting the Earth) *Environment* 33,7 (September 1991): 25(5)

Law journals

Luiki, Paul S.; and Stephenson, Dale E. "Environmental laws are stricter in 'green'-influenced Europe." *National Law Journal* 14 (September 30, 1991): 45-47 (6 p.)
"Europe has emerged as a leading force among industrial democracies in promulgating stringent environmental requirements. Growing environmental awareness among the European population has caused a 'green' philosophy to emerge across the political spectrum in many countries, acting as a catalyst for change as parties compete to have the 'greenest' platforms. Environmental law in Europe is dynamic and complex, consisting not only of regulation applicable in individual countries but also of those imposed by a supranational institution."

"Symposium on European Community Environmental law." *Boston College International and Comparative Law Review* 14 (Summer 1991): whole issue (237-480 p.)
Partial contents: "Environmental law of the European Economic Community: new powers under the Single European Act," by Christian Zacker; "The European Environment Agency and the freedom of environmental information directive: potential cornerstones of EC environmental law," by Dietrich Gorny; "Developing

a unified European environmental law and policy," by Cynthia B. Schultz and Tamara Raye Crockett; "German unification and European Community environmental policy"; "The regulation of ozone-depleting chemicals in the European Community"; "The EC's action programme for improving efficiency of electricity use"; "EC regulation of sulphur dioxide levels: directive 89/427"; "Regulating air pollution from municipal incineration plants in the EC: directives 89/369 and 89/429"; "The proposed EC directive on automobile exhaust emissions"; "Municipal waste water and sewage sludge management in the EC"; and "Report of the working group of experts from the member states on the use of economic and fiscal instruments in EC environmental policy."

Reports

1992: the environmental dimension: task force report on the environment and the internal market. Bonn (Commission of the European Communities. Task Force Environment and the Internal Market): Economica Verlag, 1990. 289 p.: ill., maps (some col.) (HC240.9.E5 C66 1990)

Alfsen, Knut H. *Natural resource accounting and analysis: the Norwegian experience, 1978-1986.* Oslo, Norway: Statistisk Sentralbyra, 1987. 71 p.: ill. (HC363.5 .A43 1987)

Environmental policies in East and West. London: Published on behalf of the European Co-ordination Centre for Research and Documentation in Social Sciences, T. Graham, 1987. 401 p.: ill. (HC240.9.E5 E6 1987)

Environmental policies in Finland. Paris: Organisation for Economic Co-operation and Development, 1988. Distributed by OECD Publications and Information Centre, Washington, D.C. 230 p.: ill. (HC340.2.Z9 E518 1988)
"Finland's economic growth in the 1970s and 1980s was stronger than that of most other European countries, and resulted in increased pressures on natural resources and the environment. The policies adopted to counter these pressures are characteristically pragmatic, but this report . . . recommends that they be implemented more effectively. It also stresses the importance of integrating environmental concerns more closely with economic development. The report concludes that other OECD countries could usefully learn from some Finnish experiences."

Environmental policies in Greece. Paris: Organisation for Economic Co-operation and Development, 1983. 140 p.
"Reducing serious air, noise, and water pollution in cities, particularly Athens and Thessaloniki, where the country's people and industry are concentrated is the main challenge. Avoiding serious problems to most of the rest of the country where the state of the environment is reasonably good is the main opportunity."

European Community and the environment. 3d ed. Luxembourg: Office for Official Publications of the European Communities, 1987. 71 p.: ill. (HC240.9.E5 E82 1987)

European environment policy: air, water, waste management. Brussels: Economic and Social Committee, 1987. Distributed by European Community Information Service, Washington, D.C. 48 p.: ill. (HC240.9.E5 E84 1987)

Flow, B. *The environmental crisis in the GDR*. Munich: Radio Free Europe, 1984. 7 p. (RAD background report/164 (Eastern Europe))
> "This paper analyzes the background of East Germany's acute ecologic problems; the regime's efforts to tackle them, both domestically and internationally; and the reasons for its failure to do so successfully."

Irish National Committee for European Year of the Environment: Ireland report on European Year of the Environment: 21st March 1987-20th March 1988. Dublin: Irish National Committee for European Year of the Environment, 1988. 44 p.: ill. (HC260.5.Z9 E55 1988)

Michalowski, Stanislaw. *Cooperation in environment protection between Eastern and Western Europe--present situation, prospects and needs*. Warsaw: Polish Institute of International Affairs, 1990. 22 p. (HC240.9.E5 M52 1990)

National report 1987. Helsinki, Finland: Ministry of the Environment, Environmental Protection Department, 1988. 361 p.: ill., maps (HC340.2.Z9 E523 1988)

OECD environmental data: compendium 1991. Paris: Organisation for Economic Co-operation and Development, 1991. 337 p.
> Contains data collected by means of the OECD/Eurostat questionnaire; supplements OECD report on the state of the environment. Discusses air; inland waters; land; forest; wildlife; solid waste; risks; pressures on the environment; energy; transport; industry; agriculture; managing the environment; responses; and general data.

Promise and performance: Irish environmental policies analysed. Dublin: Resource and Environmental Policy Centre, University College Dublin, 1983. 434 p.: ill. (HC260.5.Z9 E57 1983)

Regional strategy for environmental protection and rational use of natural resources in ECE member countries covering the period up to the year 2000 and beyond: as adopted by the Economic Commission for Europe at its forty-third session (1988) in decision E (43). New York: United Nations, 1988. 52 p. (HC240.9.E5 U55 1988)

Research on global change in the Netherlands: opportunities for research in the Netherlands in an international framework. Rijswijk: Raad voor het Milieu- en Natuuronderzoek, 1990. 90 p.: ill., maps (GF593 .R47 1990)

Strategy for industrial and urban development: plan of action for integration of environmental considerations into Danish development assistance. Copenhagen, Denmark: DANIDA, 1989. 70 p. (HT395.D44 S77 1989)

Sweden national report to UNCED 1992: United Nations Conference on Environment and Development. Stockholm, Sweden: Ministry of the Environment, 1991. 95 p.: color ill., color maps (HC380.E5 S96 1991)

Taylor, Ann. "Labour's environment protection executive." London: Fabian Society, 1991. 19 p. (HC260.E5 T39 1991)

Tendron, G. *The problems encountered by local and regional authorities as a result of their new tasks in respect of environmental protection: report*. Strasbourg, France: Council of Europe, 1986. 29 p. (HC240.9.E5 T46 1986)

Webb, Adrian Leonard. *The future for U.K. environment policy: key players, issues, and implications*. New York: Economist Intelligence Unit, 1991. 88 p. (HC260.E5 W43 1991)

Conference proceedings

Environmental protection, control, and monitoring: proceedings of the ISA International, European Region Conference, May 22-24, 1991, Birmingham, England. Research Triangle Park, N.C.: ISA International, 1991. 204 p.: ill. (TD193 .I73)

Environmental technology: proceedings of the Second European Conference on Environmental Technology, Amsterdam, the Netherlands, June 22-26, 1987. Boston: M. Nijhoff, 1987. 826 p.: ill. (TD55.A1 E97 1987)

Bibliographies

Hogg, Catherine. *National parks and other protected areas in western Europe: a selected bibliography*. Monticello, Ill. Vance Bibliographies, 1987. 27 p. (Z7405.N38 H65 1987)

Needham, Barrie. *Physical planning and environmental policy in the Netherlands: a guide to English-language publications*. Chicago, Ill.: Council of Planning Librarians, 1992.

Reference

Deziron, Mireille. *A directory of European environmental organizations*. Cambridge, Mass.: Blackwell Reference, 1991.

Milner, J. Edward. *The green index: a directory of environmental organisations in Britain and Ireland*. London, England: Cassell, 1990. 366 p. (TD171.5.G7 M55 1990)

EASTERN EUROPE AND FORMER SOVIET UNION

Books (annotated)

DeBardeleben, Joan, ed. *To breathe free: Eastern Europe's environmental crisis.* Washington, D.C.: Woodrow Wilson Center Press, 1991. 266 p. maps. (TD171.5.E86 T6 1991)
    Written just before the fall of the Communist bloc, the fourteen essays which comprise this work focus on the environmental problems, government responses, and environmental movements in Eastern Europe. First presented at a 1987 conference, these essays offer information on Bulgaria, the German Democratic Republic, Yugoslavia, Poland, Hungary, and Romania.

Feshbach, Murray; and Friendly, Alfred Jr. *Ecocide in the U.S.S.R: health and nature under siege.* New York: Basic Books/Aurum Press, 1992.
    Discusses the adversarial view the former Soviet Union had of nature. The environmental consequences of this attitude are discussed in the context of the historical, political, and social complexities of Soviet life. Analyzes the link between the environment and public health, employing statistics like a decline in life expectancy and a rise in infant mortality. Maintains that the Chernobyl nuclear accident may lead to some policy changes and greater awareness.

Johnston, R. J. *Environmental problems: nature, economy, and state.* London: Belhaven, 1989. 211 p.
    Although his argument is somewhat diffused by the changes in Eastern Europe and the Soviet Union after the book was published, Johnston nonetheless presents sound theories about why neither Capitalist countries, Socialist states, or developing nations have a true interest in "conserving the environment for the ultimate good of all. . ." His theoretical, neoMarxist approach is of interest and value to specialists.

Marcinkiewicz, Jan. *Pollution in the heart of Europe.* London: Polish Society of Arts and Sciences Abroad, 1987. 56 p.
    Examines environmental pollution in Poland and the risks it is posing to health, architecture, and the environment. Makes a strong plea for action.

Pryde, Philip R. *Environmental management in the Soviet Union.* New York: Cambridge University Press, 1991. 314 p.: ill., maps (HC340.E5 P79 1991)
    Analyzes environmental problems in the Soviet Union, which have reached crisis proportions and threaten the population. Portrays a bureaucracy that lacks the competence to deal with problems such as the shrinking of the Aral Sea, the phosphorate moonscapes of Estonia, the Chernoybl accident, and carbon monoxide pollution. The disintegration of the Soviet Union has intensified the problem, although in some areas, such as water conservation, the individual states might be better equipped to take action. An excellent overview and analysis.

Turnbull, Mildred. *Soviet environmental policies and practices: the most critical investment.* Brookfield, Vt.: Dartmouth Publishing, 1991. 240 p. (HC340.E5 T87 1990)

> Analyzes municipal wastewater treatment, the pulp and paper industry, agriculture, fossil-fueled electric power, and other heavy industries in the Soviet Union to suggest how environmental management practices could be changed. Written before the breakup of the Soviet Union, this study demonstrates how difficult it will be for the independent states to deal with a massive problem.

Yablokov, A. V.; and Ostroumov, S. A. *Conservation of living nature and resources.* Berlin: Springer-Verlag, 1991. 271 p.

> Presents conservation ideas, scientific research methods, and monitoring from a Russian perspective. Contains data not available outside the Soviet Union, such as the effects of introducing ten fish species on seven native species in Lake Balkash between 1910 and 1970.

Ziegler, Charles E. *Environmental policy in the U.S.S.R.* New York: Frances Pinter/University of Massachusetts Press, 1987. 195 p.: ill., map (HC340.E5 Z53 1987)

> Focuses on Soviet environmental law and the bodies which govern it. Examines the problem of conservation and the difficulties created by the Communist system. Discusses well-known problems such as Lake Baikal and Chernobyl.

Books (unannotated)

Astanin, L. P. *Conservation of nature.* Moscow: Progress, 1983. 148 p. (QH75 .A78 1983)

DeBardeleben, Joan. *The environment and Marxism-Leninism: the Soviet and East German experience.* Boulder, Colo.: Westview Press, 1985. 338 p. (HC710.E5 D43 1985)

Directory for the environment: organisations in Britain and Ireland, 1984-5. Boston: Routledge and Kegan Paul, 1984. 281 p. (TD169.6 .D568 1984)

Environmental problems in the Soviet Union and Eastern Europe. Boulder, Colo.: L. Rienner, 1987. 208 p.: ill. (TD171.5.S65 E58 1987)

Environmental protection and society. Moscow: USSR Academy of Sciences, 1983. 233 p. (HC79.E5 E5789 1983)

From below: independent peace and environmental movements in Eastern Europe and the USSR. New York: U.S. Helsinki Watch Committee, 1987. 263 p. (JX1952 .F868 1987)

Gustafson, Thane. *Reform in Soviet politics: lessons of recent policies on land and water.* New York: Cambridge University Press, 1981. 218 p., 1 leaf of plates: maps (JN6526 1981 .G87)

Jancar-Webster, Barbara. *Environmental management in the Soviet Union and Yugoslavia: structure and regulation in federal Communist states.* Durham, N.C.: Duke University Press, 1987. 481 p.: ill. (HC340.E5 J35 1987)

Man and biosphere: leafing through the Priroda Journal. Moscow: General Editorial Board for Foreign Publications, Nauka Publishers, 1984. 247 p.: ill. (QH75 .M343 1984)

Soviet environment: problems, policies, and politics. New York: Cambridge University Press, 1992. 245 p.: ill. (HC340.E5 S68 1991)

To breathe free: Eastern Europe's environmental crisis. Washington, D.C.: Woodrow Wilson Center Press; Baltimore: Johns Hopkins University Press, 1991. 266 p.: maps (TD171.5.E86 T6 1991)

Weiner, Douglas R. *Models of nature: ecology, conservation, and cultural revolution in Soviet Russia.* Bloomington: Indiana University Press, 1988. 312 p.: ill. (QH77.S626 W45 1988)

General interest periodicals (annotated)

Petrovsky, Vladimir. "Pure air, money, and secrecy." *New Times* 32 (August 8-14, 1989): 38-40.
> Deputy Foreign Minister Vladimir Petrovsky of the Soviet Union is interviewed by *New Times.* "The U.S.S.R. Foreign Ministry is about to launch an extensive international campaign in defence of the environment. Diplomats will have to run the campaign both abroad and at home--against those agencies that defy international standards."

Thompson, Jon. "East Europe's dark dawn: the Iron Curtain rises to reveal a land tarnished by pollution." *National geographic* 179 (June 1991): 36-37, 42-49, 52-54, 58-59, 61-65, 67-68.
> "While the world wasn't looking, Eastern Europe's regimes poisoned their environment in the name of progress. Now new leaders must assess the damage and set priorities for reversing it."

General interest periodicals (unannotated)

Balaban, Vladislav. "Danube dam plans stir unprecendented protest". (Gabcikovo-Nagymaros dam; includes related article) *Alternatives* 15,4 (Nov-December 1988): 8(2)

Bobkov, Vyacheslav; and Kaver, Yuri. "Megalopolis: an ecology round table." *Soviet Life* 4 (April 1990): 44(6)

Ellis, William S.; and Turnley, David C. "A Soviet sea lies dying: the Aral." *National Geographic* 177,2 (February 1990): 70(24)

Hofheinz, Paul. "The new Soviet threat: pollution." *Fortune* 126,2 (July 27 1992): 110(4)

Kondakov, Yevgeni. "A village inside Moscow." *Soviet Life* 8 (August 1990): 23(4)

Lang, Istvan. "Hungary: the pitfalls of growth." *UNESCO Courier* (November 1991): 25(3).

Majka, Christopher. "Environmental politics: lessons from the East bloc." *Canadian Dimension* 24,5 (July-August 1990): 36(5)

"Town without smog. (Sumy, Ukrainian industrial town) *Soviet Life* 6 (June 1988): 25(2)

Environmental and professional journals (annotated)

Bolan, Richard S. "Organizing for sustainable growth in Poland." *Journal of the American Planning Association* 58 (Summer 1992): 301-311.
> "Following the end of Communist rule in 1989, Poland has been facing the difficult challenge of responding to the legacy of a devastated environment amid an overwhelming economic crisis. This article describes how Poland has been responding to this challenge, with particular focus on the country's policy of sustainable development within the framework of a transition to a democratic market economy. The article discusses achievements and difficulties in seeking both effective environmental protection and economic restructuring since 1989 and examines the country's future needs and problems."

Budnikowski, Adam. "Foreign participation in environmental protection in Eastern Europe: the case of Poland." *Technological Forecasting and Social Change* 41 (March 1992): 147-160.
> Discusses Eastern Europe's environmental pollution and their inability to adequately protect the environment without foreign ecological assistance. Analyzes Poland's participation with other countries to clean up its environmental damage.

Charles, Daniel. "East German environment comes into the light." *Science* 247 (January 19, 1990): 274-276.
> "Environmental issues--particularly air pollution--are finally getting attention after years of political repression and neglect; an industrial complex near Leipzig indicates the extent of the change."

French, Hilary F. "Eastern Europe's clean break with the past." *Worldwatch* 4

(March-April 1991): 21-27.

"Under the assault of air pollution and acid deposition, Eastern Europe's medieval cities are blackened and crumbling, whole hillsides are deforested, and crop yields are falling. Rivers serve as open sewers, and clean drinking water is in short supply. Life expectancies in the dirtiest parts of the region are as much as five years shorter, and rates of cancer, reproductive problems, and other ailments far higher than in relatively clean areas."

French, Hilary F. "Environmental problems and policies in the Soviet Union." *Current History* 90 (October 1991): 333-337.

"The Soviet Union faces an environmental crisis that has only recently been openly acknowledged and acted on by the government. Enormous clean-up costs, inexperience with pollution control technology, and a government focused on reforming itself and the economy mean that the country 'faces a long road to environmental recovery.' "

French, Hilary F. "Green revolutions: environmental reconstruction in Eastern Europe and the Soviet Union." *Columbia Journal of World Business* 26 (Spring 1991): 28-51.

"Eastern Europe and the Soviet Union have a unique opportunity to cull from 20 years of western experience in addressing environmental problems. They could 'leapfrog' the West in both policies and technologies. By implementing the strategies that have proved most effective, Eastern Europe and the Soviet Union could realize the greatest possible return on environmental investments."

Hinrichsen, Don. "Will the sun ever shine on Budapest." *International Wildlife* 19 (September-October 1989): 18-22.

Describes the air and water pollution in Hungary and the interest of other Eastern bloc nations in Hungary's liberalized atmosphere and attempts at environmental reform.

Junkin, Elizabeth Darby. "Green cries from Red Square." *Buzzworm: The Environmental Journal* 2 (March-April 1990): 28-33.

Reports on the growth of the environmental movement in the Soviet Union and the ecological catastrophes that challenge the movement. Includes coverage of the First All-Union Conference on Agriculture and the Environment held in Moscow, November 1989.

Kabala, Stanley J. "Czechoslovakia: the reform of environmental policy." *Report on Eastern Europe* 2 (February 22, 1991): 10-14.

"Like most East European countries, Czechoslovakia inherited from its collapsed Communist regime a legacy of environmental degradation. Decades of centrally planned economic development predicated on growth in industrial output has left the country with air, water, and land polluted to such an extent that the health of the population is in danger."

Kabala, Stanley J. "EC helps Czechoslovakia pay 'debt to the environment.' " *Radio Free Europe/Radio Liberty Research Report* 1 (May 15, 1992): 54-58.

"In 1991 the EC's Polish and Hungarian Assistance for the Reconstruction of Europe program made 25.5 million ecu ($34 million) available for environmental protection projects in Czechoslovakia, which range from assessing the country's hazardous waste problem and its air quality control to upgrading safety provisions at nuclear power plants and evaluating the environmental impact of the controversial Gabcikovo-Nagymaros dam project."

Kabala, Stanley J. "Economic growth and the environment in Yugoslavia: an overview." *Ambio* 17 (no. 5, 1988): 323-329.

"Yugoslavia has not yet experienced the acute generalized effects of development-induced air and water pollution that more heavily industrialized countries in Europe are undergoing. However, economic growth, based on increased consumption of energy and materials, remains the country's goal, and if only a few of the planners' projections are realized Yugoslavia will have created for itself an environmental problem similar to that of its European neighbors." Reports on existent pollution problems and comments on prospects for environmental protection policies.

Kabala, Stanley J. "The environment and economics in Upper Silesia." *Report on Eastern Europe* 2 (August 16, 1991): 18-23.

"Upper Silesia is one of the most heavily polluted regions in the world. The extensive coal mining, steelmaking, and other heavy industries that made the region the powerhouse of Poland have also devastated the landscape and resulted in environmental deterioration so acute that the health of the population is threatened. The region is responding in a fundamental way by attempting a thorough restructuring of its pollution-prone, resource-intensive industry."

Kabala, Stanley J. "The environmental morass in Eastern Europe." *Current History* 90 (November 1991): 384-389.

"Cleaning up Eastern Europe's polluted land, air, and water will require years of effort and large infusions of aid. The cleanup must also compete with consumer demands for quality goods and services. 'The question for Eastern Europe is how to chart paths of economic redevelopment that both supply the peoples' needs and respect the environment."

Kabala, Stanley J. "The hazardous waste problem in Eastern Europe." *Report on Eastern Europe* 2 (June 21, 1991): 27-33.

Focuses on the need to clean up existing industrial waste sites and make sure that "the growing market for consumer goods and the introduction of new industrial technologies will not generate even more hazardous waste, which these economies are ill-equipped to handle."

Kabala, Stanley J. "Hungary: the environment: adjustments to a new reality." *Report*

on Eastern Europe 2 (January 11, 1991): 14-17.
> "The Hungarian environmental movement was largely responsible for mobilizing the political opposition that displaced the Communist regime. The movement now finds itself struggling with the fruits of this success."

Marshall, Patrick G. "The greening of Eastern Europe: devastated countries try to reverse communism's environmental toll." *CQ Researcher* 1 (November 15, 1991): whole issue (849-872 p.)
> The environment "suffered greatly under centrally planned economies that put a priority on production quotas and developing heavy industries while providing no incentives for conservation or pollution controls. Now two huge tasks confront Eastern Europeans as they try to bring order to their countries: They must energize crippled economies and find the resources to clean up four decades of profligate pollution." Some experts maintain that Western countries are not providing enough aid.

Mazurski, Krzyszof R. "Industrial pollution: the threat to Polish forests." *Ambio* 19 (April 1990): 70-74.
> Presents the consequences of Poland's deforestation and air pollution resulting from growing industrialization and urbanization.

"Migration to and from the Baltic States." *Report on the USSR* 2 (September 14, 1990): 20-33.
> Contains the following essays: "Migration to and from Estonia," by Riina Kionka; "Migration to and from Lithuania," by Saulius Girnius; and "Migration to and from Latvia," by Dzintra Bungs.

Miller, Marc S. "A Green wind hits the East." *Technology Review* 93,7 (October 1990): 52(12)
> Interviews with Imre Szabo, Bedrich Moldan, and Janos Vargha.

"Panel on nationalism in the USSR: environmental and territorial aspects." *Soviet Geography* 30 (June 1989): 441-509.
> "A panel of geographers, demographers, and political scientists discusses a broad range of issues related to the resurgence of nationalism in the USSR and its relationship to environmental protest and territorial disputes: the emergence of nationality politics; differential rates of nationality population growth and urbanization; various conceptions of ethnic homelands; the spatial pattern of actual and potential territorial claims."

"Policy issues environmental problems in Eastern Europe." *Forum for Applied Research and Public Policy* 6 (Winter 1991): 30-65.
> Contents: "Environmental legacy mixed in Eastern Europe," by Richard A. Liroff; "Communism and the environment," by Krzysztof R. Mazurski; "German reunification brings clean-up tasks," by Jeffrey H. Michel; "State of environment

in Czechoslovakia," by Nadja Johanisova; "Freedom, environment linked in Bulgaria," by Youli Kanev; and "Hungary seeks to lift blight of pollution," by Tamas Fleischer and Janos Vargha.

Pehe, Jiri. "A record of catastrophic environmental damage." *Report on Eastern Europe* 1 (March 23, 1990): 4-8.
"The Czechoslovak media have recently reported on some startling cases of gross ecological neglect under the ousted regime. . . it may take years before there is any visible improvement."

Rafikov, A. A. "Environmental change in the southern Aral region in connection with the drop in the Aral sea level." *Soviet geography: review and translation* 24 (May 1983): 344-353.
Describes the progressive desertification of the south shore of the Aral Sea as the inflow of water has greatly declined because of increased irrigation.

Rodgers, Lori M. "An inside look at Soviet and Czechoslovakian energy concerns." *Public Utilities Fortnightly* 126 (July 19, 1990): 37-39, 54.
Presents a first-hand report from ministers representing the Soviet Union and Czechoslovakia about energy policies and environmental concerns.

"Russia's greens: the poisoned giant wakes up." *Economist* 313 (November 4, 1989): 23-24, 26.
"The Soviet constitution says 'citizens of the U.S.S.R. are obliged to protect nature and conserve its riches.' They don't. But ever more of them think they should." Reports on efforts by the Soviet Greens.

Ryzhikov, A. I. "The size of nature reserves and costs of their maintenance." *Soviet Geography* 29 (December 1988): 917-925.
Discusses the decrease in nature reserve size in the USSR, and that many "are too small to carry out their intended functions properly. In fact, until quite recently, in spite of a dramatic expansion of the total area and number of nature reserves, average reserve size has decreased."

Tarnoff, Curt. "Eastern Europe and the environment." *Congressional Research Service Review* 11 (March-April 1990): 29-31.
Describes the relationship between the democratization process in Eastern Europe and the political concern over the deteriorating environment. Emphasizes the opportunities for western nations to help Eastern European countries through expanded trade relations and improved environmental security.

Tellegen, Egbert. "Perestroika and the rational use of materials and energy." *Environmentnal Professional* 11 (January 1989): 142-151.
Outlines "the high material and energy intensity of the Soviet economy and symptoms of waste as they are described in the Soviet press. Discusses the modest

role of prices, the lack of cooperation among different branches of the economy, the irrationality of planned targets, and the shortcomings of environmental measures of economic reform are discussed from this perspective."

Thompson, Mark. "In the laboratory: notes from Slovenia." *END Journal of European Nuclear Disarmament* 32 (February-March 1988): 17-21.
"Over recent years, peace and environmental activists in the Yugoslav republic of Slovenia have been achieving extraordinary things, providing perhaps the clearest instance in Europe of how political and social change can be worked 'from below' in authoritarian conditions."

Wolfson, Zee "Can the West help save the Soviet environment?" *Report on the USSR* 1 (August 11, 1989): 16-17.
Describes how western technology is being used to clean up and monitor environmental pollution in the Soviet Union. Questions whether the Soviet economy is capable of absorbing the sophisticated technologies it requires.

Ziegler, Charles E. "The bear's view: Soviet environmentalism." *Technology Review* 90 (April 1987): 45-51.
"In many ways the Soviet Union's philosophical perspective on the environment is similar to that of capitalist systems. Both mainstream Soviets and fiscal conservatives in the West believe that social conditions will improve through economic growth, and they therefore tend to minimize the environmental impact of development" and the Chernobyl disaster has not stopped the Soviets from finishing a larger nuclear plant with the same design 300 miles away.

Law journals

Jastrzebski, Ludwik; and Rest, Alfred. "Environmental protection in the People's Republic of Poland." *Environmental Policy and Law* 8 (January 15, 1982): 21-28.
The authors provide a brief history of environmental protection in Poland from 1918 to 1980. They outline the provisions of significant environmental legislation.

Zaharchenko, Tatiana. "The environmental movement and ecological law in the Soviet Union: the process of transformation." *Ecology Law Quarterly* 17,3 (1990): 455-475.
This article "surveys some of the changes and challenges in Soviet laws concerning the environment (referred to as ekologicheskoye pravo or ecological law)", reviewing "the Soviet Union's grave environmental problems and the new prominence of environmental issues on the Soviet political agenda."

Reports

Environmental policies in Yugoslavia. Paris: Organisation for Economic Co-operation and

Development, 1986. 160 p., 2 p. of plates: ill. (some color) (HC407.Z9 E555 1986)
This review, undertaken at the request of the Yugoslavian government, "was prompted by serious concern about the detrimental environmental effects of rapid economic development in the postwar period, by the government's determination to develop measures to improve conditions, and by its conviction that sharing experience with OECD Member countries in this area would be of considerable value to Yugoslavia."

Environmental projects in Czechoslovakia. Copenhagen, Denmark: Industriog handelsstyrelsen, 1991. 1 v. : ill., maps (TD171.5.C95 E6 1991)

Enyedi, Gyorgy. *Environmental policy in Hungary.* Pecs: Centre for Regional Studies of Hungarian Academy of Sciences, 1986. 38 p.: ill. (HT395.H8 D57 no. 2)

Evans, Sheila S. *Environmental activities in the Czech and Slovak Federal Republic, 1991.* New York (888 Seventh Ave., 19th floor New York 10106): Charter 77 Foundation, 1992. 50 p. (HC270.295.E5 E93 1992)

Expertise of the Polish team of experts, MAB 13a. Warsaw: Publishing House of the Polish Academy of Sciences, 1988. 75 p.: ill.

Fitzpatrick, Catherine; and Fleischman, Janet. *From below: independent peace and environmental movement in Eastern Europe and the USSR.* New York: U.S. Helsinki Watch Committee, 1987. 263 p.
Discusses Czechoslovakia, East Germany, Hungary, Poland, USSR, and Yugoslavia.

Impressions: essays on ecology and the public sphere in the Soviet Union. Dayton, Ohio (200 Commons Rd., Dayton 45459-2799): Kettering Foundation, 1991. 29 p.: ill. (HD340.E5 I46 1991)

Pohl, Frank. *Environmental deterioration in Czechoslovakia.* Munich Radio Free Europe, 1983. 15 p. (Research: RAD background report/95 (Czechoslovakia))
Contents include: "Legislation;" "Air pollution;" "Water pollution;" "Forests;" "Agriculture;" "Human health;" and "Conclusion."

Reforming the economies of Central and Eastern Europe. Paris, Organisation for Economic Co-operation and Development, 1992. 119 p.
"This is a comprehensive survey of the main issues facing central and eastern European countries in their transition to market economies. It provides an appraisal of economic and social reform in Poland, Hungary, the Czech and Slovak Federal Republic and, with less detail, in Bulgaria, Romania and the former USSR. Although these countries have started introducing reforms, progress has been slow in building the institutional framework of a market economy, in developing the necessary management skills, and in transforming attitudes of the

general public. The report argues that OECD governments can best contribute to these countries by ensuring greater access to OECD market and by improving the focus and co-ordination of their assistance efforts."

Sobell, Vladimir. *The ecological crisis in Eastern Europe.* Munich: Radio Free Europe, 1988. 17 p. (RAD background report/5 (Eastern Europe)
"The East European countries are much worse polluters of the environment than the West European countries in absolute terms, and they cause much more pollution than the western countries per unit of gross domestic product. This is directly or indirectly attributable to their system. The extent of ecological damage varies from country to country but it is a source of concern throughout the region."

Ziegler, Charles E. *Policy alternatives in Soviet environmental protection.* Pittsburgh, Pa.: Russian and East European Studies Program, University of Pittsburgh 1982. 28 p. (The Carl Beck Papers in Russian and East European Studies. Paper no. 102)
Describes proposals by Soviet economists to use economic incentives rather than administrative actions to encourage environmental protection. Outlines the Soviet government's response to pollution problems, and reviews environmental laws.

Conference proceedings

Ecological security for the common European home: statements by participants in the CSCE environment meeting and representatives of international and non-governmental organizations and movements, Sofia, 16 October-3 November, 1989. Sofia, Bulgaria: Sofia Press, 1990. 109 p.: ill. (HC240.9.E5 M44 1989)

Finnish-Soviet symposium on national parks and nature reserves. Helsinki: Maaja metsatalousministerio, 1983. 89 p.: ill. (QH77.F55 F59 1983)

Focus on Estonia and the Baltic Sea: research, economy, environment: proceedings from an international seminar held in Stockholm, Tallinn, and Visby, July 27-29, 1990. Uppsala, Sweden: Industrial Liaison Office, Uppsala University, 1990. 114 p. (HC340.4.Z9 E54)

Federal government reports

Eastern European telecommunications, broadcasting, and environment: report of a staff study mission to Hungary, Czechoslovakia, and Poland, November 4-17, 1990, to the Committee on Foreign Affairs, U.S. House of Representatives.. Washington, D.C.: G.P.O., 1991. 21 p. (KJC6964 .E18 1991)

Implementation of the Helsinki accords: hearing before the Commission on Security and Cooperation in Europe, One Hundredth Congress, second session, politics of pollution in the Soviet Union and Eastern Europe (on the second anniversary of the Chernobyl disaster), April 26, 1988. Washington, D.C.: G.P.O., 1988. (HC340.P55 U55 1988)

Bibliographies

Salay, Jurgen. *Literature on environment and energy in Eastern Europe and the Soviet Union: a preliminary bibliography*. Uppsala, Sweden: Uppsala University, 1991. 21 p. (Z5863.E54 S25 1991)

MIDDLE EAST

Environmental and professional journals (annotated)

Land, Thomas. "Environmental problems in the Middle East. *Middle East* 182 (December 1989): 5-10.
> "The world is slowly waking up to the enormity of the fast approaching global environmental crisis and concern is at last growing in the Middle East. Air and water pollution are already wreaking extensive damage; the Greenhouse effect will only compound the region's problems." The author "surveys what is happening to the Middle East environment and the efforts being taken to stop the rot."

Reports

Whitman, Joyce. *The environment in Israel*. 4th ed. Jerusalem: State of Israel, Ministry of the Interior, Environmental Protection Service, 1988. 294 p., 1 folded leaf of plates: ill., maps (TD171.5.I75 W47 1988)

Conference proceedings

Basic needs in the Arab region, environmental aspects, technologies, and policies: report and background papers of an expert workshop. Nairobi: United Nations Environment Programme, 1982. 216 p. (HC498.Z9 B383 1982)

Environment and development: opportunities in Africa and the Middle East: conference summary, September 25-27, 1985. New York: WEC, 1986. 119 p.: ill. (HC800 .E58 1986)

ASIA

Books (annotated)

Bruton, Henry J. *Sri Lanka and Malaysia*. Oxford: Oxford University Press, 1992. 422 p.

This World Bank study analyzes poverty, equity, and growth in these countries from 1950 to 1980. Focuses on how economic growth affects poverty and equity.

Ives, Jack D.; and Messerli, Bruno. *The Himalayan dilemma: reconciling development and conservation.* New York: Routledge, Chapman, and Hall, 1989. 302 p.
Argues against other researchers; maintains that the human role in the degradation of the Himalayas has been minimal and is not responsible for the 1988 flooding of Bangladesh. Asserts instead that a lack of understanding about complex mountain dynamics has perpetrated simplistic solutions to environmental problems, including relating environmental degradation to economic indicators.

Ji, Zhao, ed. *The natural history of China.* New York: McGraw-Hill, 1990.
An informative introduction to the geography, animals, and plants of China. Describes desert, forest, seacoast, mountain, and river habitats. The impact of human use of natural resources is discussed, as well as various conservation programs. Contains over two hundred color plates and maps. Contains appendices and an index.

Ross, Lester; and Silk, Mitchell A. *Environmental law and policy in the People's Republic of China.* Westport, Conn.: Quorum, 1987. 449 p.
Chronicles China's awakening to environmental problems and presents the viewpoints of many involved individuals, including environmentalists, scholars, jurists, and administrators. Overviews Chinese environmental law with a presentation of basic principles. Discusses environmental litigation and individually examines several specific problem areas, including air and water pollution, marine resource conservation, noise control, and toxic and solid waste disposal. The goals of Chinese conservation laws are also outlined. Presents case studies which help to delineate Chinese environmental jurisprudence.

Books (unannotated)

Bhutan and its natural resources. New Delhi: Vikas Publishing House, 1991. 149 p.: ill. (HC440.25.Z9 E53 1991)

Burnett, Alan Alexander. *The Western Pacific: challenge of sustainable growth.* Brookfield, Vt.: E. Elgar, 1992.

Chelliah, T. *Environmental training programmes and policies in ASEAN: an overview.* Konigswinter: Freidrich Naumann Stiftung, 1985. Distributed by Regional Institute of Higher Education and Development, Singapore. 165 p. (TD157.4.A785 C48 1985)

Community management: Asian experience and perspectives. West Hartford, Conn.: Kumarian, 1986. 328 p.: ill. (HD75.6 .C65 1986)

Conservation atlas of tropical forests: Asia and the Pacific. New York: Simon and Schuster, 1991. 1 atlas (256 p.): ill. (chiefly color), maps (chiefly color) (G2201.K3 C6 1991)

Conserving Indian environment. Jaipur, India: Pointer, 1992. 187 p. (S934.I4 C66 1992)

Cubitt, Gerald S. *Wild India: the wildlife and scenery of India and Nepal.* Cambridge, Mass.: MIT Press, 1991. 208 p.: chiefly color ill., maps (QH183 .C83 1991)

Danusaputro, St. Munadjat. *The marine environment of South-East Asia.* Bandung: Binacipta, 1981. 325 p.: maps (HC441.Z9 E524)

Dewan, M. L. *People's participation in Himalayan eco-system development: a plan for action.* New Delhi: Concept Publishing, 1990, 1989. 274 p.: ill., maps (HC448.H56 D49 1990)

Ecosystem degradation in India. New Delhi: Ashish Publishing House, 1990. 477 p.: ill., maps (HC440.Z9 P553 1990)

Environment planning and management in India. New Delhi: Ashish Publishing House, 1990. 2 v.: ill., maps (TD171.5.I4 E57 1990)

Environmental management in Southeast Asia: directions and current status. Singapore: Faculty of Science, National University of Singapore, 1987. 211 p.: ill., maps (HC441.Z9 E53 1987)

Environmental protection and coastal zone management in Asia and the Pacific. Tokyo: University of Tokyo Press, 1985. 244 p. (TD171.5.A7842 E58 1985)

Experiences from the Kelang Valley Region, Malaysia. Bangi: Universiti Kebangsaan Malaysia, 1987. 606 p.: ill., maps (TD883.7.M4 S53 1987)

Gian Singh. *Environmental deterioration in India: causes and control: with special reference to Punjab.* New Delhi, India: Agricole Publishing Academy, 1991. 266 p.: ill., maps (TD171.5.I42 P864 1991)

Himalayan ecology. New Delhi: Ashish Publishing House, 1989. 169 p., 7 p. of plates: ill., maps (QH193.H5 H553 1989)

Himalayas, ecology and environment. Delhi, India: Mittal Publications, 1988. 183 p. (QH193.H5 H554 1988)

Himalayas, environmental problems. New Delhi: Ashish Publishing House, 1990. 190 p. (HC437.H54 H54 1990)

Indonesia: resources, ecology, and environment. New York: Oxford University Press, 1991. 262 p.: ill. (HC447.5 .I627 1991)

Jack, D. *The Himalayan dilemma: reconciling development and conservation.* Tokyo, Japan: United Nations University; New York: Routledge, 1989. 295 p.: ill. (GF696.H55 I94 1989)

Khator, Renu. *Environment, development, and politics in India.* Lanham, Md.: University Press of America, 1991. 247 p.: ill. (HC440.E5 K43 1991)

Learning from China?: development and environment in Third World countries. Boston: Allen and Unwin, 1987. 282 p. (HC430.E5 L43 1987)

Lee, David W. *The sinking ark: environmental problems in Malaysia and Southeast Asia.* Kuala Lumpur: Heinemann, 1980. 85 p., 8 p. of plates: ill. (some color) (GF668 .L43)

Natural heritage of India: essays on environment management. Delhi: H. K. Publishers, 1989. Distributed by Deep and Deep Publications, New Delhi. 237 p.: ill., plan (QH77.I4 N36 1989)

Negi, Sharad Singh. *Environmental degradation and crisis in India.* New Delhi: Indus Publishing, 1991. 344 p. (GF661 .N45 1991)

Ross, Lester. *Environmental law and policy in the People's Republic of China.* New York: Quorum Books, 1987. 449 p.

Ross, Lester. *Environmental policy in China.* Bloomington: Indiana University Press, 1988. 240 p. (HC430.E5 R67 1988)

Rush, James R. *The last tree: reclaiming the environment in tropical Asia.* New York: Asia Society, 1991. Distributed by Westview Press, Boulder Colo. 107 p.: ill. (HC441.Z9 E57 1991)

Shiva, Vandana. *Staying alive: women, ecology, and development.* London: Zed Books, 1988. 224 p.: ill. (HQ1240.5.I4 S54 1988)

Smil, Vaclav. *China's environmental crisis: an inquiry into the limits of national development.* Armonk, N.Y.: M.E. Sharpe, 1992.

Southeast Asia regional consultation on people's participation in environmentally sustainable development. Makati, Metro Manila, Philippines: ANGOC, 1990-1991. 2 v. (HC441. S72 1990)

General interest periodicals (unannotated)

Jervis, Nancy. "Waste not, want not." (Special issue: The Endless Cycle--how a Chinese village recycles) *Natural History* 99,5 (May 1990): 70(5)

Karaosmanoglu, Atilla. "Environment, poverty, and growth: the challenge of sustainable development in Asia." *Vital Speeches of the Day* 55,13 (April 15 1989): 396(5)

Kohl, Larry. "Heavy hands on the land." (environmental impact on Himalaya Mountain region) *National Geographic* 174,5 (November 1988): 632(20)

Environmental and professional journals (annotated)

Boxer, Baruch. "China's environmental prospects." *Asian Survey* 29 (July 1989): 669-686.
> Delineates the environmental problems arising from economic development and gives a short history of the country's efforts to cope with these problems and protect its environment.

Knatz, Geraldine; and Webber, Bert. "Environmental assessment in the PRC: third phase of the Qinhuangdao coal port." *Coastal Management* 19 (July-September 1991): 343-356.
> The use of environmental impact assessment (EIA) is relatively recent in the People's Republic of China (PRC). "Today large scale harbor developments, such as that of the coal port of Qinhuangdao, are subject to the requirements of the National Environmental Protection Law (1979) and the Marine Environmental Law (1982). This article describes the environmental regulations in the PRC that affect this harbor expansion, discusses the process by which the EWC (Environment and Policy Institute, East-West Center) assisted the PRC in the preparation of the environmental assessment and evaluates the assessment product against PRC regulations and United States standards."

Matthews, William H. "Emerging strategies for Asian rural resources management." *Journal of the World Resources Institute* (1985): 17-26.
> "The emerging concepts, themes, and strategies described here cut across many traditional disciplinary, professional, sectoral, and institutional boundaries that have both helped and impeded progress on Asia's critical problems."

Stone, Roger D. "Foresight in Indonesia's Irian Jaya Province." *World Wildlife Fund Letter* 3 (1989): 1-8.
> Describes WWF's sustainable development strategies in the Irian Jaya rain forests that benefit both the people and the unique plants and animals that live there.

Environmental and professional journals (unannotated)

Greanville, David P. "Environmental causalities: the case of China." (World Economy, Part 2) *Animals' Agenda* 11,10 (December 1991): 37(2)

Law journals

Mushkat, Roda. "International environmental law in the Asia-Pacific region: recent developments." *California Western International Law Journal* 20,1 (1989-1990): 21-40.
Examines regional initiatives and cooperation in protecting the environment, including regional response to developments in international environmental law on such issues as sustainable development, protection of the ozone layer, transboundary movement of hazardous wastes, marine pollution, and state responsibility for environmental damage.

Reports

Ahmed, Sadiq. *Fiscal policy for managing Indonesia's environment.* Washington, D.C. (1818 H St., NW, Washington 20433): Country Department V, Asia Regional Office, World Bank, 1991. 34 p.: ill. (HJ1374 .A36 1991)

Cribb, R. B. *The politics of environmental protection in Indonesia.* Clayton, Vic., Australia: Monash University, 1988. 36 p.: ill., maps (HC450.E5 C75 1988)

Environmental problems of the South Asian Seas region: an overview. Nairobi, Kenya: United Nations Environment Programme, 1987. 50 p.: maps (QH77.I45 E58 1987)

Morrison, Charles E.; and Dernberger, Robert F. *Focus: China in the reform era.* Honolulu, Hawaii: East-West Center, 1989. 125 p.
Partial topics include "China's political reforms"; "China's economic reforms"; "Demographic and social change in China"; "Environmental problems and economic modernization"; and "China's role in the Asia-Pacific region."

Pakistan: a guide to conservation and development issues. Gland, Switzerland: Conservation for Development Centre, IUCN, 1986. 72 p.: color ill. (HC440.5.Z65 P35 1986)

Ross, Lester. *Public choice and public policy in China: the relationship between microincentives and macrooutcomes in environmental policy.* n.p., 1984. 36 p. Prepared for delivery at the 1984 Annual Meeting of the American Political Science Association, the Washington Hilton, August 30-September 2, 1984.
"Demonstrates that the severe problems China has suffered in terms of deforestation, pollution, and the like have been due not merely to a Stalinist bias in favor of heavy industry on the part of a dictatorial minority. Even in a dictatorship policy depends not merely on formulation at the peak of the system but also upon the effective coordination and implementation of decisions at lower levels."

Sri Lanka national conservation strategy. Colombo: Central Environmental Authority, 1988. 160 p. (HC424.Z9 E572 1988)

Conference proceedings

Coastal zone environmental planning in the Strait of Malacca: towards sustainable development of the coastal resources of the east coast of Sumatra. Bogor: Development of Environmental Study Centres Project, 1989. 241 p.: maps (HT395.I58 S888 1988)

Proceedings of the Seminar on Environmental Pollution Control in Punjab, November 26, 1990. Lahore: Environmental Protection Agency, 1990. 41 leaves: color ill. (TD187.- 5.P182 P867 1990)

Sixth National Convention of Environmental Engineers and Seminar on Environment and Ecology: Indian scenario, Ranchi, India, 8th and 9th September, 1990. Ranchi, India: Institution of Engineers, Ranchi Local Centre, 1990. 511 p.: ill., maps (TD171.5.I4 N37)

Southeast Asia Regional Consultation on People's Participation in Environmentally Sustainable Development, 1990: Puncak Pass, West Java, Indonesia. 2 vols. Makati, Metro Manila, Philippines: ANGOC, 1990-1991. (HC441. S72 1990)

Sustainable rural development in Asia: proceedings of the SUAN IV Regional Symposium on Agroecosystem Research held at Khon Kaen University, July 4-7, 1988. Khon Kaen, Thailand: KKU-USAID Farming Systems Research Project, Khon Kaen University and Southeast Asian Universities Agroecosystem Network, 1988. 246 p.: ill. (S401 .S83 1988)

Bibliographies and Reference

Directory of environmental NGOs in the Asia-Pacific region. Penang, Malaysia: SAM, 1983. 257 p. (QH77.A75 D57 1983)

Mumtaz, Khawar. *Pakistan's environment: a historical perspective and selected bibliography with annotations.* Karachi, Pakistan: JRC: IUCN, 1989. 114 p.: maps (Z5863.P6 M85 1989)

JAPAN

Books (unannotated)

Apter, David Ernest. *Against the state: politics and social protest in Japan.* Cambridge, Mass.: Harvard University Press, 1984. 271 p.: ill. (HD1265.J32 N3713 1984)

Barrett, Brendan F. D. *Environmental policy and impact assessment in Japan.* New York: Routledge, 1991. 288 p.: ill. maps. (HC465.E5 B37 1990)

Environmental policy in Japan. Berlin: Edition Sigma, 1989. 601 p.: ill. (HC465.E5 E58 1989)

Environmental protection in the industrial sector in Japan: a survey of achievement. Tokyo: Industrial Pollution Control Association of Japan, 1983. Distributed by Maruzen. 168 p.: ill.

Huddle, Norie. *Island of dreams: environmental crisis in Japan*. Cambridge, Mass.: Schenkman Books, 1987. 355 p.: ill. (HC465.E5 H82 1987)

Reed, Steven R. *Japanese prefectures and policy making*. Pittsburgh, Pa.: University of Pittsburgh Press, 1986. 197 p. (JS7373.A3 R44 1986)

Upham, Frank K. *Law and social change in postwar Japan*. Cambridge, Mass.: Harvard University Press, 1987. 269 p.

General interest periodicals (annotated)

Gross, Neil. "Charging Japan with crimes against the Earth." *Business Week* 3127 (October 9, 1989): 108, 112.
> "At home, Japan has cleaned up its environment. But abroad, the Japanese continue to earn international ire. Its companies have failed to curb industrial pollution overseas and continue to support environmentally destructive aid projects."

General interest periodicals (unannotated)

Copulos, Milton R. "The environment: a North-South conflict." *Current* 317 (November 1989): 35(5)

Environmental and professional journals (annotated)

Cross, Michael. "Japan wakes up to the environment." *New Scientist* 118 (June 23, 1988): 38-39.
> "The world's second largest economy is probably the world's second worst polluter. But the Japanese barely noticed--until now. A new government report assesses Japan's environmental responsibilities."

Kato, Kazu. "Japan and the environment." *Environment* 31 (July-August 1989): 4-5.
> In a letter to the editor, the director of Japan's Office of Planning and Research in the Environment Agency responds to criticism of the 1988 White Paper on the Environment in Japan.

Miller, Alan S. "Three reports on Japan and the global environment." *Environment* 31 (July-August 1989): 25-29.

Reviews three Japanese reports on Japan and global environmental issues: "White paper on the environment in Japan 1988: Japan's contribution toward the conservation of the global environment"; "Japan's activities to cope with global environmental problems: Japan's contribution toward a better global environment"; and "Interim report on global warming."

Sand, Peter H. Innovations in international environmental governance." *Environment* 32 (November 1990): 16-20, 40-44.

Discusses what people can expect from global institutions for environmental management or "international governance" as some have called it. The author claims that "innovative approaches are necessary and possible in three areas: transnational environmental standard setting, transnational environmental licensing, and transnational environmental auditing."

Environmental and professional journals (unannotated)

Barnard, Charles N. "Ah, Fuji, Japan's most celebrated mountain has inspired thousands of paintings and poems: now some people are calling it 'The great garbage can in the sky.' " *International Wildlife* 17 (September-October 1987): 24(4)

Reports

Japan's activities to cope with global environmental problems: Japan's contribution toward a better global environment. Tokyo: Ad Hoc Group on Global Environmental Problems, 1988. 49 p.

"The World Commission on Environment and Development made public a report entitled 'Our Common Future' in April 1987. In response to the report, the Ad Hoc Group on Global Environmental Problems set up a Special Committee in November of 1987 to study Japan's concrete approach to the objectives presented in the report." This is the report by the Special Committee, which was submitted to the Director General of Japan's Environmental Agency.

Policy recommendations concerning climate change: interim report of the advisory committee on climate change. Tokyo: Ad Hoc Group on Global Environmental Problems. Environment Agency of Japan, 1988. 30 p.

Contains the report to the Japanese government by the Panel on Global Warming of Japan, a panel organized in May 1988 by the Air Quality Bureau, Environment Agency. "The report covers the following areas: (a) review of domestic and foreign studies on global warming (b) assessment of the latest scientific knowledge on the topic (c) identification of areas for further studies, and (d) proposals for future policy selections."

Protecting Tokyo's environment. Tokyo, Japan: Tokyo Metropolitan Government, 1985. 117 p.: ill. (some color), maps (TD171.5.J32 T657 1985)

Quality of the environment in Japan, 1988. Tokyo: Tokyo Environment Agency 1989. 353 p.

English version of the "white paper on the environment" for fiscal year 1987 (April 1987-March 1988), from the Environment Agency, government of Japan. Partial contents include Part one: Global environmental problems and Japan's contribution; Trends of the global environmental problems; Global environment and Japan's position; Japan's contribution toward the conservation of the global environment. Part two: Present state of environmental problems and counter-measures; Development in environment administration; Air pollution, noise, vibration, and offensive odors; Water; Promotion of environmental health policy; Present state of natural environment and the natural environment conservation measures; Environmental research; Promotion of international cooperation; and Pollution prevention system in private sector.

Summary of annual reports of the Tokyo Metropolitan Research Institute for Environmental Protection. Tokyo: Tokyo Metropolitan Research Institute, 1988. 77 p.: ill. (TD171.5.J32 T658 1988)

PACIFIC

Books (unannotated)

Abito, Germelino Fojas. *Environmental education, training, and research in the Philippines.* Singapore: Maruzen Asia, 1982. 82 p. (HC460.E5 A25 1982)

Broad, Robin. *Plundering paradise: people, power, and the struggle for the environment in the Philippines.* Berkeley: University of California Press, 1993.

Burnett, Alan Alexander. *Western Pacific: challenge of sustainable growth.* Brookfield, Vt.: E. Elgar Publishing, 1992.

Country reports, 1980-1981. Noumea, New Caledonia: South Pacific Commission, 1983. 1 v. : ill. (HC681.Z9 E53 1983)

Johnston, Douglas M. *Environmental management in the South China Sea: legal and institutional developments.* Honolulu, Hawaii: East-West Center, 1982. 114 p.

Review of the protected areas system in Oceania. Gland, Switzerland: International Union for Conservation of Nature and Natural Resources, 1986. 239 p., 2 p. of plates: ill. (QH77.O3 R48 1986)

General interest periodicals (unannotated)

Hall, C. Michael; and Batterham, Inga. "Trouble in paradise: a special report on the state of the South Pacific environment." *Alternatives* 18,1 (June-July 1991): 14(2)

Environmental and professional journals (annotated)

Broad, Robin; and Cavanagh, John. "Marcos's ghost." *Amicus* 11 (Fall 1989): 18-29.
    "A growing grass-roots environmental movement in the Philippines discovers that the 'old politics' is back, or perhaps never left. Poverty and corruption grind on, coupled with a rate of destruction of forests that is among the fastest in the world."

Carew-Reid, Jeremy. "Conservation and protected areas on South-Pacific islands: the importance of tradition." *Environmental Conservation* 17 (Spring 1990): 29-38.
    "Innovative new approaches to the definition, establishment, and management, of conservation areas, are being tried in a number of island countries, particularly Papua New Guinea. These methods rely on local native communities to plan and manage their traditional areas according to conservation principles which reinforce appropriate traditional jurisdiction and rules."

McDowell, Mark A. "Development and the environment in ASEAN." *Pacific Affairs* 62 (Fall 1989): 307-329.
    Provides an overview of the environmental issue in the ASEAN states emphasizing the interrelationship of economic, political, legal, and geographical factors.

Myers, Norman. "Environmental degradation and some economic consequences in the Philippines." *Environmental Conservation* 15 (Autumn 1988): 205-214.
    Examines the environmental degradation in the Philippines. "The environmental degradation leads to adverse economic consequences that are pervasive and profound--as may be expected in a country where several salient sectors of development are dependent upon the natural-resource base."

Reports

Environment and resources in the Pacific. Geneva, Switzerland: United Nations Environment Programme, 1985. 294 p.: ill. (HC681.3.E5 E58 1985)

Philippines: environment and natural resource management study.. Washington, D.C.: World Bank, 1989.

Porter, Gareth. *Resources, population, and the Philippines' future: a case study.* Washington, D.C.: World Resources Institute, 1988. 68 p.: ill. (HC453.5 .P67 1988)

Segura-delos Angeles, Marian. *A Review of Philippine natural resource and environmental management, 1986-1988*. Makati, Metro Manila: Philippine Institute for Development Studies, 1990. 90 p.: ill. (HC460.E5 S44 1990)

Conference proceedings

Conference on news media coverage of Pacific environmental issues (1989: Montana State University, Bozeman) covering the Pacific environment. Missoula, Mont.: Mansfield Center for Pacific Affairs, 1989. 27 p.: ill. (PN4888.E65 C66 1989)

Convention for the protection of the natural resources and environment of the South Pacific Region and related protocols. Nairobi: United Nations Environment Programme, 1987. 41 p. (K3478.A4 C66 1986)

Hawai'i's terrestrial ecosystems: preservation and management: proceedings of a symposium held June 5-6, 1984, at Hawai'i Volcanoes National Park. Honolulu: Cooperative National Park Resources Studies Unit, University of Hawaii, 1985. 584 p., 1 leaf of plates: ill. (some color) (QH105.H3 H39 1985)

High level conference to adopt a convention for the protection of the natural resources and environment of the South Pacific Region: Noumea, New Caledonia, 17-25 November 1986: report. Noumea, New Caldeonia: South Pacific Commission, 1987. 142 p. (MLCM 92/12350 (T))

Intergovernmental meeting on the SPREP Action Plan (1988: Noumea, New Caledonia) Report. Noumea, New Caledonia: South Pacific Commission, 1988. 60 p. (HC681.3.Z9 E55 1988)

South Pacific Regional Environment Programme: report of the Regional Conference for Consideration and Review of the SPREP Work Programme for 1987-1988, Noumea, New Caledonia, 1-5 September 1986. Noumea, New Caledonia: South Pacific Commission, 1986. 55 p. (HC681.3.E5 R43 1986)

Traditional conservation in Papua New Guinea: implications for today: proceedings of a conference. Boroko, Papua New Guinea: Office of Environment and Conservation and the Institute of Applied Social and Economic Research in Port Moresby, 1982. 392 p. 20 p. of plates: ill. (S934.P26 T73 1982)

Bibliographies

Enviromental issues in the South Pacific: a preliminary bibliography. Suva: University of the South Pacific Library, Pacific Information Centre, South Pacific Regional Environment Programme PIC Bibliography, 1983. 64 p. (Z7408.S63 E58 1983)

AUSTRALIA AND NEW ZEALAND

Books (unannotated)

Battle for the Franklin: conversations with the combatants in the struggle for South West Tasmania. Melbourne: Fontana/Australia Conservation Foundation, 1981. 303 p.: ill. (QH77.A8 B37 1981)

Boardman, Robert. *Global regimes and nation-states: environmental issues in Australian politics.* Ottawa, Canada: Carleton University Press, 1990. 222 p. (HC610.E5 B63 1990)

Case studies in environmental hope. Western Australia (Western Australian State Conservation Strategy): Environmental Protection Authority, 1988. 185 p.: ill. (S934.A8 C37 1988)

Conservation in New Zealand: a citizen's guide to the law. Auckland, N.Z.: Environmental Defence Society, 1985. 180 p.: ill.

Environmental politics in Australia and New Zealand. Hobart: Board of Environmental Studies, University of Tasmania, 1989. 200 p. (HC610.E5 E58 1989)

Gilpin, Alan. *Environment policy in Australia.* St. Lucia, Qld.: University of Queensland Press, 1980. 380 p.: ill. (HC610.E5 G54 1980)

Gilpin, Alan. *The Australian environment: 12 controversial issues.* Melbourne: Sun Books, 1980. 241 p.: ill. (TD170.8.A8 G54)

Green politics in Australia: a collection of essays. North Ryde, NSW, Australia: Angus and Robertson Publishers, 1987. 245 p.: (QH77.A8 G74 1987)

Greening of Australian politics: the 1990 federal election. Melbourne, Australia: Longman Cheshire, 1990. 229 p.: ill. (JQ4094 .G73 1990)

How many more Australians?: the resource and environmental conflicts. Melbourne, Australia: Longman Cheshire, 1988. 192 p.: ill. (HB3675 .H68 1988)

Land, water, and people: geographical essays in Australian resource management. Boston: Allen and Unwin, 1988. 266 p.: ill., maps (HC610.E5 L36 1988)

Man and the Australian environment: current issues and viewpoints. New York: McGraw-Hill, 1982. 362 p.: ill., maps (DU105.2 .M36)

Quarry Australia?: social and environmental perspectives on managing the nation's resources. New York: Oxford University Press, 1982. 366 p.: ill. (HC603.5 .Q36 1982)

Smith, David. *Continent in crisis: a natural history of Australia.* New York: Penguin Books, 1990. 201 p.: ill. (QH197 .S59 1990)

Thompson, Peter. *Power in Tasmania.* Hawthorn, Victoria: Australian Conservation Foundation, 1981. 192 p.: ill. (some color) (TK1522.T3 T48 1981)

Whitelock, Derek A. *Conquest to conservation: history of human impact on the South Australian environment.* Netley, S. Australia.: Wakefield Press, 1985. 271 p.: ill. (QH77.A8 W53 1985)

Wilson, Roger. *From Manapouri to Aramoana: the battle for New Zealand's environment.* Auckland, N.Z.: Earthworks Press, 1982. 192 p.: ill. (GF805.N45 W54 1982)

Wright, Judith. *Born of the conquerors: selected essays.* Canberra: Aboriginal Studies Press, 1991. 156 p.: color ill. (GN666 .W75 1991)

Environmental and professional journals (annotated)

Dold, Catherine. "Sir Joh's last stand." *Amicus Journal* 10 (Winter 1988): 8-13.
    Reports on Australia's battle to save its vanishing rain forests. An official nomination for World Heritage listing of 3,550 square miles of remaining tropical rain forest is strongly opposed by the Queensland government.

Kennaway, Richard. "New Zealand and environmental issues: the growing international dimension." *Round Table* 317 (January 1991): 59-72.
    "In the first part of this paper, the reasons for the increased environmental emphasis in New Zealand's foreign policy are explored. In the second part, we consider the means by which New Zealand has sought to implement its goals. In this section it is noted that there have been some significant inconsistencies between domestic realities and external aspirations."

Steffen, William. "Furor over the Franklin." *Sierra* 69 (September-October 1984): 43-49.
    "Australia's conservationists not only stopped a dam, they brought down the government--and in the process changed their country's constitution."

Wilson, George R. "The development of a national conservation strategy for Australia." *Environmental Conservation* 14 (Summer 1987): 111-116.
    "The purpose of the strategy is to provide nationally-agreed guidelines for the use of living resources by Australians, so that the reasonable needs and aspirations of society can be sustained in perpetuity. Outlines the procedures adopted in developing the strategy, and indicates some of the problems encountered in arriving at a final consensus."

Reports

Dickinson, Katharine J. M. *Nokomai Ecological District: survey report for the protected natural areas programme.* Wellington, N.Z.: Department of Conservation, 1989. 139 p.: ill. (QH197.5 .D53 1989)

Fiscal measures and the environment: impacts and potential. Canberra: Australian Government Publishing Service, 1985. 25 p. (HJ1514 .F57 1985)

State conservation strategy for Western Australia: a sense of direction. Perth, W.A.: Department of Conservation and Environment, 1987. 19 p.: ill. (S934.A8 S73 1987)

Striking the balance: highlights in conservation and environmental management in Western Australia, 1983-1988. Perth: Government of Western Australia, 1988. 45 p.: color ill. (S934.A8 S77 1988)

Towards a state conservation strategy: invited Review papers. Perth, W.A.: Department of Conservation and Environment, 1986. 286 p.: ill. (QH77.A8 T69 1986)

Conference proceedings

Australian and New Zealand islands: nature conservation values and management: proceedings of a technical workshop, Barrow Island, Western Australia, 1985. Wanneroo, W.A.: Department of Conservation and Land Management, Western Australia, 1989. 233 p.: ill., maps (QH77.A8 A85 1989)

National Wilderness Conference (3rd: 1983: Katoomba, N.S.W.) Fighting for wilderness: papers from the Australian Conservation Foundation's Third National Wilderness Conference, 1983. Sydney, Australia: ACF, 1984. 256 p.: ill. (QH75.A1 N33 1983)

Proceedings of a symposium on environmental monitoring in New Zealand: with emphasis on protected natural areas. Wellington, N.Z.: Department of Conservation, 1989. 303 p.: ill. (QH197.5 .S95 1989)

Sustainable economic growth and industrial development: report of the 47th Meeting of the National Science and Industry Forum, Thredbo, February 1990. Canberra, Australia: ACT: Australian Academy of Science, 1990. 86 p.: ill., map (HC603 .A96 1990)

THIRD WORLD DEVELOPMENT

(See also *Rainforests: Sustainable Harvesting*) 2981

Books (annotated)

Ascher, William; and Healy, Robert. *Natural resource policymaking in developing countries: environment, economic growth, and income distribution.* Durham, N.C.: Duke University Press, 1990. 223 p. (HC597 .A834 1990)

> Examines the social and political aspects of Third World environmental degradation and explains how environmental-developmental projects in these countries fail to successfully promote economic growth, ensure an equitable distribution of growth, uphold environmental safety, and allow all affected parties to participate in the policy-making process. Includes case studies and makes specific recommendations for improving the policy-making process.

Conroy, Czech; and Litvinoff, Miles, eds. *The greening of aid: sustainable livelihoods in practice.* London: Earthscan with IIED, 1988.

> Analyzes the concepts and requirements of sustainable development from the perspective of the rural poor rather than urban industry. Presents thirty-four case studies from various countries.

Dankelman, Irene; and Davidson, Joan. *Women and environment in the Third World.* London: Earthscan, 1988. 210 p.

> Argues that women often are given the poorest lands to farm and as a result are most adversely affected by the loss of natural resources. Thus, women are more concerned about preserving resources and could accomplish sustainable development if given the resources.

Gupta, Avijit. *Ecology and development in the Third World.* London: Routledge, 1988. 80 p.

> Describes the ways in which development has tried to solve certain problems and ended up only causing more. Presents numerous case studies, including the Aswan Dam in Egypt, land degradation in El Salvador, and chemical pollution in Bhopal.

Hamilton, John Maxwell. *Entangling alliances: how the Third World shapes our lives.* Cabin John, Md.: Seven Locks, 1990. 212 p.

> Uses a case study approach to explore how nations are interlinked, especially in the relationship of economics and environmental problems, such as the importance of tourism in Third World countries, the destruction of the rain forests, and the introduction of modern tools to Third World countries.

Jackson, Ben. *Poverty and the planet: a question of survival.* London: Penguin, 1990.

> Focuses on how impoverished nations destroy environmental resources in order to survive while other countries seek "western-backed development schemes regardless of the environmental or human cost."

Leonard, W. Jeffrey. *Environment and the poor: development strategies for a common*

agenda. New Brunswick, N.J.: Transaction, 1989. 222 p. (HC59.72.E5 L46 1989)
"Six out of every ten of the world's poorest people are being inexorably pushed by agricultural modernization and continuing high population growth rates into ecologically vulnerable environments: tropical forests, dryland and hilly areas, and the fringes of great urban centers. Unless development strategies support their capabilities to ensure their own survival, the 470 million people living in these vulnerable areas will be forced to meet their short-term need to survive at the cost of long-term ecological sustainability and the well-being of future generations."

Pearson, Charles S., ed. *Multinational corporations, environment, and the Third World: business matters.* Durham, N.C.: Duke University Press, 1987. 295 p.
Presents eleven studies discussing effects that multinational corporations may have on the environment and natural resources of developing countries. The evidence suggests that MNC's and domestic firms receive similar legal treatment with respect to environmental law and that MNC environmental performance is probably better than that of local firms; however, developing countries need to regulate the MNC's carefully.

Redclift, Michael. *Sustainable development: exploring the contradictions.* New York: Methuen, 1987. 221 p. (HD75.6 .R43 1987)
Examines environmental problems and how they relate to the emerging global economy. Discusses "numerous aspects of development which present barriers to sustainability." Colonialism, international trade, oil prices, and the debt crisis are analyzed as environmental forces which work against sustainable development.

Schramm, Gunter; and Warford, Jeremy J. *Environmental management and economic development.* Baltimore, Md. (World Bank): Johns Hopkins University Press, 1989. 208 p.: ill. (HD75.6 .E57 1989)
These eleven essays offer information on environmental management in developing countries. Maintains that with the proper institutional structure, environmental protection and economic development can exist simultaneously and harmoniously, and concrete suggestions on how to create this institutional structure are offered. Presents specific case studies, such as forest management in Brazil and watershed management in the Philippines. Covers the role of national income accounting, incentives, and project evaluations.

Southgate, Douglas D.; and Disinger, John F., eds. *Sustainable resource development in the Third World.* Boulder, Colo: Westview, 1987. 177 p.
Covers the objectives of resource conservation programs; the requirements of documenting the economic and social effects of the projects; specific projects in Kenya, the Dominican Republic, Barbados, Costa Rica, Colombia, and Panama; and the necessity to design projects with respect to the realities of the local population.

Wildlands: their protection and management in economic development. Washington, D.C.:

World Bank, 1989.
> Addresses how World Bank policy has been applied in Third World countries to prevent damage to natural areas. Topics include afforestation, environment, public health, human ecology, and sustainability.

Books (unannotated)

Ecological development in the humid tropics : guidelines for planners. Morrilton, Ark.: Winrock International Institute for Agricultural Development, 1988.

Environmental modelling for developing countries. New York: Tycooly, 1990. 166 p.: ill. maps. (HC59.72.E5 E62 1990)

Greening of aid: sustainable livelihoods in practice. London: Earthscan in association with IIED, 1988. 302 p.: ill. (HD75.6 .G74 1988)

International justice and the Third World: studies in the philosophy of development. New York: Routledge, Chapman & Hall, 1992.

McKee, David L. *Developmental issues in small island economies.* New York: Praeger, 1990. 196 p. (HC151 .M39 1990)

Pearce, David William. *Sustainable development: economics and environment in the Third World.* Brookfield, Vt.: Gower, 1990. 217 p.: ill. (HD75.6 .P43 1990)

Resources, environment, and regional development. Hong Kong: Centre of Asian Studies, University of Hong Kong, 1989. 480 p.: (some color) ill. (HC427.92 .R47 1989)

Stone, Roger D. *The nature of development: a report from the rural tropics on the quest for sustainable economic growth.* New York: Knopf, 1992. 286 p.: ill. (HC59.72.E44 S76 1992)

Sustainable development and environmental management of small islands. Park Ridge, N.J. UNESCO: Parthenon, 1990. 419 p.: ill. (GF61 .S87 1990)

Texler, Jiri. *Environmental hazards in Third World development.* Budapest: Institute for World Economy of the Hungarian Academy of Sciences, 1986. 48 p. (TD171.5.D44 T49)

Women and the environment: a reader: crisis and development in the Third World. New York: Monthly Review, 1991. 205 p. (HQ1240.5.D44 W64 1991)

General interest periodicals (annotated)

Sachs, Ignacy. "Environment and development revisited: ten years after Stockholm conference." *Alternatives* 8 (Winter 1982-83): 369-370.

Traces the evolution of alternative development strategies from 1972 to 1982 through the course of the North-South dialogue.

General interest periodicals (unannotated)

"Making sense of Brundtland; without economic growth the Third World will ruin the environment." *Economist* 303 (May 2, 1987): 15(2).

Environmental and professional journals (annotated)

Arden-Clarke, Charles. "South-North terms of trade, environmental protection, and sustainable development." *International Environmental Affairs* 4 (Spring 1992): 122-138.
    Explores "how various trade measures might be used to alleviate the environmental impacts stemming from current South-North trade. Emphasis is placed on environmental damage incurred in developing countries, and more specifically on that relating to the production and export of commodities. Proposals are made on trade policy actions that could be taken to contribute to the objective of equitable development within the earth's carrying capacity."

Cruver, Philip C. "The trillion dollar Third World power generation market." *OPEC Bulletin* 21 (January 1990): 17-20, 86.
    "Minimizing the environmental impact of the Third World's growing demand for electricity calls for a worldwide effort to develop and market innovative technologies." Draws attention to the opportunities and the type of initiatives that should be undertaken.

Dabholkar, Uttam. "Environmental perspective to the year 2000 and beyond: a framework for world development." *Environmental Conservation* 16 (Spring 1989): 49-53, 64.
    The United Nations General Assembly's " 'Environmental Perspective to The Year 2000 and Beyond' points to the real risks of loss of economic growth and social well-being through continued environmental degradation in many developing countries, and argues for a fundamental change in modes of decision-making on economic and social issues--so as to anticipate, prevent, and mitigate, those risks."

Livernash, Robert. "The growing influence of NGOs in the developing world." *Environment* 34 (June 1992): 12-20, 41-43.
    "Nongovernmental organizations in the developing world have not achieved as much success as have their northern counterparts, mostly because of their limited funding, internal weaknesses, and governmental intolerance. Increased interaction with local governments, communication networks, international associations, and the judicial system, however, can help to reduce poverty and illuminate the links between poverty, development, and environmental degradation."

Mennerick, Lewis; and Najafizadeh, Mehrangiz. "Third World environmental health: social, technological and economic policy issues." *Journal of Environmental Health* (Spring 1991): 24-29.

> Third World nations confront numerous environmental problems, which "increasingly are byproducts of attempts to utilize 'modern' industrial and agricultural technologies and related attempts to promote economic development. Yet, such problems also are caused by or exacerbated by rapid population growth, economic and political self-interests and the ideology of consumerism."

"People speak out." *Panoscope* 24 (May 1991): 8-17.

> Examines the views of citizens of the Third World. Contents include: "The people speak out"; "Make the polluters pay"; "What they say about the 1992 conference"; "Thumbs down to trade ban"; "Lukewarm response to controls"; "Whispering in the wrong ears"; and " 'Green' is only skin-deep."

Piddington, Kenneth W. "Third World dilemma: development, environment." *Forum for Applied Research and Public Policy* 5 (Fall 1990): 26-29.

> Concludes that for sustainable development to be achieved in the Third World, both environmental protection and traditional development are necessary.

Strong, Maurice F. "The challenge of environment and development." *World Food Programme Journal* 20 (April-June 1992): 17-20.

> The views of the secretary-general regarding the 1992 Earth Summit.

"Sustainable development: no more business as usual." *Development Forum* 20 (March-April 1992): 1(4)

> Discusses the recently published book, *Environmentally Sustainable Economic Development: Building on Brundtland*, by Nobel laureates Trygve Haavelmo and Jan Tinbergen and World Bank economists Robert Goodland and Herman Day. Includes a sidebar, "Earth Summit overload," by Gillian Phillips.

Talba, Mostafa K. "The political environment for sustainable development." *Inter-parliamentary Bulletin* 1 (1988): 61-66.

> The executive director of the United Nations Environment Programme contends that sound environmental strategies should not be perceived as limits to economic growth, but pathways to sustainable development.

Zaragoza, Federico Mayor. "Challenge of sustainable development and ethical mission of UNESCO." *International Affairs* (Moscow), (April 1989): 13-21.

> UNESCO's director-general since November 1987 challenges both developed and developing nations to use intellectual cooperation and innovative diplomacy to meet future challenges in both the natural and social sciences.

Environmental and professional journals (unannotated)

Durning, Alan B. "People power and development." *Foreign Policy* 76 (Fall 1989): 66(17).

Kates, Robert W.; and Haarmann, Viola. "Where the poor live: are the assumptions correct?" (includes related articles) (cover story) *Environment* 34,4 (May 1992): 4(12).

Zaslowsky, Dyan. "Poverty, prosperity and preservation." (industrial development without environmental damage) *Sierra* 73,1 (January-February 1988): 24(3).

Law journals

"Development and the environment." *New York University Journal of International Law and Politics* 20 (Spring 1988): whole issue (p. 603-872)
> Partial contents: "Environmental protection and development in Third World countries: common destiny--common responsibility," by Gunther Handl; "The export of danger: a view from the developed world," by Robert E. Lutz; "Legal responses to the Philippine deforestation crises," by Owen J. Lynch, Jr. and Kirk Talbott; "The work of the International Law Commission relating to transfrontier environmental harm," by Stephen C. McCaffrey; "The UNEP Montreal Protocol: industrialized and developing countries sharing the responsibility for protecting the stratospheric ozone layer," by James T. B. Tripp; "Chemical exports and the age of consent: the high cost of international export control proposals," by Michael P. Walls; and "International environmental law: selective bibliography," by Blanka Kudej.

Reports

Common index and glossary to the Brandt, Palme, and Brundtland reports of the Independent Commissions on International Development, Disarmament and Security, and Environment and Development. London: Commonwealth Secretariat, 1990. 146 p. (Z7164.E17 C65)

Congressional agenda for improved resource and environmental management in the Third World: helping developing countries help themselves. Washington, D.C.: Environmental and Energy Study Institute, 1985. 50 p.
> "Strengthening the developing countries' own capability to meet their natural resource challenges is one sure and effective way for them to achieve long-term, sustainable growth, repay their debts and become effective trading partners." Proposes a new, more cost-effective approach to development aid, rather than big new expenditures. It stresses smaller scale projects which efficiently use energy, water and other resources and which are economically sustainable.

Danish government's action plan: environment and development: follow up to the recommendations in the report of the World Commission on Environment and Development and the UN environmental perspective to the year 2000. Copenhagen: Ministry of the Environment, 1988. Distributed by the State Information Service. 139 p. (HC360.E5 D36 1988)

Elkington, John; and Shopley, Jonathan. *Cleaning up: U.S. waste management technology and Third World development.* Washington, D.C.: World Resources Institute, 1989. 80 p.
"Explores the role that U.S. environmental protection and waste management industries can play in promoting environmentally sustainable economic development, particularly in the less developed and newly industrializing countries of the Third World." Profiles companies and organizations in the U.S. waste management industry, covering environmental organizations, government agencies, industry associations, companies, and industry-environment organizations.

Environment and development: institutional issues; report of the Twenty-third United Nations Issues Conference. Muscatine, Iowa: Stanley Foundation, 1992. 14 p.
"From February 26 to 28, 1992, in advance of PrepComm IV, the Stanley Foundation brought together a distinguished group of key players in the UNCED process for a frank, off-the-record exchange of what institutional arrangements would best serve the hoped-for outcome of UNCED and the demands of the future. The following summarizes their points of consensus and divergence."

Environmental assessment and development assistance. Paris: Organisation for Economic Co-operation and Development, 1986. 103 p. (Environment monographs no. 4)
Discusses the types of development assistance projects which are most in need of environmental assessment; the developing countries' constraints surrounding the use of environmental assessments and the measures taken to overcome them; the experience of AID agencies in conducting environmental assessments of development projects; and processes, procedures, organization and resources needed for the environmental assessment of development projects and programmes.

Environmental management in developing countries. Paris: Development Centre of the Organisation for Economic Co-operation and Development, 1991. Distributed by OECD Publications and Information Centre, Washington, D.C. 17 p.: ill. (HC59.72.E5 E6 1991)

Natural endowments, financing resource conservation for development: international conservation financing project report. Washington, D.C.: World Resources Institute, 1989. 33 p. (HC59.72.E44 N38 1989)

Notes for speakers: environment and development. New York: United Nations, Department of Public Information, 1991. 73 p.
"Provides an overview of the U.N. system's efforts to develop the concept of sustainable development and put it into practice."

One world or several? Paris: Development Centre of the Organisation for Economic Co-operation and Development, 1989. Distributed by OECD Publications and Information Centre, Washington, D.C. 319 p. (HC59.7 .O515 1989)

Planning for sustainable development: a resource book. Vancouver, British Columbia, Canada: UBC Centre for Human Settlements, 1989. 145 p. (HC117.B8 P57 1989)

Strengthening environmental co-operation with developing countries. Paris: Organisation for Economic Co-operation and Development, 1989. Distributed by OECD Publications and Information Centre, Washington, D.C. 147 p. (HC59.72.E5 S77 1989)

Sustainable development. Portland, Oreg.: F. Cass in association with the European Association of Development Research and Training Institutes (EADI), Geneva, 1991. 132 p.: ill. (HD75.6 .S86 1991)

Technology, development, and the global environment. Mahwah, N.J.: Institute for Environmental Studies, School of Theoretical and Applied Science, Ramapo College (505 Ramapo Valley Road, Mahwah 07430), 1991. 315 p. (TD195.E25 T43 1991)

Federal government reports

Opportunities in natural resource management for the developing world. 3d ed. Washington, D.C.: Department of Agriculture, Forest Service and Office of international Cooperation and Development, Forestry Support Program, 1990. 33 p. (HC59.7 .R467 1990)

Conference proceedings

Conservation in developing countries: problems and prospects: proceedings of the centenary seminar of the Bombay Natural History Society. New York: Oxford University Press, 1990. 656 p.: ill. (QH77.D44 C66 1990)

Energy-environment-development: proceedings of the 12th International Conference of the International Association for Energy Economics (IAEE), organised by TERI for the IAEE, from January 4-6, 1990 in New Delhi. 2 vols. New Delhi: Har-Anand Publications in association with Vikas Publishing House, 1991. (HD9502.A2 I13 1990)

CHAPTER 4

U.S. ENVIRONMENTAL POLICY AND POLITICS

(Council on Environmental Quality Assessment)
  19, 140, 168, 181, 189, 199, 206

continued

U.S. ENVIRONMENTAL POLICY AND POLITICS

(Council on Environmental Quality Assessment) 19, 140, 168, 181, 189, 199, 206

SHORT LIST OF SOURCES
FOR GENERAL RESEARCH

Books

Blueprint for the environment: a plan for federal action. Salt Lake City, Utah: Howe Brothers, 1989. 335 p.: ill. (HC110.E5 B58 1989)

This volume is a collection of multifaceted environmental policy recommendations that was originally presented to George Bush in 1988. It has since been reorganized, although it still presents the opinions of the most important environmental organizations. The book is organized around the federal agencies which control each particular issue. There is a brief historical sketch of each agency, as well as specific suggestions on how the operations of that agency can be improved. The overriding theme is that federal programs should be synthesized and given a sense of uniformity so that they can effectively combat pollution and environmental destruction. Contains the addresses and phone numbers of pertinent agencies and organizations.

Covello, Vincent T.; et al., eds. *Effective risk communication.* New York: Plenum, 1989. 370 p.

Solving environmental problems necessarily involves calculating risks and explaining them to policy makers and the public. While scientists have developed excellent risk analysis methods, they often find that the public is outraged by their findings. Maintains that ineffective communication has generated undue public reaction, and that outrage is reduced when risk is voluntarily undertaken, the extent of the risk is well known, and the public feels they have some control over the situation. Provides a full slate of suggestions for communicating with communities undergoing risk stress.

Crowfoot, James E.; and Wondolleck, Julia M. *Environmental disputes: community involvement in conflict resolution.* Washington, D.C.: Island Press, 1991. 278 p.

Offers seven case studies of local disputes: logging v. recreation; agricultural production v. soil conservation and wetland protection; oil drilling v. recreation; channel dredging v. a bird sanctuary; and water use v. aquifer protection. Describes the negotiating process and concludes that the chance of negotiating a settlement is greater if there is an experienced neutral negotiator, if the project is being planned rather than already constructed, and if scientists are involved as neutral experts.

Klapp, Merrie G. *Bargaining with uncertainty: decision-making in public health, technological safety, and environmental quality.* Dover, Mass.: Auburn House, 1992. 149p.
Describes the ways in which scientific uncertainty can influence regulatory policy—in some cases weakening it and in others strengthening it. Examples include dioxin exposure and saccharin. Contrasts U.S. experience with that of Europe. A good, advanced source on health and environmental policy making.

May, Craig L.; and Dennis, Everette E., eds. *Media and the environment.* Washington, D.C.: Island Press, 1991.
Explores how the media covers the environment and addresses problems of objectivity, advocacy, and how the subject fits into the media's concept of balanced news coverage.

Pierce, John C., et al. *Public knowledge and environment: politics in Japan and the U.S.* Boulder, Colo.: Westview Press, 1989. 229 p.
Examines the role that public knowledge plays in the environmental policy making of industrial democracies, the democratic process, the role of citizen action, the environment as a "postindustrial issue," and the role of political culture between two postindustrial nations. This comparison raises interesting questions about how economically advanced countries view the interaction of humans and nature, and the relationship between economic growth and ecological balance.

General interest periodicals

Gore, Albert, Jr. "The ecology of survival." *New Republic* 201 (November 6, 1989): 26, 28, 30, 34.
"A Strategic Environment Initiative would promote environmentally sustainable development by identifying and spreading new technologies to developing countries. In the U.S. the Strategic Environment Initiative must modernize technologies and practices in every economic sector, from more fuel-efficient cars and energy-efficient appliances to manufacturing that relies on recycled material, to a second green revolution requiring fewer fertilizers and pesticides."

Main, Jeremy. "The big cleanup gets it wrong." *Fortune* 123 (May 20, 1991): 95-96, 100-101.
"The emerging science of risk assessment says that the U.S. is spending way too much on minor threats, like asbestos, and not enough on major pollutants, like radon."

"Struggle to save our planet." Discover 11 (April 1990): 35-43, 46-49, 51-53, 55-59, 61-62, 64, 66-70, 72, 74, 76-78.

> Contains "Down in the dumps," by Dan Grossman and Seth Shulman; "The Numbers game," by David Berreby; "Ecoglasnost," by Tom Waters; "Playing dice with megadeath," by Jared Diamond; "Green giants," by Doug Stewart; "Invisible garden," by Robert Kunzig; and "The Poison eaters," by Susan Chollar.

Environmental and professional journals

"24th Environmental Quality Index. *National Wildlife* 30,2 (February-March 1992): 33(8)

> Reports on trends in wildlife, air pollution, water pollution, energy, forests, soil and quality of life. "The problems besetting America's environment have proved far tougher than anyone predicted--but the same thing is true of the resolve of the American people."

Nitz, William A. "Improving U.S. interagency coordination of international environmental policy development." *Environment* 33 (May 1991): 10-13, 31-37.

> Looks at emerging international environmental issues, examines existing interagency mechanisms for addressing these issues, and explores the boundary constraints imposed by the U.S.'s government system, political culture, and personnel selection process. Recommends that an improved interagency process should have, among other things, "a cabinet-level committee that focuses on environmental and energy issues but covers both the domestic and foreign policy aspects of those issues."

Ost, Laura. "Governing guide: cleaning up--pollution, policy and technology." *Governing* 4 (April 1991): 33-56.

> "Controlling air and water pollution is within technology's grasp. But state and local government officials trying to clean up the environment have a tough job and very little money. They must weigh factors ranging from the type of pollutants involved to the cleanup costs their community can bear."

Scarlett, Lynn. "A guide to environmental myths and realities." *Consumers' Research* 75,1 (January 1992): 11(6)

INTRODUCTORY SOURCES

Books (annotated)

Cahn, Robert, ed. *Environmental agenda for the future, by leaders of America's foremost environmental organizations.* Washington, D.C.: Agenda Press, 1985. 155 p.

Presents the results of two years' work by leaders of ten major environmental and conservation organizations in setting environmental goals. Recommends integrated, corrective action. Issues covered are protected land systems, public lands, private lands and agriculture, nuclear issues, energy strategies, wild living resources, water, population growth, cities, and international responsibilities.

Hall, Bob; and Kerr, Mary Lee. *1991-1992 green index: a state-by-state guide to the nation's environmental health.* Washington, D.C.: Island Press, 1991. 162 p.
Draws together data from dozens of government and private sources into an insightful assessment of environmental conditions in each region and state. It "is a set of 256 indicators that measure and rank each state's environmental health. Taken together, these indicators describe the condition of thing as they are, as well as the policies and political leadership in place to make things better."

Mason, Robert J.; and Mattson, Mark T. *Atlas of United States environmental issues.* New York: Macmillan, 1990. 252 p. col. ill. col. maps. 1 atlas. (G1201.G3 M3 1990)
A collection of 150 maps that illustrate the important agricultural, coastal zone management, air quality, noise and light pollution, and economic/political aspects of environmental problems. Accompanying text and tables combine to further explain these issues.

State of the Enviroment 1982. Washington, D.C.: Conservation Foundation, 1982. 439 p.
Offers a comprehensive assessment of the nation's environment. Contains chapters on trends, air quality, water resources, hazardous wastes, energy, agriculture and forestry, land, the urban environment, and the Reagan administration and institutional change.

Books (unannotated)

State of the environment: a view toward the nineties. Washington, D.C.: Conservation Foundation, 1987. 614 p.: ill. (TD171 .S73 1987)

State of the environment: an assessment at mid-decade: a report from the Conservation Foundation. Washington, D.C.: Conservation Foundation, 1984. 586 p.: ill. (TD171 .S74 1984)

Stoddard, Charles Hatch. *Looking forward: planning America's future.* New York: Macmillan, 1982. 306 p. (HC110.E5 S76 1982)

General interest periodicals (annotated)

"20th Anniversary issue on the environment." *Smithsonian* 21 (April 1990): whole issue (33-190 p.)

Commemorates Earth Day's anniversary. Partial contents include: "The past: knowing the wilderness was to love it," by Wallace Stegner; "The present: the road from Earth Day 1," by Bil Gilbert; "On the Connecticut, a sweet smell of success," by Richard Conniff; "A flight to survival for endangered raptors," by Don Moser; "Alternative agriculture is gaining ground," by Jeanne McDermott; "Rails-to-trails: an exercise in linear logic," by Judy Mills; "Down in the dumps," by Richard Wolkomir; "Trees aren't mere niceties--they're necessities," by Jon Krakauer; "Flushed with pride in Arcata, California," by Doug Stewart; and "No compromise in defense of Mother Earth," by Michael Parfit.

"Struggle to save our planet." *Discover* 11 (April 1990): 35-43, 46-49, 51-53, 55-59, 61-62, 64, 66-70, 72, 74, 76-78.

Contains "Down in the dumps," by Dan Grossman and Seth Shulman; "The Numbers game," by David Berreby; "Ecoglasnost," by Tom Waters; "Playing dice with megadeath," by Jared Diamond; "Green giants," by Doug Stewart; "Invisible garden," by Robert Kunzig; and "The Poison eaters," by Susan Chollar.

Environmental and professional journals (annotated)

"24th Environmental Quality Index. *National Wildlife* 30,2 (February-March 1992): 33(8)

Reports on trends in wildlife, air pollution, water pollution, energy, forests, soil and quality of life. "The problems besetting America's environment have proved far tougher than anyone predicted--but the same thing is true of the resolve of the American people."

"Planet strikes back." *National Wildlife* 27 (February-March 1989): 33-40.

Presents National Wildlife's annual assessment of the environmental events, gains, and losses of the past year. Rates wildlife, air, water, soil, and quality of life in the "worse" range; rates energy and forests as "same" but on the decline.

Reports

America in the 21st century: a global perspective. Washington, D.C.: Population Reference Bureau, 1989. 28 p.

Examines how demographic trends in the developed and developing countries could have future direct and indirect effects on U.S. labor markets, product markets, financial markets, environmental quality, and national security interests.

America in the 21st century: environmental concerns. Washington, D.C.: Population Reference Bureau, 1990. 24 p.

Brief report on the human dimensions of five environmental issues: air pollution; the quality and availability of water; solid waste management; the disposal of hazardous wastes; and the threat of global warming. Demographic trends often affect where and when environmental problems occur, the severity of their impact on health and jobs, and other aspects of social and economic well-being.

Caldwell, Lynton K. *U.S. interests and the global environment.* Muscatine, Iowa: Stanley Foundation, 1985. 28 p. (Occasional paper 35)
    The author "forcefully argues for policies responsive to global environmental needs. He examines six critical environmental issues, their causes and consequences, and then offers specific US policy recommendations" and concludes that "vigorous positive leadership in international environmental policy is in the national interest as well as in the interest of people everywhere."

Environment at risk: responding to growing dangers. Dayton, Ohio: National Issues Forums Institute, 1989. 48 p.
    Partial contents include: Environmental protection: choices for the 1990s;" "Planetary housekeeping: blueprint for a sustainable future;" "Balancing Act: costs and benefits of a cleaner environment;" "Carrots and sticks: putting a price on pollution;" and "Decade of decision about the global commons."

Fritsch, Albert J. *Communities at risk: environmental dangers in rural America.* Washington, D.C.: Renew America, 1989. 40 p.
    Addresses environmental dangers in rural America in six areas and highlights the dangers that particularly affect the rural poor. Each section includes policy considerations that could lead to possible solutions at the national, state, and local levels. The six areas covered are water pollution, air pollution, soil degradation, improper land use, pesticides, and waste disposal of solid and hazardous wastes.

Myers, Norman. *Not far afield: U.S. interests and the global environment.* Washington, D.C.: World Resources Institute, 1987. 73 p. : ill. (HC59.72.E5 M93 1987)
    Considers the relationship of critical issues in world environment, natural resources, and population to U.S. economic, ecological, and national security interests.

Train, Russell. *Environmental concerns in the year 2000.* Washington, D.C.: Congressional Research Service, 1989. 19 p. (Congress in the year 2000)
    Discusses the U.S.'s environmental legislative agenda and recommends four overarching environmental policy initiatives. Examines global warming, energy efficiency, solid waste, land use, and global environmental problems.

Federal government reports

Environmental quality 1990; twenty-first annual report of the Council on Environmental Quality together with the President's message to Congress. Washington, D.C. (Council on Environmental Quality): G.P.O., 1991. 388 p.
    Partial contents include: Part I: special reports; Chapter 1: Where we stand; A

national strategy for environmental quality; A federal environmental almanac for 1989-1990; Air; Budget; Department of Defense; Federal facilities; Water. Chapter 2: Making the environment count: costs; Making the environment count: benefits. Chapter 3: Technology for pollution prevention; Technology; Energy update; Existing programs; Advancing pollution prevention technology. Chapter 4: Linking ecosystems and biodiversity; Loss of biodiversity: domestic concerns; An emerging solution: ecosystem management; Diversity, ecosystems, and biological health; A strategy for conservation tomorrow. Chapter 5: The National Environmental Policy Act--integration in concept and practice; Selected NEPA cases in 1990; NEPA trends. Part II: Environmental data and trends; Population; The economy and the environment; Energy; Water; Air quality; Land, agriculture, and forestry; Protected lands, cultural and living resources; Environmental waste and hazards.

Environmental quality 1970-1990; twentieth annual report of the Council on Environmental Quality together with the President's message to Congress. Washington, D.C. (Council on Environmental Quality): G.P.O., 1990. 494 p.
    This report differs in two ways from past reports: it does not discuss in detail the environmental issues and trends of the moment, but evaluates the distance traveled since 1970; it emphasizes environmental quality and pollution control over natural resource issues, because of their historical emphasis. Partial contents include: Twenty years of change; The National Environmental Policy Act; Environmental data and trends; Environmental science and technology; Environmental enforcement; Pollution prevention; International issues; and The Great Lakes.

Environmental trends. Washington, D.C. (Council on Environmental Quality): G.P.O., 1989. 152 p.
    Updates a 1981 publication by the same name. Contains graphics on current conditions and trends in the environment in the U.S. Chapters include: minerals and energy, water, climate and air quality, land resources, wetlands and wildlife, protected areas, population, transportation, and environmental risks and hazards.

Global 2000 report to the President: entering the twenty-first century. Washington, D.C. (Council on Environmental Quality): G.P.O., 1980. 3 v.
    Concludes that if present trends continue, the world in 2000 will be more crowded, more polluted, less stable ecologically, and more vulnerable to disruption than the world we live in now. Report presents forecasts on population, income, food, fisheries, forests, water, energy, non-fuel minerals, and the environment.

List of audiovisual materials produced by the United States Government for environment and energy conservation. Washington, D.C.: General Services Administration, 1982 18 p.
    Lists audiovisual material produced by or for the federal government available for purchase or rent. Subjects covered include air pollution, alternative energy

sources, fuel economy, land, noise pollution, pesticides, solid waste recycling and water.

POLICY AND POLITICS

(Related topics in *Environmental Movement*) 419

Books (annotated)

Blueprint for the environment: a plan for federal action. Salt Lake City, Utah: Howe Brothers, 1989. 335 p.: ill. (HC110.E5 B58 1989)
 A collection of multifaceted environmental policy recommendations that was originally presented to George Bush in 1988. It has since been reorganized, although it still presents the opinions of the most important environmental organizations. Organized around the federal agencies which control each particular issue. There is a brief historical sketch of each agency, as well as specific suggestions on how the operations of that agency can be improved. The overriding theme is that federal programs should be synthesized and given a sense of uniformity so that they can effectively combat pollution and environmental destruction. Contains the addresses and phone numbers of pertinent agencies and organizations.

Caldicott, Helen. *If you love this planet*. New York: W. W. Norton, 1992.
 Part of this book on the global environmental crisis focuses on corporate interests as the greatest obstacle to solving environmental problems and poses numerous suggestions for curbing corporate power, including publicly financed elections, stronger labor unions, and a higher minimum wage. Maintains that we should strengthen Third World nations by forgiving their debt. Opposes opening up world trade because it would allow corporations greater control of resources.

Eckersley, Robyn. *Environmentalism and political theory: toward an ecocentric approach*. Albany: State University of New York, 1992. 274p. (JA75.8 .E26 1992)
 Examines the different kinds of Green politics and their relationships to traditional Western political theory. Uses this analysis to argue for an ecocentric political theory. Contains useful bibliographic notes.

Killingsworth, M. Jimmie; and Palmer, Jacqueline S. *Ecospeak: rhetoric and environmental politics in America*. Carbondale: Southern Illinois University Press, 1992. 312 p. (HC110.E5 K5 1992)
 Maintains that environmental discourse often ends in stalemate because participants often fail to link scientific or social issues to genuine environmental awareness. Discusses different philosophies and economic aspects of the relevant issues.

Klapp, Merrie G. *Bargaining with uncertainty: decision-making in public health, technological safety, and environmental quality.* Dover, Mass.: Auburn House, 1992. 149p.
    Describes the ways in which scientific uncertainty can influence regulatory policy, in some cases weakening it and in others strengthening it. Examples include dioxin exposure and saccharin. Contrasts U.S. experience with that of Europe. A good, advanced source on health and environmental policy making.

Kury, Channing. *Enclosing the environment: NEPA's transformation of conservation into environmentalism.* Albuquerque, N.M.: University of New Mexico School of Law, 1985. 180 p.
    "Natural Resources Journal 25th Anniversary Anthology." Contents include: "Conservation decision-making: a rationalization," by Ronald Beazley; "Bringing resource conservation into the main stream of American thought," by S. Blair Hutchison; "Environmental policy and the Congress," by Henry M. Jackson; "The National Environmental Policy Act: a view of intent and practice," by Daniel A. Dreyfus and Helen M. Ingram; "Citizen participation: practice in search of a theory," by Norman Wengert; "Information channels and environmental decision making," by Helen M. Ingram; "Some observation on alternative mechanisms for public involvement: the hearing, public opinion poll, the workshop and the quasi-experiment," by Thomas A. Heberlein; "Agency responses to NEPA: a comparison and implications," by Richard N. L. Andrews; "Social impacts, politics, and the environmental impact statement process," by H. Paul Friesema and Paul J. Culhane; and "NEPA litigation in the 1970s: a deluge or a dribble?" by Richard A. Liroff.

Lester, James P., ed. *Environmental politics and policy: theories and evidence.* Durham, N.C.: Duke University Press, 1990. 405 p.
    Attempts to explain the significance of public opinion, interest groups, party politics, the courts, and elite citizens in terms of socioeconomic developments in the U.S. Other forces of change include the structural reformers, reform ecologists, free market conservatives, social ecologists, and deep ecologists. In predicting the future of the environmental movement, the authors see a variety of scenarios depending on which forces dominate.

May, Craig L.; and Dennis, Everette E., eds. *Media and the environment.* Washington, D.C.: Island Press, 1991.
    Explores how the media covers the environment and addresses problems of objectivity, advocacy, and how the subject fits into the media's concept of balanced news coverage.

Piasecki, Bruce; and Asmus, Peter. *In search of environmental excellence: moving beyond blame.* New York: Simon & Schuster, 1990. 203 p.: ill., maps (TD171 .P53 1990)
    Focuses on the roles and motivations of industry, government, and the public in terms of environmental issues. Offers much information on the history of the environmental movement and the problems its has faced. Each chapter has its

own notes, which are helpful in following the development of the authors' perspectives. A good primer for environmental issues.

Pierce, John C., et al. *Public knowledge and environment: politics in Japan and the U.S.* Boulder, Colo.: Westview Press, 1989. 229 p.

Examines the role that public knowledge plays in the environment policy making of industrial democracies, the democratic process, the role of citizen action, the environment as a "postindustrial issue," and the role of political culture between two postindustrial nations. This comparison raises interesting questions about how economically advanced countries view the interaction of humans and nature, and the relationship between economic growth and ecological balance.

Romm, Joseph J. *The once and future superpower: how to restore America's economic, energy, and environmental security.* New York: Morrow, 1992. 320p. (HC106.8 .R64 1992)

Maintains that the U.S. needs to produce "a resource-efficient national and global economy that increases everyone's living standard without harming the environment." Argues for transferring funds from the military to economic security.

Science policy: the missing prerequisite for sound decision making. Westport, Conn.: Quorum, 1992. 181p.

Maintains that the public's scientific ignorance and the separation of the scientific and political communities has led to the public being disenfranchised from the development of national policies that relate to scientific matters. Uses the regulation of recombinant DNA as a case study.

Shanley, Robert A. *Presidential influence and environmental policy.* Westport, Conn.: Greenwood, 1992. 200p.

Analyzes the role of the president in formulating and implementing environmental policy. Particular emphasis is given to the areas of information flow, data collection, risk assessment, and enforcement. The Carter, Reagan, and Bush Administrations are the focus.

Rosenbaum, Walter A. *Environmental politics and policy.* 2d ed. Washington, D.C.: Congressional Quarterly, 1991. 336 p.: ill. (HC110.E5 R665 1991)

Focuses on continuity and change in the politics of American environmentalism. Discusses the the state of the environment; the politics of environmental policy; risk assessment; toxic and hazardous substances; the politics of energy; the battle for the public lands; and the political agenda for the 1990s.

Books (unannotated)

American environment. New York: H. W. Wilson, 1984. 220 p. (HC110.E5 A647 1984)

Aylesworth, Thomas G. *Government and the environment: tracking the record.* Hillside, N.J.: Enslow Publishers, 1993.

Benveniste, Guy. *Regulation and planning: the case of environmental politics.* San Francisco: Boyd & Fraser, 1981. 207 p. (HC110.E5 B44)

Beyond the individual: environmental approaches and prevention. New York: Haworth Press, 1985. 211 p.: ill. (TD145 .B49 1985)

Burford, Anne M. *Are you tough enough?* New York: McGraw-Hill, 1986. 291 p. (HC110.E5 B883 1986)

Canada-United States relationship: the politics of energy and environmental coordination. Westport, Conn.: Praeger, 1992. 222 p.: ill. (HD9502.C32 C3777 1992)

Carroll, John E. *Environmental diplomacy: an examination and a prospective of Canadian-U.S. transboundary environmental relations.* Ann Arbor: University of Michigan Press, 1983. 382 p.: maps (HC79.E5 C37 1983)

Chiras, Daniel D. *Beyond the fray: reshaping America's environmental response.* Boulder, Colo.: Johnson Books, 1990. 210 p.: ill. (HC110.E5 C48 1990)

Echard, Jo Kwong. *Protecting the environment: old rhetoric, new imperatives.* Washington, D.C. (1612 K St., NW, Suite 704, Washington 20006): Capital Research Center, 1990. 256 p. (HC110.E5 E223 1990)

Environment '90: the legislative agenda.. Washington, D.C.: Congressional Quarterly, 1990. 106 p.: ill., maps (KF3775 .E465 1990)

Environmental communication and public relations handbook. Rockville, Md.: Government Institutes, 1988. 165 p.: ill. (HD59 .E58 1988)

Environmental concepts, policies, and strategies. Philadelphia: Gordon and Breach Science Publishers, 1991. 244 p.: ill. (HC79.E5 E5757 1990)

Environmental decision making: a multidisciplinary perspective. New York: Van Nostrand Reinhold, 1991. 296 p.: ill. (TD171.8 .E58 1991)

Environmental policy formation: the impact of values, ideology, and standards. Lexington, Mass.: Lexington Books, 1981. 244 p.: ill. (HC110.E5 E4987)

Environmental policy implementation: planning and management options and their consequences. Lexington, Mass.: Lexington Books, 1982. 261 p. (HC110.E5 E49875 1982)

Environmental policy. Cambridge, Mass.: Ballinger, 1985: ill. (HC79.E5 E5785)

Environmental politics and policy: theories and evidence. Durham, N.C.: Duke University Press, 1989. 405 p.: ill. (HC110.E5 E49879 1989)

Environmental politics: lessons from the grassroots. Durham, N.C.: Institute for Southern Studies, 1988. 122 p.: ill. (HC107.N83 E514 1988)

Environmental politics: public costs, private rewards. New York: Praeger, 1992. 209 p. (HC110.E5 E498796 1992)

Environmental protection: public or private choice. Boston: Kluwer Academic, 1991. 232 p.: ill. (HC79.E5 E5793 1991)

Fernie, John. *Resources: environment and policy.* Hagerstown Md.: Harper & Row, 1985. 338 p.: ill. (HC59 .F448 1985)

Goldstein, Joan. *Demanding clean food and water: the fight for a basic human right.* New York: Plenum Press, 1990. 271 p.: ill. (TD171.7 .G66 1990)

Gorz, Andre. *Ecology as politics.* Boston: South End Press, 1980. 215 p. (HD83 .G6713)

Hammitt, James K. *Outcome and value uncertainty in environmental policy.* Santa Monica, Calif.: RAND Corp., 1988. 309 p.: ill. (HC79.E5 H326 1988)

Hardin, Garrett James. *Filters against folly: how to survive despite economists, ecologists, and the merely eloquent.* New York: Penguin, 1986, 1985. 240 p.: ill.

Harper, John L. *Mineral King: public concern with government policy.* Arcata, Calif. (6540 Fickle Hill Rd., Arcata 95521): Pacifica Publishing, 1982. 223 p.: ill., maps (TN24.C2 H37 1982)

Henning, Daniel H. *Managing the environmental crisis: incorporating competing values in natural resource administration.* Durham, N.C.: Duke University Press, 1989. 377 p. (HC79.E5 H46 1989)

Hoberg, George. *Pluralism by design: environmental policy and the American regulatory state.* New York: Praeger, 1992. 240 p. (HC110.E5 H62 1992)

House, Peter William. *The practice of policy analysis: forty years of art and technology.* Washington, D.C.: Compass Press, 1991. 188 p.: ill. (H97 .H67 1991)

Howell, Dorothy J. *Scientific literacy and environmental policy: the missing prerequisite for sound decision making.* New York: Quorum Books, 1992. 181 p. (HC110.E5 H688 1992)

McAllister, Donald M. *Evaluation in environmental planning.* Cambridge, Mass.: MIT Press, 1980. 308 p.: ill. (HC79.E5 M315)

Natural resource and environmental policy analysis: cases in applied economics. Boulder, Colo.: Westview Press, 1988. 282 p.: ill. (HC79.E5 N355 1988)

Ophuls, William. *Ecology and the politics of scarcity revisited: the unraveling of the American dream.* New York: W. H. Freeman, 1992. 379 p.: ill. (HC79.E5 O54 1992)

Ortolano, Leonard. *Environmental planning and decision making.* New York: Wiley, 1984. 431 p.: ill. (HC79.E5 O78 1984)

Petulla, Joseph M. *Environmental protection in the United States industry, agencies, environmentalists.* San Francisco, Calif.: San Francisco Study Center, 1987. 200 p.: ill. (HC110.E5 P48 1987)

Policy through impact assessment: institutionalized analysis as a policy strategy. New York: Greenwood Press, 1989. 199 p. (HC79.E5 P653 1989)

Porritt, Jonathon. *Seeing green: the politics of ecology explained.* New York: B. Blackwell, 1984, 1985. 252 p. (HC79.E5 P668 1985)

Postmodern politics for a planet in crisis: policy, process, and presidential vision. Albany, N.Y.: State University of New York Press, 1993.

Public policy and the natural environment. Greenwich, Conn.: JAI Press, 1985. 342 p.: ill. (HC110.E5 P84 1985)

Ramphal, S. S. *Our country, the planet: forging a partnership for survival.* Washington, D.C.: Island Press, 1992. 291 p. (TD195.E25 R36 1992)

Regulatory federalism, natural resources, and environmental management. Washington, D.C.: American Society for Public Administration, 1990. 203 p. (HC103.7 .R363 1990)

Restructuring the countryside: environmental policy in practice. Brookfield, Vt.: Avebury, 1992. 238 p.: ill., maps (HT395.G7 R47 1992)

Role of environmental impact assessment in the planning process. New York: Mansell, 1988. 203 p.: ill. (TD194.6 .R65 1988)

Schwarz, Michiel. *Divided we stand: redefining politics, technology, and social choice.* Philadelphia: University of Pennsylvania Press, 1990. 176 p.: ill. (HC79.E5 S296 1990)

Shanley, Robert A. *Presidential influence and environmental policy.* New York: Greenwood, 1992.

Siegmann, Heinrich. *The conflicts between labor and environmentalism in the Federal Republic of Germany and the United States.* New York: St. Martin's, 1985. 201 p.

Smith, Zachary A. *The environmental policy paradox.* Englewood Cliffs, N.J.: Prentice Hall, 1992. 266 p. (HC110.E5 S57 1992)

Tensions at the border: energy and environmental concerns in Canada and the United States. New York: Praeger, 1992. 176 p.: ill., map (HD9502.C32 T46 1992)

Thobaben, Robert G. *Issues in American political life: money, violence, and biology.* Englewood Cliffs, N.J.: Prentice Hall, 1991. 242 p.: ill. (JK271 .T46 1991)

Uusitalo, Liisa. *Environmental impacts of consumption patterns.* New York: St. Martin's Press, 1986. 184 p.: ill. (HF5415.3 .U87 1986)

Wann, David. *Biologic: environmental protection by design.* Boulder, Colo.: Johnson Books, 1990. 284 p.: ill. (TD170 .W36 1990)

Waterfield, Larry W. *Conflict and crisis in rural America.* New York: Praeger, 1986. 235 p.: ill. (HD256 .W35 1986)

Zube, Ervin H. *Environmental evaluation: perception and public policy.* New York: Cambridge University Press, 1984. 148 p.: ill.

Technical books

Cohen, Steven; and Kamieniecki, Sheldon. *Environmental regulation through strategic planning.* Boulder, Colo.: Westview, 1991. 194 p.

> Discusses how researches designed a strategic regulatory program that would increase the probability of compliance for the underground tank provisions of the Hazardous Solid Waste Amendments of 1984. Focuses on three variables: capability, motivation, and feasibility and maintains that these variable must be in harmony for maximum compliance.

General interest periodicals (annotated)

Downey, Terrence J. "Understanding policy-making: a necessary first step for environmentalists." *Alternatives* 14 (May-June 1987): 30-34.

> "In the absence of a grasp of the broad range of factors which impinge on policy, any attempt to comprehend the ways of politicians and bureaucracy--the techniques and subtleties of decision making and the realities of the bureaucratic world--is fruitless and may lead to nothing more than frustration and, finally resignation."

"Environmental policy: special issue." *Financier* 9 (December 1985): whole issue (45 p.)

> Contents include "Focus now on risk assessment: editorial;" "Intellectual

underpinning required: interview with William D. Ruckelshaus," by J. Berry; "Environment gains against budget cuts," by J. Dineen; "Rising political issue in U.K.," by F. Cairncross; "Ten principles for action," by R. Anderson; "Put consensus-building first," by L. Fernandez; "Ozone threat, not acid rain," by G. Weyerhaeuser; "Superfund creates insurance problems," by R. Kilpatrick; "How to manage the wastes?" by D. Buntrock; "Man as predator wrecks development," by B. Walker; and "EPA has not cut industrial air pollution," by P. MacAvoy.

Gore, Albert, Jr. "The ecology of survival." *New Republic* 201 (November 6, 1989): 26, 28, 30, 34.
"A Strategic Environment Initiative would promote environmentally sustainable development by identifying and spreading new technologies to developing countries, where 95 percent of world population growth will take place during the next century. Here in the United States, the Strategic Environment Initiative must modernize technologies and practices in every economic sector, from more fuel-efficient cars and energy-efficient appliances to manufacturing that relies on recycled material, to a second green revolution requiring fewer fertilizers and pesticides."

Hayward, Steven. "The big green monster." *Reason* 22 (June 1990): 32-36.
California's proposed "Environmental Protection Act of 1990," known popularly as 'Big Green', would impose huge costs on consumers and taxpayers and create a new and very powerful state bureaucracy. It is not clear that Big Green will do much for the environment."

Lee, Robert W. "Conservatives consider the crushing cost of environmental extremism." *American Opinion* 26 (October 1983): 7-12, 99 100, 102-103, 106, 108, 110.
Cites examples of the "financial burden which American industry has suffered from overzealous, often malicious, environmentalism during the past decade. The EPA should be abolished."

Manning, Edward W. "Prophets and profits: a critique of benefit/cost analysis for natural resource decisions." *Alternatives* 15 (December 1987-January 1988): 36-41.
"Benefit/cost analysis often is used to address entirely the wrong questions, and even if done properly is ill-suited to the types of decisions politicians must really make. Yet such is the strength of the myth surrounding benefit/cost analysis that it will be extremely difficult to dislodge it as a touchstone for most public investment decisions."

"Politics of the environment, 1970-1987." *Wilson Quarterly* 11 (Autumn 1987): 50-83.
The four articles include "A big agenda" by David Vogel, which analyzes the rise of the environmental movement; "Cleaning up the Chesapeake; Learning the lessons" by Robert W. Crandall, discussing the complexities of environmental regulation; and "Background books."

General interest periodicals (unannotated)

Dentzer, Susan. "The mega-issues of the '90s." (political and social issues) *U.S. New & World Report* 105,9 (August 29 1988): 56(4)

Felten, Eric. "Close candidates, tepid issues and House districts in balance." (1990 elections; includes related articles on California's "Big Green" initiative, Proposition 128) *Insight* 6,46 (November 12 1990): 18(3)

Horowitz, David. "Making the Green one red: environmental politics." *National Review* 42,5 (March 19 1990): 39(2)

Mano, D. Keith. "Environmental Trojan horse." (Learning Alliance seminar) (column) *National Review* 41,14 (August 4 1989): 50(2)

Reed, Susan. "Hollywood heavyweights turn out in force to urge California voters to give the green light to Big Green." *People* 34,18 (November 5 1990): 125(3)

Repetto, Robert. "Environmental productivity and why it is so important." *Challenge* 33,5 (September-October 1990): 33(6)

Sancton, Thomas A. "The fight to save the planet: as concern for the environment grows, and some promising international initiatives take shape, the U.S. must do its share." *Time* 134,25 (December 18 1989): 60(2)

Scarlett, Lynn. "A guide to environmental myths and realities." (includes related article on legislation based on bad advice) *Consumers' Research* 75,1 (January 1992): 11(6)

Schechner, Richard. "Ploughshares or perish II." (need to redirect government spending from defense to the environment) (editorial) *The Drama Review* 34,2 (Summer 1990): 4(4)

Webb, Kernaghan. "Between the rocks and hard places: bureaucrats, the law, and pollution control." *Alternatives* 14 (May-June 1987): 4(10)

Wysocki, Annette. "Read our lips: an election-year primer on how to put a real environmental precedent in office." *Backpacker* 20,5 (August 1992): 28(4)

Environmental and professional journals (annotated)

Brockett, Patrick L.; Golden, Linda L.; and Aird, Paul R. "How public policy can define the marketplace: the case of pollution liability insurance in the 1980s." *Journal of Public Policy and Marketing* 9 (1990): 211-226.

"Using the case of pollution liability insurance, this paper describes how public policy can both stimulate and negate marketplace functions. The paper focuses on the role of marketing in public policy and illustrates that if public policy decision makers do not incorporate explicit consideration of both buyer and seller motivations, as well as other marketplace realities, they can work against their own objectives and stifle exchange."

Brown, Lester R.; Flavin, Christopher; and Postel, Sandra. "No time to waste: a global environmental agenda for the Bush administration." *World Watch* 2 (January-February 1989): 10-19.
Suggests the Bush administration pursue a climate-sensitive energy policy, adopt forest conservation and reforestation policies, reverse the global decline in food security, and slow population growth.

Caldwell, Lynton K. "Environmental impact analysis (EIA): orgins, evolution, and future directions." *Policy Studies Review* 8 (Autumn 1988): 75-83.
"The technique of environmental impact analysis (EIA) was developed initially in the United States in response to a requirement of the National Environmental Policy Act of 1969. Now adopted with variations in at least 30 countries and by the European Community, EIA has proved to be a valuable component of a group of related techniques for discovering and projecting the probable consequences of proposed action."

Chase, Anne. "Privatization: who, what and how." *Governing* 5 (May 1992): 51, 53-55, 57-60.
"Examines four different areas of privatization--environmental services, corrections, infrastructure, and facilities management--all points at which the government's need to get things done has intersected with business's desire to make money."

Costanza, Robert "Social traps and environmental policy." *BioScience* 37 (June 1987): 407-412.
Argues that environmental problems persist despite the availability of technical solutions because people fall into social traps, that is, situations in which short-term, local reinforcements guiding individual behavior are inconsistent with global and long-term individual and societal interests. Discusses methods by which traps could be seen and avoided.

Crotty, Patricia McGee. "Assessing the role of federal administrative regions: an exploratory analysis." *Public Administration Review* 48 (March-April 1988): 642-648.
"Proposes a framework to judge the success of national administrative regions in convincing states to accept responsibility for implementing federal programs. Successful policy supervision is seen as the ability to use communication, treasury, authority, and organization tools. Reveals why some federal regions are more successful than others."

Davis, Charles E.; and Lester, James P. "Decentralizing federal environmental policy: a research note." *Western Political Quarterly* 40 (September 1987): 555-565.
> "Examines the extent to which states have replaced federal aid reductions in the environmental area with their own-source funds and tests the explanatory power of several hypothetical explanations of state replacement behavior. The findings suggest that most of the states surveyed have not replaced federal aid cuts with their own-source funds."

"Energy-environment tradeoff." *Canada-U.S. Outlook* 1 (April 1990): whole issue (107 p.)
> Special issue on Canadian-U.S. environmental policies includes: "Implementing sustainable development, by Colin F. W. Isaacs;" "Through a glass darkly: the environment-energy interface, by Gilles Paquet;" "Power from the North: the benefits of the project, by Robert Bourassa;" "Will free trade release a flood of Canadian water?, by Kendall Moll and Ned Rosenbrook;" "The greening of U.S. and Canadian electricity trade, by Amy Abel and Larry B. Parker;" "Our changing climate: challenges and opportunities in a warming world, by Irving M. Mintzer;" and "Toward cooperation in North American hazardous waste facility siting: the case of Alberta," by Barry G. Rabe.

Fitzgerald, Michael R.; McCabe, Amy Snyder; and Folz, David H. "Federalism and the environment: the view from the states." *State and Local Government Review* 20 (Fall 1988): 98-104.
> Reports on a survey of state legislators and senior state executives in Tennessee, Massachusetts, and Texas on whether states should regulate and manage hazardous waste sites and treatment facilities. Finds a substantial antifederal government sentiment among the state officials but says further research is needed concerning the capacity of states to manage hazardous waste problems.

"Four men and a planet." *Sierra* 77 (September-October 1992): 46-48.
> Examines how the presidential candidates stack up on the critical environmental issues of the day.

"If I were President." *Environmental Action* 24 (Spring 1992): 31-36.
> "Six progressive leaders outline their platforms for environmental change." The leaders are: Barry Commoner, Dr. Benjamin F. Chavis, Jr., Heather Booth, Ralph Nader, Hazel Henderson, and Jim Hightower.

Jamail, Milton H.; and Mumme, Stephen P. "The International Boundary and Water Commission as a conflict management agency in the United States-Mexico border-lands." *Social Science Journal* 19 (January 1982): 45-62.
> Traces the history of the Commission since 1889, when it was instituted to delineate and preserve the boundary, until today when its function is both as a water engineering agency and a diplomatic agency.

Kagan, Robert A. "Adversarial legalism and American government." *Journal of Policy Analysis and Management* 10 (Summer 1991): 369-406.
    "The United States is uniquely prone to adversarial, legalistic modes of policy formulation and implementation, shaped by the prospect of judicial review. While adversarial legalism facilitates the expression of justice-claims and challenges to official dogma, its costs are often neglected or minimized."

Meyerson, Adam. "The vision thing, continued: a conservative research agenda for the '90s." *Policy Review* 53 (Summer 1990): 2-5.
    The editor of *Policy Review* discusses "six major strategic goals of the conservative movement that he believes require major scholarly work." The issues include developing a conservative environmentalism.

Nitz, William A. "Improving U.S. interagency coordination of international environmental policy development." *Environment* 33 (May 1991): 10-13, 31-37.
    Looks at emerging international environmental issues, examines existing interagency mechanisms for addressing these issues, and explores the boundary constraints imposed by the U.S.'s government system, political culture, and personnel selection process. Recommends that an improved interagency process should have, among other things, "a cabinet-level committee that focuses on environmental and energy issues but covers both the domestic and foreign policy aspects of those issues."

"Organizing for analysis." *Journal of Policy Analysis and Management* 8 (Summer 1989): 373-410.
    Contents include "Policy analysis in the bureaucracy: how new? How effective?" by Laurence E. Lynn, Jr.; "Policy analysis at the Department of State: the Policy Planning Staff," by Lucian Pugliaresi and Diane T. Berliner; "The Office of Policy Analysis in the Department of the Interior," by Robert H. Nelson.

Ost, Laura. "Governing guide: cleaning up--pollution, policy and technology." *Governing* 4 (April 1991): 33-56.
    "Controlling air and water pollution is within technology's grasp. But state and local government officials trying to clean up the environment have a tough job and very little money. They must weigh factors ranging from the type of pollutants involved to the cleanup costs their community can bear."

Peach, J. Dexter; and Steinhardt, Bernice. "What we've learned since Earth Day." *G.A.O. Journal* 11 (Winter 1990-91): 29-35.
    "With too many problems and too few resources, we need to target the environmental problems that threaten us most."

"Power of the earth." *Congressional Quarterly Weekly Report* 48 (January 20, 1990): 137-191.
    Looks at environment and energy-related issues that the 101st Congress will

debate. Articles cover topics such as clean air, agriculture, timber, solid wastes, nuclear weapons, energy policy, global warming, fuel economy, and lobbyists and key players.

Russell, Dick. "Big no." *Amicus Journal* 13 (Winter 1991): 4-6.
"California's rejection of Big Green is a mandate for even more fundamental change."

Russell, Milton. "Environmental protection for the 1990s and beyond." *Environment* 29 (September 1987): 12-15, 34-38.
"We will be faced with explicit tradeoffs between reducing toxic risks still further and reducing yet more personal freedom and the consumption of other goods and services--including risk reductions elsewhere. For the future, much of the remaining pollution that will cause the most risk is from widely dispersed sources whose control will depend on changing the behavior of individual citizens."

Sabatier, Paul; Hunter, Susan; and McLaughlin, Susan. "The devil shift: perceptions and misperceptions of opponents." *Western Political Quarterly* 40 (September 1987): 449-476.
"Examines the argument for a 'devil shift,' i.e., that political elites in relatively high conflict situations perceive their opponents to be more 'evil' and more powerful than do most other actors in their policy community. The extent of distortion is largely a function of ideological distance. Data from a study of land use conflict at Lake Tahoe largely support this argument."

"Saving America." *Buzzworm: The Environmental Journal* 4 (September-October 1992): 28-36, 89.
This Special EcoReport on the 1992 election offers the presidential candidates' statements on ten environmental issues. It also provides an analysis of their records by environmental reporters Keith Schneider of the *New York Times* and Bobbi Ridlehoover of the Arkansas Democrat Gazette; Schneider also presents a brief essay on their running mates. Governor Jerry Brown and Senator Al Gore give thoughtful suggestions about their visions of an environmental president. Finally, departing Senator Tim Wirth gives an election year interview after his return from the Earth Summit.

Schwab, Robert M. "Environmental federalism." *Resources* 92 (Summer 1988): 6-9.
"While it is true that we should apply the same principle in every region, we should not set the same environmental standard everywhere unless the costs and benefits of pollution control are the same in each region. There are excellent reasons to believe that this is not the case." Argues for decentralization of environmental policy making.

"Six issues for a campaign." *National Journal* 20 (September 24, 1988): 2370-2372, 2375-2378, 2383-2387.

Lists six issues the two 1988 presidential candidates "have merely danced around the edges of," but will likely have to be confronted by the next administration. Rochell L. Stanfield identifies urgent environmental problems, and Bruce Stokes foresees global disasters spreading from the Third World.

Smith, Kerry. "Resource evaluation at the crossroads." *Resources* 90 (Winter 1988): 2-5.
Contends that the framework of benefit-cost analysis, devised originally to cope with traditional public investment projects, needs to be modified in theory and practice. Proposes a broader term, "resource evaluation," and describes the modifications to benefit-cost methods that are necessary to evaluate current environmental and natural resource policy issues.

Stanfield, Rochelle L. "The green blueprint." *National Journal* 20 (July 2, 1988): 1735-1737.
"The environmental community, having learned some important lessons from the Heritage Foundation, is determined to greet the new president-elect in November with a detailed program for change."

Stoel, Thomas B., Jr. "Global tomorrow." *Amicus Journal* 10 (Fall 1988): 21-27.
Suggests actions that President Bush and his administration should take to solve the following global environmental problems: ozone depletion; global warming; and, in the Third World, the interrelated problems of rapid population growth, desertification, deforestation, and loss of wildlife species.

Environmental and professional journals (unannotated)

"102nd Congress off to active start." *Parks & Recreation* 26,4 (April 1991): 7(3)

Dunlap, Riley E.; and Scarce, Rik. "The polls--poll trends: environmental problems and protection. " *Public Opinion Quarterly* 55,4 (Winter 1991): 651(22)

O'Riordan, Timothy. "The politics of environmental regulation in Great Britain." *Environment* 30,8 (October 1988): 4(12)

"Picking up the environmental check." (local governments, from an Environmental Protection Agency report) *American City & County* 105,9 (September 1990): 107(3)

"Public's capital: a forum on infrastructure issues." (Special Report) *Governing* 5,4 (January 1992): 45(12)

Rauber, Paul. "Loose lips sink careers." (federal government silencing of whistle-blowers) *Sierra* 77,2 (March-April 1992): 33(3)

Rauber, Paul. "Losing the initiative? Lessons from the 1990 election: keep it simple;

keep it cheap; and keep it off the ballot if there's going to be a war." (planning environmental initiatives) *Sierra* 76,3 (May-June 1991): 20(3)

Sampson, Neil. "Budget battleground." *American Forests* 95,11-12 (November-December 1989): 17(4)

Schmidt, David D. "Voting on the environment." *Technology Review* 90 (August-September 1987): 15(2)

Law journals

Kimmel, James Paul, Jr. "Disclosing the environmental impact of human activities: how a federal pollution control program based on individual decision making and consumer demand might accomplish the environmental goals of the 1970s in the 1990s." *University of Pennsylvania Law Review* 138 (December 1989): 505-548.
> "Proposes an 'environmental impact index' (EII), that would provide consumers with simple, concise, at-a-glance numerical information concerning the environmental impact of a product or package during its production, use, and disposal." Linked to this index, an environmental impact tax (EIT) would effectively charge consumers, not producers, for the environmental costs of the goods they consume. "The overall program would achieve environmental goals by reducing the demand for environmentally inefficient goods rather than by altering the supply."

Trubatch, Sheldon L. "Informed judicial decisionmaking: a suggestion for a judicial office for understanding science and technology." *Columbia Journal of Environmental Law* 10,2 (1985): 255-270.
> "This article proposes a modification to the existing process of judicial review of agency decisions. The proposed modification should improve the accuracy of review of those agency actions that affect the public health or the environment, and that are based upon uncertain or controversial scientific conclusions."

Reports

Baden, John. *Destroying the environment: government mismanagement of our natural resources.* Dallas, Tex. (7701 N. Stemmons, Dallas 75247): National Center for Policy Analysis, 1986. 45 leaves: ill. (HC103.7 .B24 1986)

Environmental protection: a strategic issue for the 1990s: a report prepared for Chubu Electric Power Company. Cambridge, Mass. (20 University Rd., Cambridge 02138): The Associates, 1990. 30, 23 leaves: ill. (HC110.E5 E498797 1990)

Hahn, Robert W.; and Stavins, Robert N. *Incentive-based environmental regulation: a new era from an old idea?* Cambridge, Mass.: Energy and Environmental Policy Center, John

F. Kennedy School of Government, Harvard University, 1990. 40 p. (Discussion paper no. E-90-13)

This paper begins with a brief overview of conventional and alternative approaches to environmental regulation, and a review of previous U.S. experience with incentive-based policies. We chronicle how a shift in attitudes among influential interest groups is leading to new consideration of market-based proposals at the federal level, and we seek to explain why these changes are occurring. In this context, we identify key factors that have affected the emergence of market-based approaches throughout the world. Finally, we draw some conclusions regarding these policy mechanisms and their likely future role in addressing environmental issues.

Irwin, Frances H. *Making decisions on cumulative environmental impacts: a conceptual framework.* Washington, D.C.: World Wildlife Fund, 1992. 54 p.: ill. (HC79.E5 I78 1991)

Jester, John. *Searching for success: environmental success index.* Washington, D.C. (1400 16th St., NW, Suite 710, Washington 20036): Renew America, 1991. 154 p.: ill. (TD171 .J47 1991)

Public policies for environmental protection. Washington, D.C.: Resources for the Future, 1990. 308 p. (HC110.E5 P83 1990)

Addresses the economic issues, including benefits and costs, involved in major areas of environmental regulation and policy. Partial contents include: "The evolution of federal regulation," by Paul R. Portney; "Air pollution policy," by Paul R. Portney;" "Water pollution policy," by A. Myrick Freeman; "Hazardous wastes," by Roger C. Dower; "Toxic substances policy," by Michael Shapiro; "Monitoring and enforcement," by Clifford S. Russell; and "Overall assessment and future directions," by Paul R. Portney.

Smith, Kerry. *On separating defensible benefits transfers from "smoke and mirrors".* Washington, D.C.: Resources for the Future, Quality of the Environment Division, 1991. 29 p. (Discussion paper QE91-10)

"Benefits transfer methods increasingly are being applied to value nonmarketed resources for both policy evaluation and natural resource damage litigation. This paper illustrates the need for guidelines for deciding when benefits transfer methods can be used to efficiently value changes in environmental resources."

Federal government reports

Our changing planet: the FY 1990 research plan; executive summary. Washington, D.C.: U.S. Office of Science and Technology Policy, Federal Coordinating Council on Science, Engineering, and Technology. Committee on Earth Sciences, 1989. 43 p.

The goal of the research program is to provide a sound scientific basis for policymakers to make decisions on global- change issues such as ozone depletion, climate warming, sea level change, desertification, droughts, and acid rain. This plan defines the key scientific questions, details gaps and the best agencies to fill them, and coordinates research and budgetary activities.

Conference proceedings

Conference on the Environment (19th: 1990: Airlie House) Federal versus state environmental protection standards: can a national policy be implemented locally? Washington, D.C.: Division for Public Services, American Bar Association, 1991. 35 p. (KF3775.A75 C66)

Bibliographies

Bibliography of documents issued by the GAO on matters related to environmental protection. Washington, D.C.: General Accounting Office, 1985. 81, 63 p. (Z5863.P7 B5 1985)

Environmental protection: bibliography of GAO documents, August 1988-April 1990; Resources, Community, and Economic Development Division, September 1990. Washington, D.C.: General Accounting Office, 1990. 223 p. (GAO/RCED-90-228)
    Includes information on G.A.O. documents issued between August 1988 and April 1990 that directly or indirectly discuss environmental protection. A previous G.A.O. environmental bibliography GAO/RCED-89-23, included documents issued between January 1985 and August 1988.

Goehlert, Robert. *Policy studies environmental affairs: a selected bibliography.* Monticello, Ill.: Vance Bibliographies, 1984. 7 p. (Z5863.P6 G62 1984)

Pearson, Arn H.; and Buttel, Frederick H. *Environmental politics: an annotated bibliography.* Monticello, Ill.: Vance Bibliographies, 1984. 65 p. (Public administration series: bibliography P-1369)

Van Liere, Kent D. *Public perception of environmental trade-offs: a preliminary bibliography.* Monticello, Ill.: Vance Bibliographies, 1980. 10 p. (Z5863.P6 V36)

REAGAN AND BUSH ADMINISTRATION POLICIES

(See also *Global Environmental Policy: 1992 Earth Summit*) 278

Books (annotated)

Vig, Norman J., and Kraft, Michael E., eds. *Environmental policy in the 1990's.*

Washington, D.C.: Congressional Quarterly Press, 1990. 418 p.

> A comprehensive compilation of essays written by prominent environmentalists and political scientists that reviews the environmental policies of the last twenty years and examines some of the problems facing the environmental policies of the next ten. Discusses alternative methods of resolving environmental conflicts and the international aspects of environmental problems.

Vig, Norman J.; and Kraft, Michael E., eds. *Environmental policy in the 1980s: Reagan's new agenda.* Washington, D.C.: Congressional Quarterly Press, 1984. 377 p. (HC110.E5 E49865 1984)

> "This collection by seventeen environmental experts explores the major changes in environmental policy that took place during the first three years of the Reagan presidency. By focusing on the process the administration used to shape its environmental program, the book attempts to answer this question: to what extent can an incoming presidential administration strongly committed to reorienting national priorities achieve substantial policy changes."

Books (unannotated)

Arnold, Ron. *At the eye of the storm: James Watt and the environmentalists.* Chicago: Regnery Gateway, 1982. 282 p. (HC110.E5 A76 1982)

Environmental policy under Reagan's executive order: the role of benefit-cost analysis. Chapel Hill: University of North Carolina Press, 1984. 266 p.: ill. (HC110.E5 E49878 1984)

Lash, Jonathan. *A season of spoils: the Reagan administration's attack on the environment.* New York: Pantheon Books, 1984. 385 p. (HC110.E5 L34 1984)

Natural resources and the environment: the Reagan approach. Washington, D.C.: Urban Institute Press, 1984. 144 p.: ill. (HC110.E5 N4 1984)

Ronald Reagan and the American environment: an indictment. San Francisco: Friends of the Earth, 1982. 144 p.: ill. (HC110.E5 R658 1982)

General interest periodicals (annotated)

Roberts, Steven; Walsh, Kenneth T.; and Satchell, Michael. "Is Bush in nature's way?" *U.S. New & World Report* 108 (March 19, 1990): 20-22.

> "A clean-air bill is coming, but the President's environmental policies are getting low marks from both conservationists and the business community. *U.S. News* asked leaders of ten conservation groups and ten business and trade associations to grade the Bush administration's environmental efforts in twelve key areas."

Environmental and professional journals (annotated)

Gray, C. Boyden; and Rivkin, David B., Jr. " 'No regrets' environmental policy."
Foreign Policy 83 (Summer 1991): 47-65.
> Argues that a policy that balances environmental and economic considerations is
> appropriate because of the uncertainties underlying the global warming debate
> and the limits on available resources.

Hager, Mary. "How goes the environmental presidency?" *National Wildlife* 29
(December 1990-January 1991): 14, 16-17.
> "After two years, the Bush administration's conservation record is clouded; critics
> wonder who really is setting policy."

Pope, Carl. "The politics of plunder." *Sierra* 73 (November-December 1988): 49-55.
> Presents the Sierra Club's deputy conservation director's assessment of the
> impacts of the Reagan years on environmental protection. Calls Reagan appointees
> and policy initiatives "antienvironmentalist" and presents data on environmental
> funding, staffing, and activities, 1980-1987.

Stanfield, Rochelle L. "Global guardian." *National Journal* 19 (December 12, 1987):
3138-3142.
> "Environmentalists give the administration higher marks on global environmental
> issues than on domestic ones and acknowledge dramatic improvement in efforts
> on the world front," but still see problem areas, including ozone depletion,
> international development projects, and the export of pesticides banned in the U.S.

"Tough new winds blowing from the Bush administration, Congress." *Chemical Times
& Trends* 12 (April 1989): 56-60.
> Discusses the Bush administration's environmental policy as it might affect the
> chemical industry. Looks at state environmental regulation, industry taxes, user
> fees, Clear Air Act amendments, California's Proposition 65, and waste disposal.

Environmental and professional journals (unannotated)

Dunlap, Riley E. "Polls, pollution, and politics revisited: public opinion on the
environment in the Reagan era." *Environment* 29 (July-August 1987): 6(12)

Sampson, Neil. "Mr. Bush's lasting legacy." *American Forests* 96, 3-4 (March-April
1990): 6(2)

Law journals

Atcheson, John. "The department of risk reduction or risky business." *Environmental
Law* 21,4 (1991): 1375-1412.

"This essay explores what sets the environmental agenda of the next twenty years apart from the set of problems we faced in the first twenty years, examines some of the shortcomings of a risk-dominated policy framework for dealing with the new generation of environmental problems, and briefly outlines elements of a policy framework that can help us set the right course. It is offered as part of the national dialogue called for by William Reilly and Hank Habicht."

Brown, William Y. "Environmental leadership: the search for priorities and power." *Environmental Law* 21,4 (1991): 1413-1423.

"The U.S. Environmental Protection Agency (E.P.A.) is re-evaluating its role in environmental risk assessment. It is changing from a reactive posture toward a more active role in anticipating environmental problems and reducing the likelihood of deregulation. The author observes that subjective variables complicate efforts to assess risk. This essay addresses the E.P.A., industry, and environmental organizations, and suggests cooperative strategies to achieve risk reduction."

Reilly, William K. "Taking aim toward 2000: rethinking the nation's environmental agenda." *Environmental Law* 21,4 (1991): 1359-1374.

"To meet the environmental challenges of the twenty-first century, the U.S. Environmental Protection Agency (E.P.A.) must focus on setting priorities. Risk analysis, supported by sound science and properly communicated to the public will provide a rational and important means of setting those priorities. After it has determined the order in which the problems will be solved, E.P.A. must address these problems by moving beyond the traditional media-specific, command-and-control regulations. The Clean Air Act Amendments of 1990 show the promise of market-based incentives, and efforts like the 33/50 Project demonstrate the effectiveness of integrated, voluntary programs that focus on pollution prevention."

Reports

Blueprint for the environment: advice to the president-elect from America's environmental community. Washington, D.C.: Council on the Environment, 1988. 32 p.

Gillman, Katherine, ed. *Hitting home: the effects of the Reagan environmental policies on communities across America.* Washington, D.C.: Natural Resources Defense Council, 1982. 66 p.

"The stories we report here cover only a few problems in a few dozen communities, but they are the symptoms of a much larger malady. It may be several years before all of the consequences of the Reagan administration policies are apparent--it has taken years, after all, to begin to see the results from our efforts to control pollution--but even now we are seeing some of the damage. And we can predict much more."

Global environmental change: recommendations for President-Elect George Bush. Prepared by the National Academies of Sciences and Engineering and the Institute of Medicine. Washington, D.C.: National Academy Press, 1988. 24 p.

> Contains recommendations to the Bush administration on U.S. policy regarding ozone depletion, climate warming, tropical deforestation and acid deposition.

Kloepfer, Deanne. *The Watt/Clark record: environmental policies of the Reagan administration: the National Wildlife Refuge System.* Washington, D.C. (1400 I Street, N.W., Washington, D.C. 20006): Wilderness Society, 1984. 48 p.: ill. (QH76 .K57 1984)

Kloepfer, Deanne. *The Watt record: James Watt and wilderness: the environmental policies of the Reagan administration.* Washington, D.C.: Wilderness Society, 1983. 28 p.: ill. (QH76 .K576 1983)

Skilbred, Amy. *The Watt/Clark record: environmental policies of the Reagan administration: Alaska.* Washington, D.C. (1400 I Street, N.W., Washington 20006): Wilderness Society, 1984. 63 p.: ill. (HC107.A47 E57 1984)

Smith, Fred L., Jr. *What environmental policy? In Assessing the Reagan years.* Washington, D.C.: Cato Institute, 1988.

> Concludes that the Reagan Administration failed to gain control over environmental policy. Calls for "free-market environmentalism."

Federal government reports

Bush, George. "Message to the Congress on environmental goals." *Weekly Compilation of Presidential Documents* 44 (March 24, 1992): 537-541.

> The President outlines his environmental strategy, which seeks to merge economic and environmental goals and discusses the Administration's goals and accomplishments for the global environment and for the domestic environment.

Bush, George. *U.S. committed to safe environment.* Washington, D.C.: Department of State, Bureau of Public Affairs, 1990. 3 p. (Current policy no. 1249)

> In an address to the Intergovernmental Panel on Climate Change (IPCC) at Georgetown University, Washington, D.C., on February 5, 1990, President Bush outlined U.S. efforts to protect the environment and combat global climate change. He emphasized reconciling environmental protection with the continued benefits of economic development.

Bibliographies

Gilbert, Janice Dee. *Secretary James G. Watt's Department of the Interior.* Monticello, Ill., Vance Bibliographies, 1983. 12 p. (Public administration series: bibliography P-1256)

Reference

Reducing risk: Appendix A: the report of the Ecology and Welfare Subcommittee; Relative Risk Reducton Project. Washington, D.C.: Environmental Protection Agency, Science Advisory Board, 1990. 77 p. (EPA SAB-EC-90-021A)
> Suggests an alternative approach to defining environmental problems, ranking them from an ecological perspective. Identifies a need to more accurately reflect ecological concerns in economic/welfare considerations.

Unfinished business: a comparative assessment of environmental problems. Washington, D.C.: Environmental Protection Agency, 1987. 5 v.
> Considers four different types of risks, namely cancer risks, non-cancer health risks, ecological effects, and welfare effects, of thirty-one environmental problem areas. Observes that the final rankings by risk do not correspond well to E.P.A.'s current program priorities.

ENVIRONMENTAL LAW

Books (annotated)

Buck, Susan J. *Understanding environmental administration and law.* Washington, D.C.: Island Press, 1991.
> Offers a synopsis of U.S. environmental law and how it relates to judicial policy-making. Meant to be an aid to those who deal with the law on a daily basis, it contains a section on environmental law in Britain and a chapter on international environmental law, covering such issues as ozone depletion and acid rain.

Caldwell, Lynton K. *Science and the National Environmental Policy Act: redirecting policy through procedural reform.* University, Ala.: University of Alabama Press, 1982 178 p.
> A series of six essays describes how a procedural invention, the environmental impact analysis, was enlisted to effect major reorientation of public policy and administration.

Carpenter, David A., et al., eds. *Environmental dispute handbook: liability and claims.* New York: Wiley, 1991. 2 v. (KF1299 .H39 E575 1991)
> Sorts out the types and classes of legal actions related to federal statutes and state court cases. Explains in broad, lay terms the claims and litigations, and identifies partics potentially at risk, such as property owners and transporters. Especially valuable as a supplement to government publications related to environmental regulations.

DiMento, Joseph F. *Environmental law and American business: dilemmas of compliance.*

New York: Plenum, 1986. 228 p.

> Outlines the difficulties involved in formulating, communicating, and enforcing environmental law. Identifies the factors that result in businesses being unable or unwilling to comply. Analyzes the methodologies agencies have at their disposal to enforce compliance, including their intrinsic problems--such as policy changes associated with a new administration. Offers suggestions for simplifying governmental rulemaking.

Miller, Jeffrey G. *Citizen suits: private enforcement of federal pollution control laws.* Washington, D.C.: Environmental Law Institute, 1987. 306 p.

> Aims to delineate the procedures and limitations of the public's ability to file citizen suits designed to enforce environmental regulation when government has failed to do so. Offers a great deal of practical information concerning procedures.

Paul, Ellen Frankel. *Property rights and eminent domain.* New Brunswick, N.J.: Transaction Books, 1987. 276 p.

> The Wisconsin Supreme Court upheld a 1972 decision prohibiting a property owner from using his coastal marshlands as a landfill. Raises the question as to whether the state can force a property owner to maintain open land for the scenic, ecological, or moral benefit of other citizens, and if the same principle of eminent domain applies when the state takes land from a citizen for public use.

Tietenberg, T. H., ed. *Innovation in environmental policy: economic and legal aspects of recent developments in environmental enforcement and liability.* Brookfield, Vt.: E. Elgar, 1992. 269p.

> Evaluates regulatory sanctions, tort law, and private enforcement to determine which elements of U.S. environmental policy are the most effective. Argues that fines are often more efficient and less expensive than jail terms and highlights the recent increase in law suits filed by environmental organizations. Maintains that currently U.S. environmental policy is somewhat inconsistent and that a more cohesive package of reformed tort law, economic incentives, and fines would be more conducive to environmental protection.

Yeager, Peter C. *The limits of the law: the public regulation of private pollution.* New York: Cambridge University Press, 1991.

> Examines how government legislation protects the environment, often despite great inconvenience. Focusing on water pollution, maintains that the passage of environmental laws has been impeded by biased politicians and polluters. Analyzes the subsequent green movement and the influence of consumers on such things as drinking water quality.

Books (unannotated)

Adams, Sherman. *The Weeks Act: a 75th anniversary appraisal.* New York: Newcomen Society of the United States, 1987. 24 p.: ill. (SD412 .A33 1987)

Byrne, Noel. *Legal protection of plant varieties.* New York: Stockton Press, 1993.

Davis, T. R. *Successor liability for environmental damage.* Calgary, Alberta, Canada: University of Calgary Press, 1989. 46p.

Kernes, Steven T. *Law enforcement reference: selected statutes and rules.* Port Angeles, Wash.: Justice Systems Press, 1989. 49 p. (KFW452 .A3 1989)

Laitos, Jan. *Energy and natural resources law in a nutshell.* St. Paul, Minn.: West Publishing, 1992. 554 p.: ill. (KF2120.Z9 L27 1992)

Law, policy, and the environment. Cambridge, Mass.: Blackwell, 1991. 173 p. (KD3372 .L39 1991)

Natural resources statutes. Rockville, Md.: Government Institutes, 1991. 539 p. (KF5505 .A3 1991)

Plater, Zygmunt J. B. *Environmental law and policy: a coursebook on nature, law, and society.* St. Paul, Minn.: West Publishing, 1992. 1039, 33 p.: ill. (KF3775.A7 P55 1991)

Environmental and professional journals (annotated)

Bear, Dinah. "NEPA at 19: a primer on an old law with solutions to new problems." *Environmental Law Reporter* 19 (February 1989): 10060-10069 (20 p.)
> "The general counsel of the Council on Environmental Quality, outlines N.E.P.A.'s purposes, scope, and implementation procedures. She describes current issues in N.E.P.A. practice and policy, and observes that N.E.P.A. has continuing vitality in the context of a new generation of environmental concerns that could only have been guessed at by its original supporters."

"Enforcing environmental laws." *EPA Journal* 13 (March 1987): 2-27.
> Contents include: "Enforcement today: an interview with Thomas L. Adams, Jr.;" "Pollution doesn't pay: a landmark case, by Matthew Coco;" "Penalties on the rise, by Carol Hudson Jones;" "Tools to deter violators, by Terrell E. Hunt;" "Enforcement in the year 2000, by Richard H. Mays and Julie C. Becker;" "The Justice Department: when the polluter meets the judge, by F. Henry Habicht, II;" "EPA's regional offices: a case of being on the front line, by Victor J. Janosik;" "The States: innovative ways to enforce the cleanup, by LeRoy C. Paddock;" "Local government: the pollution didn't wash, by Carol Panasewich;" "Initiatives by EPA's enforcement offices;" and "The sludgebusters," by Thomas Graf.

"FACTA: Food, Agriculture, Conservation, and Trade Act 1990: conservation and environmental highlights." *Journal of Soil and Water Conservation* 46 (January-February 1991): 20-22.

"The 101st Congress not only reauthorized and modified the landmark 1985 Conservation Title, it also enacted new provisions dealing with water quality, pesticides, organic food, sustainable agriculture research, global warming, and other environmental topics."

Parenteau, Patrick A. "NEPA at twenty: great disappointment or whopping success?" *Audubon Magazine* 92 (March 1990): 104, 106-107.

"N.E.P.A. took an old ethic and made it a national policy. The ethic is steward-ship, which has not always occupied a high place in American values. To be good stewards means accepting the principle that we belong to the land, not the other way around. Stewardship means putting something back into the land, and sometimes leaving it alone. It means wise use, with the emphasis on wise."

Law journals

"1990 Ninth Circuit environmental review." *Environmental Law* 21 (1991): 1117-1319. First annual survey of environmental cases decided by the U.S. Court of Appeals for the Ninth Circuit. Discusses air and water quality; hazardous waste; the National Environmental Policy Act; natural resources; the E.S.A. and the spotted owl; fish and wildlife; public lands; reclamation law; wetlands and scenic rivers; land use; and native rights.

Benfield, F. Kaid; Ward, Justin R.; and Kinsinger, Anne E. "Conservation gains in the Tax Reform Act: an analysis of the implications of tax reform for farmers and natural resources in rural America, with a policy agenda for the future." *Harvard Environmental Law Review* 11,2 (1987): 415-435.

"This article considers four general topics. First, it presents an overview of what ramifications tax reform might have for economic conditions in the nation's farming regions. Second, the analysis focuses on reform provisions of special significance for natural resources. The third section introduces issues likely to be faced by conservationists as the new law is implemented, and the article con-cludes with some thoughts on a future agenda for conservation tax legislation."

Blumm, Michael C. "A primer on environmental law and some directions for the future." *Virginia Environmental Law Journal* 11 (Spring 1992): 381-399.

Discusses a number of characteristics of modern American environmental law, including the process for assessing environmental impact, the legislation for pollution control, the program of environmental cleanup, and the direction in which environmental legislation is heading.

Brady, Timothy Patrick. "But most of it belongs to those yet to be born: the public trust doctrine, NEPA, and the stewardship ethic." *Boston College Environmental Affairs Law Review* 17 (1990): 621-646.

"Proposes that the future of the public trust doctrine lies in its ability to change the way society considers the needs and rights of future generations"; examines the history of the doctrine; argues that its legitimate purpose is as a catalyst for social change; examines N.E.P.A. as the situs for application of the doctrine; "calls for a renewed appreciation of the public trust doctrine as a legal device that challenges legal perceptions and assumptions about land ownership."

"Symposium on NEPA at twenty: the past, present and future of the National Environmental Policy Act." *Environmental Law* 20 (1990): whole issue (447-810 p.)
    Contents include the following essays: "The National Environmental Policy Act at Twenty: a preface," by Michael C. Blumm; "Perspectives on NEPA in the courts," by Donald N. Zillman and Peggy Gentles; "NEPA's promise--partially fulfilled," by Nicholas C. Yost; "Judicial enforcement of NEPA-inspired promises," by Thomas O. McGarity; "NEPA and private rights in public mineral resources: the fee complex relative?," by George Cameron Coggins and Jane Elizabeth Van Dyke; "NEPA's impacts on federal agencies, anticipated and unanticipated," by Paul J. Culhane; and "The myth of mitigation under NEPA and SEPA," by Peter J. Eglick and Henryk J. Hiller.

"Ten years of environmental law." *Virginia Environmental Law Journal* 10 (Fall 1990)
    "The Journal has asked five leading environmental scholars and practitioners to reflect on the last ten years of environmental law, with special emphasis on their particular areas of interest." Contents include: "Regulation of pollution: is the system mature or senile?," by Samuel A. Bleicher; "The emergence of environmental mediation," by Richard C. Collins; "Legislative approaches to solving natural resource, energy, and environmental problems," by D. Michael Harvey; "Regulating environmental carcinogens: evolving paradigms," by Richard A. Merrill; and "Natural resources law," by Ronald J. Tipton.

Webster, Timothy K. "Protecting environmental consent decrees from third party challenges." *Virginia Environmental Law Journal* 10 (Fall 1990): 137-164.
    "Focuses on the assailability of environmental consent decrees by third parties. Part I examines the history and use of consent decrees. Part II reviews the methods by which such attacks could be avoided. Part III suggests congressional strategies for encouraging the use of consent decrees by uniformly discouraging collateral challenges, and Part IV proffers the conclusion that some circumstances may pose problems that cannot be avoided."

REGULATION AND THE EPA

(Related topics in *Air Pollution*) 1607
(See also *Groundwater Pollution: Legislation and Regulation*) 1961
(See also *Hazardous Waste Management: Superfund*) 2288

Books (annotated)

Landy, Marc K.; Roberts, Marc J.; and Thomas, Stephen R., eds. *The Environmental Protection Agency: Asking the wrong questions*. New York: Oxford University Press, 1990. 309 p. (TD171.L36 1990)

> This critical evaluation of the E.P.A. examines four areas--responsiveness to the public, fidelity to technical merits, civic education, and capacity building--and concludes that education is the E.P.A.'s weakest link. Maintains that the E.P.A. has consistently underrated the complexity of environmental issues when presenting them to the public. Replete with details of political conflict within the organization. While the primary focus is from the E.P.A.'s inception in 1970 to the end of the Carter administration, there is also a scathing review of the early Reagan years.

Legal aspects of recent developments in environmental enforcement and liability. Brookfield, Vt.: Ashgate, 1992. 269p.

> Focuses on the different types of environmental policy, including tort law, private enforcement, and regulatory sanctions. Examines these different types of legislation to determine which are the most effective in attaining environmental policy goals. Maintains that fines are often as effective (as well as less costly) than imprisonment. Suggests a combined strategy of fines, economic incentives, and reformed tort law.

Luken, Ralph A. *Efficiency in environmental regulation: a benefit-cost analysis of alternative approaches*. Boston: Kluwer Academic, 1990. 364 p. (HD9826.L84 1990)

> With the narrow focus of the Environmental Protection Agency's regulation of air and water pollutants emitted by the pulp and paper industry from 1973 to 1984, this study concludes that ambient-based standards for air pollution are more successful, and more efficient, than technology-based standards used to regulate water pollution. Different methods for regulating pollution in the future are also discussed, including benefit tailoring, ambient concentration limits, and technology specification. Contains a plethora of data.

Vogel, David. *National styles of regulation: environmental policy in Great Britain and the United States*. Ithaca, N.Y.: Cornell University Press, 1986.

> Compares the environmental policies of Britain, which are piecemeal, practical, and centered around the idea of acceptable risk, and the environmental policies of the United States, which are impractical, unrealistic, and catalysts for lawsuits. On a deeper level, this book is examining the two very different ways these democratic countries have tried to control the actions of private businesses.

Books (unannotated)

Environmental protection: regulating for results. Boulder, Colo.: Westview Press, 1991. 190 p.: ill. (HC110.E5 E498798 1991)

Russell, Clifford S. *Enforcing pollution control laws.* Washington, D.C. Resources for the Future, 1986. Distributed by Johns Hopkins University Press. 231 p.: ill. (HC110.E5 R87 1986)

Yandle, Bruce. *The political limits of environmental regulation: tracking the unicorn.* New York: Quorum Books, 1989. 180 p. (HC110.E5 Y36 1989)

General interest periodicals (annotated)

Ramirez, Anthony. "EPA should clean up its own act." *Fortune* 120 (November 6, 1989): 139-140, 142.
> "Even the E.P.A.'s staunchest supporters consider it poorly managed, balkanized, and slow to carry out a complex agenda legislated over the past two decades." Author recommends what the Administration, Congress, business, and environmentalists can do to unsnarl the E.P.A..

General interest periodicals (unannotated)

Mathews, Jay. "Big Green's big question mark." (Environmental Protection Act of 1990) *California Magazine* 15,6 (June 1990): 98(2).

Powell, Monica. "Throwing the book at polluters: environmental law enforcement." *Insight* 6,19 (May 7 1990): 52(2)

Environmental and professional journals (annotated)

Crotty, Patricia McGee. "The new federalism game: primacy implementation of environmental policy." *Publius* 17 (Spring 1987): 53-67.
> "Federal environmental laws have substantially preempted state powers over pollution control. Many of these laws contain a unique implementation scheme called 'primacy,' which offers a state the opportunity to become the primary enforcement agent for federal policies. Primacy relieves the federal government from enforcing laws within state boundaries even while it retains ultimate control over the policies involved and sets minimum standards. By and large, states have chosen to accept primacy."

"Environmental permitting." *Natural Resources & Environment* 2 (Winter 1987): whole issue (50 p.)
> Contents include: "The role of in-house personnel," by D. Stafford Johnson; "The role of outside counsel," by Robert T. Connery; "Precious metals mine and processing facility permitting," by Brian R. Hanson and Sheila Bush; "Environmental permitting: expediting the N.E.P.A. process," by Albert I. Herson; "Clean

Water Act permitting: the NPDES program at fifteen," by Kristy A. Niehaus; "Section 404 of the Clean Water Act: an E.P.A. perspective," by Bruce D. Ray; "Public input in the permitting process: the Section 404 example," by James T. B. Tripp; "RCRA permitting, by Caroline Wehling;" and "Complying with environmental permits after issuance," by Stephen D. Bundy.

Gaynor, Kevin; and Starr, Judson W. "Too many cooks." *Environmental Forum* 6 (January-February 1989): 9-14.
"President George Bush has promised stricter enforcement of environmental regulations. How can the Justice Department and E.P.A. work together to more effectively achieve this goal? Two seasoned experts offer criminal and civil enforcement strategies to new E.P.A. Administrator Bill Reilly."

Kadec, Sarah T. "The Environmental Protection Agency: a profile of its information collection and dissemination." *Government Information Quarterly* 6,3 (1989): 295-309.
"This article provides an overview of the agency and its publication distribution programs. The article illustrates the E.P.A.'s relationship to both the Government Printing Office and the National Technical Information Service."

Rademaker, Ken. "ECO-COPS." *Occupational Hazards* 53 (September 1991): 125-127. (E.P.A.'s criminal investigators are the point men in the agency's drive to bust the environmental bad guys.)
Fully authorized to carry guns, executive search warrants, and make arrests, "the criminal investigators employed by E.P.A. track down crimes that are committed against the environment, seek out those who commit them, and work to get the responsible parties convicted."

Roberts, Leslie. "Counting on science at E.P.A.." *Science* 249 (August 10, 1990): 616-618.
"William Reilly is trying to give science a bigger role in E.P.A. policy" allowing it "to focus the agency's resources on the environmental problems that pose the biggest risks rather than those that have attracted the most political attention."

Wann, David. "Environmental crime: putting offenders behind bars." *Environment* 29 (October 1987): 5, 44-45.
In past years some U.S. companies found that it was less expensive to pay a fine than to clean up toxic effluent or emission. Now special investigators for the E.P.A. "are deputized with the full powers of U.S. deputy marshals to stalk individual offenders--company presidents--and bring them to justice. Many states are also using law enforcement officers in this effort."

Weidenbaum, Murray. "Return of the "R" word: the regulatory assault on the economy." *Policy Review* 59 (Winter 1992): 40-43.
Argues that "those who wonder why the economy has not responded more swiftly to successive doses of monetary stimulus should consider a development

overlooked by most analysts of the macro-economy: any U.S. company brave enough to consider embarking on a new capital investment faces a thicket of obstacles in the form of expanded environmental and other social regulations."

Environmental and professional journals (unannotated)

Crawford, Mark. "R&D eroding at EPA." *Science* 236 (May 22 1987): 904(2)

Law journals

Andreen, William L. "In pursuit of N.E.P.A.'s promise: the role of executive oversight in the implementation of environmental policy." *Indiana Law Journal* 64 (Spring 1989): 205-261.

Argues for expanded surveillance of N.E.P.A. implementation. Expansion would require E.P.A.'s watchdog role to extend beyond mere monitoring of actions proposed in final impact statements to include "records of decisions and the implementation of the commitments found in those decisions. Furthermore, to enhance the effectiveness of this new role, agencies will have to provide E.P.A. with more information on their progress in implementing their decisions, including the mitigation measures to which they are committed. Armed with these new tools, E.P.A. would be able for the first time to refer truly 'unsatisfactory' final actions to C.E.Q., rather than just proposals as presented, perhaps in unduly flattering terms, in an impact statement."

Carlson, J.; and Braden, John B. "Implications of Executive Order 12,291 for discretion in environmental regulation." *Boston College Environmental Affairs Law Review* 12 (Winter 1985): 313-340.

"The purpose of this article is to show that Executive Order 12,291 (which addresses the economic burden generated by federal regulation) provides insufficient guidance to ensure that inefficient regulations will be avoided. Due to vagueness in the order, administrative agencies are forced to make important decisions concerning when and how to conduct R.I.A.s (regulatory impact analyses). Considerations other than efficiency may easily enter into these decisions and undermine the intent of the order. These issues are explored with reference to regulations issued by the U.S. Environmental Protection Agency (E.P.A.)."

Hempel, Lamont C. "EPA in the year 2000: perspectives and priorities." *Environmental Law* 21,4 (1991): 1493-1508.

"There are at least five driving forces that may strongly shape and direct E.P.A.'s mission as it enters the next century: (1) the growing importance of transnational environmental dilemmas; (2) the intensifying struggle between science and politics in environmental policy making; (3) the shift in emphasis from 'end of pipe,' single medium regulation (for instance, air, land, or water) to integrated, and more preventive, multimedia approaches; (4) the increasing tension between human health-based approaches and holistic ecology approaches to environmental risk reduction; and (5) the growing reliance on markets and economic incentives for environmental protection."

Mazmanian, Daniel A.; and Morell, David L. "EPA: coping with the new political economic order." *Environmental Law* 21,4 (1991): 1477-1491.

> While the "command-and-control E.P.A. has had some success with the goals of the 1970s, to meet the challenges of the twenty-first century, E.P.A. must adopt a new attitude. EPA must embrace a true cross-media focus and shift from centralized regulatory control to reliance on market-based solutions. Further, E.P.A. must rely on technology standards and delegate as much regulation to the state and local level as possible."

Paddock, LeRoy C. "Environmental enforcement at the turn of the century." *Environmental Law* 21,4 (1991): 1509-1525.

> "Since the mid-1980s, Congress and state legislatures have passed numerous environmental protection statutes. Agencies will have to cope with the increased workload created by these statutes by developing new enforcement techniques and strategies. A comparison with the Dutch enforcement system shows that increasing the role of local governments will help promote new enforcement schemes. Other methods for coping with the new workload will include increased reliance on criminal penalties, citizen enforcement efforts, field citations, strategic planning, and administrative penalty orders."

Reports

Magat, Weley A.; Krupnick, Alan J.; and Harrington, Winston. *Rules in the making: a statistical analysis of regulatory agency behavior.* Washington, D.C.: Resources for the Future, 1986. 182 p.

> Using multivariate statistics as a foundation, this work examines regulatory decision making. Uses the Environmental Protection Agency's Clean Water Act as a case study to demonstrate how guidelines are set. Suggests that rule making should be reformed by minimizing the industrial monopoly on relevant information, by increasing cost-effectiveness, and by cutting down on agency turnover.

Quirmbach, Herman. *Environmental regulation and R&D incentives.* Santa Monica, Calif.: Rand Corp., 1982. 126 p.: ill. (HC79.E5 Q57 1982)

Stroup, Richard L.; and Goodman, John C. *Making the world less safe: the unhealthy trend in health, safety, and environmental regulation.* Dallas, Tex.: National Center for Policy Analysis, 1989. 33 p. (NCPA policy report no. 137)

> Argues that when federal and state agencies ban suspected carcinogenic products,

the public often turns to other products that are less safe. Cites bans on EDB, DDT, and California well water as examples of hazards for which replacements were more hazardous. Condemns California's Proposition 65. Emphasizes that risks must be viewed in perspective.

Federal government reports

Environmental enforcement: alternative enforcement organizations for EPA; report to congressional requesters. April 14, 1992. Washington, D.C.: General Accounting Office, 1992. 36 p. (GAO/RCED-92-107, B-247208)

This report discusses the advantages and disadvantages of the current and alternative organizational structures for enforcement responsibilities at the Environmental Protection Agency."

Environmental enforcement: EPA needs a better strategy to manage its cross-media information; report to the Chairman, Committee on Governmental Affairs, U.S. Senate. April 2, 1992. Washington, D.C.: General Accounting Office, 1992. 20 p. (GAO/IMTEC-9214, B-246338)

Deficiencies in E.P.A.'s information systems are impeding its ability to enforce environmental laws and regulations. E.P.A. cannot readily bring together and correlate data from its various programs--such as air, water, hazardous waste, and pesticides--in order to comprehensively assess environmental risks or identify and target the most important enforcement priorities. These shortcomings are likely to persist unless E.P.A. develops an information management strategy that directly addresses the need for better integrating the program offices' data and systems."

Environmental enforcement: penalties may not recover economic benefits gained by violators; report to congressional requesters. June 17, 1991. Washington, D.C.: General Accounting Office, 1991. 31 p. (GAO/RCED-91-166, B-243879)

Examines E.P.A.'s penalty policies and practices. "Because penalties should serve as a deterrent to violators and should ensure that regulated entities are treated fairly and consistently, it has been E.P.A.'s policy since 1984 that penalties for significant violations of environmental regulations be at least as great as the amount by which a company would benefit by not being in compliance." However, this has not always been the case.

Environmental progress and challenges: EPA's update. Washington, D.C.: Environmental Protection Agency, 1988. 140 p.: ill., maps (EPA-230-07-88-033) (TD171 .E567 1988)

Summarizes E.P.A.'s accomplishments and agenda for environmental protection. Discusses problems that relate to air, water, land, and toxic chemicals and for each problem, outlines efforts to date, immediate challenges, and the E.P.A. agenda for the problem.

EPA's review of progress and new directions in environmental protection. Washington, D.C.: Environmental Protection Agency, 1990. 26 p.: ill., maps.

Meeting the environmental challenge: EPA's review of progress and new directions in environmental protection. Washington, D.C.: Environmental Protection Agency, 1990. 26 p.: ill., maps.

ENVIRONMENTAL MANAGEMENT

Books (annotated)

Hansen, P. E.; and Jørgensen, S. E., eds. *Introduction to environmental management.* New York: Elsevier, 1991. 403 p.: ill. (TD170 .I584 1990)
> Examines, using a multidisciplinary approach, a wide variety of issues, such as different kinds of pollution and the effects of law and economics on the planning process. Contains numerous European case studies.

Books (unannotated)

Annotated reader in environmental planning and management. New York: Pergamon Press, 1983. 460 p.: ill. (TD170.3 .A56 1983)

Applied social science for environmental planning. Boulder, Colo.: Westview Press, 1984. 278 p.: ill. (HC79.E5 A64 1984)

Buckley, Ralf. *Perspectives in environmental management.* New York: Springer-Verlag, 1991. 276 p. (TD170 .B83 1991)

Friedman, Frank. *Practical guide to environmental management.* Washington, D.C.: Environmental Law Institute, 1990. 208 p.: ill. (KF3775 .F77 1990)

Gilpin, Alan. *Environmental planning: a condensed encyclopedia.* Park Ridge, N.J.: Noyes Publications, 1986. 348 p.: ill. (HC79.E5 G52 1986)

Greenland, David. *Guidelines for modern resource management: soil, land, water, air.* Columbus: C.E. Merrill Publishing, 1983. 224 p.: ill. (S938 .G74 1983)

Integrated environmental management. Chelsea, Mich.: Lewis Publishers, 1991. 214 p.: ill. (HC107.G42 S285 1991)

Natural resources for the 21st century. Washington, D.C.: Island Press in cooperation with American Forestry Association, 1990. 349 p.: ill. (HC103.7 .N296 1990)

Selman, Paul H. *Ecology and planning: an introductory study*. London: G. Godwin, 1981. 174 p.: ill. (HD111 .S44 1981)

Young, Oran R. *Resource regimes: natural resources and social institutions*. Berkeley: University of California Press, 1982. 276 p.: ill. (HC59 .Y69 1982)

Technical books (unannotated)

Cooke, Ronald U. *Geomorphology in environmental management: a new introduction*. 2nd ed. New York: Oxford University Press, 1990. 410 p.: ill, maps (GB406 .C64 1990)

Dorney, Robert S. *The professional practice of environmental management*. New York: Springer-Verlag, 1989. 228 p.: ill. (TD159 .D67 1989)

Goldberg, Michael A. *On systemic balance: flexibility and stability in social, economic, and environmental systems*. New York: Praeger, 1989. 180 p.: ill. (HC79.E5 G63 1989)

Janssen, Ron. *Multiobjective decision support for environmental management*. Boston: Kluwer Academic, 1992.

Reports

Environmental resource management. Arlington, Va. (P.O. Box 12208, 1401 Wilson Blvd., Suite 600, Arlington 22209): Management Institute for Environment and Business, 1990. 1 v. (HD69.P6 E594 1990)

Conference proceedings

Strategic planning in energy and natural resources: proceedings of the Second Symposium on Analytic Techniques for Energy, Natural Resources, and Environmental Planning. New York: Elsevier Science Publishers, 1987. 339 p.: ill. (TJ163.15 .S94 1986)

Bibliographies

Griffith, C. R. *The theory and practice of alternative mechanisms for environmental quality management: a bibliography*. Chicago, Ill.: CPL Bibliographies, 1981. 25 p.

Selected annotated bibliography on ecological planning resources. Rev. Monticello, Ill.: Vance Bibliographies, 1983. 84 p. (Z5863.P6 S44 1983)

RISK ASSESSMENT
▬▬▬▬▬▬▬▬▬▬

Books (annotated)

Bartell, Steven M. *Ecological risk estimation.* Chelsea, Mich.: Lewis Publishers, 1992. 252 p.

> Provides a background for the rapidly evolving field of ecological risk assessment directed to estimate the impact of development and/or chemical pollution on natural ecosystems and their member populations. Describing experimental approaches in ecosystem mesocosms, and the legislative framework for making such investigations, there are discussions of theoretical approaches to modeling, including risk forecasting. Finally, future research needs are identified in the chapter on future directions.

Bromley, Daniel W.; and Segerson, Kathleen, eds. *The social response to environmental risk: policy formulation in an age of uncertainty.* Boston: Kluwer Academic, 1992. 216 p.: ill. (HC79.E5 S57 1992)

> These eight essays examine the growing field of environmental risk management. Discusses environmental rights and perceptions, the role of the media, how to value risks and establish priorities, and the role of the government. Presents three relevant case studies.

Clarke, Lee. *Acceptable risk?: Making decisions in a toxic environment.* Berkeley: University of California Press, 1989. 229 p. (TD171 .C57 1989)

> Takes the perspective of organizational sociology and maintains that acceptable risk is determined by organizational or political interests rather than the public good or scientific understanding. As an example, the author demonstrates how these types of factors influenced decisions relating to the release of PCBs in an office building in New York State.

Covello, Vincent T.; et al., eds. *Effective risk communication.* New York: Plenum, 1989. 370 p.

> Solving environmental problems necessarily involves calculating risks and explaining them to policy makers and the public. While scientists have developed excellent risk analysis methods, they often find that the public is outraged by their findings. Maintains that ineffective communication has generated undue public reaction, and that outrage is reduced when risk is voluntarily undertaken, the extent of the risk is well known, and the public feels they have some control over the situation. Provides a full slate of suggestions for communicating with communities undergoing risk stress.

Fiksel, Joseph; and Covello, Vincent T. *Biotechnology risk assessment: issues and methods for environmental introductions.* New York: Pergamon, 1986.

Written as a report for the U.S. Office of Science and Technology Policy Executive, this technical work focuses on the methodology of risk assessment. Aspects such as the modelling of risks; how to assess risks from bacteria, fungi, and viruses; the spread of microorganisms; and risk monitoring are all discussed. Contains a detailed list of items to be considered before a recombinant micro-organism is released into the environment.

Greenberg, Harris R.; and Cramer, Joseph J. *Risk assessment and risk management for the chemical process industry*. New York: Van Nostrand Reinhold, 1991. 369 p.
   Overviews the techniques and requirement of risk assessment and management in the chemical industry. Discusses different types of analysis and auditing, as well as topics such as emergency planning and financing.

Lewis, H. W. *Technological risk*. New York: W. W. Norton, 1990. 353 p. (T174.5.L48 1990)
   Written with the aim of supporting technological research. Maintains that risks and benefits should be weighed against the dangers of alternative choices. There is a general discussion of the perception, assessment, and uncertainty of risk, as well as case studies and an introduction to probability.

Smith, Keith. *Environmental hazards: assessing risk and reducing disaster*. Boston, Mass.: Routledge, Chapman and Hall, 1992. 324 p.
   Focuses on hazard topology, risk assessment and perception, and adjustment to hazard. Details hazard experiences. Although most of the hazards discussed are natural, there is also a brief discussion of human-induced hazards such as nuclear power.

Waterstone, Marvin, ed. *Risk and society: the interaction of science, technology, and public policy*. Hingham, Mass.: Kluwer Academic, 1992. 178 p.
   Contains twelve essays from a 1988-1989 symposium held at the University of Arizona. Discusses risk, science, and public policy. Individual topics include the nuclear power industry, medical resources and the rationing of medical care, waste disposal, and air quality, among others.

Books (unannotated)

Aldrich, Tim. *Environmental epidemiology and risk assessment*. New York: Van Nostrand Reinhold, 1992.

Calabrese, Edward J. *Risk assessment and environmental fate methodologies*. Boca Raton, Fla.: Lewis Publishers, 1992. 150 p. (TD879.P4 E58 1992)

Davis, W. S. *Biological risk assessment: tools for risk based decision*. Chelsea, Mich.: Lewis, 1993.

Hammitt, James K. *Probability is all we have: uncertainties, delays, and environmental policy making.* New York: Garland, 1990. 291 p.: ill. (HC79.E5 H327 1990)

Risk assessment of environmental and human health hazards: a textbook of case studies. New York: Wiley, 1989. 1155 p.: ill. (RA427.3 .R57 1989)

Social and cultural construction of risk: essays on risk selection and perception. Boston: D. Reidel Publishing, 1987. Distributed by Kluwer Academic Publishers. 403 p.: ill. (HD61 .S59 1987)

Social response to environmental risk: policy formulation in an age of uncertainty. Boston: Kluwer Academic Publishers, 1992. 216 p.: ill. (HC79.E5 S57 1992)

General interest periodicals (annotated)

Main, Jeremy. "The big cleanup gets it wrong." *Fortune* 123 (May 20, 1991): 95-96, 100-101.
> "The emerging science of risk assessment says that the U.S. is spending way too much on minor threats, like asbestos, and not enough on major pollutants, like radon."

Scroggin, Don G. "EPA health risk policy will have broad impact." *Legal Times* 8 (July 15, 1985): 19, 24-25.
> "Perhaps no subject prompts more concern from industry, environmental activist groups, government regulators, and the general public than the question of how environmental risk to public health is to be determined and what level of risk is sufficient to justify government intervention."

Environmental and professional journals (annotated)

Baram, Michael S. "Risk communication and the law for chronic health and environmental hazards." *Environmental Professional* 8 (1986): 165-178.
> "This paper provides an overview of the complex subject of risk communication and the law, discussing several areas of the law that provide rights of access and duties to disclose risk and information and which restrict such rights and duties under certain circumstances. It focuses on chronic health and environmental hazards arising from exposure to chemical substances under 'routine' industrial operations."

Fiksel, Joseph; and Covello, Vincent T. "Expert systems and knowledge system for risk communication." *Environmental Professional* 9,2 (1987): 144-152.
> "This paper suggests that knowledge technology can make a significant contribution to the risk communication process by providing a structured,

interactive environment in which qualitative knowledge about specific risk situations can be examined and interpreted. Knowledge systems can help decisionmakers (1) focus their attention on the specific concerns of different stakeholders (2) systematically review alternative courses of action (3) explain the reasoning and factual basis for risk assessments or control decisions, and (4) show the effects of assumptions and uncertainties."

Freeman, A. Myrick, III; and Portney, Paul R. "Economics clarifies choices about managing risks." *Resources* 95 (Spring 1989): 1-4.

"Like it or not, environmental risks cannot be completely eradicated. Difficult choices must be made about how best to control particular risks using the limited resources available. These choices invariably involve tradeoffs. Economics can help with these decisions by providing information on the pros and cons of particular courses of action."

Harvey, Pamela D. "Educated guesses: health risk assessment in environmental impact statements." *American Journal of Law and Medicine* 16,3 (1990): 399-427.

Recommends that the Council of Environmental Quality directly address risk assessment in the environmental impact statement process.

Hattis, Dale; and Kennedy, David. "Assessing risks from health hazards: an imperfect science." *Technology Review* 89 (May-June 1986): 60-67, 71.

"Quantifying people's risks is a process rife with uncertainty. Analysts should make public those uncertainties so we can make more informed decisions about controlling hazards."

"Informed choice or regulated risk?" *Environment* 30 (May 1988): 12-15, 30-35.

Reports on a social experiment in which the same information about radon risk was presented in six different ways to New York State homeowners. Discusses how the content, format, and tone of informative messages affect how people understand and use information, and comments on the experiment's ethical and practical implications for E.P.A. as it communicates risk information.

Lave, Lester B. "Health and safety risk analyses: information for better decisions." *Science* 236 (April 17, 1987): 291-295.

Knowing the nature and magnitude of health and safety risks is helpful in setting priorities as well as in making decisions about pursuing recreational activities, foods, jobs, and other aspects of everyday living. Emphasizes importance of information flow between risk assessors and risk managers.

Sorenson, Jay B. "The assurance of reasonable toxic risk?" *Natural Resources Journal* 24 (July 1984): 549-569.

"The objective of this paper is to examine the criteria employed to assure 'reasonable' risk in the context of the 1980-1982 NMWQCC (New Mexico Water Quality Control Commission) regulations on the control of organic toxic contaminants in groundwater." It emphasizes the need for better environment, safety, and health assurance by which high and low risk choices can be made.

Environmental and professional journals (unannotated)

Paehlke, Robert. "The criteria of net benefit." (risk, safety and capitalism) *Society* 27,1 (November-December 1989): 13(2)

Segal, Marian;Sandman, Peter M. "Is it worth the worry? Determining risk." (health concerns) (includes related information) *FDA Consumer* 24,5 (June 1990): 7(5)

Law journals

Atcheson, John. "The department of risk reduction or risky business." *Environmental Law* 21,4 (1991): 1375-1412.
"Explores what sets the environmental agenda of the next twenty years apart from the set of problems we faced in the first twenty years, examines some of the shortcomings of a risk-dominated policy framework for dealing with the new generation of environmental problems, and briefly outlines elements of a policy framework that can help us set the right course. It is offered as part of the national dialogue called for by William Reilly and Hank Habicht."

Brown, William Y. "Environmental leadership: the search for priorities and power." *Environmental Law* 21,4 (1991): 1413-1423.
"The U.S. Environmental Protection Agency (E.P.A.) is re-evaluating its role in environmental risk assessment. It is changing from a reactive posture toward a more active role in anticipating environmental problems and reducing the likelihood of deregulation. The author observes that subjective variables complicate efforts to assess risk. Addresses the E.P.A., industry, and environmental organizations, and suggests cooperative strategies to achieve risk reduction."

Reilly, William K. "Taking aim toward 2000: rethinking the nation's environmental agenda." *Environmental Law* 21,4 (1991): 1359-1374.
"To meet the environmental challenges of the twenty-first century, the U.S. Environmental Protection Agency (E.P.A.) must focus on setting priorities. Risk analysis, supported by sound science and properly communicated to the public, will provide a rational and important means of setting those priorities. After it has determined the order in which the problems will be solved, E.P.A. must address these problems by moving beyond the traditional media-specific, command-and-control regulations. The Clean Air Act Amendments of 1990 show the promise of market-based incentives, and efforts like the 33/50 Project demonstrate the effectiveness of integrated, voluntary programs that focus on pollution prevention."

Reports

Dietz, Thomas. *The risk professionals*. New York: Russell Sage Foundation, 1987.

Risk assessment and risk control.. Washington, D.C.: Conservation Foundation, 1985. 69 p.: ill. (TD171 .R56 1985)

Federal government reports

Cohrssen, John J. *Risk analysis: a guide to principles and methods for analyzing health and environmental risks*. Washington, D.C.: Council on Environmental Quality, 1989. Available from the National Technical Information Service, Springfield, Va. 407 p.: ill. (RA566.3 .C64 1989)

Reducing risk: Appendix A: the report of the Ecology and Welfare Subcommittee; Relative Risk Reducton Project. Washington, D.C.: Environmental Protection Agency, Science Advisory Board, 1990. 77 p. (EPA SAB-EC-90-021A)
> Suggests an alternative approach to defining environmental problems, ranking them from an ecological perspective. Identifies a need to more accurately reflect ecological concerns in economic/welfare considerations.

ENVIRONMENTAL CONFLICT MANAGEMENT

(See also *Environmental Movement*) 419

Books (annotated)

Amy, Douglas J. *The politics of environmental mediation*. New York: Columbia University Press, 1987. 255 p. (KF3775 .A949 1987)
> Argues that mediating may not be as fast or economical as litigating issues with environmental groups.

Crowfoot, James E.; and Wondolleck, Julia M. *Environmental disputes: community involvement in conflict resolution*. Washington, D.C.: Island Press, 1991. 278 p.
> Offers seven case studies of local disputes: logging v. recreation; agricultural production v. soil conservation and wetland protection; oil drilling v. recreation; channel dredging v. a bird sanctuary; and water use v. aquifer protection. Describes the negotiating process and concludes that the chance of negotiating a settlement is greater if there is an experienced neutral negotiator, if the project is being planned rather than already constructed, and if scientists are involved as neutral experts.

Killingsworth, M. Jimmie; and Palmer, Jacqueline S. *Ecospeak: rhetoric and environmental politics in America*. Carbondale, Ill.: Southern Illinois University Press, 1992. 312 p.
   Maintains that discussions of environmental issues almost invariably never reach consensuses. Argues that such discussion are characterized by the unwillingness of participants to allow the scientific and social reform areas to be linked in the name of environmental awareness.

Books (unannotated)

Gorczynski, Dale M. *Insider's guide to environmental negotiation*. Chelsea, Mich.: Lewis Publishers, 1991. 242 p. (HC79.E5 G654 1991)

Environmental and professional journals (annotated)

Blomquist, William; and Ostrom, Elinor. "Institutional capacity and the resolution of a commons dilemma." *Policy Studies Review* 5 (November 1985): 383-393.
   "Examines the dynamic process of resolving a commons dilemma without an externally imposed solution. A description of the case of West Basin in Southern California offers an example of the interaction of institutional capacity with participants' actions to produce a successful resolution of a commons dilemma."

"Environmental dispute resolutions." *Natural Resources Journal* 28 (Winter 1988): 1-170.
   Partial contents include "Natural resources dispute resolution: an overview," by Lawrence J. MacDonnell; "The Denver metropolitan water roundtable: a case study in researching agreements," by Susan L. Carpenter and W. J. D. Kennedy; "Conflict in Alaska," by Thomas R. Berger; "The use of negotiated agreements to resolve water disputes involving Indian rights," by John A. Folk-Williams; "Negotiating the cleanup of toxic groundwater contamination: strategy and legitimacy," and by Lloyd Burton; "The future of environmental dispute resolution," by An Painter.

"Symposium on alternative dispute resolution and public policy." *Policy Studies Journal* 16 (Spring 1988): 493-626.
   Contents include: "Environmental mediation as an alternative to litigation," by J. Walton Blackburn; "Environmental dispute resolution and policy making," by Alfred Levinson;" and "The politics of environmental dispute resolution," by Barry G. Rabe.

Environmental and professional journals (unannotated)

"Year of unfortunate conflict." (24th Environmental Quality Index) *National wildlife* 30,2 (February-March 1992): 33(8)

Bibliographies

Ehrmann, John R. *A bibliography on natural resource and environmental conflict: management strategies and processes: (a complementary volume to CPL bibliography no. 64).* Chicago, Ill.: Council of Planning Librarians, 1982. 35 p. (Z7164.S66 E47 1982)

Jacobs, Harvey M.; and Rubino, Richard. *Environmental mediation: an annotated bibliography.* Chicago: Council of Planning Librarians, 1987. 24 p. (Bibliography 189)
Deals with environmental mediation including regulatory issues and the role of the mediator. It also includes a section on mediation, negotiation and bargaining.

Lesnick, Michael T. *A bibliography for the study of natural resource and environmental conflict.* Chicago, Ill.: CPL Bibliographies, 1981. 49 p. (Z5863.P6 L47)

NORTH AMERICAN FREE TRADE AGREEMENT: ENVIRONMENTAL IMPACT

Books (unannotated)

Dynamics of North American trade and investment: Canada, Mexico, and the United States. Stanford, Calif.: Stanford University Press, 1991. 281 p.: ill. (HF3211 .D96 1991)

Environmental hazards and bioresource management in the United States-Mexico borderlands. Los Angeles, Calif.: UCLA Latin American Center Publications, 1990. 483 p.: ill. (TD180 .E47 1989)

Investment in the North American free trade area: opportunities and challenges. Provo, Utah: David M. Kennedy Center for International Studies, Brigham Young University; Toronto: Ontario Centre for International Business, University of Toronto, 1991. 219 p. (HF1455 .I66 1991)

Negotiating and implementing a North American free trade agreement. Toronto: Centre for International Studies, 1992. 152 p.(HF1746 .N44 1992)

North American free trade: assessing the impact. Washington, D.C.: Brookings Institution, 1992. 274 p.: ill. (HF3211 .N667 1992)

Glick, Leslie Alan. *Understanding the North American Free Trade Agreement: legal and business consequences of NAFTA.* Boston: Kluwer, 1993.

Hosten-Craig, Jennifer. *The effect of a North American Free Trade Agreement on the Commonwealth Caribbean.* Lewiston, N.Y.: E. Mellen Press, 1992. 142 p.(HF1506.5.Z4 N674 1992)

Hufbauer, Gary Clyde. *NAFTA: an assessment.* Washington, D.C.: Institute for International Economics, 1993.

Hufbauer, Gary Clyde. *North American free trade: issues and recommendations.* Washington, D.C.: Institute for International Economics, 1992. 369 p.: map (HF3211 .H84 1992)

Johnson, Jon Ragnar. *The Free Trade Agreement: a comprehensive guide.* Aurora, Ontario: Canada Law Book, 1988. 352 p.(KF6668.C321988 J64 1988)

General interest periodicals (unannotated)

"From fast track to back burner." (North American Free Trade Agreement) *Economist* 322,7748 (February 29 1992): A25(2)

Mahony, Rhona. "Dirty deal?" (environmental impact of the proposed North American Free Trade Agreement) *Reason* 24,1 (May 1992): 50(2)

"Rio Grande illusions." (free trade with Mexico) *The New Republic* 204,21 (May 27 1991): 7(2)

Satchell, Michael. "Poisoning the border; many American-owned factories in Mexico are fouling the environment, and their workers aren't prospering." *U.S. News & World Report* 110,17 (May 6 1991): 32(7)

Environmental and professional journals (annotated)

"Social charter implications of the NAFTA." *Canada-U.S. Outlook* 3 (August 1992): whole issue (60 p.)
> Contents include: "Implications of a social charter for the North American Free Trade Agreement, by Peter Morici;" "Trade negotiations and social charters: the case of the North American Free Agreement, by Elizabeth C. DeBoer and Gilbert R. Winham;" and "The social side to free trade, by Sidney Weintraub."

Environmental and professional journals (unannotated)

"Environment and free trade with Mexico." (includes related fact sheets) (George Bush address) *US Department of State Dispatch* 3,9 (March 2 1992): 182(4)

Cloud, David S. "Environmental groups look for ways to ensure a 'green' trade agreement." (North American Free Trade Agreement) *Congressional Quarterly Weekly Report* 50,47 (November 28 1992): 3712(2)

Fisher, Jonathan. "We are talking about our children." (interview with Mexican president Carlos Salinas) *International Wildlife* 22,5 (September-October 1992): 48(4)

Nikiforuk, Andrew. "Seeking a voice at the free trade talks." (includes related article on Grand Canal water-for-free-trade plan) *Nature Canada* 15 (Summer 1986): 38(7)

Pope, Carl. "Borderline issues." (environmental protection and trade policy) *Sierra* 76,5 (September-October 1991): 22(2)

Rauber, Paul. "Trading away the environment." *Sierra* 77,1 (Jan-February 1992): 24(3)

Ward, Justin; and Prickett, Glenn T. "Prospects for a green trade agreement." (North American Free Trade Agreement) *Environment* 34,4 (May 1992): 2(4)

Law journals

Barr, Michael S., et al. "Labor and environmental rights in the proposed Mexico-U.S. free trade agreement." *Houston Journal of International Law* 14 (Fall 1991): 1-114.
  Discusses "whether labor and environmental rights should be incorporated into a F.T.A., to minimize any adverse impact on workers and the environment, and to promote a more desirable form of economic development."

Bucholtz, Barbara K. "Coase and the control of transboundary pollution: the sale of hydroelectricity under the United States-Canada Free Trade Agreement of 1988." *Boston College Environmental Affairs Law Review* 18,2 (1991): 279-317.
  "Proposes that, in the absence of requisite domestic political consensus, market forces -through the auspices of free trade agreements--may be employed as a substitute for traditional international regulation. Proposes that Hardin's 'commons' may occasionally be saved from 'tragedy' by applying a variant of Coase's theorem: in the absence of 'hard law,' the bargaining of private parties in free trade arenas may result in a curtailment of transboundary pollution."

"Competition and Dispute Resolution in the North American Context (1987: Cleveland, Ohio): Proceedings." *Canada-United States Law Journal* 12 (1987): whole issue (377 p.)
  Partial contents include: "The free trade negotiations: Canadian and U.S. perspectives," by Leonard Legault; "The anticipated economic effect of a North American free trade area on business in the North American context," by Carl Beigie; "The North American political outlook for the future," by Ken Freed; "Environmental protection, unfair competition and the U.S.-Canada free trade discussions," by Van Carson; "The competitive edge: impact of taxes on a North American free trade area," by Robert Brown; "Comments on dispute resolution under a North American free trade agreement," by T. B. Smith; and "A look at the future: Canada and the United States in the future world economic context--can we be competitive?" by James D. Fleck.

Reports

Binational statement on environmental safeguards that should be included in the North American Free Trade Agreement (NAFTA). Washington, D.C.: National Wildlife Federation; Toronto: Pollution Probe, 1992. 4 p.

> This document is supported by eleven major conservation organizations in the U.S. and Canada and is designed to serve as a guide for those negotiating the N.A.F.T.A.

Canada and the United States in the 1990s: an emerging partnership. Cambridge, Mass.: Institute for Foreign Policy Analysis, 1991. 96 p.

> Contents include: "The Canada-United States science and technology relationship in a globalized economy: challenges for the 1990s, by William C. Winegard;" "Environmental issues and future Canadian policy, by John E. Carroll;" "U.S.-Canadian trade and technology cooperation: risks and opportunities, by David Leyton- Brown and Michael Clark;" "North American Defense and Security Cooperation: sustaining a limited partnership, by Joel J. Sokolsky;" "Canada and the United States: issues for the 1990s, by Robert L. Pfaltzgraff, Jr.

Grossman, Gene M. *Environmental impacts of a North American Free Trade Agreement.* Cambridge, Mass.: National Bureau of Economic Research, 1991. 39 p. (NBER working paper no. 3914)

> "We use comparable measures of three air pollutants in a cross-section of urban areas located in 42 countries to study the relationship between air quality and economic growth. We find for two pollutants (sulfur dioxide and 'smoke') that concentrations increase with per capita GDP at low levels of income. We investigate whether the size of pollution abatement costs in the U.S. industry influences the pattern of international trade and investment."

Hart, Michael. *A North American free trade agreement: the strategic implications for Canada.* Ottawa: Centre for Trade Policy and Law; Halifax, N.S.: Institute for Research on Public Policy, 1990. 142 p.: ill. (HF1479 .H377 1990)

Kelly, Mary E.; Kamp, Dick; and Gregory, Michael. *Mexico-U.S. free trade negotiations and the environment: exploring the issues.* Austin: Texas Center for Policy Studies, 1991. 33 p.

> "Explores some of the environmental and public health issues that are associated with increased economic integration between the United States and Mexico. Because of the perspective and location of the organizations involved in preparing this discussion paper, there is a substantial focus on the U.S./Mexico border area."

Minimal environmental safeguards to be included in the North American Free Trade Agreement. Washington, D.C.: National Wildlife Federation; Toronto: Pollution Probe, 1992. 11 p.

Details what must be included in the N.A.F.T.A. if it is to live up to the assurances of environmental protection made by the governments of Canada, Mexico and the U.S." Contents include: Sustainable development as a goal for the N.A.F.T.A.; General exceptions for environmental protection; Harmonization of environmental standards; International environmental agreements; Dispute resolution process, enforcement and public participation; and Investment.

Rich, Jan Gilbreath. *Planning for the border's future: the Mexican-U.S. integrated border environmental plan.* Austin: U.S.-Mexican Policy Studies Program, LBJ School of Public Affairs, University of Texas, 1992. 48 p. (U.S.-Mexican occasional paper no. 1)
 "Begins with a brief history of the economic trends that led to creation of the Integrated Environmental Plan for the Mexico-U.S. Border Area. A second section analyzes the Mexican and U.S. border community response to the first publicly released draft of this plan in August 1991. The final section discusses the revised version of the plan and whether these modifications adequately address the concerns of border communities affected by the plan." Covers such issues as hazardous waste, water quality, air quality, housing, wetlands, and colonias.

Smith, Wesley R. *Protecting the environment in North America with free trade.* Washington, D.C.: Heritage Foundation, 1992. 12 p. (Backgrounder no. 889)
 Rejects the notion that North American free trade will generate greater pollution as U.S. industries move to Mexico to take advantage of less strict environmental laws. Says that "a N.A.F.T.A. would improve Mexico's chances of protecting its environment by increasing the cooperation and interaction between Mexican and American companies and the U.S. and Mexican governments. As Mexican companies merge with American ones, or receive U.S. investors, they will be more inclined to adopt the clean environment policies of the U.S. companies."

Federal government reports

Agriculture in a North American free trade agreement: analysis of liberalizing trade between the United States and Mexico. Washington, D.C.: Department of Agriculture, 1992. 167 p.: ill. (HF2651.F27 U532 1992)

Environmental plan for the Mexican-U.S. border area: first stage (1992-1994). Washington, D.C.: Environmental Protection Agency, 1992. 34 p.
 Summarizes the first stage of a binational border environmental protection program prepared jointly by the E.P.A. and Mexico's Secretariat for Urban Development and Ecology. Topics covered include wastewater treatment, hazardous waste disposal, air quality, and cooperative enforcement actions.

Integrated environmental plan for the Mexican-U.S. border area: (first stage, 1992-1994). Washington, D.C.: Environmental Protection Agency, 1992. 135 p.
 Discusses the present status of environmental issues and describes "plans to mobilize the cooperative efforts of governments at all levels, in seeking solutions."

Report of the Administration on the North American Free Trade Agreement and actions taken in fulfillment of the May 1, 1991 commitments. Washington, D.C.: G.P.O., 1992. 177 p.
> "On August 12, 1992, the president announced that the U.S., Canada, and Mexico had completed negotiation of the North American Free Trade Agreement. Report describes the actions the Administration has taken in fulfillment of the President's May 1, 1991 commitments to Congress on worker adjustment, labor rights, environmental protection, and thorough consultation."

Review of U.S.-Mexico environmental issues. Washington, D.C.: U.S. Trade Representative, 1992. 231 p.
> Reviews the possible environmental effects of a free trade agreement. Includes analysis on air and water quality, hazardous wastes, chemical emergencies, and endangered species.

U.S.-Mexico trade: information on environmental regulations and enforcement; report to the Chairman, Committee on Commerce, Science and Transportation, U.S. Senate. May 13, 1991. Washington, D.C.: General Accounting Office, 1991. 16 p.
> Evaluates Mexico's environmental problems, laws and regulations, and resources and efforts to implement and enforce its environmental protection program. Discusses environmental protection cooperation between the U.S. and Mexico.

Conference proceedings

Free trade within North America: expanding trade for prosperity: proceedings of the 1991 Conference on the Southwest Economy sponsored by the Federal Reserve Bank of Dallas. Boston: Kluwer Academic Publishers, 1993. 224 p.: ill. (HF1746 .C66 1993)

Region North America: Canada, United States, Mexico: proceedings. Waco, Tex.: Regional Studies, Baylor University, 1990. 175 p.(HF1766 .R44 1990)

Second U.S.-Mexico Border Governors Finance Summit, November 9-10, 1989: final report. San Diego, Calif.: Institute for Regional Studies of the Californias, San Diego State University, 1989. 1 v. : ill., maps (HC135 .U17 1989)

U.S./RUSSIA ENVIRONMENTAL RELATIONS

General interest periodicals (unannotated)

Bobkov, Vyacheslav; and Kurbatov, Alexander. "Working together for the sake of our planet." (includes excerpts of speeches of Mikhail Gorbachev, Alexei Yablokov, Albert Gore, Jr., Nikolai Vorontsov, Carl Sagan and Pitirim, Metropolitan of Volokolamsk and Yuriev, People's Deputy of the USSR) *Soviet Life* 5 (May 1990): 22(7)

Bolshakov, Valeri. "From Lake Okeechobee to Lake Ladoga." (USSR-USA) *Soviet Life* 8 (August 1990): 42(3)

Environmental and professional journals (annotated)

"U.S. and Soviets spur scientific collaboration." *Conservation Foundation Letter* 1 (January-February 1988): 1-8.

> States that since "relations between the two nations have thawed, a greater openness in the Soviet Union has improved the climate for bilateral exchanges, and both countries have become increasingly concerned about changes in the atmosphere and climate they share."

Environmental and professional journals (unannotated)

Bear, Dinah; and Elkind, Jonathan. "Soviet-U.S. cooperation." (Soviet environmental protection laws) *Environment* 32,3 (April 1990): 5(4)

Graham, Frank, Jr. "U.S. and Soviet environmentalists join forces across the Bering Strait." (includes related articles on joint efforts) (Cover Story) *Audubon* 93,4 (July-August 1991): 42(18)

Sweet, William. "US and Soviet academies reach global ecology agreement." *Physics Today* 42,2 (February 1989): 101(2)

Federal government reports

Fifteen years of cooperation toward solving environmental problems in the USSR and USA. Bethesda, Md. (Environmental Protection Agency): American Fisheries Society, 1988. 178 p.: ill. (TD365 .P79 1988)

CHAPTER 5

THE ENVIRONMENTAL MOVEMENT

(Council on Environmental Quality Assessment) 1
(See also *Public Lands: Conservation Movement*) 3108
(See also *Animal rights: Movement and Leaders*) 598

continued

THE ENVIRONMENTAL MOVEMENT

(Council on Environmental Quality Assessment) 1
(See also *Public Lands: Conservation Movement*) 3108
(See also *Animal Rights: Movement and Leaders*) 598

SHORT LIST OF SOURCES
FOR GENERAL RESEARCH

Books

Allaby, Michael, ed. *Thinking green: an anthology of essential ecological writing.* n.p.: Barrie and Jenkins, 1989.
> Anthologizes extracts from the principal literature of the Green movement, as well as writing from their supporters, and a few of their opponents. Everything from *Silent Spring*, to Malthus, to *Spaceship Earth* is included.

Cairncross, Frances. *Costing the Earth: the challenge for government, the opportunities for business.* Cambridge, Mass.: Harvard Business School Press, 1992. 341 p.
> Maintains that environmental protection will only come about after governments decide to take an active role and business recognizes the economic potential of ecologically sound practices. Focuses on the needs of the green consumer, recycling, energy efficiency, taxing polluters, government failures, and the challenge to business.

Chiras, Daniel D. *Beyond the fray: reshaping America's environmental response.* Boulder, Colo.: Johnson, 1990. (HC110.E5 C48 1990)
> Recommends a change in environmental tactics from reactive to proactive. Maintains that large, money wasting environmental organizations that rally supporters to change laws and enforce regulations cannot achieve long-term goals. Recommends making environmentalism part of the mainstream body politic rather than part of the fringe.

Hadden, Susan G. *A citizen's right to know.* Boulder, Colo.: Westview, 1989.
> Hadden believes that the U.S. has led the way in informing citizens about environmental problems, but she has no illusions about the technical and political difficulties that restrict public knowledge. She analyzes the obstacles to citizen involvement and power sharing, including the public's low level of scientific understanding and the absence of trust in government and mass political activity.

Maintains that as environmental problems increase, government must lead the way to informing its citizens.

Hollender, Jeffrey A. *How to make the world a better place: a beginner's guide to doing good*. New York: Morrow, 1990. 303 p.: ill. (HN90.V64 H65 1990)
    Maintaining that the fate of humankind and the planet can be shaped by social behavior, this book offers a wide range of small suggestions on how we should change our behavior. Issues such as the environment, human rights, and world hunger are the targets of the authors suggestions.

McCormick, John. *Reclaiming paradise: the global environmental movement*. Bloomington: Indiana University Press, 1990. 274 p.
    Traces the origins and accomplishments of the environmental movement beginning with nineteenth-century naturalists. Argues that the environmental prophets of doom, spurned by Rachael Carson's *Silent Spring* in 1962, set the quarrelsome tone for modern environmentalists. The 1972 Stockholm conference solidified the global concern about the environment and "marked a transition from the emotional and occasionally naive New Environmentalism of the 1960s to the more rational, political, and global perspectives of the 1970s." Devotes much attention to United Nations efforts.

Scheffer, Victor B. *The shaping of environmentalism in America*. Seattle: University of Washington Press, 1991 249 p. (HC110.E5 S39 1991)
    Outlines the public awakening to air, land, water, and wildlife abuse by unbridled human consumption and activities. Provides an overview of the conservation activities in the nineteenth and first half of the twentieth century, then focuses on the formative years of the environmental movement (1960-1980) before it encountered the Reagan administration (1981-1989). Argues for striking the middle ground between conservation natural resources use.

Schmidheiny, Stephan. *Changing courses*. Cambridge, Mass.: MIT Press, 1992. 374 p.
    Large corporations argue that as long as they are profitable, they can afford to pay for many environmental programs, and therefore it is not in the best interest of the environment to hamper their operations. Presents success stories of large companies contributing to the environmental good, but ignores the blatant offenders and the willingness of the world's corporate community to police itself.

Steger, Will, and Bowermaster, Jon. *Saving the Earth: a citizen's guide to environmental action*. New York: Knopf, 1990. 306 p.: ill. (TD171.7 .S74 1990)
    This guide is divided into four general areas requiring citizen involvement: atmosphere, land, water, and people, with subdivisions into individual problems. At the end of each chapter there are sections defining what can be done by individuals and government, listing organizations to contact, and including a bibliography.

General interest periodicals

Bailey, Ronald. "The apocalypse boosters: raining in their hearts." *National Review* 42 (December 3, 1990): 32-36.
"Nuclear winter, population explosion, non-renewable resources, vanishing species--some of them are genuine problems, others are made up out of whole cloth; all are easily exploited for fun and left-wing profit." This atmosphere of catastrophe has been sustained by a cadre of professional doomsters, intellectuals and policy-makers whose apocalyptic predictions are meant to frighten the public into adopting their leftist policy prescriptions. Includes sidebar by Bruce Ames: Misconceptions about pollution and cancer.

Berry, Wendell. "Out of your car, off your horse." *Atlantic* 267 (February 1991): 61-63.
"A noted environmentalist warns that those who would save the planet may sometimes be as dangerous as those who recklessly exploit its resources. Ecological good sense, he argues, begins not in thinking globally but in acting locally." Author offers 26 propositions about global thinking and the sustainability of cities.

"Cleaner environment." *Changing Times* (February 1990): 28-31.
Discusses "environment friendly" products and tradeoffs for careful shoppers. An "Environmental Shopper's Guide" accompanies the article which comments on how to be environmentally friendly with various categories of products.

Cross, Frank B. "The weaning of the green: environmentalism comes of age in the 1990s." *Business Horizons* 33 (September-October 1990): 40-46.
"Traditional environmental regulation, unfortunately, has been neither particularly efficient nor effective. Environmental concerns have been marked by considerable antagonism between the environmental and industrial communities. Environmental activists have demanded the most extreme levels of environmental protection and refused to consider the costs of such protection. Industrial representatives have too often fought even efficient environmental protection measures and employed spurious science on behalf of their position."

Eberhart, Annie. "If you can't stand the heat . . . Read on for a citizen's guide to some sensible steps for cooling off the Earth." *Public Citizen* 8 (November-December 1988): 13-17.
Outlines steps which individuals and groups can take to reduce atmospheric pollution and to pressure local, regional, national and international policymakers into taking action.

Gifford, Bill. "Inside the environmental groups." *Outside* 15 (September 1990): 69-78, 80, 82, 84.
A user's guide to the top twenty-five environmental groups, describing "who they are and what issues they most heartily embrace," their leadership, membership,

current activities, funding and financial information, celebrity spokesperson, etc. Also included is an honor roll list of "eight of the finest grass-roots efforts" for protecting the environment.

Holt, Thomas Harvey. "Growing up green: are schools turning our kids into eco-activists?" *Reason* 23 (October 1991): 37-41.

"Ten years ago, one unit of a social-studies or science class or one chapter of a textbook might have been dedicated to ecology. Today, entire classes and textbooks focus on the environment, and dozens of activist organizations are working to shape environmental curricula nationwide. Environmental education has become a growth industry."

Kirkpatrick, David. "Environmentalism: the new crusade." *Fortune* 121 (February 12, 1990): 44-48, 50, 54-55.

"It may be the biggest business issue of the 1990s. Here's how some smart companies are tackling it."

Kittredge, William. "White people in paradise." *Esquire* 116 (December 1991): 152-156, 158, 161-162, 193-195.

"Our paradise is played out. The old order in the West is right to fear the environmentalists: They are going to prevail; they represent the will of the nation (except maybe in the matter of drilling for petroleum); they have a great crisis on their side, and the votes."

Radelat, Ana. "Whoooo says saving the Earth costs jobs?" *Public Citizen* 13 (May-June 1992): 10-14.

"It looks like tough times for environmentalists. Not only have environmental groups lost members and money, but they're facing a backlash fueled by industries that insist the greens are blocking the nation's economic recovery by putting environmental concerns ahead of jobs and development."

Smith, Emily. "The greening of corporate America." *Business Week* 3156 (April 23, 1990): 96-103.

Describes how tougher laws, stepped-up enforcement and grass-roots environmental movements are causing corporate America to become more involved in controlling environmental pollution.

Stegner, Wallace. "It all began with conservation; our first response to the New World's wild Eden was to break it - but seeds of a land ethic were planted early on." (special issue on the environment) *Smithsonian* 21,1 (April 1990): 34(10)

Zuckerman, Seth. "Environmentalism turns 16." Nation 243 (October 18, 1986): 368-370.

Discusses changes in the environmental movement since Earth Day in 1970. Many national groups have adopted sophisticated tactics, including economic analysis, demonstration projects, and sizable Washington offices. Another brand of environmentalism has also developed; it relys more on monkey wrench and grassroots tactics and on bioregionalism.

Environmental and professional journals

Borrelli, Peter. "Environmentalism at a crossroads." *Amicus Journal* 9 (Summer 1987): 24-37.

> Reviews developments in environmentalism in the U.S., discussing key environmental legislation, key pressure groups, and key players in the movement. Considers differences among environmental protection groups and identifies a few general trends; these include increased willingness by industry to sit down with environmentalists, and increasingly global interests of environmentalists.

"Kid crusaders: children and the environment." *Environmental Action* 23 (September-October 1991): 19-26.

> "From lead poisoning to pesticide exposure, children are especially vulnerable to environmental risks. At the same time, young activists are championing environmental causes in the classroom and in Congress."

Meyer, Eugene L. "Environmental racism: why is it always dumped in our backyard? Minority groups take a stand."*Audubon* 94,1 (January February 1992). 30(3)

Postrel, Virginia I. "The environmental movement: a skeptical view." *Chemtech* 21 (August 1991): 457-461.

> For the green ideology, the core value is stasis, or 'sustainability: the ideal is of an Earth that shows little effects of human activity. This static view "leads environmentalists to advocate policies that will make growth hard on people, as a way of discouraging further development. Second, the static view leads environmentalists to misunderstand how real environmental problems can be solved."

Steinhart, Peter. "Respecting the law: there must be limits to environmental protest." *Audubon* 89 (November 1987): 10, 12-13.

> "We can't applaud spiking trees or scuttling whaling ships any more than we can applaud official abandonment of environmental law and policy. For one form of lawlessness tends to invite the other. We cannot really expect business to temper greed with understanding, or individuals to take their best ethics into the woods. In the end, conservation depends upon an awareness that we are all in this thing together."

INTRODUCTORY SOURCES

Books (annotated)

Allaby, Michael, ed. *Thinking green: an anthology of essential ecological writing.* N.p.: Barrie and Jenkins, 1989.
> Anthologizes extracts from the principal literature of the Green movement, as well as writing from their supporters, and a few of their opponents. Everything from *Silent Spring*, to Malthus, to *Spaceship Earth* is included.

Chiras, Daniel D. *Beyond the fray: reshaping America's environmental response.* Boulder, Colo.: Johnson, 1990. (HC110.E5 C48 1990)
> Recommends a change in environmental tactics from reactive to proactive. Maintains that large, money wasting environmental organizations that rally supporters to change laws and enforce regulations cannot achieve long-term goals. Recommends making environmentalism part of the mainstream body politic rather than part of the fringe.

Day, David. *The environmental wars: reports from the front lines.* New York: St. Martin's, 1990.
> Beginning with a litany of eight well-known people who have died fighting for various environmental causes, including Chico Mendes, Dian Fossey, Karen Silkwood, and Joy Adamson, this work uses the war metaphor to describe the conflict between environmental and political/economic interests. Discusses animal rights especially in terms of endangered species, pollution, chemical waste, and chemical and biological warfare.

Dyer, Ken; and Young, J., eds. *Changing directions.* Adelaide, Australia: University of Adelaide, 1990. 667 p.
> The product of a 1989 conference, records the current trends in Green philosophy and practice. While the emphasis is on Australia, there are also articles on Antarctica and the Indian-Pacific area.

Elkington, John. *A year in the greenhouse: an environmental diary.* North Pomfret, Vt.: Victor Gollancz, 1990.
> Offers a firsthand perspective of the Green movement. Discusses how world leaders are discovering the political benefits of environmental protection.

Mills, Stephanie. *Whatever happened to ecology?* San Francisco: Sierra Club, 1989.
> Traces the history of the environmental movement from the personal perspective of one who chose to forgo motherhood because of overpopulation and who currently resides in a bioregional community--a group of people dedicated to reinhabiting the land for safe and productive use. Discusses society's failure to handle environmental problems, her commitment to the future of the planet, and the issue of personal choice. Contains a list of environmental publications and organizations.

Norton, Bryan G. *Toward unity among environmentalists.* New York: Oxford, 1991. 287 p.

Focuses on the political and social facets of environmental policy. Maintains that contextualism and pluralism provide a framework in which the various views of environmentalists can be examined. Written from a philosophical perspective that includes ethics and spirituality.

Paehlke, Robert C. *Environmentalism and the future of progressive politics.* New Haven, Conn.: Yale University Press, 1989.

Maintains that environmentalism is a political ideology equal to liberalism or conservatism and that it will only take hold as a political force when it is able to be seen as nonpartisan. Traces the history of the modern environmental movement and examines its underlying values. Concludes that modern conservatism and environmentalism do not mix, and the environmental movement is therefore obliged to join the moderate Left.

Rosenbaum, Walter A. *Environmental politics and policy.* 2d ed. Washington, D.C.: CQ Press, 1991. 336 p.

Focuses on continuity and change in the politics of American environmentalism. Discusses the environmental era; the state of the environment; the politics of environmental policy; risk assessment; toxic and hazardous substances; the politics of energy; the battle for the public lands; and an agenda for the 1990s.

Schaefer, Paul. *Defending the wilderness: The Adirondack Writings of Paul Schaefer.* Syracuse, N.Y.: Syracuse University Press, 1989. 250 p.: ill. (QH76.5.N7 S33 1989)

This conservation-oriented book focuses on the Adirondack region of New York. Spanning over forty years, these essays offer much information on the history of the conservation movement and political history of the area. Contains accounts of the Higley and Panther mountain reservoir proposals.

Snow, Donald. *Inside the environmental movement: meeting the leadership challenge.* Washington, D.C.: Island Press, 1992. 295 p.: ill. (S944.5.L42 I57 1991)

Presents the findings of a two-year study by the Conservation Leadership Project of nonprofit environmental and conservation organizations. Discusses leadership issues, including leadership of volunteers, and the academic backgrounds of persons involved in such organizations. Also discusses issues at state, local, and regional levels. Maintains that these organizations are often poorly organized and underfunded. Concludes with some specific recommendations.

Snow, Donald, ed. *Voices from the environmental movement: perspectives for a new era.* Washington, D.C.: Island Press, 1992. 237 p. (S930 .V65 1991)

The product of the Conservation Leadership Project, this work addresses a wide range of issues relating to the environmental movement, including the lack of minorities involved, political shortcomings, the need for education, and the difficulties of international cooperation, as well as the movement's leadership and goals. Based on a public discussion among conservationists, business leaders, and academics, this work offers much insightful information.

Yandle, Bruce. *The political limits of environmental regulation: tracking the unicorn*. New York: Quorum Books, 1989. 180 p. (HC110.E5 Y36 1989)

In analyzing the tensions between environmentalists and those who seek to balance economic efficiency and productivity with pollution control, the author not only sides with the latter group, but also maintains that the environmental movement is responsible for a massive, unnecessary diversion of funds to the issue of pollution control. Claims that these tensions have endured because of competing congressional pressures and unusual coalitions among environmental interest groups.

Books (unannotated)

American environmentalism: the U.S. environmental movement, 1970-1990. Philadelphia, Pa.: Taylor & Francis, 1992. 121 p. (HC110.E5 A648 1992)

Arnold, Ron. *Ecology wars: environmentalism as if people mattered*. Bellevue, Wash.: Free Enterprise Press, 1987. Distributed by Merril Press, Bellevue, Wash. 182 p. (HC110.E5 A77 1987)

Bahro, Rudolf. *Building the Green movement*. London: GMP, 1986. 219 p. (JN3971.A98 G72313 1986)

Berle, Gustav. *The green entrepreneur: business opportunities that can save the Earth and make you money*. n.p.: Liberty Hall, 1991. 242 p. (HD69.P6 B47 1991)

Blowers, Andrew. *Something in the air: corporate power and the environment*. Cambridge, Mass.: Harper & Row, 1984. 359 p.: ill.

Bookchin, Murray. *Defending the Earth: a dialogue between Murray Bookchin and Dave Foreman*. Boston, Mass.: South End Press, 1991. 147 p. (JA75.8 .B66 1990)

Bryant, Bunyan I. *Environmental advocacy: concepts, issues, and dilemmas*. Ann Arbor, Mich.: Caddo Gap Press, 1990. 140 p. (HC110.E5 B79 1990)

Buenfil, Alberto Ruz. *Rainbow nation without borders: toward an ecotopian millennium*. Santa Fe, N.M.: Bear, 1991. 226 p.: (some color) ill. (HN17.5 .B8313 1991)

Dobson, Andrew. *Green political thought: an introduction*. Boston: Unwin Hyman, 1990. 224 p. (JA75.8 .D63 1990)

Eckersley, Robyn. *Environmentalism and political theory: toward an ecocentric approach*. Albany: State University of New York Press, 1992. 274 p. (JA75.8 .E26 1992)

Goodin, Robert E. *Green political theory*. Cambridge, Mass.: Polity Press, 1992.

Green fuse: the Schumacher lectures 1983-88. New York: Quartet Books, 1990. 198 p. (JA75.8 .G73 1990)

Green reader: essays toward a sustainable society. San Francisco: Mercury House, 1991. 280 p. (JA75.8 .G74 1991)

Green reporting: accountancy and the challenge of the nineties. New York: Chapman & Hall, 1992. 299 p.: ill. (HF5686.C7 G56 1992)

Greening of Australian politics: the 1990 federal election. Melbourne, Australia: Longman Cheshire, 1990. 229 p.: ill. (JQ4094 .G73 1990)

Institutions for the Earth: sources of effective international environmental protection. Cambridge, Mass.: MIT Press, 1993.

Into the twenty-first century: an agenda for political re-alignment. Basingstoke, Hants, UK: Green Print, 1988. 203 p.: ill. (JA75.8 .I58 1988)

Irvine, Sandy. *A Green manifesto: policies for a green future*. London: Optima, 1988. 178 p.: ill.

Keenan, Charles J. *Environmental anarchy: the insidious destruction of social order: a legacy of the sixties*. British Columbia, Canada: Cappis Press, 1984. 206 p. (HC117.B8 K43 1984)

Kelly, Petra Karin. *Nonviolence speaks to power*. Honolulu: Center for Global Nonviolence Planning Project, Spark M. Matsunaga Institute for Peace, University of Hawaii, 1992.

Korten, David C. *Getting to the 21st century: voluntary action and the global agenda*. West Hartford, Conn.: Kumarian Press, 1990. 253 p.: ill. (HC60 .K67 1990)

Lowe, Philip. *Environmental groups in politics*. Boston: Allen & Unwin, 1983. 208 p.: ill. (HC260.E5 L68 1983)

Milbrath, Lester W. *Environmentalists: vanguard for a new society*. Albany: State University of New York Press, 1984. 180 p.: ill. (HC79.E5 M47 1984)

Nicholson, Max. *The new environmental age*. New York: Cambridge University Press, 1987. 232 p.: ill. (QH75 .N53 1987)

Petulla, Joseph M. *American environmentalism: values, tactics, priorities*. College Station: Texas A&M University Press, 1980. 239 p. (GF21 .P43)

Pilat, Joseph F. *Ecological politics: the rise of the Green movement*. Beverly Hills: Sage Publications, 1980. 96 p. (HC110.E5 P54)

Porritt, Jonathon. *Seeing green: the politics of ecology explained*. New York: B. Blackwell, 1985. 252 p. (HC79.E5 P668 1985)

Prokop, Marian K. *Managing to be green*. San Diego, Calif.: Pfeiffer, 1992.

Sale, Kirkpatrick. *The green revolution: the environmental movement since 1962*. New York: Hill and Wang, 1993.

Seabrook, Jeremy. *The race for riches: the human cost of wealth*. Basingstoke, Hants, U.K.: Marshall Pickering, 1988. 182 p. (HC79.W4 S4 1988)

Shabecoff, Philip. *A fierce green fire: the American environmental movement*. New York: Hill & Wang, 1992.

Spretnak, Charlene. *The spiritual dimension of green politics*. Sante Fe, N.M.: Bear, 1986. 95 p. (JN3971.A98 G72376 1986)

Tucker, William. *Progress and privilege: America in the age of environmentalism*. Garden City, N.Y.: Anchor Press/Doubleday, 1982. 314 (TD170 .T82 1982)

Voices from the environmental movement: perspectives for a new era. Washington, D.C.: Island Press, 1992. 237 p. (S930 .V65 1991)

Wachtel, Paul Spencer. *Eco-bluff your way to greenism: the guide to instant environmental credibility*. Chicago: Bonus Books, 1991. 112 p.: ill. (HC79.E5 W26 1991)

Wenner, Lettie McSpadden. *U.S. energy and environmental interest groups: institutional profiles*. New York: Greenwood Press, 1990. 358 p. (HD9502.U52 W45 1990)

Wilds, Leah J. *Understanding who wins: organizational behavior and environmental politics*. New York: Garland, 1990. 207 p.: ill. (HC110.E5 W535 1990)

Yearley, Steven. *The green case: a sociology of environmental issues, arguments, and politics*. Boston: HarperCollinsAcademic, 1991. 197 p. (JA75.8 .Y43 1991)

Young, John. *Post environmentalism*. London: Belhaven Press, 1990. 225 p.: ill. (HC79.E5 Y67 1990)

General interest periodicals (annotated)

Atkinson, Carla. "A white 'green' movement: where do minorities fit into the environmental picture?" *Public Citizen* 9 (September-October 1990): 16-20.
"Many national minority groups are just beginning to turn their attention towards environmental concerns . . . but there is a gaping social and economic gap between minorities and the mainly white, middle-class environmental movement. Most minorities must address what they feel are more pressing problems and don't have the time or energy to protest acid rain and global warming."

Berry, Wendell. "Out of your car, off your horse." *Atlantic* 267 (February 1991): 61-63.
"A noted environmentalist warns that those who would save the planet may sometimes be as dangerous as those who recklessly exploit its resources. Ecological good sense, he argues, begins not in thinking globally but in acting locally." Author offers 26 propositions about global thinking and the sustainability of cities.

Brooks, David; Smith, Fred L; and Kushner, Kathy H. "A sustainable environmentalism: saving the Earth from its friends." *National Review* 42 (April 1, 1990): 28-33, 59.
"Does protecting the environment require collectivism and an end to economic growth? David Brooks examines the full agenda of the environmental activists. Fred L. Smith and Kathy H. Kushner present a conservative alternative: restoring property rights and individual responsibilities so that free markets will protect natural resources."

Commoner, Barry. "The environment." *New Yorker* 63 (June 15, 1987): 46-47, 50-54, 56-71.
Reviews trends in the state of the environment over the last fifteen years and discusses the environmental movement.

Cordes, Colleen. "The NIMBY complex." *American Politics* 1 (February 1986): 11-15, 23, 31.
"Some NIMBY (not-in-my-backyard) watchers applaud the movement as a vital sign of democracy in action, protecting the only patch of Earth we've got. Others indict it as evidence of a fearful nation, whose will to grow is increasingly paralyzed by a belated reckoning with the risks of technology." Describes neighborhood reaction to LULU's (locally unwanted land uses)

Cross, Frank B. "The weaning of the green: environmentalism comes of age in the 1990s." *Business Horizons* 33 (September-October 1990): 40-46.
"Traditional environmental regulation, unfortunately, has been neither particularly efficient nor effective. Environmental concerns have been marked by considerable antagonism between the environmental and industrial communities. Environmental activists have demanded the most extreme levels of environmental protection and refused to consider the costs of such protection. Industrial representatives

have too often fought even efficient environmental protection measures and employed spurious science on behalf of their position. The recent shift in attitudes of both these political combatants offers promise for improved environmental policies."

Day, Barbara; Knight, Kimberly; and Donloe, Darlene. "The rain forest in our backyard." *Essence* 21,9 (January 1991): 75(6)
    Afro-Americans most concerned with local environmental issues in their neighborhoods. Includes related articles on the neighborhood organization, Take It Back.

Ettinger, Christian; and Block, A. G. "New party has Democrats seeing green." *California Journal* 23 (February 1992): 97-98.
    "California voters will have another slate of candidates to evaluate in 1992 after the Green Party qualified for the ballot. And Democrats are none too happy about it."

"Going green." *World Press Review* 37 (August 1990): 11-18, 20.
    Contents include: "Green diplomacy takes off"; "Ecology vs. economics," by Ricardo Bayon; "Rewriting the rules of business," by Robyn Allen; and "The highjack of reason? Wanted: plain speech on ecology," by John Horsfall.

Johnson, Bob. "Grassroots growth control." *Golden State Report* 5 (June 1989): 22-24.
    "Will growth management be the driving political force of the '90s? A Santa Cruz supervisor (Gary A. Patton) is forming a statewide coalition (called Grassroots California), and he predicts environmentalists will have a candidate for governor by 1994."

Kittredge, William. "White people in paradise." *Esquire* 116 (December 1991): 152-156, 158, 161-162, 193-195.
    "Our paradise is played out. The old order in the West is right to fear the environmentalists: They are going to prevail; they represent the will of the nation (except maybe in the matter of drilling for petroleum); they have a great crisis on their side, and the votes."

Levinson, Marc. "The green gangs." *Newsweek* 120 (August 3, 1992): 58-59.
    "Environmental activists are reshaping the staid diplomacy of trade."

MacLeish, William H. "Ways to beat the noxious '90s." *World Monitor* 3 (June 1990): 16-18, 20, 22.
    "The East bloc's wasteland. California's super-bowl of smog. Everywhere the pollution forecasts are horrendous. One gleam in the cloud: Industry, environmentalists, and other old adversaries are getting together to help save the planet."

Paehlke, Robert. "Participation in environmental administration: closing the open door?" *Alternatives* 14 (May-June 1987): 43-48.

"Environmentalist demands for open access to administrative decision processes are, then, neither an incidental characteristic of the movement, nor something likely to be easily achieved in modern bureaucratic societies."

Pastrel, Virginia I. "The green road to serfdom." *Reason* 21 (April 1990): 22-28.
   Describes the ideology of "greens", environmental activists, and why the idea is "dangerous".

"Politics of the environment, 1970-1987." *Wilson Quarterly* 11 (Autumn 1987): 50-83.
   Presents four articles: "A Big Agenda" by David Vogel, which analyzes the rise of the environmental movement; "Cleaning Up the Chesapeake"; "Learning the Lessons" by Robert W. Crandall, discussing the complexities of environmental regulation; and "Background Books."

Radelat, Ana. "Whoooo says saving the Earth costs jobs?" *Public Citizen* 13 (May-June 1992): 10-14.
   "It looks like tough times for environmentalists. Not only have environmental groups lost members and money, but they're facing a backlash fueled by industries that insist the greens are blocking the nation's economic recovery by putting environmental concerns ahead of jobs and development."

Wallis, L. R. "Endgame ecology." *Vital Speeches of the Day* 58 (June 15, 1992): 529-533.
   This speech, delivered at the Cal Rad Forum Annual Conference, San Francisco, California, May 3-4, 1992, by the executive secretary, examines the growth of the environmental movement, discussing its negative attributes.

Weber, Thomas. "Is there still a Chipko andolan?" *Pacific Affairs* 60 (Winter 1987-88): 615-628.
   "About fifteen years ago India experienced the large-scale emergence of action groups which drew conventionally apolitical citizens into expressing and fighting for their basic right to participate in the decision making process. The Chipko andolan--formed to save trees--is perhaps India's most celebrated action group."

Zuckerman, Seth. "Environmentalism turns 16." Nation 243 (October 18, 1986): 368-370.
   Discusses changes in the environmental movement since Earth Day in 1970. Many national groups have adopted sophisticated tactics, including economic analysis, demonstration projects, Washington offices, and monkey wrench grassroots tactics.

General interest periodicals (unannotated)

Ahlberg, Brian. "Green around the blue collar?" (new alliance between labor and environmentalists raises political hopes) *Utne Reader* 46 (July-August 1991): 44(2)

Aiyar, Shahnaz Anklesaria. "Going green: joining forces to clean up the planet." (cover story: the environment) *World Press Review* 37,8 (August 1990): 11(2)

Alexander, Charles P. "Gunning for the greens." (environmental movement) *Time* 139,5 (February 3, 1992): 50(3)

Alexander, Don. "Is there a green future in progressive politics?" (interview with Robert Paehlke, editor of *Alternatives*) *Alternatives* 17,2 (July-August 1990): 56(3)

Ananyev, Andrei. "Vasili Peskov: on man and nature." *Soviet Life* 11 (November 1988): 18(3)

"Apocalypse now." (criticism of environmental movement) (editorial) *National Review* 42,9 (May 14, 1990): 14(2)

Berlet, Chip. "Activists face increased harassment; protestors are perceived as security risks." *Utne Reader* 49 (January-February 1992): 85(4)

Berlet, Chip. "Hunting the 'green menace': the environmental movement has become the target of domestic covert operations." *The Humanist* 51,4 (July-August 1991): 24(9)

"Black marks for Greens." (National People of Color Environmental Leadership Summit) *Economist* 321,7730 (October 26, 1991): A29(2)

Bookchin, Murray; and Tyrrell, R. Emmett, Jr. "We can't heal the environment without remaking our society." (includes related articles) *Utne Reader* 36 (November-December 1989): 80(4)

Brown, Linda J. "Saving the land-one piece at a time." (greenway conservation) *Good Housekeeping* 214,5 (May 1992): 163(2)

Buckley, William F., Jr. "Ecological misgivings." (criticism of environmental movement) (column) *National Review* 42,10 (May 28, 1990): 54(2)

Carter, Clay. "Integration 101 for environmentalists." (how to make green groups a rainbow) *Utne Reader* 41 (September-October 1990): 116(2)

Charles, Prince of Wales. "Green watch; what's happening in the fight to save our environment; a message from Prince Charles." *Good Housekeeping* 212,5 (May 1991): 146(2)

Delattre, Pierre. "Beauty and the aesthetics of survival." (includes related article on visionary arts movement) *Utne Reader* 34 (July-August 1989): 64(5)

"Earthen renewal begins at home." *People* 33,SPEISS (Spring 1990): 42(2)

Easterbrook, Gregg. "Green Cassandras." (environmentalists) *The New Republic* 207,2 (July 6, 1992): 23(3)

Easterbrook, Gregg. "Saving the Earth—what's really important?" (includes related information on setting priorities) *Good Housekeeping* 212,4 (April 1991): 73(2)

Eckersley, Robyn. "Green politics: a practice in search of a theory?" (international environmental activism) *Alternatives* 15,4 (November-December 1988): 52(10)

Ehrlich, Paul; and Ornstein, Robert. "New World, New Mind." (toward species consciousness) *New Perspectives Quarterly: NPQ* 6,1 (Spring 1989): 26(8)

Gottlieb, Robert. "The hazards of eco-chic: would the real greens please stand up?" *Utne Reader* 39 (May-June 1990): 109(2)

Gray, Elizabeth Dodson. "The parable of the sandhill cranes: women, men and the Earth." *The Catholic World* 233,1396 (July-August 1990): 182(5)

"Green diplomacy takes off: sea and air know no boundaries." (cover story: the environment) *World Press Review* 37,8 (August 1990): 12(2)

"Greening of the invisible hand." (environmentalists) *Economist* 309,7582 (December 24 1988): 107(2)

Grover, Ronald. "Fighting back: the resurgence of social activism." *Business Week* 3106 (May 22, 1989): 34(2)

Hutchings, Vicky. "The good life." (the green movement) *New Statesman Society* 3,104 (June 8, 1990): 10(2)

Irani, Ray R. "Environmental literacy; building a coalition for the 1990's." *Vital Speeches of the Day* 57,17 (June 15, 1991): 542(3)

Irvine, Sandy. "No growth in a finite world." (essay on ecology and politics) *New Statesman Society* 3,128 (November 23, 1990): 16(3)

Iverem, Esther. "By Earth obsessed." (environmental protection) *Essence* 22,5 (September 1991): 37(2)

Johnson, William Oscar. "Back on track: Earth Day success story; the Chattanooga choo-choo no longer spews foul air." *Sports Illustrated* 72,18 (April 30, 1990): 26(4)

Kay, Jane Holtz. "Applying the brakes." (environmental movement and the automobile) Nation 251,8 (September 17, 1990): 280(5)

Kihn, Martin. "SDS Jr.; wrapping itself in the cloak of environmentalism, a new movement is spreading fast in the universities. But it seems less interested in cleaning the air than in bashing capitalism". (Students for a Democratic Society, Student Environmental Action Coalition) *Forbes* 147,12 (June 10, 1991): 51(2)

Klein Joe. "Cleaning up: how to save the Earth the American way." *New York* 23,15 (April 16, 1990): 57(2)

Miller, William H. "Environmental activism is back." *Industry Week* 212 (March 8, 1982): 40(6)

Morris, David. "The four stages of environmentalism." (column) *Utne Reader* 50 (March-April 1992): 159(2)

Muwakkil, Salim. "Black and white and green all over. (environment) *Utne Reader* 39 (May-June 1990): 18(2)

Nash, J. Madeleine. "Learning how to revive the wilds of Eden." (environmentalism) *Time* 138,15 (October 14, 1991): 62(3)

Porrit, Jonathan. "Seeing green: how we can create a more satisfying society." (environmentalism) *Utne Reader* 36 (November-December 1989): 70(7)

Postrel, Virginia I. "Environmental movement; a skeptical view." *Vital Speeches of the Day* 56,23 (September 15, 1990): 729(4)

Postrel, Virginia I. "The green road to serfdom." (Green Party) *Reason* 21,11 (April 1990): 22(7)

Rensenbrink, John; and Smith, Sam. "Protest too much: the greening of America means new political strategies." (includes related article) *Utne Reader* 48 (November-December 1991): 93(4)

Satchell, Michael. "Any color but green." (political opposition to environmental movement) *U.S. News & World Report* 111,17 (October 21, 1991): 74(3)

Scarlett, Lynn. "Don't buy these environmental myths." *Reader's Digest* 140,841 (May 1992): 100(3)

Seabrook, Jeremy. "The nature of nature." *New Statesman Society* 3,128 (November 23, 1990): 23(2)

Shankland, Alexander. "Green roots and red ends." *New Statesman Society* 3,128 (November 23, 1990): 14(2)

Sinclair, A. J.; Symington, D.; and Winn, S. N. "Eco rock: environmentalism in music." *Alternatives* 16,4 (March-April 1990): 22(2)

Trinkl, John. "The environment movement." *Canadian Dimension* 24,5 (July-August 1990): 25(3)

Vare, Robert. "Save the Earth now!" *Rolling Stone* 577 (May 3, 1990): 44(2)

Walljasper, Jay. "Can Green politics take root in the U.S.?" *Utne Reader* 35 (September-October 1989): 140(4)

Weir, David. "10 years left." (special section on the environment) (cover story) *Mother Jones* 17,2 (March-April 1992): 29(3)

Wooster, Martin Morse. "Eco-logic." (environmentalist attitudes) (editorial) *Reason* 23,2 (June 1991): 48(2)

Yandle, Bruce. "Environment and efficiency lovers."*Society* 29,3 (March-April 1992): 23(10)

Environmental and professional journals (annotated)

Bailey, Ronald. "The apocalypse boosters: raining in their hearts." *National Review* 42 (December 3, 1990): 32-36.
   "Nuclear winter, population explosion, non-renewable resources, vanishing species--some of them are genuine problems, others are made up out of whole cloth; all are easily exploited for fun and left-wing profit." This atmosphere of catastrophe has been sustained by a cadre of professional doomsters, intellectuals and policy-makers whose apocalyptic predictions are meant to frighten the public into adopting their leftist policy prescriptions. Includes sidebar by Bruce Ames: Misconceptions about pollution and cancer.

Baugh, Joyce A. "African-Americans and the environment: a review essay." *Policy Studies Journal* 19 (Spring 1991): 182-191.
   "Until recently, the participation of African-Americans in environmental causes was negligible. Reasons often cited for this absence include a lack of concern for the environment by African-Americans, a lack of attention by mainstream environmentalists to issues affecting the quality of life for African-Americans, and racism in environmental organizations. Whatever the reasons for this in the past, there now is increasing activism by African-Americans in environmental issues."

Borrelli, Peter. "Environmentalism at a crossroads." *Amicus Journal* 9 (Summer 1987): 24-37.

Reviews developments in environmentalism in the U.S., discussing key environmental legislation, key pressure groups, and key players in the movement. Considers differences among environmental protection groups and identifies a few general trends; these include increased willingness by industry to sit down with environmentalists, and increasingly global interests of environmentalists.

Bosso, Christopher J. "Transforming adversaries into collaborators." *Policy Sciences* 1 (1988): 3-22.
"After more than a decade of sharp conflict, environmental and chemical industry lobbyists sat down in direct, private negotiations on reforming pesticide regulatory policy in the United States. This article examines the dynamics that transformed longtime adversaries into temporary collaborators during 1985-1986. It also examines why the efforts of this ad hoc coalition eventually failed." Includes discussion of current U.S. interest group politics.

Broad, Robin; and Cavanagh, John. "Marcos's ghost." *Amicus Journal* 11 (Fall 1989): 18-29.
"A growing grass-roots environmental movement in the Philippines discovers that the 'old politics' is back, or perhaps never left. Poverty and corruption grind on, coupled with a rate of destruction of forests and other natural resources that is among the fastest in the world."

Darst, Robert G., Jr. "Environmentalism in the U.S.S.R.: the opposition to the river diversion projects." *Soviet Economy* 4 (July-September 1988): 223-252.
"The opposition to the Soviet river diversion projects is analyzed in order to explore the nature of environmentalism in the U.S.S.R." Divides the opposition into three groups: utilitarians, populists, and Russian nationalists.

Eckersley, Robyn. "Green politics and the new class: selfishness or virtue?" *Political Studies* 37 (June 1989): 205-223.
"The predominantly new middle-class social composition of the green movement has become a matter of increasing interest. This paper considers the concept of the 'new class' in relation to two explanations for the social composition of the green movement."

"Environmental protection--has it been fair?" *EPA Journal* 18 (March-April 1992): whole issue (64 p.)
"Is it time that we broadened the definition of environmental protection in this country? The physical environment of America's minorities--Hispanics, Native Americans, Asians, African Americans, the poor of any color--has in one way or another been left out of the environmental cleanup of the past two years."

Fowler, Linda L.; and Shaiko, Ronald G. "The grass roots connection: environmental activists and Senate roll calls." *American Journal of Political Science* 31 (August 1987): 484-510.

"Interest groups resort to grass roots lobbying tactics with increasing frequency, but they are unsure when and how such a strategy affects congressional decision making. In analyzing citizen activities who belong to environmental organizations and their influence on 1978 Senate roll calls, this paper defines the limits and possibilities of such grass roots pressure. In general, the statistical results indicate that grass roots mobilization has a modest value as a lobbying tool."

Guskind, Robert. "Big Green light." *National Journal* 22 (October 6, 1990): 2400-2404.
"Backers of Big Green, California's major environmental ballot initiative, say that approval would give the go signal to more-aggressive environmental politics around the country." Includes a sidebar entitled "Warning: litigation can be hazardous," which discusses California's Safe Drinking Water and Toxic Enforcement Act.

Jain, Purnendra C. "Green politics and citizen power in Japan: the Zushi movement." *Asian Survey* 31 (June 1991): 559-575.
"Examines an ongoing movement in the city of Zushi, which has been opposing since 1982 a central government plan to construct a housing complex for U.S. naval personnel on a 290-hectare forest area of the Ikego Hills. The Zushi movement is related to the conservation of the forest, its flora and fauna, and can be regarded as the first major 'green' movement in Japan. The movement does not oppose the government's housing plan, only the site of the plan."

"Kid crusaders: children and the environment." *Environmental Action* 23 (September-October 1991): 19-26.
"From lead poisoning to pesticide exposure, children are especially vulnerable to environmental risks. At the same time, young activists are championing environmental causes in the classroom and in Congress."

Klapp, Merrie. "Challenges from smart publics." *Environmental Professional* 3 (1988): 189-197.
"Examines the problem that 'smart publics' pose for environmental decision makers. Smart publics are public groups that hire or attract scientists to challenge government decisions proposed on their behalf. Reviews the scientific controversies and roles of smart publics in two cases of decision making: one about cancer risks from dioxin emissions of a municipal waste incinerator, the other about fire risks from the transport of liquefied natural gas (LNG)." Suggests how decision makers can cooperate with smart publics.

Kowalewski, David; and Porter, Karen L. "Ecoprotest: alienation, deprivation, or resources?" *Social Science Quarterly* 73 (September 1992): 523-534.
Protest is said to be driven by political alienation, relative deprivation, and resource availability. Yet debate surrounds the meanings of the three perspectives, their applicability to moderate and militant protest, and their relevance to post-industrial policies. This study tests the three perspectives with data on protest

against a nuclear waste facility. All three perspectives help explain both moderate and militant participation."

Lovrich, Nicholas P., Jr.; and Pierce, John C. "The good guys and bad guys in natural resource politics: content and structure of perceptions of interests among general and attentive publics." *Social Science Journal* 23,3 (1986): 309-326.
    Using a survey of Idahoans about water resource policy, the authors examine public attitudes regarding the degree of preferred influence to be given interest groups on environmental issues.

MacAdam, Gil. "Environmental protection through citizen action." *Parks and Recreation* 25 (October 1990): 46-51.
    "In spite of the known difficulties, a few concerned residents of Broward County, Florida initiated the effort, rallied public support, and persevered over an extended time to ultimately succeed in preserving 1,300 acres of environmentally sensitive land for the public."

Nownes, Anthony J. "Interest groups and the regulation of pesticides: Congress, coalitions, and closure." *Policy Sciences* 24 (February 1991): 1-18.
    "This article is essentially a rejoinder to Christopher Bosso's piece, 'Transforming Adversaries Into Collaborators: Interest groups and the regulation of chemical pesticides,' which appeared in *Policy Sciences* (21: 3-22). The case of pesticides regulation is re-examined and some new insights are offered. At the center of Bosso's argument is the contention that Congress is passive. John Kingdon's agenda/alternative distinction is utilized to arrive at an alternative way to think about the role of Congress in today's 'permeable pressure system.' "

Ohnuma, Keiko. "New kids on the Earth." *Sierra* 76 (January-February 1991): 34-36.
    "Not content to prepare only for a high-income future, students are working today to improve the world. Among the largest components of this activist mechanism is the Student Environmental Action Coalition, known as 'Seek.' "

Overby, L. Marin; and Ritchie, Sarah J. "Mobilized masses and strategic opponents: a resource mobilization analysis of the clean air and nuclear freeze movements." *Western Political Quarterly* 44 (June 1991): 329-351.
    "We extend previous research on social movements by arguing that the strategic decisions of status quo opponents, as well as those of the mobilizing masses, are important determinants of the success of such movements. Reanalyzing case studies of the clean air and nuclear freeze movements, we contend that the ineffective mobilization of public opinion by interests opposed to the clean air movement accounts for the resulting 'policy spiral,' while effective mobilization of public opinion by interests resistant to the nuclear freeze precluded a non-incremental outcome in that case."

Pierce, John C.; Lovrich, Nicholas P.; and Matsuoka, Masahiko. "Support for citizen

participation: a comparison of American and Japanese citizens, activists and elites."
Western Political Quarterly 43 (March 1990): 39-59.

"This paper examines the dynamics underlying contemporary demands for broader citizen involvement in environmental policy formation processes in two postindustrial nations. The subjects of study are Japanese and American citizens, activists, and elected/administrative elites in two comparable locations of recent high public activism concerning environmental issues."

Postrel, Virginia I. "The environmental movement: a skeptical view." *Chemtech* 21 (August 1991): 457-461.

For the green ideology, the core value is stasis, or sustainability: the ideal is of an Earth that shows little effects of human activity. This static view "leads environmentalists to advocate policies that will make growth hard on people, as a way of discouraging further development. Second, the static view leads environmentalists to misunderstand how real environmental problems can be solved."

Rauber, Paul. "No second warning." *Sierra* 76 (January-February 1991): 24, 26, 28, 30.

"Defenders of the ancient forests were ready for a summer of nonviolent confrontation. Then a car bomb went off."

Rohrschneider, Robert. "The roots of public opinion toward new social movements: an empirical test of competing explanations." *American Journal of Political Science* 34 (February 1990): 1-30.

"Analyzing public opinion toward environmental groups in four nations, this research employs a Eurobarometer (1986) to test and synthesize various models. Postmaterial values, left policy orientations, and the perceptions of national pollution problems have the strongest direct influence on public evaluations of environmental organizations."

Ruben, Barbara. "Root rot." *Environmental Action* 24 (Spring 1992): 25-30.

"Commandeering traditional grassroots organizing strategies, the anti-environmental movement is rallying its ranks to gut years of environmental progress."

Rudig, Wolfgang. "Green party politics around the world." *Environment* 33 (October 1991): 6-9, 25-31.

"Green political parties have been established in many countries during the last 20 years, yet very few have experienced significant electoral success. Here, the author discusses the development of green parties, identifies the factors behind their sporadic successes and recurring defeats, and evaluates the ability of the Greens to attract voters, given the growing environmental zeal of many major political parties."

Russell, Dick. "Environmental racism." *Amicus Journal* 11 (Spring 1989): 22-35.

Describes the efforts of grass-roots minority groups in their battle against pollution, focusing on the Los Angeles area.

Stanfield, Rochelle L. "Out-standing in court." *National Journal* 20 (February 13, 1988): 388-391.
> "Environmentalists trying to sue federal agencies and private businesses are being challenged at the courthouse door by the Justice Department and Reagan-appointed judges."

Stanfield, Rochelle L. "The green blueprint." *National Journal* 20 (July 2, 1988): 1735-1737.
> "The environmental community, having learned some important lessons from the Heritage Foundation, is determined to greet the new President-elect in November with a detailed program for change."

Taylor, Jerry. "Environmental lobby suffers crushing defeat at hands of voters." *F.Y.I.* (American Legislative Exchange Council), (November 2, 1990): 1-5.
> In last Tuesday's election, "the across-the-board rout of environmentalist causes has left politicians stunned, environmentalists reeling, and analysts convinced that the 'green lobby,' once so feared and appeased, is a political paper-tiger that registers more noise than clout with the American electorate. Every single major environmental ballot initiative not only lost, but generally lost big."

Thompson, Grant P. "New faces, new opportunities." *Environment* 27 (May 1985): 6-11, 30.
> While polls show that the public is concerned with environmental issues, the environmental movement seems to have lost momentum in the last few years. Article discusses environmental organizations' needs to have financially savvy managers in leadership positions and to recapture the interest and energy of the American public in environmental protection.

Truax, Hawley. "Labor and the environment: workers and environmental protection." *Environmental Action* 24 (Spring 1992): 11-24.
> Contents include: "When progress bears a bitter price," by Hawley Truax; "Viewpoint: the case for unity," by Ruth Caplan; "Viewpoint: confronting corporate power," by Richard Grossman; "The growth dilemma," by Andy Feeney; Feeney; and "Easing a painful transition," by Denny Scott.

Whelan, Tensie. "A tree falls in Central America." *Amicus Journal* 10 (Fall 1988): 28-38.
> Describes the growth of environmentalism in the countries of Central America. "Central Americans have taken the goal of sustainable development (the notion that conservation is integral to economic development) and made it theirs." Countries discussed include: Costa Rica, Panama, Nicaragua, El Salvador, Honduras, Belize and Guatemala.

Wolok, Miriam S. "Standing for environmental groups: procedural injury as injury-in-fact." *Natural Resources Journal* 32 (Winter 1992): 163-193.
> "Explores (1) the changing standing requirements generally under the United

States Constitution and specifically for environmental organizations, (2) the evolution and use of procedural injury to satisfy standing under Article III through its adoption by the Eighth Circuit in Defenders of Wildlife vs. Lujan, and (3) possible future applications of procedural injury. It also examines the pitfalls that attend the use of procedural injury."

Yearley, Steven. "Greens and science: a doomed affair?" *New Scientist* 131 (July 13, 1991): 37-40.
"Environmentalists have turned to science to give weight to their campaigns to save the planet. But science can prove an unreliable ally."

Zuckerman, Seth. "Living there: call it bioregion, a watershed, or a life-place, your backyard by any other name is just as sweet." *Sierra* 72 (March-April 1987): 61-67.
"The conceptual roots of bioregionalism go back to indigenous peoples who lived as members rather than rulers of their ecosystems." Describes areas of North America where the residents take the growing body of knowledge about ecology and apply it to real-life situations. Residents of the Mattole River Valley work to restore salmon, residents of the Kansas Area Watershed Council, British Columbia and New York State also have established bioregional councils.

Environmental and professional journals (unannotated)

Arrandale, Tom. "The mid-life crisis of the environmental lobby: can it stay pure and still succeed in politics?" (cover story) *Governing* 5,7 (April 1992): 32(5)

Boerner-Ein, Deborah. "The tree gangs of glittertown." (environmental groups in Los Angeles, California) *American Forests* 97,11-12 (November-December 1991): 32(5)

Covault, Craig. "Major space effort mobilized to blunt environmental threat." (Mission to Planet Earth special report) *Aviation Week & Space Technology* 130,11 (March 13 1989): 36(4)

Czapski, Silvia. "Grassroots environmentalism in Brazil." *Conservationist* 46,1 (July-August 1991): 42(6)

Grove, Noel. "Which way to the revolution?" (environmentalism) *American Forests* 96,3-4 (March-April 1990): 21(4)

Hair, Jay D., et al. "Beyond the 1990s." (exchange about "Environmental Protection for the 1990s - and Beyond") *Environment* 30,2 (March 1988): 3(3)

Junkin, Elizabeth Darby. "Losses and gains." (man's efforts to save the environment) *Buzzworm: The Environmental Journal* 4,3 (May-June 1992): 35(7)

Letto, Jay. "One hundred years of compromise."(the changing environmental movement) *Buzzworm: The Environmental Journal* 4,2 (March-April 1992): 26(6)

Lewis, Thomas A. "A new day must dawn." (environmental progress) *National Wildlife* 28,2 (February-March 1990): 4(4)

McFadden, Jeff. "How far do we go with environmentalism?" (protecting the environment may require giving up technology and the comfortable lifestyle it has created) *Countryside & Small Stock Journal* 75,4 (July-August 1991): 54(3)

Meyer, Eugene L. "Environmental racism: why is it always dumped in our backyard? Minority groups take a stand."*Audubon* 94,1 (January-February 1992): 30(3)

Polsgrove, Carol. "Unbroken circle." (activist training in environmental protection at Highlander Center) *Sierra* 77,1 (January-February 1992): 130(5)

Quinnett, Paul. "Memories of the Green Guard." (Oregon Green Guard, an environmental group for children) *American Forests* 96,9-10 (September-October 1990): 9(2)

Rauber, Paul. "Greenhorns." (Green Party) *Sierra* 77,3 (May-June 1992): 45(3)

Wells, Malcolm. "A jigsaw puzzle." (racism, environmentalism, and the social conscience of an architect) *Whole Earth Review* 62 (Spring 1989): 22(4)

Worcman, Nira Broner. "Brazil's thriving environmental movement." *Technology Review* 93,7 (October 1990): 42(10)

Conference proceedings

Proceedings of the World Environment Day, June 5, 1988, Lahore. Lahore: Environmental Protection Agency, 1988. 85 p.: ill. (TD187.5.P18 W67)

Bibliographies

Duensing, Edward; and Chilik, Yvonne. *The negotiations of environmental conflicts: techniques, case studies, and analyses.* Monticello, Ill.: Vance Bibliographies, 1986. 10 p. (Public administration series: bibliography P 1942)

Gray, Dorothy A. "Voices for wilderness: conservation society serials." *Serials Review* 15 (Summer 1989): 23-33.
  Provides a bibliography and brief description of periodicals and magazines published by organizations that support the conservation and preservation of nature.

Linden, Eugene. "Literary guides to turning green: to respect nature, people need a whole new way of thinking." (books dealing with environmental concerns) *Time* 136,24 (December 3 1990): 104(2)

Pearson, Arn H.; and Buttel, Frederick H. *Environmental politics: an annotated bibliography.* Monticello Ill.: Vance Bibliographies, 1984. 65 p. (Public administration series: bibliography P-1369)

Vaz, Mark. "Leaves of green." (environmentalists choose most important books) *Sierra* 71 (May-June 1986): 56(6)

Books for young adults

50 simple things kids can do to save the Earth. Kansas City, Mo.: Andrews and McMeel, 1990. 156 p.: ill. (TD171.7 .A16 1990)

Bellamy, David J. *How green are you?* New York: C. Potter, 1991. Distributed by Crown Publishers. 31 p.: color ill. (TD171.7 .B45 1991)

Brown, Laurene Krasny. *Dinosaurs to the rescue!: a guide to protecting our planet.* Boston: Joy Street Books, 1992. 1 v. color ill. (TD171.7 .B76 1992)

Doney, Meryl. *The green activity book.* Batavia, Ill.: Lion Publishing, 1991.

Elkington, John. *The young green consumer guide.* London: V. Gollancz, 1990. 96 p.: color ill. (TD171.7 .E44 1990)

Fry-Miller, Kathleen M. *Peace works: young peacemakers project book 2.* Elgin, Ill.: Brethren Press, 1989. 116 p.: ill. (JX1963 .F87 1988)

Gay, Kathlyn. *Caretakers of the Earth.* Hillside, N.J.: Enslow, 1993.

Going green: a kid's handbook to saving the planet. New York: Viking, 1990. 111 p.: color ill. (TD171.7 .G64 1990)

Goodman, Billy. *A kid's guide to how to save the planet.* New York: Avon Books, 1990. 137 p.: ill.

Holmes, Anita. *I can save the Earth: a kids' handbook for rescuing life on Earth.* Englewood Cliffs, N.J.: Julian Messner, 1992.

Jakobson, Cathryn. *Think about the environment.* New York: Walker, 1992. 143 p.: ill. (TD176 .J35 1992)

Katz, Bobbi. *The Care Bears and the big cleanup.* New York: Random House, 1991. 1 v.: color ill. (PZ7.K157 Car 1991)

Krull, Kathleen. *It's my Earth, too: how I can help the Earth stay alive.* New York: Delacorte Press, 1992. color ill. (TD171.7 .K78 1992)

Langone, John. *Our endangered Earth: what we can do to save it.* Boston: Little, Brown, 1992. 197 p. (TD176 .L36 1992)

Levine, Shar. *Projects for a healthy planet: simple environmental experiments for kids.* New York: Wiley, 1992. 95 p.: ill. (TD171.7 .L48 1992)

Lowery, Linda. *Earthwise at play: a guide to the care & feeding of your planet.* Minneapolis, Minn.: Carolrhoda Books, 1992.

Lowery, Linda. *Earthwise at school: a guide to the care & feeding of your planet.* Minneapolis, Minn.: Carolrhoda Books, 1993.

Markle, Sandra. *The kids' Earth handbook.* New York: Atheneum, 1991. 58 p. : ill. (TD171.7 .M37 1991)

McVey, Vicki. *The Sierra Club kid's guide to planet care and repair.* San Francisco: Sierra Club Books, 1992.

Newton, David E. *Taking a stand against environmental pollution.* New York: Watts, 1990. 157 p.: ill. (TD171 .N59 1990)

Our poisoned planet: can we save it?. New York: Facts on File, 1989. 216 p.: ill. (TD176 .O97 1989)

Ross, Anna. *Grover's 10 terrific ways to help our wonderful world.* New York: Random House, 1992. 1 v.: color ill. (PZ7.R71962 Gr 1992)

Wheeler, Jill C. *The food we eat.* Edina, Minn.: Abdo & Daughters, 1991. Distributed by Rockbottom Books, Minneapolis, Minn.

Wheeler, Jill C. *The throw-away generation.* Edina, Minn.: Abdo & Daughters, 1991. Distributed by Rockbottom Books, Minneapolis, Minn.

Videos

Environmental dog. Produced and directed by Randolph Wright. Pyramid Film and Video (Box 1048, Santa Monica, CA 90406). 1990. 20 min.
    Narrated by Ralphie the Airedale, this video opens with Ralphie accidentally

swallowing toilet bowel cleaner. His subsequent crusade to make his owners' home more environmentally sound is the focus of the rest of the story. Includes examples of water and electricity conservation, as well as recycling.

One second before sunrise: a search for solutions. Bullfrog Films (Oley, PA 19547). 1989. 60 min.
    Presents examples of individuals making small-scale effort in an attempt to help remedy gigantic environmental problems. Examples from inner-city New York, Amazonia, Puerto Rico, Holland, and southern California.

HISTORY

Books (annotated)

Bailes, Kenneth E., ed. *Environmental history: critical issues in comparative perspective.* Lanham, Md: University Press of America, 1985. 697 p.
    This volume is comprised of papers presented at the 1982 American Society for Environmental History conference, and covers topics such as geographical determinism, overly enthusiastic advocacy, the role of institutions, views of nature in the nineteenth century, conservation history, women as conservationists, urban environmentalism, world climate, the role of scientists in shaping public values, deforestation, pesticides, and the importance of enviornmental history to other historical views.

Bramwell, Anna. *Ecology in the Twentieth Century: A History.* New Haven, Ct.: Yale University Press, 1989. 292 p. (QH 541.B7 1989)
    This work traces the history of the ecological movement from its beginnings in the ideas of Ernst Haeckel to modern day economic activism. Other aspects, such as the green movement, organic farming, and conservation, are also addressed. The focus is European because the history of the movement in North America has been covered in other works. Well researched, with many footnotes and a good bibliography.

Conzen, Michael. *The making of the American landscape.* Boston: U. Hyman, 1990. 432 p.
    Explains the impact of colonization, ethnicity, industrialization and urbanization on the American landscape, then analyzes the effect of the automobile and recreation in an examination of the interaction between human activities and the natural landscape.

Engle, Ed. *Seasonal: a life outside.* Boulder, Colo.: Pruett, 1989. (SD129.E5 A3 1989)
    The author has been a seasonal employee of the Forest Service in Idaho, Colorado, and California. He relates his personal experiences during the environmental

awakening of the 1970s and offers his unique perspective on the state of the Earth's health. Presents a good description of his love for nature.

Fox, Stephen. *The American Conservation Movement: John Muir and His Legacy.* Madison: University of Wisconsin, 1986.
Traces the history of the American conservation movement from 1890 to 1975, with particular emphasis given to the life of John Muir, who is often regarded as the father to the movement.

Hays, Samuel P. *Beauty, health, and permanence: environmental politics in the United States, 1955-1985.* New York: Cambridge University Press, 1987. 630 p.: ill. (HC110.E5 H39 1987)
Argues that environmental objectives have been part of the core of changing social values since World War II, and that the rigorous antienvironmental revolution of the Reagan administration tested, and thus revealed, the vitality and breath of popular support for environmental values and objectives.

Huth, Hans. *Nature and the American: three centuries of changing attitudes.* 1957. New ed. Lincoln: University of Nebraska Press, 1991.
Chronicles the history of Americans' attitudinal changes toward nature over three centuries, with emphasis on the changes which led to the conservation movement. This reissue includes a new introduction by historian Douglas H. Strong.

Mannion, Antoinette M. *Global environmental change: a natural and cultural environmental history.* New York: Wiley, 1991. 410 p.
Provides a synopsis of environmental change for three million years and examines the environmental changes caused by industrial and agricultural developments during the past two hundred years. Topics in this section include crop chemicals, salinization, desertification, deforestation, recreation, tourism, and biotechnology.

McCormick, John. *Reclaiming paradise: the global environmental movement.* Bloomington: Indiana University Press, 1990. 274 p.
Traces the origins and accomplishments of the environmental movement beginning with nineteenth-century naturalists. Argues that the environmental prophets of doom, spurned by Rachael Carson's *Silent Spring* in 1962, set the quarrelsome tone for modern environmentalists. The 1972 Stockholm conference solidified the global concern about the environment and "marked a transition from the emotional and occasionally naive New Environmentalism of the 1960s to the more rational, political, and global perspectives of the 1970s." Devotes much attention to United Nations efforts.

Paehlke, Robert C. *Environmentalism and the future of progressive politics.* New Haven, Conn.: Yale University Press, 1989. 325 p. (HC110.E5 P34 1989)
Begins with the background of the nineteenth century conservation movement, Reviews the scientific, ethical, and ideological structure of environmentalism, and

concludes with an ecological program which aims to be acceptable to moderate politics. A case against consumerism and waste.

Scheffer, Victor B. *The shaping of environmentalism in America*. Seattle: University of Washington Press, 1991. 249 p. (HC110.E5 S39 1991)
Outlines the public awakening to air, land, water, and wildlife abuse by unbridled human consumption and activities. Provides an overview of the conservation activities in the nineteenth and first half of the twentieth century, then focuses on the formative years of the environmental movement (1960-1980) before it encountered the Reagan administration (1981-1989). Concludes with an argument for striking the middle ground that balances conservation with the reality that people must use natural resources to survive.

Smith, Michael L. *Pacific visions: California scientists and the environment 1850-1915*. New Haven, Conn.: Yale University Press, 1987. 243 p.: ill. (Q127.U6 S62 1987)
Explains how after the gold rush California scientists became involved in predicting ecological problems because of attitudes that regarded the environment as "a storehouse of commodities to fuel the ever expanding engines of progress" and because of the rapid expansion of technological innovation. By 1892 these scientists had organized the Sierra Club, and in 1916 the Panama-Pacific Exposition dramatized the enormous changes technology had forced on the California landscape. Argues that the involvement of California scientists with the environment was different from scientific interest and involvement in the East.

Tolba, Mostafa Kamal, ed. *Evolving environmental perceptions: from Stockholm to Nairobi*. Boston: Butterworths, 1988. 458 p. (HC79.E5 E96 1988)
Analyzing the attitudinal changes which have come about over a period of fifteen year--from the 1972 U.N. Conference on the Human Environment in Stockholm, to the 1982 Nairobi conference, and the 1987 U.N. General Assembly meeting--this work demonstrates how awareness of environmental problems has greatly increased. Maintains that much more international cooperation is still needed.

Weiner, Douglas R. *Models of nature: ecology, conservation, and cultural revolution in Soviet Russia*. Bloomington: Indiana University Press, 1988. 312 p.
Chronicles the ecological movement in Russia during the two decades after the Bolshevik Revolution. It wasn't until Stalin forced industrialization that the Soviet Union adopted the "utilitarian" approach that led to massive destruction of the environment. Maintains that some of the same preservation spirit is alive in modern Russia and that the Russian people are interested in restoring their losses.

Worster, Donald, ed. *The ends of the Earth: perspectives on modern environmental history*. New York: Cambridge University Press, 1988. 341 p.: ill. (GF13 .E52 1988)
Analyzes the historical relationship between climate, food supplies, population, technology, social change, and imperialism to provide background for environmental concerns.

Books (unannotated)

American environmentalism: readings in conservation history. 3rd ed. New York: Knopf, 1990.

Brant, Irving. *Adventures in conservation with Franklin D. Roosevelt.* Flagstaff, Ariz.: Northland Publishing, 1989. 348 p.: ill. (E807 .B74 1989)

Environmental history: critical issues in comparative perspective. Lanham, Md.: University Press of America; Denver, Colo.: American Society for Environmental History, 1985. 697 p. (GF13 .E58 1985)

Evolving environmental perceptions: from Stockholm to Nairobi. Boston: Butterworth, 1988. 458 p. (HC79.E5 E96 1988)

Gould, Peter C. *Early Green politics: back to nature, back to the land, and socialism in Britain, 1880-1900.* New York: St. Martin's Press, 1988. 225 p. (DA550 .G68 1988)

Government and environmental politics: essays on historical developments since World War Two. Washington, D.C.: Wilson Center Press, 1989. Distributed by the University Press of America. 325 p. (HC110.E5 G686 1989)

Mitchell, Lee Clark. *Witnesses to a vanishing America: the nineteenth-century response.* Princeton, N.J.: Princeton University Press, 1981. 320 p.: ill. (QH76 .M54)

Pepper, David. *The roots of modern environmentalism.* Dover, N.H.: Croom Helm, 1984. 246 p.: ill. (QH540.5 .P47 1984)

Scheffer, Victor B. *The shaping of environmentalism in America.* Seattle: University of Washington Press, 1991. 249 p.: ill. (HC110.E5 S39 1991)

Shabecoff, Philip. *A fierce green fire: the American environmental movement.* New York: Hill & Wang, 1992.

Wild, Peter. *Pioneer conservationists of eastern America.* Missoula, Mont.: Mountain Press, 1985. 280 p.: ill. (S926.A2 W53 1985)

Zahniser, Howard. *Where wilderness preservation began: Adirondack wilderness writings of Howard Zahniser.* Utica, N.Y.: North Country Books, 1992.

General interest periodicals (unannotated)

Paehlke, Robert. "Eco-history: two waves in the evolution of environmentalism." (includes related article) (cover story) *Alternatives* 19,1 (September-October 1992): 18(6)

Stegner, Wallace. "It all began with conservation; our first response to the New World's wild Eden was to break it--but seeds of a land ethic were planted early on." (special issue on the environment) *Smithsonian* 21,1 (April 1990): 34(10)

Environmental and professional journals (annotated)

Grove, Richard. "The origins of environmentalism." *Nature* 345 (May 3, 1990): 11-14. "There's nothing new in worries about man's effect on the natural world. Indeed, that history runs deeper (and wider) than latter-day environmentalists would have it."

Environmental and professional journals (unannotated)

Grove, Richard H. "Origins of western environmentalism." *Scientific American* 267,1 (July 1992): 42(6)

Madson, John. "A legacy of love and anger." (environmental heritage) *National Wildlife* 28,2 (February-March 1990): 54(6)

Bibliographies

Casper, Dale E. *The nation's response to its environment at risk: historical perspectives, 1974-1984.* Monticello, Ill.: Vance Bibliographies, 1985. 11 p. (Z5863.P6 C38 1985)

Elbers, Joan S., comp. *Changing wilderness values, 1930-1990: an annotated bibliography.* New York: Greenwood, 1991. 138 p. (Z5861.E4 1991)
> The purpose of this bibliography is to identify works that "encompass the different values Americans have sought in or attributed to the wilderness," and includes political, scientific, historical, philosophical, or personal accounts of wilderness experience. Each citation contains a brief but useful annotation.

LEADERS

(See also *Radical Environmentalism*, below) 466

Books (annotated)

Brower, David R. *For Earth's sake: the life and times of David Brower.* Salt Lake City, Utah: Peregrine Smith, 1990. 556 p. (QH31.B859 A3 1990)

David Brower has been one of the cornerstones of the American environmental movement that has emerged in the last few decades. Traces many of his important accomplishments, includes passages from the Sierra and other environmental publications, and profiles of people who have influenced his life. Discusses the flooding of Glen Canyon, which is often regarded as Brower's greatest defeat.

Cohen, Michael P. *The pathless way: John Muir and American wilderness*. Madison: University of Wisconsin Press, 1984. 408 p. (QH31.M9 C64 1984)

This is a biography of Muir's spiritual journey toward nature. Cohen explains Muir's repudiation of American progress, his disgust for man, disappointment with mankind, and his disillusionment with secular humanism and Christian humanitarianism. Cohen describes Muir's intellectual shallowness in his early years and his decline into a genteel political activist in his later years. Nonetheless, Muir's hopes and visions have greatly influenced generations of naturalists, and Cohen believes that he will continue to be revered for his inspiration.

Eisenbud, Merril. *An environmental odyssey: people, pollution, and politics in the life of a practical scientist*. Seattle: University of Washington Press, 1990. (TD140.E35 A3 1990)

This professional autobiography covers a career which included expertise in industrial hygiene, occupational safety, air pollution control, and radiation. Eisenbud taught at New York University, was affiliated with the Atomic Energy Commission, and served in the New York City Environmental Protection Administration. Relating his fifty-four years of experience, Eisenbud discusses the environmental challenges he faced and analyzes environmental hazards, including political problems and the harmful effects of technology on human health.

Glover, James M. *A wilderness original: the life of Bob Marshall*. Seattle, Wash.: Mountaineers Books, 1986. 323 p.

Chronicles Bob Marshall's role in the wilderness movement, which included well-known publications, involvement with federal policies, and the founding of the Wilderness Society.

Loeffler, Jack. *Headed upstream: interviews with iconoclasts*. Tucson, Ariz.: Harbinger, 1989. 194 p.: ill. (GF21 .L63 1989)

This work takes a bioethical perspective as it interviews a wide range of individuals, from poets Philip Whalen and Gary Snyder to Dave Foreman, the founder of Earth First!, and others. The interviewees all share a belief that cultural values must change in order for environmental degradation to be allayed.

Lyon, Thomas J.; and Stine, Peter, eds. *On nature's terms: contemporary voices*. College Station: Texas A&M University Press, 1992. 224 p.

Focuses on nature writing as literature. Contains essays by Bass and Terry Tempest, Barry Lopez, Gary Snyder, William Kittredge, and Charles Bowden. Attempts to bridge the gap between urban and wild environments and to allow the reader to once again understand the wild.

Meine, Kurt. *Aldo Leopold: his life and work.* Madison: University of Wisconsin Press, 1988. 638 p.

> A complete biography that covers every aspect of Leopold's professional and private life. Explains Leopold's transition from forester to the Forest Service and his break from bureaucratic culture. Discusses the importance of his changing view of wildlife from recreational to biotic.

Merrow, Susan; and Rickerby, Wanda. *One for the Earth: journal of a Sierra Club President.* Champaign, Ill.: Sagamore, 1992.

> As well as discussing some of the Earth's more pressing environmental concerns, this work offers personal information on Merrow's year as president of the Sierra Club (all Sierra Club officers are limited to one-year terms). In that year, March 1990 to March 1991, Merrow traveled extensively abroad, worked as a lobbyist and political insider, and tried to work out ecological solutions.

Montgomery, Sy. *Walking with the great apes: Jane Goodall, Dian Fossey, Biruté Galdikas.* Boston: Houghton Mifflin, 1991. 280 p. (QL26.M66 1991)

> Focusing more on three prominent women who studied apes than the actual animals, this work offers many interesting details on their personal lives, and their relationships with each other, Louis Leakey, and the scientific community. Contains some photographs and some suggested readings.

Murray, John A., ed. *Nature's new voices.* Golden, Colo.: Fulcrum, 1992. 256p.

> Introduces modern nature writers such as Gerard Gormley, Conger Beasley, and Jan DeBlieu. These essayists discuss natural history and wildlife in the tradition of Annie Dillard, Ted Hoagland, and Edward Abbey.

Paul, Sherman. *For the love of the world: essays on nature writers.* Iowa City: University of Iowa Press, 1992. 224 p.

> Analyzes the literary nature writing of several prominent authors, including Henry David Thoreau, John Muir, Barry Lopez, and Aldo Leopold.

Pinchot, Gifford. *Breaking new ground.* Washington, D.C.: Island Press, 1988. 522 p.

> Discusses the relationship between Pinchot's political career and the establishment of the Forest Service. His bitter fight with President Taft over the course of conservation led to his dismissal, and thus set the precedent for resisting government control of public lands.

Stineman, Esther Lanigan. *Mary Austin: song of a maverick.* New Haven, Conn.: Yale University Press, 1990.

> Chronicles the life of one of the first women to join the environmental movement. Mary Austin was one of the first to point out the intrinsic spiritual values of desert landscapes, fought against industrialization, and opposed the damming of the Colorado River.

Tanner, Thomas, ed. *Aldo Leopold: the man and his legacy*. Ankeny, Iowa: Soil Conservation Society of America, 1989.
> Discusses how Leopold's ideas are making their way into the modern world and the pertinency of his concepts of sustainable use and management.

Vale, Thomas R., ed. *Progress against growth: Daniel B. Luten on the American landscape*. New York: Guilford Press, 1986. 366 p.
> Daniel B. Luten left a twenty-five year career as a research chemist for Shell Oil Company so that he could study population and resource issues. Begins with a brief biographical sketch, followed by topically arranged groupings of Luten's articles (which are further grouped chronologically) on population, food and agriculture, energy, water, nature, conservation, and the future.

Watkins, Tom H. *Righteous pilgrim: the life and times of Harold L. Ickes, 1874-1952*. New York: Henry Holt, 1990. 1010 p. (E748.I28 W37 1990)
> Chronicles the career of Harold L. Ickes, Secretary of the Interior from 1933 to 1946. Maintains that Ickes truly understood that wilderness preservation is vital to the spiritual and ecological health of the country. Traces Ickes's efforts to create a cabinet-level Department of Conservation.

Books (unannotated)

Brooks, Paul. *Speaking for nature: how literary naturalists from Henry Thoreau to Rachel Carson have shaped America*. San Franciso: Sierra Club Books, 1983. 304 p.: ill.

Carty, Winthrop P. *The rhino man and other uncommon environmentalists*. Washington, D.C.: Seven Locks Press, 1992.

Cutright, Paul Russell. *Theodore Roosevelt, the making of a conservationist*. Urbana: University of Illinois Press, 1985. 285 p., 7 p. of plates: ill. (E757 .C985 1985)

DeStefano, Susan. *Theodore Roosevelt, conservation president*. New York: Twenty-First Century Books, 1993.

Fox, Stephen R. *The American conservation movement: John Muir and his legacy*. Madison: University of Wisconsin Press, 1985. 436 p.: ill. (QH31.M9 F68 1985)

Gould, Lewis L. *Lady Bird Johnson and the environment*. Lawrence: University Press of Kansas, 1988. 312 p.:ill. (E848.J64 G68 1988)

Ho'i Ho'i Hou, a tribute to George Helm and Kimo Mitchell. Honolulu, Hawaii: Bamboo Ridge Press, 1984. 113 p.: ill., maps, ports. (DU627.83.H45 H65 1984)

Mowat, Farley. *Rescue the Earth!: conversations with the green crusaders.* Toronto, Canada: McClelland & Stewart, 1990. 282 p. (HC120.E5 M69 1990)

Muir, John. *John Muir, in his own words: a book of quotations.* Lafayette, Calif.: Great West Books, 1988. 98 p. (QH31.M9 A3 1988)

Raymond, Jack. *Robert O. Anderson: oilman/environmentalist and his leading role in the international environmentalist movement: a monograph.* Lanham, Md. (Aspen Institute for Humanistic Studies): University Press of America, 1988. 49 p.: ill. (HC110.E5 A6617 1988)

General interest periodicals (annotated)

McKibben, Bill. "David Brower: the *Rolling Stone* interview." *Rolling Stone* 581 (June 28, 1990): 59-62, 88.
> The man who has been America's most militant and effective environmentalist in the years since World War II talks about his past--about his climbs, about the birth of the environmental movement, about battles won and lost--and about what lies ahead for the movement.

General interest periodicals (unannotated)

Abrahams, Andrew. "In a race against time, Michael Stewart takes wing on an aerial mission to save America's vanishing forests." *People* 34,15 (October 15 1990): 71(3)

Altany, David. "Andy and Goliath: activist Andy Kerr battles the timber industry to save America's last ancient trees from the saw." *Industry Week* 239,22 (November 19, 1990): 50(4)

Blyskal, Jeff; and Hodge, Marie. "The old man and the bay." (Lou Figurelli and the New York Bay) *New York* 23,15 (April 16 1990): 48(4)

Bonnett, Margie. "Finding the planet sicker than ever, eco-watchman Lester Brown gives us 40 years to clean up our act." (interview) *People* 33,23 (June 11 1990): 113(3)

Brower, Kenneth. "Grey Owl: he became famous as a half-Scot, half-Apache defender of wildlife, and some believe he should rank with John Muir and Rachel Carson in the environmentalists' pantheon." *The Atlantic* 265,1 (January 1990): 74(11)

Chu, Dan. "Ed and Betsy Marston's biweekly may be small, but it is still a mighty voice in the wilderness." *People* 33,14 (April 9 1990): 117(3)

Chu, Daniel. "Toxin avenger; Robert F. Kennedy Jr., naturalist and environmental legal eagle, stalks 'bad guys' who pollute." *People* 36,2 (July 22 1991): 72(3)

Denver, John. "John Denver's faith in the environment." *Sunset* 187,5 (November 1991): 40(2)

Elnadi, Bahgat; and Rifaat, Adel. "Jacques-Yves Cousteau." (interview) *UNESCO Courier* (November 1991): 8(6)

Fincher, Jack. "One man against a wasteland." (Paul Rokich's crusade to save Utah's Oquirrh Mountains) *Reader's Digest* 137,820 (August 1990): 99(5)

"For their love of children . . . the 'As They Grow' Awards." (Parents Magazine awards for people who improve the world for children) *Parents* 65,2 (February 1990): 127(6)

"For Tonys, movies and Mother Earth, the stars party hearty during a busy week of bashes." (environmental fund-raising) *People* 33,24 (June 18 1990): 46(3)

Ginsburg, Ina. "Otto Schily." (interview) *Interview* 16 (November 1986): 77(3) "

Gross, Linden. "Stars who take care of the Earth." (Ted Danson, Jane Fonda, Jeff and Beau Bridges, Olivia Newton-John, Ed Begley Jr.) *Redbook* 176,1 (November 1990) 28(3)

Harlin, John; Chase, Jim; Begley, Marita; and Agurcia, Ana G. "Environmental leaders." *Backpacker* 16,1 (January 1988): 69(4)

Kerasote, Ted. "Bob Marshal: outdoorsman with a vision." (column) *Sports Afield* 195 (February 1986): 48(2)

Kiernan, Michael. "A step forward, a step back." (a day in the life of an environmentalist) *U.S. News & World Report* 108,13 (April 2 1990): 62(2)

McRae, Michael. "Meet 'Grandmother Nature.' " (Elizabeth Terwilliger teaches children to love the environment) *50 Plus* 27 (July 1987): 25(4)

Picker, Lauren. "On top of the heap." (Patricia Poore's magazine *Garbage*) *House Beautiful* 134,4 (April 1992): 50(3)

Reed, Susan K. "Siberian writer Valentin Rasputin fears for the planet's fate." (includes excerpt from 'The Fire') *People* 27 (April 6 1987): 127(2)

Reed, Susan. "Environmentalist Fred Krupp helps crush the ubiquitous fast-food clamshell." *People* 35,14 (April 15 1991): 61(2)

Reed, Susan. "Hollywood heavyweights turn out in force to urge California voters to give the green light to Big Green." *People* 34,18 (November 5 1990): 125(3)

Searle, Rick. "Following the path of the heart." (environmental activist Andy Russell) *Nature Canada* 18,3 (Summer 1989): 45(2)

Sadruddin Aga Khan." (interview) *UNESCO Courier* (May 1991): 4(6)

Tropkin Alexander. "Man of the noosphere." (Alexander Ianshin) *Soviet Life* 7 (July 1990): 36(4)

" 'We still want no stragglers'; a valedictory conversation with The Wilderness Society's president, William A. Turnage." *Wilderness* 48 (Summer 1985): 34(11).

Weill, Jeanne. "The king of 'do-goodship.' " (Jim Bob Moffett) *New Orleans* 22,7 (April 1988): 38(9)

Weiss, Michael J. "Protect the ozone layer, warns NASA's Robert Watson, or prepare to do a slow burn." (National Aeronautics and Space Administration) (interview) *People* 29,24 (June 20 1988): 53(3)

Environmental and professional journals (annotated)

Bohlen, Janet Trowbridge. "Three faces of Eve." *Buzzworm: The Environmental Journal* 1 (Summer 1989): 65-69.
> Profiles three high-ranking women with strong influence in the areas of environmental protection and conservation: Kathryn Fuller, President, World Wildlife Fund; Joan Martin-Brown, Washington representative and senior liaison officer, United Nations Environment Programme; D. Jane Pratt, Chief, World Banks' Division of Environmental Operations and Strategy.

Graham, Frank, Jr. "Audubon prepares for U.N. conference." *Audubon* 93 (JulyAugust 1991): 124, 126-127.
> "The State Department has appointed Fran Spivy-Weber (director of National Audubon Society's International Program) to represent U.S. private environmental organizations on the Preparational Committee for the U.N. Conference on Environment and Development." Discusses some of her current activities relating to UNCED.

Kraus, Susan J. "Those women of Oregon: they've lead the way to pesticide reform." *Environmental Action* 17 (May/June 1986): 28-31.
> Presents profiles of Oregonians Therese Ohlson, who supervises sheep-grazing and tansy-beetle programs rather than herbicide application in Siuslaw National Forest; Lynn Coody, organic farmer; and Mary O'Brien, Barbara Kelley, and Carol

van Strum, who were instrumental in securing a ban on chemical spraying on federal lands.

McLarney, William. "An interview with Alvaro Umana." *Amicus Journal* 10 (Fall 1988): 39-41.
    Costa Rica's Minister of Natural Resources, Energy and Mines gives "a candid assessment of Central America's conservation and development challenges and the role of developed nations."

Revkin, Andrew. "The lonely prophet of the world environmental movement." *Audubon* 93 (November-December 1991): 97-100.
    "Brazil's first environmental czar, Jose Lutzenberger struggles to make the transition from maverick visionary to government insider." Lutzenberger will play an important role when more than 70 percent of the world's heads of state convene for UNCED.

Wolfson, Elissa. "Interview--Dr. Noel Brown--uniting nations for the environment." *E: the Environmental Magazine* 91 (July-August 1991): 13-14, 60- 62.
    "If mother Earth could choose a spokesperson, Dr. Noel Brown, UNEP's Regional Director for North America, might well be the one."

Environmental and professional journals (unannotated)

Beasley, Conger, Jr. "Moore takes on all." (Richard Moore campaigns for the rights of people of color) *Buzzworm: The Environmental Journal* 3,3 (May-June 1991): 38(6)

Bourne, Juliet, et al. "People who make a difference." (personal efforts in wildlife conservation) *National Wildlife* 25 (April-May 1987): 4(10)

Fersko-Weiss, Henry. "A man, a ship, and a dream to sail by." (Peter Willcox of Greenpeace) *Sierra* 73,3 (May-June 1988): 66(7)

Frankel, Bruce. "Report on the commissioner: how New York's environmental chief thinks he can change the world." (Albert F. Appleton) *Audubon* 94,1 (January-February 1992): 94(3)

Jackson, Donald Dale; and Ernst, Tim. "Every state should have a Leo Drey." *Audubon* 90,4 (July 1988): 78(6)

Klinkenborg, Verlyn. "The making of a biopolitician." (Russell A. Mittermeier) *Audubon* 94,1 (January-February 1992): 90(4)

Line, Les. "Travelin' man. (Audubon field editor John G. Mitchell) *Audubon* 88 (November 1986): 6(2)

Meine, Curt; and Watkins, T. H. "How Leopold learned to think like a mountain." (excerpt from *Aldo Leopold: His Life and Work*; includes editor's notes) *Wilderness* 51,179 (Winter 1987): 57(5)

"Next half-century: a golden opportunity for conservation leadership." (transcript) *National Wildlife* 24 (June-July 1986): 25(2)

"Paul Wachtel." (interview) *Whole Earth Review* 61 (Winter 1988): 128(2)

"Peter Berg." (interview) *Whole Earth Review* 61 (Winter 1988): 12(2)

Quammen, David. "A murder in Madagascar." (Bedo Jaosolo, naturalist murdered in Andasibe) *Audubon* 93,1 (January 1991): 48(11)

"Rockefeller up for congressional medal." (Laurance Rockefeller) *National Parks* 64,5-6 (May-June 1990): 9(2)

Stutz, Bruce. "Ted Turner turns it on." (views on the environment) (interview) *Audubon* 93,6 (November-December 1991): 110(6)

Sunquist, Fiona. "The private passion of Tomas Blohm." (protecting the Orinoco crocodile) *International Wildlife* 18,1 (January-February 1988): 20(5)

Wray, Elizabeth. "Travelers between two worlds." (environmental activists) *Sierra* 77,3 (May-June 1992): 106(8)

Videos

For Earth's sake: the life and times of David Brower. Produced by John DeGraaf. Bullfrog Films (Oley, PA 19547), 1989. 58 min.
> David Brower is one of the founding fathers of the environmental movement. This video traces his life from his early days working in Yosemite, his World War II mountaineering experiences, his transformation of the Sierra Club from a hiking group to a leading environmental advocacy organization, his founding of the Friends of the Earth, and his falling out with both organizations, as well as his recent activities toward reviving damaged ecosystems. Provides a good introduction to the twentieth-century environmental movement.

Muir Woods: for generations to come. Chip Taylor Communications (15 Spollett Dr., Derry, NH 03038). 15 min.
> Focuses on the Muir Woods National Monument near San Francisco. Reflects Muir's conservationists philosophy and maintains that humankind can observer and appreciate the natural world without damaging it. Contains excellent footage of the Muir Woods, including its giant redwoods.

Wild by law. Produced and directed by Lawrence Hott and Diane Garey for Florentine Films. (Direct Cinema, P.O. Box 10003, Santa Monica, CA 90410). 1991. 60 min.

     Focuses on the three founders of the Wilderness Society, Aldo Leopold, Bob Marshall, and Howard Zahniser, and chronicles their efforts at nature preservation. Combines archival stills with nature footage and interviews with historians.

Wilderness idea. Direct Cinema (P.O. Box 10003, Santa Monica, CA 90410). 1989. 58 min.

     Assesses the early twentieth-century conservationists John Muir and Gifford Pinchot. Explains the schism over preservation of undeveloped wilderness or multi-use management.

ORGANIZATIONS

(See also *Reference*, below) 512
(See also *Radical Environmentalism*, below) 466

Books (annotated)

Graham, Frank. *The Audubon ark: a history of the National Audubon Society*. New York: Knopf, 1990.

     Traces the history of the Audubon Society from its beginnings as a late nineteenth century group formed with the aim of saving "pretty birds" to its current role as an important conservation organization that sponsors annual bird counts, plays an important part in related legislation, and is involved in preventing environmental calamities. Discusses how this evolution took place and outlines conservation efforts by private citizens.

Turner, Tom. *Sierra Club: one hundred years of protecting nature*. New York: Abrams, 1991. 288 p.: (some color) ill. (some color) maps. (QH76 .T87 1991)

     A commemoration of the Sierra Club's one-hundredth anniversary. Contains almost three hundred excellent photographs and an introduction by Frederick Turner, who wrote a biography of John Muir, founder of the club. Chronicles the history of the Sierra Club, its goals, successes, and failures, and descriptions of many of the natural areas the club has worked so hard to preserve through publishing, public relations, and congressional lobbying.

Turner, Tom. *Wild by law: The Sierra Club Legal Defense Fund and the places it has saved*. New York: Random House, 1990.

     Traces the history of environmental law, discussing landmark cases and containing excellent photographs. Presents examples of successful litigation, including Redwood National Park, the Colorado Plateau, and Admiralty Island.

Wenner, Lettie McSpadden. *U.S. energy and environmental interest groups: institutional profiles.* New York: Greenwood, 1990. 358 p.

"All three categories of organizations that lobby in the energy and environmental policy areas are included in this volume; trade associations, public interest groups, and professional/governmental organizations. All profiles are alphabetized together, although it should be clear from reading each entry which category the group represents."

Books (unannotated)

Allen, Thomas B. *Guardian of the wild: the story of the National Wildlife Federation, 1936-1986.* Bloomington: Indiana University Press, 1987. 212 p., 81 p. of plates: ill. (QH76 .A43 1987)

Citizens, political communication, and interest groups: environmental organizations in Canada and the United States. New York: Praeger, 1992.

Davies, J. Clarence. *Training for environmental groups.* Washington, D.C.: Conservation Foundation, 1984. 124 p. (HC110.E5 D36 1984)

Earthworks, ten years on the environmental front. San Francisco: Friends of the Earth, 1980. 255 p.: ill. (QH541.145 .E253)

Grove, Noel. *Preserving Eden: the Nature Conservancy.* New York: Abrams, 1992. 176 p.: color ill. (QH76 .K73 1992)

Parkin, Sara. *Green parties: an international guide.* London: Heretic Books, 1989. 335 p.: ill. (HC79.E5 P35 1989)

United States and the global environment: a guide to American organizations concerned with international environmental issues. Claremont, Calif.: California Institute of Public Affairs, 1983. 72 p. (TD169.6 .U49 1983)

Wise use agenda: the citizen's policy guide to environmental resource issues: a task force report. Bellevue, Wash.: Free Enterprise Press, 1989. Distributed by Merrill Press. 167 p. (HC110.E5 W57 1989)

General interest periodicals (annotated)

Desky, Joanne. "Wolves in sheep's clothing." *Public Citizen* 11 (May-June 1991): 14-15, 21-22.

Identifies alleged "bogus" consumer groups that are in fact supported by corporate interests, it is charged, to "mislead the public about their agendas and backers. These groups conduct direct-mail campaigns, buy advertising space from major media outlets and promote themselves in slick, glossy packages--not to give the public insight and information on the issues, critics charge, but to promote an 'objective' stance on policy issues from corporate America."

Gifford, Bill. "Inside the environmental groups." *Outside* 15 (September 1990): 69-78, 80, 82, 84.
> A user's guide to the top twenty-five environmental groups, describing "who they are and what issues they most heartily embrace," their leadership, membership, current activities, funding and financial information, celebrity spokesperson, etc. Also included is an honor roll list of "eight of the finest grass-roots efforts" for protecting the environment.

Krakauer, Jon. "Brownfellas." *Outside* (December 1991): 69-72, 114-116.
> Ron Arnold and Alan Gottlieb are the principal strategists for an anti-environmental movement known as Wise Use. Their followers are "convinced that creeping environmentalism is an insidious crusade to cripple the American economy." Their goal: the destruction of environmentalism.

Miller, James Nathan. "The happy landgrabbers." *Reader's Digest* 119 (July 1981): 148-157.
> Describes the Nature Conservancy which since 1954 has been preserving ecologically unique places and is today the world's largest privately owned system of nature refuges.

Mundy, Alicia. "Conflicts of interest: acres and pains." *Regardie's* 9 (March 1989): 57-66.
> Describes controversy over a tract of land in Loudoun County which Claude Moore, a multimillionaire in his nineties, donated to the National Wildlife Federation for a nature center. The Federation sold the land to developers, raising the wrath of both Moore and the county. Includes an inside look at the National Wildlife Federation and a biography of NWF president Jay D. Hair.

Russell, Dick. "We are all losing the war." *Nation* 248 (March 27, 1989): 403-408.
> Predicts future tensions between local grass-roots environmental activist organizations and the larger, more established conservation and environmental protection groups. Includes a sidebar article by Viveca Novak, "Hazardous Choice" on EPA administrator William Reilly.

General interest periodicals (unannotated)

Gilbert, Bill. "The Nature Conservancy game." *Sports Illustrated* 65 (October 20 1986): 86(11)

"Institute for 'Earth keeping.' (AuSable Institute of Environmental Studies)" *The Christian Century* 105,26 (September 14, 1988): 808(3)

Reiger, George. "Unnatural developments." (The Nature Conservancy makes deals with land developers) (column) *Field & Stream* 95,4 (August 1990): 12(3)

Environmental and professional journals (annotated)

Fernandez, Lisa. "Private conservation groups on the rise in Latin America and the Caribbean." *World Wildlife Fund Letter* 1 (1989): 1-8.
  As a result of the growing magnitude of the environmental crisis, well over two hundred private nonprofit conservation organizations (NCOs) have emerged in Latin America and the Caribbean. This article describes their activities and the challenges they face, such as limited traditions of philanthropy and discouraging tax and labor laws.

Gills, Lisa. "Who's who among environmental organizations." *Waste Age* (January 1989): 124-142.
  Overviews public and private organizations with a focus on affecting environmental policy, especially waste management issues concerning both solid and hazardous wastes. Arranged in a directory with three classifications: public interest groups; industry/professional associations; government/public organizations. Examines selected organizations in greater detail.

Hammond, Gary. "The Nature Conservancy: protecting the rare and the beautiful." *Western Wildlands* 10 (Spring 1984): 28-31.
  "Since 1953, the conservancy has been responsible for protecting more than two million acres of significant wildlands using a variety of means, including fee simple acquisition, conservation easements, leases and government cooperative management agreements."

Knox, Margaret L. "The wise use guys." *Buzzworm: The Environmental Journal* 2 (November-December 1990): 30-36.
  "A whole new breed of special interest groups in entering the public land use fray. Their goal: to make use of the land, and maybe derail the environmental movement in the process."

Martin, Lawrence R. "Demanding waste reduction: the roles of public interest organizations in promoting the institutionalization of waste and toxics reductions." *Environmental Professional* 11 (January-March 1989): 132-141.
  Public interest groups included are the Environmental Defense Fund, the National Campaign Against Toxic Hazards, the Citizen Clearinghouse for Hazardous Wastes, Inform, the Environment and Energy Study Institute, and the Local Government Commission.

McIntosh, Phyllis. "Embracing dilemmas the world over." *International Wildlife* 20 (November-December 1990): 12-15.

> "A success story in itself, the International Union for Conservation of Nature and Natural Resources was the brainchild forty-two years ago of conservationists, mostly European, concerned with preserving wildlife. In the years since, it has evolved from an insulated think-tank churning out reports, technical data, and endangered species lists to a real-world force for solving tough environmental problems that often go far beyond wildlife."

Russell, Dick. "The rise of the grass-roots toxics movement." *Amicus Journal* 12 (Winter 1990): 18-21.

> Describes the rise of the Citizen's Clearinghouse for Hazardous Wastes (CCHW), the National Toxics Campaign, Greenspeace, and other citizen actions groups united to clean up toxic substances.

Stanfield, Rochelle L. "Environmental lobby's changing of the guard is part of movement's evolution." *National Journal* 17 (June 8, 1985): 1350-1353.

> Includes a list of the "Group of 10," the ten largest and oldest organizations in the environmental movement, and gives their budgets, and incoming and outgoing leaders.

Wood, Daniel. "Lighthawk: exposing park threats from the air." *National Parks* 63 (January-February 1989): 27-31.

> "Opening people's eyes to the ravages of clearcutting is typical of Project Lighthawk. Aerial crusaders in the cause of conservation, Lighthawk is a Santa Fe-based group of pilots and volunteers who have named themselves after a mythical bird 'whose purpose is to shed light.' "

Environmental and professional journals (unannotated)

Cahn, Robert. "International Union for Conservation of Nature and Natural Resources." (assembly in San Jose, Costa Rica) *Environment* 30,2 (March 1988): 44(2)

Eidsvik, Harold. "International defense: IUCN--a union that ranks protecting wildlands with world peace." (International Union of Conservation of Nature and Natural Resources) *National Parks* 61 (September-October 1987): 12(2).

Fri, Robert W. "Resources for the Future." (non-profit research organization) *Environment* 30,1 (January-February 1988): 5(3)

McIntosh, Phyllis. "Embracing dilemmas the world over." (International Union for Conservation of Nature.) *International Wildlife* 20,6 (November-December 1990): 12(4)

McLarney, Bill. "ANAI: idealism survives in the humid tropics." (nonprofit Costa Rican environmental group) *Whole Earth Review* 62 (Spring 1989): 44(4)

Pierce, Jim. "Environmental Task Force." (new organization to help citizen groups fight environmental dangers) *Environment* 28 (December 1986): 3(2)

Watkins, T. H. "Systems management." (Wilderness Society is now active in New England) *Wilderness* 53,186 (Fall 1989): 16(2)

Reference

Carey, Ann. "The 1990 directory to environmental organizations." *Buzzworm: The Environmental Journal* 2 (May-June 1990): 65-68, 70-72, 74-77.
> Provides information on over 150 organizations, including "their purpose, current emphasis, membership, funding and expenditures. The focus of these groups ranges from hydropower to hiking trails, from primate protection to preservation of public lands to careers in conservation."

Directory of non-governmental environment and development organisations in OECD member countries. Paris: Environment and Development in the Third World; Organisation for Economic Co-operation and Development, Development Center, 1992. 409 p.
> "This specialised Directory provides information on 649 non-governmental organisations (NGOs) in OECD Member countries that focus on environment and development. Profiles of the NGOs describe their aims, education work and actions in developing countries. Cross-referenced indexes provide access to information on 'who is doing what and where' in this field."

Lanier-Graham, Susan D. *The nature directory: a guide to environmental organizations.* New York: Walker, 1991. 190 p.
> "Gives an overview of environmental organizations. From the large, highly organized to the small grassroots organizations, the directory will provide you with a basic understanding of each organization." Discusses the environmental problems addressed by the environmental groups, provides a sampling of 120 environmental groups, as well as a guide to personal involvement in the environmental issues. There is also a selected resource list including a special guide to children's books on the environment.

Trzyna, Thaddeus C.; and Childers, Robert A., eds. *World directory of environmental organizations.* 4th ed. Claremont: California Institute for Public Affairs, 1992. 231 p.
> Opens with a listing of acronyms and key events. Lists organizations according to specific topics and geographic regions. Provides the full addresses, and telephone and fax numbers for these organizations. There is an appendix of directories and databases that provide further information on environmental organizations. The fourth edition adds important new citizens groups.

Your resource guide to environmental organizations: includes the purposes, programs, accomplishments, volunteer opportunities, publications, and membership benefits of 150 environmental organizations. Irvine, Calif.: Smiling Dolphins Press, 1991. 514 p. (TD171 .Y68 1991)

RADICAL ENVIRONMENTALISM

(See also *Europe,* below)   471

Books (annotated)

Bosso, Christopher J. "Adaptation and change in the environmental movement." In *Interest group politics.* 3d ed. Washington, D.C.: CQ Press, 1991. p. 151-176.
"Details the evolution of environmental politics from 1970 to 1990, and notes a universe of groups that differs substantially from that of the mid-1970s. Not only have more radical groups split off from the major or 'Big Ten' organizations, but a third set of groups has become an important force in environmental politics: narrow or local interests that emphasize specific projects or issues rather than broad objectives."

Lewis, Martin W. *Green delusions: an environmentalist critique of radical environmentalism.* Durham, N.C.: Duke University Press, 1992. 290 p.
Maintains that if the agenda of the radical environmentalists were enacted, ecological catastrophe would result. Argues that capitalism, large-scale government, advanced technology, and big cities have much to offer the environment.

Manes, Christopher. *Green rage: radical environmentalism and the unmaking of civilization.* Boston, Mass.: Little, Brown, 1990. 291 p. (HC79.E5 M353 1990)
In recent years, environmental activism has taken a new direction, away from the sedate practices of organizations like the Sierra Club and the Wilderness Society, and toward guerrilla tactics. Written by a former writer and editor for Earth First!, a radical environmental organization, this work presents an insider's perspective which includes discussions of the new philosophy of "deep ecology," civil disobedience, and the goal within the radical environmental movement of creating a civilization more in harmony with nature. Criticizes mainline environmental organizations.

Scarce, Rik. *Eco-warriors: understanding the radical environmental movement.* Chicago: Noble Press, 1990. 291 p. (HC110.E5 S385 1990)
Traces the history of the radical environmental movement and reviews the actions of Greenpeace and Earth First!, as well as other, lower profile organizations. Discusses their beliefs, aims, and actions, offering accounts of both well-known and obscure incidents--mostly tales of ecological sabotage, such as the sinking of whaling ships and raiding animal research facilities. Offers information which

cannot be found elsewhere on animal liberation groups and also reviews the influx of environmentalism in the arts.

Books (unannotated)

Brown, Michael Harold. *The Greenpeace story.* Rev. ed. New York: Dorling Kindersley, 1991. 192 p.: (some color) ill.

Earth first! reader: ten years of radical environmentalism. Salt Lake City: Peregrine Smith Books, 1991. 272 p. (HC110.E5 E22 1991)

Ecodefense: a field guide to monkeywrenching. 2d ed. Tucson, Ariz.: N. Ludd, 1987. 311 p., 2 leaves of plates: ill. (HV6431 .E36 1987)

Foreman, Dave. *Confessions of an eco-warrior.* New York: Harmony Books, 1991. 228 p.

Keenan, Charles J. *Environmental anarchy: the insidious destruction of social order: a legacy of the sixties.* British Columbia, Canada: Cappis Press, 1984. 206 p. (HC117.B8 K43)

Pearce, Fred. *Green warriors: the people and the politics behind the environmental revolution.* London: Bodley Head, 1991. 331 p.: ill. (HC79.E5 P386 1991)

Radical environmentalism: philosophy and tactics. Belmont, Calif.: Wadsworth, 1992.

Scarce, Rik. *Eco-warriors: understanding the radical environmental movement.* Chicago: Noble Press, 1990. 291 p. (HC110.E5 S385 1990)

General interest periodicals (annotated)

Cooper, Marc. "Redwood bummer: if an Earth First! protest falls in the forest and no one hears it, did it happen?" *Village Voice* 35 (September 18, 1990): 19-20, 22-25.
  Discusses the lengthy standoff between redwood loggers and California environmentalists. Earth First! "announced that militant 'direct action' had to be taken this past summer to save the trees. Court fights, initiatives, ballot measures, and polite lobbying were not enough to save the forests, said the Earth First!ers; last spring they issued a dramatic call to bring thousands of activists into the woods to place themselves between the chain saws and the trees."

Foreman, Dave. "Earth First!" *Progressive* 45 (October 1981): 39-42.
  A self-proclaimed environmental extremist describes "Earth First" an organization dedicated to environmental activism.

Gabriel, Trip. "If a tree falls in the forest, they hear it." *New York Times Magazine* (November 4, 1990): 34, 58-59, 62-64.

"Earth First is a group that in no way resembles other environmental organizations. Frustrated with the political system, its followers take direct, often unlawful, actions aimed at timber, mining and other development companies. To some mainstream environmentalists, the passion of Earth First is an inspiration. But others find it counterproductive, tarring radicals as irresponsible or, in some cases, eco-terrorists."

Convey, Kevin R. "Soldiers in mother nature's army." *Boston Magazine* (February 1989): 123-125, 198-205.
   Profiles environmentalist Marco Kaltofen, consultant to Greenpeace.

Hoffmann-Martinot, Vincent. "Grune and Verts: two faces of European ecologism." *West European Politics* 14 (October 1991): 70-95.
   "Using recent Eurobarometer data, this article aims to verify, differentiation of European Ecologism in two poles: the New Left radicalism of the German Grune, and the more pragmatic way of the newly more visible French Verts."

Ostertag, Bob. "Greenpeace takes over the world: it's rich, powerful, and looking for new lands to conquer. But can the most successful eco-force keep its seal of approval?" (includes related articles) *Mother Jones* 16,2 (March-April 1991): 32(8)

Parfit, Michael. "Earth First!ers wield a mean monkey wrench; convinced that genteel approaches won't win the war, these activists go to extremes to slow the demise of the natural world." *Smithsonian* 21,1 (April 1990): 184(16)

Rosen, Rebecca. "This bomb had my name on it: Judi Barr talks about Redwood Summer." *Progressive* 54 (September 1990): 29-33.
   "Earth First! activist Judi Barr was the architect of Redwood Summer, but she has to watch the action from the hospital bed where she is recovering from a bomb that exploded in her car. The blast slowed her down, but it hasn't stopped her campaign to save the forest."

Royte, Elizabeth. "Extreme in nature." *New England Monthly* 7 (March 1990): 28-34.
   "Describes the beginnings of Earth First!, a hard hitting environmental movement, its growth and activities. Although the movement started in the Western U.S., the next battleground for the group seems to be New England.

Spencer, Leslie; Bollwerk, January; and Morais, Richard C. "The not so peaceful world of Greenpeace." *Forbes* 148 (November 11, 1991): 174-179.
   "A band of scrappy protesters in rubber rafts, saving whales. That's the Greenpeace image. The reality: a multinational organization accountable only to itself, with large revenues and a brilliant ability to manipulate the public."

Stang, Alan. "A polemical environment." *American Opinion* 27 (May 1984): 7-12, 85, 87, 89, 91, 93, 95-96.

Criticizes environmentalists and the press for "environmental terrorism" in warnings about nuclear power, ozone, acid rain, leaded gasoline, EDB, and aspartame. Questions the research validity on which environmentalists make their cases.

Talbot, Stephen. "Earth First! What next?" *Mother Jones* 15 (November-December 1990): 47-49, 76-81.
Examines the recent activities of this environmental group; discusses the bombing of Judi Bari, the defection of Dave Foreman, Redwood Summer, and nonviolent demonstrations aimed at saving ancient forests in California. Includes sidebar by Dave Foreman: "Ecotage updated."

Waller, Michael. "West European communism--red for 'stop', green for 'go'." *World Today* 44 (March 1988): 43-46.
"The Communist parties of western Europe have been passing through a period of crisis, marked in the late 1970s and early 1980s by sharp falls in electoral performance and in membership. On the other hand, the same period saw an upsurge in the political presence of radical movements mobilizing around the issues of peace and the protection of the environment."

General interest periodicals (unannotated)

Cockburn, Alexander. "The trees strike back." (Redwood summer in the press) *Nation* 251,19 (December 3 1990): 670(2)

Foote, Jennifer. "Trying to take back the planet; radical environmentalists are honing their militant tactics and gaining followers." *Newsweek* 115,6 (February 5 1990): 24(2)

Foreman, Dave. "Ecotage updated." (book excerpt: 'Confessions of an Eco-Warrior's use of sabotage in environmental protests') *Mother Jones* 15,7 (November-December 1990): 49(3)

Harwood, Michael. "Daredevils for the environment." (Greenpeace and attention-grabbing stunts) *The New York Times Magazine* 138 (October 2 1988): 72

Kuipers, Dean. "Earth First." (interview) *Interview* 21,9 (September 1991): 140(4)

Environmental and professional journals (annotated)

Brown, Michael. "The zeal of disapproval." *Oceans* 20,3 (1987): 36-41.
Greenpeace International, under David McTaggart's leadership practices a unique brand of environmentalism "non-violent direct action." Among the targets are whaling, nuclear fallout, and toxic waste pollution.

Cherfas, Jeremy. "Greenpeace and science: oil and water?" *Science* 247 (March 16, 1990): 1288-1290.
> "Greenpeace U.K. has hired a director of science and begun an ambitious program of funding scientific research, raising questions about what role advocacy groups should play in science."

Russell, Dick. "The monkeywrenchers." *Amicus Journal* 9 (Fall 1987): 28, 30-42.
> Describes two groups, Earth First! and the Sea Shepherd Conservation Society, "a new extreme of the environmental movement." Greenpeace, in "one of the ironies of eighties environmentalism, is acquiring a mainstream image by comparison."

Steinhart, Peter. "Respecting the law: there must be limits to environmental protest." *Audubon* 89 (November 1987): 10, 12-13.
> "We can't applaud spiking trees or scuttling whaling ships any more than we can applaud official abandonment of environmental law and policy. For one form of lawlessness tends to invite the other. We cannot really expect business to temper greed with understanding, or individuals to take their best ethics into the woods. In the end, conservation depends upon an awareness that we help each other."

Environmental and professional journals (unannotated)

Barnes, Donald J.; Spring, Dona; Pacelle, Wayne; and Spira, Henry. "The "terrorist" label: how to neutralize it." (tactics for animal rights activists) *The Animals' Agenda* 9,7 (September 1989): 39(4)

Bookchin, Murray; Foreman, Dave; and Nilsen, Richard. "Defending the Earth and burying the hatchet." (includes article on Earth First!) *Whole Earth Review* 69 (Winter 1990): 108(7)

Bowden, Charles. "At sea with the shepherd: a shooting script." (Ecoessay; life on board the Sea Shepherd, a Greenpeace boat) *Buzzworm: The Environmental Journal* 3,2 (March-April 1991): 38(10).

Franklin, Karen E.; and Sowell, Janet. "The timber terrorists." (activist activities to save forests) (Special Section: Forests Under Siege) *American Forests* 93 (March-April 1987): 41(2).

Reports

Bandow, Doug. *Ecoterrorism: the dangerous fringe of the environmental movement.* Washington, D.C.: Heritage Foundation, 1990. 10 p. (Backgrounder no. 764)
> "Individuals and scattered bands of environmental or ecological radicals, usually called ecoterrorists, have been sabotaging industrial facilities, logging operations,

construction projects, and other economic targets around the country. They have inflicted millions of dollars in damage and have maimed innocent people."

"Planet in peril: a view from the campus: an environmental opinion survey." Washington, D.C.: National Wildlife Federation, 1989. 55, 14 p.: ill. (HC110.E5 P57)

Bibliographies

Buttel, Frederick H. *Labor and the environment: an analysis of and annotated bibliography on workplace environmental quality in the United States.* Westport, Conn.: Greenwood Press, 1984. 148 p. (Z5863.P6 B87 1984)

Morales, Leslie Anderson. *The impact of radical environmentalism on policy and practice: a bibliography.* Monticello, Ill.: Vance Bibliographies, 1991. 8 p. (Z5863.P6 M67 1991)

Videos

Rainbow Warrior Affair. Wombat Film and Video (930 Pitner Ave., Evanston, IL 60201). 1986. 48 min.
  Chronicles the events which lead to the bombing of the environmental organization Greenpeace's flagship because it was protesting the French government's policy of nuclear testing in the South Pacific.

EUROPE

Books (annotated)

Allison, Lincoln. *Ecology and utility: the philosophical dilemmas of planetary management.* Madison, N.J.: Fairleigh Dickinson University Press, 1992. 185 p.
  Focuses on the English environmental movement and offers both a philosophical analysis and personal recollections. Maintains that modern environmental holism is not incompatible with more scientific world views. Presents a moderate, aesthetic, hygienic perspective.

Capra, Fritjof; and Spretnak, Charlene. *Green politics.* New York: Dutton, 1984. 244 p. (JN3971.A98 G7232 1984)
  Focuses upon the emergence of the Green Party in West Germany. Explains the philosophy behind the Green Party and calls for "the application of the Green perspective to the whole Earth and the entire human family."

Crosby, Alfred. *Ecological imperialism: the biological expansion of Europe, 900-1900.* Cambridge: Cambridge University Press, 1986.

    Traces the role European explorers and settlers took in spreading crops, weeds, animals, and diseases around the world.

Evans, D. *A history of nature conservation in Britain.* London: Routledge, 1992. 274 p.

    A historical survey and assessment of the conservation movement in Britain, which is often cited as having the most advanced system of conservation in the world. Evans explains how this system developed and analyzes its deficiencies.

Greens of West Germany: origins, strategies, and transatlantic implications. Cambridge, Mass.: Institute for Foreign Policy Analysis, 1983. 105 p.

    Contents include: "The Greens: implications for the United States," by R. Pfaltzgraff, Jr.; "The origins, development, and composition of the Green movement," by K. Holmes; "The Green Program for German society and international affairs," by C. Clemens; and "The Greens/alternatives and the peace movement: a challenge to the German Party system," by W. Kaltefleiter.

Parkin, Sara. *Green parties: an international guide.* London: Heretic Books, 1989. 335 p.: ill. (HC79.E5 P35 1989)

    Describes the history, growth, and performance of the green parties of Europe and demonstrates that the environment would not be safe in their hands. They are disorganized, politically confused, parochial, and distrustful of fellow members. Explains their influence on other political parties and provides specific examples of how they have raised public awareness on particular issues.

Pilat, Joseph F. *Ecological politics: the rise of the Green movement.* Beverly Hills, Calif.: Sage Publications, 1980. 96 p. (HC110.E5 P54)

    "An ecological movement arose in the 1970s in the industrial West in some ways similar to the environmental movement of the previous decade. It has developed its own political force and has often worked outside of the law rather than through it. Ecological, or Green, parties have already influenced local and regional politics in Western Europe and may influence national politics in the near future."

Salzman, James. *Environmental labelling in OECD countries.* Paris: Organisation for Economic Co-operation and Development, 1991. 133 p.

    "This report describes the origin and aims of existing government-sponsored labeling programs, the selection of criteria, how products are assessed, the financing of labeling programs, ways of measuring their success, legal issues and potential implications for international trade in 'green' products. The report reviews labeling programs already operating in Canada, Germany, and Japan and looks at plans for environmental labels in other OECD countries and the European Community."

Spretnak, Charlene; and Caproa, Fritjof. *Green politics.* Rev. ed. Santa Fe, N.M.: Bear, 1986. 255 p. (JN3971.A98 G7232 1986)

Discusses green politics in West Germany; global green politics; and green politics in the United States; includes addresses of Green parties.

Books (unannotated)

Cannon, Beryl. *The European Green consumer market*. Menlo Park, Calif. (333 Ravenswood Ave., Menlo Park 94025-3476): SRI International, Business Intelligence Program, 1990. 25 leaves (HF1040.9.E78 C36 1990)

Dominick, Raymond H. *The environmental movement in Germany: prophets and pioneers, 1871-1971*. Bloomington: Indiana University Press, 1992.

Green light on Europe. London: Heretic Books, 1991. Distributed by Inbook, East Haven, Conn. 367 p.: ill.

Jamison, Andrew. *The making of the new environmental consciousness: a comparative study of the environmental movements in Sweden, Denmark, and the Netherlands*. Edinburgh: Edinburgh University Press, 1990. 216 p. (HC380.E5 J36 1990)

Kitschelt, Herbert. *Beyond the European left: ideology and political action in the Belgian ecology parties*. Durham, N.C.: Duke University Press, 1990. 262 p.: ill. (JN6371.A42 K58 1990)

Mallinckrodt, Anita M. *The environmental dialogue in the GDR: literature, church, party, and interest groups in their socio-political context: a research concept and case study*. Lanham, Md.: University Press of America, 1987. 198 p.: ill. (JA76 .M335 1987)

New politics in Western Europe: the rise and success of green parties and alternative lists. Boulder, Colo.: Westview, 1989. 230 p. (JN94.A979 N49 1989)

Papadakis, Elim. *The Green Movement in West Germany*. New York: St. Martin's Press, 1984. 230 p. (JN3971.A98 G7236 1984)

Porritt, Jonathon. *The coming of the Greens*. London: Fontana/Collins, 1989, 1988. 287 p. (JN1129.G74 P67 1989)

Rawlinson, Roger. *Communities of resistance: nuclear and chemical pollution cross frontiers, and so did the protestors of the Upper Rhine*. London: Quaker Peace & Service, 1986. 55 p.

Robinson, Mike. *The greening of British party politics*. Manchester: Manchester University Press, 1992. Distributed by St. Martin's Press, New York. 246 p.: ill. (HC260.E5 R63 1992)

Webb, Adrian Leonard. *German politics and the Green challenge*. New York: Economist Intelligence Unit, 1990. 44 p.: color map (HC290.5.E5 W43 1990)

General interest periodicals (annotated)

Brumlik, Micha. "Fear of the father figure: Judaeophobic tendencies in the new social movements in West Germany." *Patterns of Prejudice* 21 (Winter 1987): 19-37.
> "Argues that the feminist, ecology and peace movements in West Germany manifest worrying Judaeophobic tendencies. This tendency is an ideology which can only be understood as a successor to National Socialism. It seems that Jews, and the influence of 'patriarchal Judaism' on modern civilization, are held responsible for man's rush to nuclear and ecological self-destruction, and that this turning of Jews into culprits is a way of dealing with the guilt of the Nazi past."

Daniels, Nomsa. "Guardian of Eden." *Africa Report* 36 (September-October 1991): 13-17.
> "The perennial conflict between European environmentalists and Africans the author argues, should be resolved in favor of the indigenous population, whose environmentalism grows out of a need for survival and has spurred homegrown grassroots organizations. In addition, the effort to impose Western priorities and values on Africans denies them the right to determine which conservation policies best suit their needs and conditions."

Hanninen, Sakari. "How to combat pollution by words." *Alternatives* 17 (Spring 1992): 209-229.
> Examines the direct action environmental movement in Finland aimed at closing down the Kylasaari incineration plant; this historical approach considers social factors and risk perception.

Kelley, Kevin J. "A green fringe." *Progressive* 54 (April 1990): 30-33.
> "Britain's Green Party achieved a startling electoral breakthrough in last year's elections to the European Parliament. The nation's gray politics take on a new hue."

Kreuzer, Markus. "New politics: just post-materialist? The case of the Austrian and Swiss Greens." *West European Politics* 13 (January 1990): 12-30.
> Using examples from the Swiss and Austrian Green Parties, this article examines the issue of what mobilizes voter participation in Green and New Left parties.

Pehe, Jiri. "The green movements in Eastern Europe." *Report on Eastern Europe* 1 (March 16, 1990): 35-37.
> Reports on the rise of "green" parties and other ecological groups formed throughout Eastern Europe and the part these groups are likely to play in future politics.

Rudig, Wolfgang. "Peace and ecology movements in Western Europe." *West European Politics* 11 (January 1988): 26-39.

"The article seeks to provide a comparative analysis of peace and ecology movements in Western Europe. Common features and differences across national boundaries are analyzed in terms of cycles of protest, their social basis, the structures of the movements, their forms of action and their political impact."

Rudig, Wolfgang; and Lowe, Philip D. "The withered 'greening' of British politics: a study of the Ecology Party." *Political Studies* 34 (June 1986): 262-284.

"Britain appears to be largely removed from the new political tide of 'green' parties that is currently sweeping over West European countries. This article will put forward some explanations for this 'stillborn' character of 'green' party politics in Britain. A detailed scrutiny of the history of the Ecology Party will be provided. It will be argued that the relative weakness of the party is mainly due to its failure to attract the support of 'new' social movements."

General interest periodicals (unannotated)

"Fresh Greens." (environment party in Sweden) *Economist* 304 (July 25 1987): 40(2)

Hayden, Tom. "All quiet on the Western front?" (special feature analysis on the spate of demonstrations across Western Europe) *Tikkun* 5,4 (July-August 1990): 51(2)

Johnson, R. W. "Green harvest." (Stroud, England is the first Green district council in Great Britain) *New Statesman Society* 4,155 (June 14 1991): 14(2)

Environmental and professional journals (annotated)

Burke, Tom. "The year of the Greens." *Environment* 31 (November 1989): 18-20, 41-44.

"The most significant consequence of the Euro-elections has been to place the environment right at the top of Britain's political agenda."

Hay, P. R.; and Haward, M. G. "Comparative Green politics: beyond the European context?" *Political Studies* 36 (September 1988): 433-448.

Attempts to "demonstrate the existence of significant deep-seated differences within the international Green movement and its attendant national political manifestations. The emergence of Green politics has been an important factor in the history of recent electoral politics in Western Europe. The nature of environmental political practice is the factor towards determining differences within the European Greens."

Kelly, Petra. "The need for eco-justice." *Fletcher Forum of World Affairs* 14 (Summer

1990): 327-331.
> "The co-founder of Die Grunen, the German Green Party, Petra Kelly has ignited a grassroots movement felt throughout Europe and, increasingly, beyond Europe's borders. Here, Petra Kelly recounts the guiding principles and tenets of the party, which calls for a complete redrafting of the international political economy."

MacKenzie, Debora; and Milne, Roger. "A fresh Green tinge to Europe's research." *New Scientist* 123 (August 19, 1989): 23-25.
> "Politicians across Europe are wondering how important the Green parties will become. But laboratories and institutes throughout the Community are already feeling the influence."

Mewes, Horst. "The Green Party comes of age." *Environment* 27 (June 1985): 13-17, 33-38.
> "Radical environmental politics in Germany is a highly complex and problematic phenomenon that may carry the seeds of its own destruction."

Redclift, Michael. "Turning nightmares in dreams: the Green Movement in Eastern Europe." *Ecologist* 19 (September-October 1989): 177-183.
> "Local concern over environmental degradation, together with increasing demands for political reform, has led to the emergence of green groups throughout the countries of the Eastern Bloc. Such movements have exposed problems in orthodox Marxist analysis and remind us that ecological politics cannot simply be understood in terms of Western capitalism. They illustrate how the ecological crisis is bound up with questions of human rights, freedom of information and democracy."

Schmid, Carol L. "The Green Movement in West Germany: resource mobilization and institutionalization." *Journal of Political and Military Sociology* 15 (Spring 1987): 33-46.
> "Briefly examines the general social, economic, and generational conditions which created the potential for a new social movement in West Germany. The second section analyzes the resources and constraints that have influenced each step of the evolution of the Green movement, leading to the electoral victory of the Green party in 1983 and again in 1987."

Veen, Hans-Joachim. "From student movement to Ecopax: the Greens." *Washington Quarterly* 10 (Winter 1987): 29-39.
> "Starting as a movement against the established party system, the Greens, since the end of the 1970s, have built an increasingly strong position as the fifth party in the Federal Republic of Germany."

Waller, Michael. "The ecology issue in Eastern Europe: protest and movements." *Journal of Communist Studies* 5 (September 1989): 303-328.
> "Two developments have favoured the emergence of an ecology movement in Eastern Europe. The first is a shift in the strategy of the major dissenting groups

in the region, which have broadened their emphasis on human rights to include ecology and peace issues. The second, which affects Hungary and Poland in particular, are the new possibilities for autonomous organization that have opened up in the wake of Gorbachev's reforms in the Soviet Union."

Law journals

Luiki, Paul S.; and Stephenson, Dale E. "Environmental laws are stricter in 'green'-influenced Europe." *National Law Journal* 14 (September 30, 1991): 45-47 (6 p.)
> "Growing environmental awareness among the European population has caused a 'green' philosophy to emerge across the political spectrum in many countries, acting as a catalyst for change as parties compete to have the 'greenest' platforms. Environmental law in Europe is complex, consisting of regulations applicable in individual countries and those imposed by a supranational institution."

Reports

Langguth, Gerd. *The Green Movement in Germany*. Washington, D.C.: Konrad Adenauer Stiftung, 1986. 9 p.
> "In the federal elections on March 6, 1983, the Greens polled 5.6 percent of the vote. Their series of successes continued with the elections to the European Parliament in June 1984. Presently the Greens are represented in half of the state parliaments of the Federal Republic of Germany."

ENVIRONMENTAL EDUCATION

Books (annotated)

Orr, David. *Ecological literacy: education and the transition to a postmodern world*. Albany: State University of New York Press, 1992. 219 p.
> Maintains that the only solution to the environmental crisis is education and presents numerous suggestions for improving environmental awareness through education. Argues that environmental education "ought to change the way people live, not just how they talk."

Walker, Richard. *Nature projects on file*. New York: Facts on File, 1992. 1 v.: ill. (QH55 .N38 1992)
> Presents nature projects for students ranging from plant and animal life to pollution measurement and environmental quality. Each project contains a list of materials needed, safety precautions to be taken, and step by step instructions. Contains appendices on grade levels for each project (most range from sixth to

twelfth grade), which projects require adult supervision, which projects can be done at home, and which should be done at school.

Books (unannotated)

Environment and science and technology education. New York: Pergamon Press, 1987. 430 p.: ill. (TD170.6 .E58 1987)

Environment in engineering education. Paris: Unesco, 1980. 111 p.: ill. (TD170 .E57)

Environmental training: an overview. Lanham, Md.: UNIPUB, 1985. 150 p. (GF26 .E586 1985)

Harrison, James D. *Community and environmental simulations: annotated guide to over 200 games for college and community education.* Monticello, Ill.: Vance Bibliographies, 1981. 139 p. (LB1029.S53 H37)

Hunter, Lisa. *Into adolescence. Caring for our planet and our health: a curriculum for grades 5-8.* Santa Cruz, Calif.: Network Publications, 1991. 104 p.: ill. (TD171.7 .H86 1991)

Issues in environmental education in ASEAN. Singapore: Regional Institute of Higher Education and Development, 1985. 138 p.: maps (TD170.8.A785 I87 1985)

Lloyd-Kolkin, Donna. *Entering adulthood. Creating a healthy environment: a curriculum for grades 9-12.* Santa Cruz, Calif.: Network, 1991. 146 p.: ill. (QH541.26 .L62 1991)

Metzger, Mary. *This planet is mine: teaching environmental awareness and appreciation to children.* New York: Simon & Schuster, 1991. 224 p. (TD171.7 .M48 1991)

New ideas in environmental education. New York: Croom Helm, 1988. 219 p. (HC79.E5 N45 1988)

Project W.I.Z.E.: Diversity of Lifestyles, Module I; *Survival Strategies,* Module II (Wildlife Inquiry through Zoo Education) Washington, D.C. (New York Zoological Society): Beacham Publishing, 1989.
    Two complete curriculum kits for grades 5-6 and 7-11, covering all aspects of wildlife management and environmental responsibility. A holistic, multidisciplinary approach to the relationship between individuals, society, and nature.

General interest periodicals (unannotated)

Irani, Ray R. "Environmental literacy; building a coalition for the 1990's." *Vital Speeches of the Day* 57,17 (June 15 1991): 542(3)

Robinson, Michael G. "Teaching integrated science around community and global problems." (includes bibliography) *The Clearing House* 64,3 (January-February 1991): 171(2)

Scott, Sylvia. "Teaching tots to save the Earth: even the youngest members of society can learn how to be kind to their planet." (includes some ecological issues and how to describe them for young children) *American Baby* 53,4 (April 1991): A28(4)

Stone, Jody M. "Preparing teachers as environmental educators." (Is education ready for the 1990s?) *The Education Digest* 55,5 (January 1990): 43(3)

Whittaker, Michael. "Heart o' Texas Council's down-to-Earth conservation college." *Scouting* 73 (October 1985): 40(5)

Environmental and professional journals (annotated)

Brennan, Matthew J. "A curriculum for the conservation of people and their environment." *Journal of Environmental Education* 17 (Summer 1986): 1-12.
> Outlines the structure of "Total Education for the Total Environment" an international curriculum developed at the United Nations International School in New York and endorsed by ISA-the International Schools Association.

Holt, Thomas Harvey. "Growing up green: are schools turning our kids into eco-activists?" *Reason* 23 (October 1991): 37-41.
> "Ten years ago ecology was rarely taught. Today, entire classes and textbooks focus on the environment, and dozens of activist organizations are working to shape environmental curricula nationwide. Environmental education has become a growth industry."

Lewis, Barbara A. "The children's cleanup crusade." *Sierra* 74 (March-April 1989): 62-66.
> Describes how a sixth-grade class in Salt Lake City waged a campaign to clean up a toxic waste site and drafted Utah House Bill 199, a state contributory Superfund, which passed unanimously.

Stenger, Richard S. "The corporate classroom." *Environmental Action* 23 (September-October 1991): 27-30.
> "Corporations are moving into classrooms with a host of new environmental curricula. Just what are they teaching our kids?"

Environmental and professional journals (unannotated)

Benenati, Frank. "Fun for a week: environmental awareness for a lifetime." (New

York State Department of Environmental Conservation's annual summer Rogers Ecology Workshop at Pack Demonstration Forest in the Adirondack Mountains) *Conservationist* 46,6 (May-June 1992): 26(4)

Brawer, Jennifer. "A+ teachers' toolbox: two programs that involve students in environmental issues." (Audubon Wildlife Adventures, Grizzly Bears; Adventures with Charts and Graphs) *A+* 7,4 (April 1989): 88(4)

Chepesiuk, Ron. "The Green Library: making an environmental difference". *Wilson Library Bulletin* 66,7 (March 1992): 36(4)

Gibbons, Ann. "Conservation biology in the fast lane: funding for this new kind of study has zoomed up - along with the number of academic programs - but critics think it's more a fad than a new scientific discipline." *Science* 255,5040 (January 3 1992): 20(3)

Seideman, David. "Wading into the fight." (innovative school programs help teenage activists in conservation) *National Wildlife* 29,1 (December-January 1990): 34(4)

Thompson, Craig D. "Project WILD earns straight A's in the three R's." (Wildlife in Learning Design; includes related article on teaching in the project) *Conservationist* 41 (November-December 1986): 14(6)

Waxman, Don. "Teaching restoration to kids." (special issue: environmental restoration) *Whole Earth Review* 66 (Spring 1990): 64(3)

White, Christopher. "Our classroom is the whole outdoors." (environmental conservation department of the Community College of the Finger Lakes) *Conservationist* 41 (May-June 1987): 42(6)

Reports

E.E.C. 2000: a study of environmental education centers. St. Paul, Minn.: Department of Natural Resources, 1992. 3 v.: ill., maps

Universities and environmental education. Paris: UNESCO and the International Association of Universities, 1986. 127 p.: ill. (TD178 .U55 1986)

Conference proceedings

Teaching conservation: proceedings of a Seminar on Teaching Conservation Overseas. Cambridge: African Studies Centre, University of Cambridge, 1987. 106 p.: ill. (S946 .S46 1986)

Reference

Earthwatching III: an environmental reader with teacher's guide. Madison, Wis.: Institute for Environmental Studies, University of Wisconsin-Madison and the University of Wisconsin Sea Grant Institute, 1990.

Education for the Earth: a guide to top environmental studies programs. Princeton, N.J.: Peterson's Guides, 1993. 175 p.
> Profiles over 100 colleges and universities that offer programs in environmental studies. Each institution has a one page description that includes general information about the school and specific information about the environmental program and the type of career at which it is aimed. Contains essays which examine the future of environmental careers in business, labor, health, education, and environmental areas.

Environmental and natural resource mathematics. Providence, R.I.: American Mathematical Society, 1985. 143 p.: ill. (HC79.E5 E575 1985)

Jenkins, Neil W. *Instructor reference manual: a glossary of selected terms of conservation, ecology, and resource use.* 3d ed. Jefferson City, Mo. (P.O. Box 180, Jefferson City 65102-0180): Missouri Department of Conservation, 1989. 77 p. (S922 .J45 1989)

ENVIRONMENTAL BUSINESS

Books (annotated)

Cairncross, Frances. *Costing the Earth: the challenge for government, the opportunities for business.* Cambridge, Mass.: Harvard Business School Press, 1992. 341 p.
> Maintains that environmental protection will only come about after governments decide to take an active role and business recognizes the economic potential of ecologically sound practices. Focuses on the needs of the green consumer, recycling, energy efficiency, taxing polluters, government failures, and the challenge to business.

Gilbreath, Kend, ed. *Business and the environment toward common ground.* 2d ed. Washington, D.C.: Conservation Foundation, 1984. 533 p. (HC110.E5 B885 1984)
> "Advocates no policies and proposes no specific solutions. It illustrates the diversity of opinions held by some of the leading thinkers and actors in the field. It offers evidence of a growing cooperative spirit--one that recognizes the indisputable link between economic and environmental goals. And it testifies to the increasing acceptance of a healthy environment and a productive economy as inseparable--each necessary for the other to thrive."

Hoffman, W. Michael, et al., eds. *Business, ethics, and the environment.* New York: Quorum, 1990.

> Addresses the responsibility of business to deal with environmental problems in the absence of law or regulation. Some contributors argue that business must assume responsibility with no tradeoff in gain, while others argue that "Business does not have an obligation to protect the environment over and above what is required by law; however, it does have a moral obligation to avoid intervening in the political arena in order to defeat or weaken environmental legislation."

Leonard, H. Jeffrey. *Polluting and the struggle for the world product.* New York: Cambridge University Press, 1988.

> Examines the conventional idea that multinational companies gravitate to countries that have the fewest controls on environmental pollution. For case studies he looks at U.S. companies that moved to Mexico, Rumania, Spain, and Ireland, countries that openly recruited business as pollution havens. Concludes that companies with a vital market share prefer cleaning up to moving.

Nanney, Donald. *Real estate transactions and environmental risks: a practical guide.* New York: Executive Enterprises, 1989. 451 p.

> Explains how to assess environmental risk and minimize personal liability to avoid civil or criminal penalities associated with real estate sales.

Pennell, Allison A., et al., eds. *Business and the environment: a resource guide.* Washington, D.C.: Island Press, 1992. 364 p.

> Aims to help the business educator integrate environmental issues into management research, education, and practices. Covers accounting and finance; business, government, and society; management; production and operations management; and strategic management. Contains a guide to resources, including journals, books, reports, and videos. Lists 185 business/environment educators.

Schmidheiny, Stephan. *Changing courses.* Cambridge, Mass.: MIT Press, 1992. 374 p. (HD75.6 .S35 1992)

> Large corporations argue that as long as they are profitable, they can afford to pay for many environmental programs, and therefore it is not in the best interest of the environment to hamper their operations. Half of this volume is devoted to success stories of large companies contributing to the environmental good, but Schmidheiny ignores the blatant offenders and seems naive in presenting the world's corporate community as willing to police itself.

Smart, Bruce, ed. *Beyond compliance: a new industry view of the environment.* Washington, D.C.: World Resources Institute, 1992. 285 p.

> Focuses on changes business has undergone, and is undergoing, in order to meet environmental standards. Discusses the incentives, process, difficulties, and rewards of such change. Presents reports from company officials, along with commentary and summaries. A good source for research into particular com-

panies and for information on how business has responded to particular environmental problems.

Sullivan, Thomas F. P., ed. *Greening of American business: making bottom-line sense of environmental responsibility.* Rockville, Md.: Government Institutes, 1992. 372 p.
Focuses on the legal aspects of environmental responsibility. Discusses such things as liability for environmental cleanup and agency regulations, as well as business opportunities that can alleviate the high cost of pollution control by creating jobs and sales.

Wann, David. *Biologic: environmental protection by design.* Boulder, Colo: Johnson Books, 1990. 248 p. (TD170 .W36 1990)
Argues that engineers should study nature's systems and create harmonious designs, which are efficient and sustainable, rather than dominating and destructive. Maintains that the building, manufacturing, and packaging industries should lead the way to prevent, rather than solve, environmental problems.

Wasik, John F. *The green company resource guide: a reference for any organization facing environmental concerns.* Kansas City, Mo.: Quality Books, 1992.
This guide to environmental resources--including everything from environmental lawyers to environmental conferences to recycled paper suppliers--is aimed at business with an interest in environmental marketing and at companies with an interest in making the work place environmentally sound.

Books (unannotated)

Blowers, Andrew. *Something in the air: corporate power and the environment.* Cambridge, Mass.: Harper & Row, 1984. 359 p.: ill.

Cahill, Lawrence B. *Environmental audits.* 5th ed. Rockville, Md.: Government Institutes, 1987. 656 p.: ill. (HC110.E5 C26 1987)

Cohn, Susan. *Green at work: finding a business career that works for the environment.* Washington, D.C.: Island Press, 1992.

Egan, James T. *Regulatory management: a guide to conducting environmental affairs and minimizing liability.* Chelsea, Mich.: Lewis, 1991. 231 p.: ill. (HC79.E5 E34 1991)

Eicher, George J. *The environmental control department in industry and government: its organization and operation.* Beaverton, Oreg.: Words Press, 1981. 144 p. (HD69.P6 E38 1981)

Elkington, John. *The green capitalists: industry's search for environmental excellence.* London: V. Gollancz, 1987. 258 p. (HD75.6 .E45 1987)

Kazis, Richard. *Fear at work: job blackmail, labor, and the environment*. New York: Pilgrim Press, 1982. 306 p. (HC110.E5 K39 1982)

Khozin, Grigorii Sergeevich. *Big business against nature*. Moscow: Novosti Press Agency Pub. House, 1984. 78 p. (HD75.6 .K49 1984)

Silverstein, Michael. *The environmental factor: its impact on the future of the world economy and your investments*. Chicago, Ill.: Longman Financial Services Pub., 1990. 227 p. (HG4521 .S567 1990)

General interest periodicals (annotated)

Beers, David; and Capellaro, Catherine. "Greenwash!" *Mother Jones* 16 (April 1991): 38-41, 88.
> "The advertising and public-relations industries have long been paid to jujitsu potential anticorporate movements into image builders for their clients. But their attempted co-opting of environmentalism has been so frenzied, so nakedly nineties as to make Orwell's imagination seem stunted. And yet, because most of the media rely on the greenwashers for ad revenue, it's been hard to find anyone who is willing to shout that the emperor has dropped trow." Includes sibebar, "Oiling the works," by Eve Pell, describing how Chevron bought its way into environmentalism's power circle.

"Corporations and the environment: greening or preening?" *Business and Society Review* 75 (Fall 1990): whole issue (96 p.)
> Partial contents include: "Are corporations playing clean with green?" by Jonathan Schorsch; "Do World Bank loans yield deforested zones?" by Bruce Rich; "Kodak's picture-perfect PR (chemical spills)," by Karen Paul; "The super morass of superfund," by Rock Grundman; "Can capitalists be environmentalists?" by Barry Commoner; "Death watch on the Rhine (by Aaron Schwabach; "Can the free market clean up our air?" by Anne Sholtz-Vogt and Kenneth Chilton; "Plum Creek's chainsaw massacre," by Tom E. Thomas; "Let's clean up clear air legislation," by Richard L. Klimisch; "Bhopal aftermath: Union Carbide rethinks safety," by Cornelius C. Smith; "How to fill the holes in ozone policy," by Paul Shrivastava; "Stemming the tide of chemical waste," by Joanna D. Underwood and Mark Dorfman; "A solid program for solid waste (Polystyrene recycling)," by Thomas A. Hemphill; "Corporations captialize on environmentalism," by Ulrich Steger; "Environmentalists lock up canal development," by Robert H. Hogner; "Will oil spills sink Exxon's bottom line?" by Ramesh Gehani; and "Canada's solution to pollution," by Jean Pasquero.

Johnson, Roberta Ann; and Kraft, Michael E. "Bureaucratic whistleblowing and policy change." *Western Political Quarterly* 43 (December 1990): 849-874.

Presents two case studies of whistleblowing--the EPA's Hugh Kaufman on hazardous waste sites and HHS's Hal Freeman on the Office for Civil Rights' failure to protect AIDS victims--and suggests that both instances had policy impacts "attributable to characteristics of the whistleblower, the nature of the issues raised, and especially the existence of several critical conditions: extensive and sympathetic media coverage, active support by politically influential groups within the agency and outside of it, and strong interest on the part of members of Congress in a position to conduct oversight investigations and promote policy change."

Kirkpatrick, David. "Environmentalism: the new crusade." *Fortune* 121 (February 12, 1990): 44-48, 50, 54-55.
"It may be the biggest business issue of the 1990s. Here's how some smart companies are tackling it."

Norton, Robert E. "Can business win in Washington?" *Fortune* 122 (December 3, 1990): 75-76, 80, 84.
"Yes, but it's increasingly on the defensive against environmentalists, consumer advocates, and resurgent regulators. Its main weapon: high-tech grass-roots lobbying."

Schrecker, Ted. "Resisting regulation: environmental policy and corporate power." *Alternatives* 13 (December 1985): 9-21.
Discusses the deep and fundamental conflict of priorities which underlies environmental regulation and some of the ways in which business firms are able to resist regulation.

Smith, Emily. "The greening of corporate America." *Business Week* 3156 (April 23, 1990): 96-103.
Describes how tougher laws, stepped-up enforcement and grass-roots environmental movements are causing corporate America to become more involved in controling environmental pollution.

Staehelin-Witt, Elke; and Blochliger, Hansjorg. "Proposed: a central environment bank." *Swiss Review of World Affairs* 40 (January 1991): 24-25.
"Analogous to the status of Switzerland's National Bank, the Central Environment Bank which we are proposing would be structured as an institutionally independent organization." The CEB would receive no direct instructions from the government or any of the government's departments. Regarding decision-making powers, it would be legally obligated to achieve certain environmental objectives, and it would be provided with a set of tools (prescribed in CEB Law) to accomplish its aims.

General interest periodicals (unannotated)

Bowles, Jerry. "Environment and industry: harnessing the power of the marketplace." *Fortune* 125,12 (June 15 1992): 17(10)

Cairncross, Frances. "Cleaning up." (survey of industry and the environment) *Economist* 316,7671 (September 8 1990): S1(16)

"Environment & industry: cooperating to create market-driven solutions." *Fortune* 124,12 (November 18 1991): 205(7)

Ferguson, Anne. "Good to be green." (environmental ethics) *Management Today* (February 1989): 46(6)

Gaskins, Richard H. "Extending the search for safety." (symposium: risk, safety and capitalism) *Society* 27,1 (November-December 1989): 19(3)

Grondin, James. "Business tackles an environmental agenda." (includes related article on environmental fund investments) *Financial World* 160,17 (August 20 1991): 40(3)

Kagan, Daniel. "The greening of environmental PR." (corporate environmental public relations) *Insight* 7,11 (March 18 1991): 38(3)

Kleiner, Art. "What does it mean to be green? Environmentalism, like the quality movement, challenges companies to do what's good for them: perfect their manufacturing processes." (includes information on business and the environment) *Harvard Business Review*: 38(8)

Mann, Eric. "Environmentalism in the corporate climate." *Tikkun* 5,2 (March-April 1990): 60(6)

McKee, Bradford. "From the ground up: firms that serve the nation's increasing environmental demands are at the core of a growing, recession-resistant industry." *Nation's Business* 79,1 (January 1991): 39(3)

"When green is profitable." *Economist* 312,7618 (September 2 1989): S16(2)

Wilsher, Peter. "The feeling grows that going green is good for business." *Management Today* (October 1991): 30(2)

Environmental and professional journals (annotated)

MacKerron, Conrad B.; and Chynoweth, Emma. "Europe's CPI plays catch-up in environmental cleanup." *Chemical Week* 146 (March 7, 1990): 24-28.

"It is difficult to tell if companies are serious or merely responding to tighter government rules and increasingly bold citizen demands for pollution control. But the question may be moot because two things seem certain on the environmental front. First, things will never be the same for the industry after the 1986 Sandoz spill that devastated the Rhine River. Second, the high cost of environmental compliance is forcing the industry to deal with tough issues like waste reduction and recycling even if it does not want to."

Russell, Dick; and Delong, Owen. "Can business save the environment?" *E: The Environmental Magazine* 2 (November-December 1991): 28-37, 57.
"From professionals all over the country these days, the word is that a huge new wave of 'greening' has broken over the profit-oriented shores of American business. But the overriding question remains: do business leaders realize that the wave is here to stay and that the public will no longer tolerate products or processes that harm the environment?"

Law journals

Pink, Daniel H. "The Valdez Principles: is what's good for America good for General Motors?" *Yale Law & Policy Review* 8,1 (1990): 180-195.
Argues "that the Valdez Principles are a trend toward extra-legal, market-driven enforcement of corporate responsibility, and that to succeed they must exploit market forces and appeal directly to companies' economic self-interests."

Reports

Environmental aspects of the activities of transnational corporations: a survey. New York: United Nations, 1985. 114 p.: ill. (HD62.4 .E58 1985)

Gentry, Bradford S. *Global environmental issues and international business: a manager's guide to trends, risks, and opportunities.* Washington, D.C.: Bureau of National Affairs, 1990. 1 v.: ill. (HD2755.5 .G458 1990)

Healy, Robert G. *America's industrial future: an environmental perspective.* Washington, D.C.: Conservation Foundation, 1982. 49 p. (HC110.I52 H4 1982)

Sims, William Allen. *The impact of environmental regulation on productivity growth.* Ottawa, Canada: Economic Council of Canada, 1983. 101 p.: ill. (HC120.I52 S55 1983)

Conference proceedings

Business, ethics, and the environment: the public policy debate. (Eighth National

Conference on Business Ethics, 1989: Bentley College) New York: Quorum Books, 1990. 253 p.: ill. (HC110.E5 N32 1989)

Freight transport and the environment. Paris: European Conference of Ministers of Transport, OECD Publications Service, 1991. 172 p.
> "Freight transport is vital for trade and for the economy. However, the movement of goods causes pollution and other environmental nuisances. This publication is an account of an international seminar held in Paris in May 1991, which examined ways to reduce the environmental damage caused by the movement of goods without undermining the important economic role of freight transport. Recommendations for action at the international level are also made."

Bibliographies

Buttel, Frederick H. *Labor and the environment: an analysis of and annotated bibliography on workplace environmental quality in the United States.* Westport, Conn.: Greenwood Press, 1984. 148 p. (Z5863.P6 B87 1984)

Kazis, Richard. *Jobs and the environment: a selected and annotated bibliography.* Chicago, Ill.: CPL Bibliographies, 1983. 13 p. (Z5863.P6 K39 1983)

ENVIRONMENTAL MARKETING

(See also *Food Safety: Labeling*) 1198

Books (annotated)

Meech, Karen N., ed. *Environmental industries marketplace.* Detroit, Mich.: Gale, 1992. Presents information on a large number of firms that provide environmental products and services.

Books (unannotated)

Carson, Patrick. *Green is gold: business talking to business about the environmental revolution.* New York: HarperBusiness, 1991. 216 p. (HF5413 .C37 1991)

Makower, Joel. *The E-factor: turning environmental responsibility into good, green profits.* New York: Times Books, 1993.

Ottman, Jacquelyn. *Green marketing: responding to environmental consumerism.* Lincolnwood, Ill.: NTC Business Books, 1992.

Samli, A. Coskun. *Social responsibility in marketing: a proactive and profitable marketing management strategy*. New York: Quorum Books, 1992.

General interest periodicals (annotated)

"Cleaner environment." *Changing Times* (February 1990): 28-31.
   Discusses "environment friendly" products and tradeoffs for careful shoppers. An "Environmental Shopper's Guide" accompanies the article which comments on how to be environmentally friendly with various categories of products.

McRae, Michael. "Bad wood, good wood." *Harrowsmith Country Life* 6 (November--December 1991): 29-39.
   Describes programs that encourage consumers to buy products made of "good wood" (wood from non-endangered trees harvested by sustainable forestry practices) and reject "bad wood" products (wood from endangered trees whose harvesting may contribute to the loss of biological diversity or the dislocation of forest dwellers); such programs use marketplace forces to support rain forest conservation.

General interest periodicals (unannotated)

Barrier, Michael. "How Richard Thalheimer is trying to sharpen an out-of-date Image." (The Sharper Image) Nation's Business 79,11 (November 1991): 16(2)

Bauerlein, Monika. "Green labels aim to aid eco-conscious shoppers." (environmental/ecological advertising) *Utne Reader* 44 (March-April 1991): 38(2)

Bowles, Jerry. "Environment and industry: harnessing the power of the marketplace." *Fortune* 125,12 (June 15 1992): 17(10)

"Earth-conscious beauty." (ecologically sound grooming) *Seventeen* 49,9 (September 1990): 174(2)

Ellis, Januarye. "Green ambition." (standards for labeling environmentally friendly products) *House Beautiful* 133,8 (August 1991): 26(2)

Fierman, Jaclyn. "The big muddle in green marketing: from ketchup bottles to garbage bags, companies are making products they claim will spare the environment. Trouble is, the definition of what's good keeps changing." (includes related articles on McDonald's wrap, aseptic juice boxes, and making mulch of disposable diapers) *Fortune* 123,11 (June 3 1991): 91(5)

Foster, Anna. "Decent clean and true." (sale of environmentally safe products and the effect on profits) *Management Today* (February 1989): 56(5)

Gallon, Gary; and Reid, Susanna. "The green product endorsement controversy: lessons from the Pollution Probe/Loblaw experience. (Pollution Probe endorsement of Loblaw products in 1989) (includes related articles) (cover story) *Alternatives* 18,3 (January-February 1992): 16(10)

Gifford, Bill. "The greening of the golden arches." (McDonald's environmental marketing) *Rolling Stone* 611 (August 22 1991): 34(3)

Irani, Ray R. "Environmental literacy; building a coalition for the 1990's." *Vital Speeches of the Day* 57,17 (June 15 1991): 542(3)

Irvine, Sandy. "The limits of green consumerism." *Canadian Dimension* 23,7 (October 1989): 22(3)

Kohl, Helen. "Earthly goods." (environmental product marketing) (includes related articles) *Canadian Consumer* 20,7-8 (July-August 1990): 9(10)

Landler, Mark. "Suddenly, green marketers are seeing red flags." *Business Week* 3201 (February 25 1991): 74(2)

Lord, Shirley. "Beauty's new nature." (natural beauty products and save-the-planet ideas) *Vogue* 180,10 (October 1990): 394(6)

Schannon, Mark L. "One businessperson's view of the ecological crisis; the restoration of trust." *Vital Speeches of the Day* 57,6 (January 1, 1991): 174(5)

"Selling green." (environmental marketing) *Consumer Reports* 56,10 (October 1991): 687(6)

Shaw, Januarye S.; and Stroup, Richard L. "Can consumers save the environment?" (includes related information) (cover story) *Consumers' Research* 73,9 (September 1990): 11(5)

Smith, Michael A. "Keeping shipshape and free of fines: can cleaning your boat kill the environment?" *Yachting* 171,3 (March 1992): 30(2)

Solheim, Mark K.; Wilcox, Melynda Dovel; and Young, Sarah. "A cleaner environment: what to buy." (includes environmental shopper's guide insert) *Changing Times* 44,2 (February 1990): 28(4)

Vogel, Carol. "Environmentally yours." (marketing of environmentally safe products for the home) *The New York Times Magazine* 140 (March 17, 1991): 70

Voss, Bristol. "The green marketplace." *Sales and Marketing Management* 143,8 (July 1991): 74(3)

Walsh, S. Kirk. "Green tech: new gadgets promise users an eco-conscious hot time." *Rolling Stone* 608-9 (July 11, 1991): 118(2)

Wasik, John. "Put your purchasing power to work." (becoming an environmentally friendly consumer) *Better Homes and Gardens* 68,11 (November 1990): 114(2)

Weber, Lynette; and Walker, Maura. "Behind the green veil." (marketing environmentally friendly products) *Alternatives* 17,4 (March-April 1991): 11(2)

Zinn, Laura. "Whales, human rights, rain forests--and the heady smell of profits." (Anita Roddick) *Business Week* 3222 (July 15, 1991): 114(2)

Environmental and professional journals (annotated)

Hayes, Denis. "Harnessing market forces to protect the Earth." *Issues in Science and Technology* 7 (Winter 1990-91): 46-51.
"A credible environmental seal for products and packaging could inform consumers and help prompt manufacturers to do the right thing." Environmental labeling programs in the U.S., Germany, Canada, and Japan are discussed.

Lefferts, Lisa Y.; and Blobaum, Roger. "Eating as if the Earth mattered." *E: The Environmental Magazine* 3 (January-February 1992): 31-37.
"Environmentally savvy consumers steer clear of toxic cleaners, bleached coffee filters and plastic bags at the supermarket, and fret about the recyclability of containers. But most of us barely give the environment a second thought when it comes to choosing food, the product we buy most often at the grocery store." But besides profoundly affecting our health, our food choices greatly affect the environment via the impact of farming practices.

Marinelli, January. "Packaging." *Garbage* 2 (May-June 1990): 28-33.
"In the past few years, as garbologists have poked around in our trashcans, they've discovered that packaging accounts for more than sixty-five percent of what we throw away every day. Pollsters say American shoppers are willing to pay up to five percent more for 'environmental-friendly' packaging."

Schwartz, Joe. "Shopping for a model community." *Garbage* 2 (May-June 1990): 35-38.
"Developed by Champaign-based Central States Education Center, Model Community uses education and financial incentives to help local towns reduce the volume and toxicity of their trash. Model supermarkets use shelf labels, posters, and fliers to educate consumers about less voluminous and more recyclable packaging, as well as non-toxic products. The Model Community also includes

garbage haulers who give discounts for recycling, as well as waste-reducing schools, newspapers, churches, and copy shops."

Vandervoort, Susan Schaefer. "Big 'green brother' is watching: new directions in environmental public affairs challenge business." *Public Relations Journal* 47 (April 1991): 14-19, 26.

> "With environmental awareness at an all-time high, companies are adopting proactive strategies that address consumer concerns. Everyone is talking 'ecospeak,' the language of accuracy in describing environmental problems and solutions. The Big 'Green Brother' monitoring industry represents a collection of activist, regulatory, media and consumer interests. Keeping the dialogue flowing is public relations'challenge for 1991 and beyond."

Environmental and professional journals (unannotated)

Banashek, Mary Ellen. "Back to the garden." *Health* 22,9 (October 1990): 58(6)

Frankel, Carl. "The trouble with green products." *Buzzworm: The Environmental Journal* 3,6 (November-December 1991): 36(5)

O'Brien, Kathyleen A. "Green marketing: Can it be harmful to your health?" (environmental advertising) *Industry Week* 241,8 (April 20 1992): 56(4)

Law journals

Howett, Ciannat M. "The 'green labeling' phenomenon: problems and trends in the regulation of environmental product claims." *Virginia Environmental Law Journal* 11 (Spring 1992): 401-461.

> "Provides an overview of and insight into the problem and trends in environmental claims regulation;" explores the regulatory approach at the state and federal levels; discusses efforts by state attorneys general to crack down on illegitimate claims; examines the private litigation approach to control illegitimate claims; describes methods of industry self-regulation; examines controling claims through third party regulation; and overviews international claims regulation, looking at Canada and the European Community.

Kimmel, James Paul, Jr. "Disclosing the environmental impact of human activities: how a Federal pollution control program based on individual decision making and consumer demand might accomplish the environmental goals of the 1970s in the 1990s." *University of Pennsylvania Law Review* 138 (December 1989): 505- 548.

> "Proposes an 'Environmental Impact Index' (EII), that would provide consumers with simple, concise, at-a-glance numerical information concerning the environmental impact of a product or package during its production, use, and disposal."

Linked to this index, an environmental impact tax (EIT) would effectively charge consumers, not producers, for the environmental costs of the goods they consume. "The overall program would achieve environmental goals by reducing the demand for environmentally inefficient goods rather than by altering the supply."

Reports

Green Report II: recommendations for responsible environmental advertising. n.p.: 1991. 30 p. "Attempts to take advantage of consumers' increasing interest in the environment have led some companies to make environmental advertising claims that are trivial, confusing and misleading." In November, 1990, an ad hoc task force issued "The Green Report," which made a series of preliminary recommendations for responsible environmental advertising to guide industry until national standards are effected. Green Report II is a revision incorporating suggestions and criticisms made by industry, environmental groups, and consumers.

Schorsch, Jonathan. *Can our economy come clean?* New York: Council on Economic Priorities, 1990. 6 p. (CEP research report)
Not all products that claim to be green are green. Addresses unanswered questions regarding potential for false labeling and advertising of green products, and what constitutes environmental soundness.

Bibliographies

Morales, Leslie Anderson. *The economics of "green consumerism": a bibliography.* Monticello, Ill.: Vance Bibliographies, 1991. 7 p. (Z7164.M18 M67 1991)

ENVIRONMENTAL INVESTMENT

Books (unannotated)

Domini, Amy L. *Ethical investing.* Reading, Mass.: Addison-Wesley, 1984. 288 p.: ill. (HG4528 .D65 1984)

Harrington, John C. *Investing with your conscience: how to achieve high returns using socially responsible investing.* New York: Wiley, 1992. 276 p.: ill. (HG4528 .H37 1992)

How to invest with your conscience. New York: New York Institute of Finance, 1991. 367 p. (HG4528 .M55 1991)

Meeker-Lowry, Susan. *Economics as if the Earth really mattered: a catalyst guide to socially conscious investing*. Santa Cruz, Calif.: New Society Publishers, 1988. 282 p.

Sethi, S. Prakash. *Interfaith Center on Corporate Responsibility (ICCR)*. Richardson, Tex. (P.O. Box 688, Richardson 75080): School of Management and Administration, University of Texas at Dallas, 1980. 42 leaves (HD60.5.U5 S465 1980)

General interest periodicals (unannotated)

Antilla, Susan. "Socially responsible investments; a new breed of investor is betting that today's ethics will affect tomorrow's balance sheet." *Working Woman* 10 (April 1985): 38(3)

Cohn, Jordan. "Interest on principles." (investing conscientiously) *Ms.* 17,5 (November 1988): 56(4)

Edgerton, Jerry. "Money and morals: despite the dilemmas, you can profit by following your conscience." *Money* 14 (December 1985): 153(6)

Katz, Donald R. "Cashing in with a conscience." (ethical investments) *Esquire* 106 (November 1986): 83(2)

Kilgo, Edith Flowers. "Concerned about future finances?" (Christian money management) *Christian Herald* 109 (January 1986): 41(4)

Kohn, Alfie. "Fund from the heart; new market funds let you put your money where your morals are." *Savvy* 5 (December 1984): 48(2)

Louthan, Shirley. "The ethical investor." *Ms.* 13 (August 1984): 20(11)

Rentoul, John. "The ethical sector borrowing requirement." (ethics of investments) *New Statesman* 112 (August 15 1986): 13(3)

Rothchild, John. "I'm irresponsible!" (managing a morally proper portfolio) *Mother Jones* 14,7 (September 1989): 42(2)

Rowland, Mary. "Putting your money where your heart is." (mutual funds for 'ethical investors') *50 Plus* 29,12 (December 1989): 57(4)

Siwolop, Sana. "Ethical investing?" (socially responsible investment portfolios) *Financial World* 158,13 (June 27 1989): 86(2)

Stern, Linda; and Goldwasser, Joan. "These investors take a stand." (on social objectives as well as financial) *Changing Times* 41,11 (November 1987): 134(3)

Stovall, Robert. "When do-gooders do good." (Domini 400 Social Index invests in environmentally responsible companies) *Financial World* 161,17 (September 1 1992): 68(2)

Stroup, Margaret A., et al. "Doing good, doing better: two views of social responsibility. (corporate applications) *Business Horizons* 30 (March-April 1987): 22(4)

Tritch, Teresa. "How to put your principles behind your principal." *Savvy* 10,12 (December 1989): 27(3)

Wasik, John F. "Environmentally conscious investing; you can help clean up the the world and profit at the same time." *Consumers Digest* 29,5 (September-October 1990): 27(6)

Environmental and professional journals (unannotated)

"Ethics and investments." *Public Management* 69 (March 1987): 20(2)

EARTH DAY

Books (annotated)

Baden, John, ed. *Earth Day reconsidered*. Washington, D.C.: Heritage, 1980. 108 p.
Calls for a reassessment of environmental policy, arguing that the current policy uses expensive mechanisms to buy increments in environmental quality. Evaluates several environmental programs including timber management in the Rockies, federal water development policies, and public land management.

Books (unannotated)

Balasubramaniam, Arun. *Ecodevelopment: towards a philosophy of environmental education.* Konigswinter, Germany: Friedrich Naumann Stiftung; Singapore: Regional Institute of Higher Education and Development, 1984. 82 p. (GF75 .B36 1984)

New ideas in environmental education. New York: Croom Helm, 1988. 219 p. (HC79.E5 N45 1988)

General interest periodicals (annotated)

Hayes, Denis. "Earth Day 1990: threshold of the green decade." *Natural History* 99

(April 1990): 55-58, 67-70.
> Twenty years after Earth Day 1970, the world finds itself in worse environmental shape. In this essay, the chairman of Earth Day 1990 analyzes the mistakes that have led to our current dilemma and suggests ways to make the present environmental movement more effective.

Painton, Priscilla. "Earth Day: greening from the roots up." *Time* 135 (April 26, 1990): 76-80.
> "The fanfare masks a quiet revolution: millions of ordinary Americans are leading the environmental movement from their homes and town halls." Includes brief descriptions of the efforts of ten environmental activists, including the first six winners of the Goldman Environmental Prize.

General interest periodicals (unannotated)

"After Earth Day." (environmental awareness) *Nation* 250,17 (April 30 1990): 583(2)

Anderson, Walter. "Green politics now comes in four distinct shades." (beyond Earth Day: 10 views on where to go from here) *Utne Reader* 40 (July-August 1990): 52(2)

"Beyond Earth Day, 1990." (editorial) *The Progressive* 54,5 (May 1990): 8(2)

Burchill, Julie. "Good reasons to distrust and despise the greens." (beyond Earth Day: 10 views on where to go from here) *Utne Reader* 40 (July-August 1990): 54(2)

Carpenter, Betsy. "Living with our legacy." (Earth Day and the environment) *U.S. News & World Report* 108,16 (April 23 1990): 60(5)

Dane, Abe. "Earth Day, then and now." (environmental protection) *Popular Mechanics* 167,4 (April 1990): 54(2)

"Defenders of the planet." (Earth Day: profiles of Goldman Environmental Prize winners) *Time* 135,17 (April 23 1990): 78(2)

"Global festival." (Earth Day 1990) *Time* 135,17 (April 23 1990): 84(2)

Gottlieb, Robert. "Earth Day revisited." (ecology and social meaning) *Tikkun* 5,2 (March-April 1990): 55(5)

Lerner, Michael. "Critical support for Earth Day: an editorial." (ecology and social meaning) *Tikkun* 5,2 (March-April 1990): 48(3)

Minucci, Mary Beth Spann. "Earth Day extravaganza." (pupils celebrate Earth Day) *Instructor* 100,7 (March 1991): 75(2)

Phoenix, River; and Hendle, Jayne. "We are the world." (Earth Day, 1990; includes related article on history of Earth Day) *Seventeen* 49,4 (April 1990): 220(4)

Postrel, Virginia I. "Forget left and right, the politics of the future will be growth vs. green." (beyond Earth Day: 10 views on where to go from here) *Utne Reader* 40 (July-August 1990): 57(2)

Scheer, Robert. "Up to here with green; spare us the niceties of Earth Day, which deflected attention from the true perils faced by our planet." *Playboy* 37,8 (August 1990): 50(2)

Twiest, Linda; and Egan, Lorraine Hopping. "Suddenly endangered: a book-based Earth Day unit about animals on the brink." *Instructor* 101,8 (April 1992): 14(3)

Walljasper, Jay. "Who are the greens and what do they believe?" (beyond Earth Day: 10 views on where to go from here) *Utne Reader* 40 (July-August 1990): 58(3)

Environmental and professional journals (annotated)

Cahn, Robert; and Cahn, Patricia. "Did Earth Day change the world?" *Environment* 32 (September 1990): 16-20, 36-43.
   "The 200 million people who created the largest grassroots demonstration in history by celebrating Earth Day 1990 have generated a global environmental current that is quietly galvanizing voters, and politicians into action. More people are assuming individual responsibility for changing environmentally harmful practices and are calling for global cooperation in service to a common cause."

"Earth Day and the future of our planet." *National Parks* 64 (March-April 1990): 16-28, 42-43.
   Contains "Global prescription: leading conservationists look to the future and speak their mind--Earth Day," by Elizabeth Hedstrom; and "Planet at the crossroads: eight issues that will determine Earth's future: species loss, global warming, acid rain, ozone depletion, rain forests, protected areas, marine management, and land abuse," by John Kenney.

Mitchell, John G. "A perfect day for Earth." *Audubon* 92 (March 1990): 109-110, 112, 114, 116, 118, 120, 122-123.
   "A long, thin trail with many branches led to Earth Day I. What really happened on April 22, 1970? What did it all mean? And as Earth Day XX approaches, have we lost the path?"

Ross, David M.; and Driver, B. L. "Importance of appraising responses of subgroups in program evaluations: the Youth Conservation Corps." *Journal of Environmental Education* 17 (Spring 1986): 16-23.

"The overall conclusion drawn from this research is that the participants, parents of participants, and program administrators studied perceive that these programs provide a wide variety of benefits to both the participants and to society in general."

Environmental and professional journals (unannotated)

Autobahn, J. J. "Earth Day 1990: T-shirts and environmental justice." (Earth Day exhibition area, Washington, D.C.) *National Review* 42,10 (May 28 1990): 23(2)

Barrell, Jeannie. "Earth Day - one year later." *Parks & Recreation* 26,4 (April 1991): 36(6)

Comp, T. Allan. "Earth Day and beyond. (includes related information) *American Forests* 96,3-4 (March-April 1990): 54(5)

Desmond, Mary Beck. "Global environment - Earth Day 1990." *Earth Science* 43,1 (Spring 1990): 6(2)

Ehrlich, Paul. "Ecovoice - Earth Day 1990." (column) *Buzzworm: The Environmental Journal* 2,2 (March-April 1990): 14(2)

Hedstrom, Elizabeth. "Earth Day; the green movement kicks off the environmental decade". (includes related articles) *National Parks* 64,3-4 (March-April 1990): 18(6)

Polsby, Emily. "A new environmental launching point." (1990 Earth Day festivities) Nation's Cities Weekly 13,15 (April 16 1990): 1(2)

Rome, Linda. "Celebrating Earth Day all year. (libraries and the environment; includes related information on Earth Day ideas) *Wilson Library Bulletin* 65,6 (February 1991): 40(4)

"Seeking a world conservation ethic; Earth Day twenty years later." *Wilderness* 53,188(Spring 1990): 3(3)

Steinhart, Peter. "Bridging the gap: can Earth Day 1990 bring together Greens and mainline conservationists?" (clash between environmental values and politics underscores problem of achieving consensus on ecological issues) *Audubon* 92,1 (January 1990): 20(4)

Bibliographies

Levine, Beth. "Earth Day anniversary celebrated with bumper crop of books." *Publishers Weekly* 237,11 (March 16 1990): 41(4)

Books for young adults

Gardner, Robert. *Celebrating Earth Day: a sourcebook of activities and experiments.* Brookfield, Conn.: Millbrook Press, 1992. 96 p.: (some color) ill. (TD170.2 .G37 1992)

Gore, Willma Willis. *Earth Day.* Hillside, N.J.: Enslow, 1992. 48 p.: ill. (some color) (HC110.E5 G66 1992)

Lowery, Linda. *Earth Day.* Minneapolis, Minn.: Carolrhoda Books, 1991. 45 p.: color ill. (HC110.E5 L69 1991)

Wheeler, Jill C. *Earth Day every day.* Edina, Minn.: Abdo & Daughters, 1991. Distributed by Rockbottom Books, Minneapolis, Minn.

CITIZEN PARTICIPATION

(See also *Nuclear Energy: Movement Against Nuclear Power*) 1513
(See also *Organizations*, above) 460
(Related topics in *Recycling: Citizen Participation*) 2112

Books (annotated)

50 simple things you can do to save the Earth. Berkeley, Calif.: EarthWorks, 1989. 96 p.
    This introductory work offers a wide range of suggestions on things individuals can do to stop environmental degradation, ranging from car pooling to stopping junk mail. Contains the addresses of environmental organizations and lists publications which offer further information.

Caplan, Ruth, et al. *Our Earth, Ourselves: the action-oriented guide to help you protect and preserve our planet.* New York: Bantam, 1990. 340 p.: ill. (TD175 .C37 1990)
    This work is a compilation of statistical facts, activist biographies, expert opinions, and suggestions on individual environmental awareness. It is the product of Environmental Action, the group which founded Earth Day.

CEIP Fund. *Complete guide to environmental careers.* Washington, D.C.: Island Press, 1989. 328 p. (TD170.2 C66 1989)
    The increased environmental awareness which has pervaded recent years has spawned a whole range of environmental careers--from toxic waste management to outdoor recreation and environmental education. This work offers a variety of information on such careers, including requirements, salaries, outlook, job listings,

and professional organizations. Also addressed in terms of these careers are key issues, developments, and trends. Includes institutional case studies.

Environmental politics: lessons from the grassroots. Durham, N.C.: Institute for Southern Studies, 1988. 122 p. (HC107.N83 E514 1988)
> Focuses on the efforts of activists in North Carolina to cut across racial, class, and cultural barriers in their efforts to promote a progressive environmental agenda.

Erickson, Brad, ed. *Call to action: handbook for ecology, peace and justice.* San Francisco: Sierra Club Books, 1990.
> This politically centered collection contains essays by luminaries from Carl Sagan and Jacques Cousteau to Mike Roselle and John Conyers. The introduction by Jesse Jackson sets the tone. Chapters consist of an explanation of the particular topic, a detailed plan for what individuals can do to change the situation, and a list of resources. Not all topics are environmentally related and include such broad subjects as world hunger and Third World debt.

Hadden, Susan G. *A citizen's right to know.* Boulder, Colo.: Westview, 1989.
> Hadden believes that the U.S. has led the way in informing citizens about environmental problems, but she has no illusions about the technical and political difficulties that restrict public knowledge. She analyzes the obstacles to citizen involvement and power sharing, including the public's low level of scientific understanding and the absence of trust in government and mass political activity. Maintains that as environmental problems increase, government must lead the way to informing its citizens.

Harris, D. Mark. *Embracing the Earth: choices for environmentally sound living.* Chicago: Noble Press, 1990. 162 p.: ill. (TD171.7 .H37 1990)
> Focusing on practical solutions to environmental problems, this work examines the choices that we make every day such as what we buy, and how we heat our homes, and offers alternatives that are more environmentally sound. Includes a bibliography of environmental publications and information on EcoNet, a computer network for environmentalists.

Hollender, Jeffrey A. *How to make the world a better place: a beginner's guide to doing good.* New York: Morrow, 1990. 303 p.: ill. (HN90.V64 H65 1990)
> Maintaining that the fate of humankind and the planet can be shaped by social behavior, this book offers a wide range of small suggestions on how we should change our behavior. Issues such as the environment, human rights, and world hunger are the targets of the authors suggestions.

Hynes, H. Patricia. *Earthright: every citizen's guide; or, what you can do in your home, workplace, and community to save our environment.* Rocklin, Calif.: Prima, 1990. 236 p. (TD171.7 N96 1990)
> The author presents scientific statistics on environmental crisis and follows them

up with specific examples of how individuals have succeeded in making a difference. Offers a plethora of suggestions on everything from how to organize a beach cleanup to how to put pressure on toxic polluters. Contains a bibliography and lists resources.

Naar, Jon. *Design for a livable planet*. New York: Harper, 1990. 338 p.: ill. (TD171.7 .N33 1990)

Naar's approach is to define the primary forms of pollution and then to provide advice as to what citizens can do to improve the situation. Organizations and phone numbers are listed, but for individual citizen involvement there are tips on such things as what to buy, how to save energy, and how to influence legislation. Argues that only when we become a nation of practicing environmentalists will conditions change at the governmental level.

Rifkin, Jeremy; and Rifkin, Carol Grunewald. *Voting green: your complete environmental guide to making political choices in the 1990s*. New York: Doubleday, 1992. 390 p.: ill. (HC110.E5 R53 1992)

In examining the mass of "green" legislation which has been introduced in the last two Congresses, this work evaluates the performance of individual members of Congress and ranks them according to their votes on environmental issues. Critically examines the Bush administration, concluding that, despite proenvironment rhetoric, the administration is decidedly anti-environment. Suggests an alternative political framework that emphasizes sustainability over wealth.

Rifkin, Jeremy, ed. *The green lifestyle handbook*. New York: Henry Holt, 1990. (TD170.2 .G74 1990)

Focusing on the theme of "Think globally, act locally," this work is an attempt to make people change their lifestyles so that the ecological consequences of every act are considered. Philosophically analyzes society as being greedy and centered on short-term gain. Includes sections on life-style choices and ways to organize environmental groups. Includes a bibliography.

Sargent, Frederic, et al. *Rural environmental planning for sustainable communities*. Washington, D.C.: Island Press, 1991. 254 p.: ill. (HT392 .R873 1991)

Outlines a method and process for citizens in small towns and rural areas to plan the future of their own communities and suggests growth plans that are environmentally sound, politcally acceptable, and economically feasible.

Steger, Will; and Bowermaster, Jon. *Saving the Earth: a citizen's guide to environmental action*. New York: Knopf, 1990. 306 p.: ill. (TD171.7 .S74 1990)

This guide is divided into four general areas requiring citizen involvement: atmosphere, land, water, and people, with subdivisions into individual problems. At the end of each chapter there are sections defining what can be done by individuals and government, listing organizations to contact, and including a bibliography.

Books (unannotated)

Abzug, Malcolm J. *Palisades oil: a community battles over oil drilling*. Pacific Palisades, Calif.: M. J. Abzug, 1991. 315 p.: ill. (TN871.2 .A216 1991)

Alternative environmental conflict management approaches: a citizens' manual. Ann Arbor, Mich.: Environmental Conflict Project, School of Natural Resources (Dana Bldg., Ann Arbor 48109-1115), University of Michigan, 1986. 252 p.: ill. (HC110.E5 A643 1986)

Bellamy, David J. *How green are you?*. New York: C. Potter, 1991. Distributed by Crown Publishers. 31 p.: color ill. (TD171.7 .B45 1991)

Call to action: handbook for ecology, peace, and justice. San Francisco: Sierra Club Books, 1990. 250 p.: ill. (H97 .C35 1990)

Camino Velozo, Ronnie de. *Incentives for community involvement in conservation programmes*. Rome: Food and Agricultural Organization of the United Nations, 1987. 159 p.: ill. (S934.D44 C35 1987)

Campolo, Anthony. *50 ways you can help save the planet*. Downers Grove, Ill.: InterVarsity Press, 1992. 144 p.: ill. (TD171.7 .A47 1992)

Capone, Lisa. *The Conservationworks book: practical conservation tips for the home and outdoors*. Boston, Mass.: Appalachian Mountain Club Books, 1992. Distributed by Talman. 96 p.: ill. (TD171.7 .C37 1992)

Christensen, Karen. *Home ecology: simple and practical ways to green your home*. Golden, Colo.: Fulcrum, 1990. 334 p.: ill. (TD171.7 .C47 1990)

Crampton, Norman. *Complete trash: the best way to get rid of practically everything around the house*. New York: M. Evans, 1989. 136 p.: ill. (TX324 .C72 1989)

Dadd, Debra Lynn. *Nontoxic, natural & Earthwise: how to protect yourself and your family from harmful products and live in harmony with the Earth*. Rev. ed. Los Angeles: Jeremy P. Tarcher, 1990. Distributed by St. Martin's Press. 360 p. (RA565 .D34 1990)

Davenport, Marc. *Dear Mr. President: 100 Earth-saving letters*. Secaucus, N.J.: Carol Publishing Group, 1992. 260 p. (TD170.2 .D38 1991)

Donahoe, Sydney L. *Earth keeping: making it a family habit*. Grand Rapids, Mich.: Zondervan, 1990. 144 p. (TD171.7 .D64 1991)

Earthkeeping, Christian stewardship of natural resources. Grand Rapids, Mich.: Eerdmans, 1980. 317 p. (HC55 .E27)

Earthkeeping in the nineties: stewardship of creation. Rev. ed. Grand Rapids, Mich.: Eerdmans, 1991. 391 p.: ill. (HC55 .E27 1991)

Ecologue: the environmental catalogue and consumer's guide for a safe Earth. New York: Prentice Hall Press, 1990. 255 p.: ill. (TD171.7 .A53 1990)

Environmental crisis: a handbook for all friends of the Earth. Exeter, N. H.: Heinemann Educational Books, 1984. 196 p. (TD170.3 .E54 1984)

Freudenberg, Nicholas. *Not in our backyards!: community action for health and the environment.* New York: Monthly Review Press, 1984. 304 p. (RA566.3 .F74 1984)

Gay, Kathlyn. *Caretakers of the Earth.* Hillside, N.J.: Enslow, 1993.

Getis, Judith. *You can make a difference: help protect the Earth.* Dubuque, Iowa: Wm. C. Brown, 1991. 88 p. (TD171.7 .G48 1991)

Global ecology handbook: what you can do about the environmental crisis. Boston: Beacon Press, 1990. 414 p.: ill. (TD171.7 .G56 1990)

Gorder, Cheryl. *Green Earth resource guide: a comprehensive guide about environmentally-friendly services and products.* Tempe, Ariz.: Blue Bird Publishing, 1991. 256 p.: ill. (TD171.7 .G68 1990)

Green alternative: guide to good living. London: Methuen, 1987. 368 p.: ill. (HC260.E5 G736 1987)

Green lifestyle handbook. New York: Henry Holt, 1990. 198 p. (TD170.2 .G74 1990)

Green pages: your everyday shopping guide to environmentally safe products. New York: Random House, 1990. 237 p.: ill. (TD171.7 .G74 1990)

Heloise. *Heloise, hints for a healthy planet.* New York: Perigee Books, 1990. 160 p.: ill. (TD171.7 .H44 1990)

Howell, Robert E. *Designing a citizen involvement program: a guidebook for involving citizens in the resolution of environmental issues.* Corvallis, Oreg.: Western Rural Development Center, Oregon State University, 1987. 178 p.: map (HC110.E5 H69 1987)

King, Angela. *Holding your ground: an action guide to local conservation.* London: Maurice Temple Smith, 1985. 326 p.: ill. (QH75 .K56 1985)

Krull, Kathleen. *It's my Earth, too: how I can help the Earth stay alive.* New York: Delacorte Press, 1992. 1 v.: color ill. (TD171.7 .K78 1992)

Levine, Michael. *The environmental address book: how to reach the environment's greatest champions and worst offenders.* New York: Perigee, 1991. 252 p.: ill. (TD169.6 .L48 1991)

Lind, Brenda. *The conservation easement stewardship guide : designing, monitoring, and enforcing easements.* Washington, D.C.: Land Trust Alliance; Concord, N.H.: Trust for New Hampshire Lands, 1991. 107 p.: ill., maps (KF658.C65 L56 1991)

Love, Ann. *Take action.* New York: Tambourine Books, 1993.

Lowery, Linda. *Earthwise at home: a guide to the care & feeding of your planet.* Minneapolis, Minn.: Carolrhoda Books Inc., 1993.

MacEachern, Diane. *Save our planet: 750 everyday ways you can help clean up the Earth.* New York: Dell, 1990. 210 p.: ill. (QH541 .M2 1990)

Makower, Joel. *The green commuter.* Washington, D.C.: National Press Books, 1992. 171 p. (TD195.T7 M35 1992)

Meeker-Lowry, Susan. *Economics as if the Earth really mattered: a catalyst guide to socially conscious investing.* Philadelphia, Pa: New Society Publishers, 1988. 282 p.

McKean, Margaret A. *Environmental protest and citizen politics in Japan.* Berkeley: University of California Press, 1981. 291 p. (HC465.E5 M3)

Moore, Roger L. *Organizing outdoor volunteers.* 2d ed. Boston, Mass.: Appalachian Mountain Club, 1992.

Mother Earth handbook: what you need to know and do--at home, in your community, and through your church--to help heal our planet now. New York: Continuum, 1991. 320 p.: ill. (TD171.7 .M68 1991)

Naar, Jon. *Design for a livable planet: how you can help clean up the environment.* New York: Perennial Library, 1990. 338 p.: ill. (TD171.7 .N33 1990)

Next step: 50 more things you can do to save the Earth. Kansas City, Mo.: Andrews and McMeel, 1991. 120 p. (TD171.7 .N49 1991)

Pick, Maritza. *How to save your neighborhood, city, or town: the Sierra Club guide to community organizing.* San Francisco: Sierra Club Books, 1993.

Progress as if survival mattered: a handbook for a conserver society. San Francisco: FOE, 1981. 456 p.: ill. (HC79.E5 P733 1981)

Purcell, Arthur H. *The waste watchers: a citizen's handbook for conserving energy and resources.* Garden City, N.Y.: Anchor Books, 1980. 286 p.: ill. (TD793 .P87)

Rees, Andrew. *The pocket green book.* Atlantic Highlands, N.J.: Zed Books, 1991. 154 p. (TD175 .R44 1991)

Richman, Beth. *Creating a healthy world: 101 practical tips for home and work: everyday chemicals.* Snowmass, Colo.: Windstar Foundation, 1989. 46 p.: ill. (TD170 .R53 1989)

Save the Earth at work!: how you can create a waste-free, non-polluting, non-toxic workplace. Holbrook, Mass.: Bob Adams Publishers, 1991. 144 p. TD171.7 .S28 1991)

Seymour, John. *Blueprint for a green planet: your practical guide to restoring the world's environment.* New York: Prentice Hall, 1987. 192 p.: ill. (QH545.A1 S49 1987)

Skills for simple living. Point Roberts, Wash.: Hartley & Marks, 1991. 218 p.: ill. (TX147 .S56 1991)

Stepaniak, Joanne. *Ecological cooking: recipes to save the planet.* Summertown, Tenn.: Book Publishing, 1991. 228 p.: ill. (TX837 .S74 1991)

Tasaday, Laurence. *Shopping for a better environment: a brand name guide to environmentally responsible shopping.* Deephaven, Minn.: Meadowbrook, 1991. Distributed by Simon and Schuster. 341 p. (HF1040.8 .T37 1991)

Teitel, Martin. *Rain forest in your kitchen: the hidden connection between extinction and your supermarket.* Washington, D.C.: Island Press, 1992. 112 p. (TX335 .T388 1992)

Vallely, Bernadette. *1001 ways to save the planet.* New York: Penguin Books, 1990. 336 p.: ill. (TD171.7 .V35 1990)

Viner, Michael. *365 ways for you and your children to save the Earth one day at a time.* New York: Warner Books, 1991. 99 p.: ill. (TD171.7 .H55 1991)

Wachtel, Paul Spencer. *Eco-bluff your way to greenism: the guide to instant environmental credibility.* Chicago: Bonus Books, 1991. 112 p.: ill. (HC79.E5 W26 1991)

Wald, Michael. *What you can do for the environment.* New York: Chelsea House, 1993.

Wilkinson, Loren. *Caring for creation in your own backyard: over 100 things Christian families can do to help the Earth.* Ann Arbor, Mich.: Vine Books, 1992. 268 p.: ill. (TD171.7 .W55 1992)

Winn, Chris. *Legal daisy spacing: the Build-a Planet manual of official world improvements.* New York: Random House, 1985. 123 p.: ill. (NC1479.W53 A4 1985)

Zisk, Betty H. *The politics of transformation: local activism in the peace and environmental movements.* New York: Praeger, 1992.

General interest periodicals (annotated)

Berry, Wendell. "Out of your car, off your horse." *Atlantic* 267 (February 1991): 61-63.
Offers twenty-six propositions about global thinking and the sustainability of cities and "warns that those who would save the planet may sometimes be as dangerous as those who recklessly exploit its resources. Argues that ecological good sense . . . begins not in thinking globally but in acting locally."

"Cleaner environment." *Changing Times* (February 1990): 28-31.
Discusses "environment friendly" products and tradeoffs for careful shoppers. An "Environmental Shopper's Guide" accompanies the article which comments on how to be environmentally friendly with various categories of products.

Eberhart, Annie. "If you can't stand the heat . . . Read on for a citizen's guide to some sensible steps for cooling off the Earth." *Public Citizen* 8 (November-December 1988): 13-17.
Outlines steps which individuals can take to reduce atmospheric pollution and to pressure local, regional, national and international policymakers into taking action.

Moore, Thomas. "For whom the corps toils." *U.S. News & World report* 106 (February 13, 1989): 24-25.

Describes the California Conservation Corps, "the oldest and most successful state-run public-service youth corps in the country."

General interest periodicals (unannotated)

"133 ways to save the Earth and improve your life at the same time" (environmentalism) *Utne Reader* 36 (November-December 1989): 71(10)

Cox, Yvonne. "Citizens to save the environment." *Chatelaine* 63,4 (April 1990): 106(4)
Hager, Mary. "What you can do to save the Earth." (includes related information) *Consumers Digest* 29,2 (March-April 1990): 62(4)

Dobb, Edwin. "Catch the spirit" (environmental cleanup program by community volunteers) *Reader's Digest* 138,829 (May 1991): 94(5)

Dranov, Paula. "Last chance to save the planet." (includes information on celebrity-endorsed environmental causes) *Cosmopolitan* 208,2 (February 1990): 204(5)

Durning, Alan B. "Grass-roots groups are our best hope for global prosperity and ecology." *Utne Reader* 34 (July-August 1989): 40(8)

"Earth-conscious beauty." (ecologically sound grooming) *Seventeen* 49,9 (September 1990): 174(2)

"Earthen renewal begins at home." *People* 33 (Spring 1990): 42(2)

Gorney, Cynthia. "Green and sober: no sweat--this family was ready to kick its environmentally incorrect habits. Then withdrawal set in." *Mother Jones* 15,3 (April-May 1990): 44(7)

Hall, Doug. "How you can help improve the health of planet Earth." (plans and plants for the planet) (cover story) *Flower and Garden* 35,4 (August-September 1991): 27(3)

Halter, Jon C. "Low-impact camping; you can enjoy nature without ruining it for others." (Scout program: forestry) *Boys' Life* 80,3 (March 1990): 44(2)

Hartig, John H.; and Hartig, Patricia D. "Remedial action plans: an opportunity to implement sustainable development at the grassroots level in the Great Lakes Basin." *Alternatives* 17,3 (November-December 1990): 26(6)

Heiligman, Deborah. "How I helped save the planet; one mom describes her family's efforts to make a healthier place." *Parents* 66,4 (April 1991): 63(2)

Henderson, Karla A. "50 ways camps and campers can save the Earth." *The Camping Magazine* 63,5 (March 1991): 17(5)

Howells, Bob. "We can save the Earth." *Trailer Life* 50,4 (April 1990): 55(5)

Jackson, Donald Dale. "Confessions of a do-gooder." (practicing ecological responsibility) *Reader's Digest* 140,837 (January 1992): 41(2)

Jackson, Tom. "Earth-friendly family." (how one suburban family practices an environmental lifestyle) *Better Homes and Gardens* 69,4 (April 1991): 170(3)

Jackson, Tom. "What's right for our environment?" *Better Homes and Gardens* 69,8 (August 1991): 126(2)

Joseph, Lawrence E.; and Holing, Dwight. "Saving the Earth; the time has come for each of us to help. There is no alternative." *50 Plus* 30,4 (April 1990): 48(7)

Jubak, Jim; and D'Amico, Marie. "9 things you can do in the '90s to save the planet." (includes related information) *McCall's* 117,7 (April 1990): 34(4)

Kanner, Bernice. "Friends of the Earth." (marketing ecologically sound products) *New York* 23,2 (January 15 1990): 19(2)

Karlsberg, Elizabeth. "Rx for Earth ills: be part of the cure." (includes related articles on helping to clean up the Earth) *Teen* 34,8 (August 1990): 18(5)

Kerasote, Ted. "How to be a lobbyist; getting involved in a conservation movement is easier than you might think." (column) *Sports Afield* 202,3 (September 1989): 20(3)

King, Chris Savage. "Shopping for salvation." (environmentally friendly consumer goods) *New Statesman Society* 3,103 (June 1 1990): 26(3)

Laycock, George. "How to help save our home planet." *Boys' Life* 81,5 (May 1991): 24(7)

Loomis, Christine. "What you can do to save the Earth." (environmental protection activities for the whole family) *Parents* 65,8 (August 1990): 98(3)

Martin, Brian. "What's your problem?" (selecting environmental issues) *Alternatives* 16,4 (March-April 1990): 88(5)

"Mary Ellen says 'It's possible to clean up without making the environment dirtier.' " *Woman's Day* (January 15 1991): 93(2)

Moser, Penny Ward. "How I saved the Earth in one week." *Life* 11,14 (December 1988): 165(5)

Reed, Susan. "Want an ecologically correct house? Architect Michael Reynolds builds Earthships out of beer cans and tires." *People* 35,1 (January 14 1991): 105(3)

Schubert, John. "Trail skills." (the Student Conservation Association is helping the U.S. Forest Service to maintain wilderness trails) *Backpacker* 17,5 (August 1989): 8(2)

"Simple ways you can help save the Earth." (tips for the concerned citizen) *Reader's Digest* 136,818 (June 1990): 135(4)

Sobol, Ken; Sobol, Julie Macfie; and Butterill, John. "Chalk up one small victory." (students clean the Moira river) *Canadian Geographic* 111,2 (April-May 1991): 30(5)

Sombke, Laurence. "What you can do: how to be a friend to the Earth." (practical tips on protecting the environment) *New York* 23,15 (April 16 1990): 44(3)

Stern, Januarye; and Stern, Michael. "It's hard to be green." (environmentally safe habits) (includes bibliography) *Metropolitan Home* 21,5 (May 1990): 93(3)

Textor, Ken. "Are you pitching in? Environmental management is everyone's responsibility. Here are a dozen ways cruising sailors can pitch in for a cleaner Earth." *Cruising World* 17,4 (April 1991): 97(2)

Villeneuve, Claude. "The citizen and the environment." (environment and development: a global responsibility and commitment) *UNESCO Courier* (November 1991): 16(3)

Warshaw, Robin. "Save money--and the environment too." *Woman's Day* (November 5, 1991): 28(2)

Wasik, John. "Put your purchasing power to work." (becoming an environmentally friendly consumer) *Better Homes and Gardens* 68,11 (November 1990): 114(2)

"Waste not! 32 little things you can do at home to save the Earth."*Redbook* 175,2 (June 1990): 130(4)

Watts, Patti. "Rising star in social marketing." (cleaning the environment with a woman's touch) *Executive Female* 13,6 (November-December 1990): 8(2)

Williamson, Lonnie. "Look, listen, and litigate." (Hudson River Fishermen's Association) *Outdoor Life* 175 (January 1985): 50(2)

Environmental and professional journals (annotated)

Desai, Uday. "Public participation in environmental policy implementation: Case of the Surface Mining Control and Reclamation Act." *American Review of Public Administration* 19 (March 1989): 49-65.
   "Describes the use of various mechanisms by participants in one major environmental policy, the Surface Mining Control and Reclamation Act. Our data indicate that the enforcement stage of policy implementation generates most individual citizen involvement, and citizen complaints are widely used mechanisms for citizen participation. The paper concludes with thoughts on why extensive legislative provisions for citizen participation are nevertheless justified."

Gelpe, Marcia R. "Citizen boards as regulatory agencies." *Urban Lawyer* 22 (Summer 1990): 451-483.
   Concludes that citizen boards (a group of lay people who are not full time government employees, who constitute the body responsible for some or all decisions of an administrative agency) "provide significant advantages in making government decisions clearer to the public and giving the public greater access to the decision-making process. The advantages presented by citizen boards should be preserved while their disadvantages are reduced."

Pierce, John C.; Lovrich, Nicholas P.; and Matsuoka, Masahiko. "Support for citizen participation: a comparison of American and Japanese citizens, activists and elites." *Western Political Quarterly* 43 (March 1990): 39-59.
   "Examines the dynamics underlying contemporary demands for broader citizen

involvement in environmental policy formation processes in two postindustrial nations. The subjects studied are Japanese and American citizens, activists, and elected/administrative elites in two comparable locations."

Ruben, Barbara. "Good intentions." *Environmental Action* 24 (Fall 1992): 28-33. "Are intentional communities and eco-villages the route to ecotopia or escapism? They are about how you can very nearly extricate yourself from your car for good. How you and your neighbors can grow much of your own food without pesticides, fertilizers or a big-bucks factory farm."Though a major reshuffling of the world's population into rural eco-villages is unfeasible, these ecotopias may teach us about making our own communities more sustainable.

Environmental and professional journals (unannotated)

Fisher, David. "Living conservatively: the greening of America begins at home." (column) *Health* 22,1 (January 1990): 30(2)

Hannum, Kristen; and Juniper, Christopher. "Shades of green; consumer shopping as if the Earth counted." *Buzzworm: The Environmental Journal* 2,1 (January-February 1990): 36(6)

Hutchison, Sue. "How to protect yourself from your environment." (steps to safeguard oneself) *National Wildlife* 28,5 (August-September 1990): 30(13)

Lawren, Bill. "How safe is your world? You have a right to know." (Emergency Planning and Community-Right-To-Know Act of 1986) *National Wildlife* 28,2 (February-March 1990): 18(2)

Lewis, Sanford J. "Turning industrial polluters into good neighbors." (includes a short list of resource organizations for grassroots environmental activists and environmental restoration) *Whole Earth Review* 66 (Spring 1990): 116(4)

Loomis, Laura. "Taking action; a guide to help you get involved in park protection from the roots up." *National Parks* 62,1-2 (January-February 1988): 36(4)

MacAdam, Gil. "Environmental protection through citizen action." (Broward County, Florida) *Parks & Recreation* 25,10 (October 1990): 46(6)

Moll, Gary. "The best way to plant trees." *American Forests* 96,3-4 (March-April 1990): 61(4)

Pierce, Jim. "Environmental Task Force." (new organization to help citizen groups fight environmental dangers) *Environment* 28 (December 1986): 3(2)

Schmidt, David D. "Voting on the environment." *Technology Review* 90 (August-September 1987): 15(2)

Steinhart, Peter. "Waterway watchdogs: citizen keepers enforce laws the government can't." *Audubon* 92,6 (November 1990): 26(5)

Stranahan, Susan Q. "Putting the heat on polluters; increasingly, frustrated citizens are banding together to keep their neighborhoods free of chemical contamination." *National Wildlife* 23 (August-September 1985): 30(4)

Udall, Stewart L.; and Dunlap, Riley E. "Local electoral clout." (public opinion on the environment) (letter to the editor) *Environment* 29,9 (November 1987): 2(2)

Van Putten, Mark, et al. "People who make a difference: Americans from all walks of life are taking extraordinary steps to safeguard wildlife and the environment." (Frederick L. Brown, Marcy Benstock, Dale Shields, Sam King, Dorothy Green, Marcy Golde) *National Wildlife* 28,6 (October-November 1990): 42(7)

Varnes, Paul R.; and Holland, Steve. "Looking out for the environment." *Parks & Recreation* 26,9 (September 1991): 84(4)

"Volunteer opportunities and environmental jobs." *Buzzworm: The Environmental Journal* 2,2 (March-April 1990): 23(5)

Wann, David. "Designing the environment." (developing technologies and lifestyles to protect and live in a healthier world) *Buzzworm: The Environmental Journal* 2,6 (November-December 1990): 18(2)

Zuckerman, Seth. "Living there." (bioregionalism) *Sierra* 72 (March-April 1987): 61(7)

Reports

U.S. Citizens Network on the United Nations Conference on Environment and Development. *U.S. Citizens Network on UNCED brochure.* Washington, D.C.: The Network 1991.

> Describes UNCED and the important role that citizen groups have to play in the 1992 U.N. conference. Summarizes the five major functions of the U.S. Citizens Network on UNCED: clearinghouse; outreach; policy development; policy input; and international cooperation.

Bibliographies

Frankena, Frederick. *Citizen participation in public administration: a bibliography.* Monticello, Ill.: Vance Bibliographies, 1987. 25 p. (Z7164.P79 F73 1987)

Howell, Robert E. *Who will decide?: the role of citizen participation in controversial natural resource and energy decisions*. Monticello, Ill.: Vance Bibliographies, 1981. 30 p. (Z7164.N3 H68)

Jaffray, B. *Public involvement in environmental decisionmaking: an annotated bibliography*. Chicago: Council of Planning Librarians, 1981. 48 p. (CPL bibliography no. 61)

Sells, Jennifer. "1,851 (and counting) ways to save the Earth." (selection of the best and brightest books) *Utne Reader* 39 (May-June 1990): 93(5)

Woodwell, George M. "Do the right thing." (books to help the environmentally concerned consumer) *Natural History* 99,5 (May 1990): 84(3)

REFERENCE WORKS ON THE ENVIRONMENTAL MOVEMENT

(See also *Global Environmental Crisis: Reference*) 242

Books (annotated)

Crump, Andy. *Dictionary of environment and development: people, places, ideas and organizations*. London: Earthscan, 1991. 272 p. (GE10 .C78 1991)
>   Provides definitions, acronyms, organizations, events, and people associated with the environment worldwide. It includes statistics and cross references to make each concept as clear as possible.

Deziron, Mireille; and Bailey, Leigh. *A directory of European environmental organizations*. New York: B. Blackwell Reference, 1992. 177 p.
>   An international directory to European governmental and nongovernmental organizations. Highlights the independent and cooperative efforts of these organizations, including the OECD. Presents flowcharts to illustrate the interrelationships and offers addresses and background information for each organization.

Dictionary of environmental quotations. Compiled by Barbara K. Rodes and Rice Odell. New York: Simon and Schuster, 1992. 335 p.
>   A collection of almost four thousand quotations relating to the environment. There is an alphabetical arrangement of 143 categories that are further subdivided chronologically. The sources range from current leaders in the environmental movement to classical Greece.

Education for the Earth: a guide to top environmental studies programs. Princeton, N.J.: Peterson's Guides, 1993. 175 p.

Profiles over 100 colleges and universities that offer programs in environmental studies. Includes general information about the institution and specific information about the environmental program and the type of career at which it is aimed. Contains essays which examine the future of environmental careers in business, labor, health, and education.

Elkington, John. *The green consumer*. New York: Penguin Books, 1990. 342 p.: ill (HC110.C6 E44 1990)
Lists products and services that work toward environmental preservation.

Ferguson, Marilyn. *Who is in service to the Earth: people, projects, organizations, key words and forty-one visions of a positive future*. Waynesville, N.C.: Visionlink Educational Foundation, 1991.
A compilation of forty-one essays by environmental activists. Provides a list of programs and projects individuals can join or start in their own communities.

Hill, Karen; and Piccirelli, Annette, eds. *Gale environmental sourcebook: a guide to organizations, agencies, and publications*. Detroit, Mich.: Gale Research, 1991.
Lists information concerning national and international organizations, government agencies, research facilities and educational programs, publications and information services, scholarships and awards, and sources for green consumers. Contains a list of environmental contacts at U.S. businesses and another list of environmentally sound products.

Information please environmental almanac 1992. Compiled by World Resources Institute. Boston, Mass.: Houghton Mifflin, 1992. 544 p.
Discusses the issue which relate to the environmental crisis on both a national and global scale. Outlines activities and attitudes that relate to environment, possible future problems, and possible solutions. Offers governmental statistics on a variety of issues, including food, energy, water, pollution, and hazardous materials. Sixty-four U.S. cities are ranked on an assortment of environmental issues. The international section focuses on global warming and ozone depletion and ranks 164 countries.

Lanier-Graham, Susan D. *The nature directory: a guide to environmental organizations*. New York: Walker and Company, 1991. 190 p. (TD169 .L36 1991)
"Gives an overview of environmental organizations. From the large, highly organized to the small grassroots organizations, the directory provides basic understanding of each organization. Part I discusses the environmental problems addressed by the environmental groups. Part II is a sampling of 120 environmental groups. Part III is a guide to personal involvement in the environmental issues. A selected resource list follows Part III including a special guide to children's books on the environment."

Meech, Karen Napoleone. *Environmental industries marketplace: a guide to U.S. companies providing environmental regulatory compliance products and services.* Detroit, Mich.: Gale, 1992. 779 p.

> Presents pertinent information for approximately 11,000 companies providing products or services to aid in environmental compliance. Information includes addresses, telephone numbers, and brief descriptions. The companies are listed alphabetically, but there is also a geographic index.

Seredich, John, ed. *Your resource guide to environmental organizations.* Irving, Calif.: Smiling Dolphins Press, 1991. 514 p. (TD171 .Y68 1991)

> A directory to 150 environmentally sensitive U.S. organizations. Lists the purpose, program, accomplishments, publications, member benefits, history, address, and telephone and fax numbers for each organization. Includes members of EcoNet, an electronic-mail service for the environmental community. Contains biographies of fourteen leaders in the environmental movement and numerous indexes.

Stein, Edith Carol. *The environmental sourcebook.* New York: Lyons and Burford, 1992.

> Focuses on eleven environmental issues: global warming, climate and atmosphere, oceans, biodiversity, water, solid waste, hazardous substances, endangered lands, development, population, agriculture, and energy. Each issue is discussed and followed by a listing of relevant literature and organizations, including addresses and telephone and fax numbers. Offers information for funding and grant-making foundations.

Books (unannotated)

Advances in technology provide environmental solutions: a user-friendly guide to the latest technology. Dubuque, Iowa: Kendall/Hunt, 1990. 71 p. (TD170.2 .A38 1990)

Basta, Nicholas. *Environmental jobs for scientists and engineers.* New York: Wiley, 1992. 228 p.: ill. (TD170.2 .B37 1992)

Basta, Nicholas. *The environmental career guide: job opportunities with the Earth in mind.* New York: Wiley, 1991. 195 p.: ill. (TD170 .B38 1991)

Miller, Louise. *Careers for nature lovers & other outdoor types.* Lincolnwood, Ill.: VGM Career Horizons, 1992. 123 p. (S945 .M55 1992)

Shapiro, Stanley Jay. *Exploring environmental careers.* Rev. ed. New York: Rosen Publishing Group, 1985. 195 p.: ill. (TD170.2 .S45 1985)

Warner, David J. *Environmental careers: a practical guide to opportunities in the 90s.* Boca Raton: Lewis Publishers, 1992. 267 10 p. (TD170 .W37 1992)

Environmental and professional journals (unannotated)

Jones, David L.; and Brennan, Januaryet F. "Careers in environmental conservation."
Conservationist 46,2 (September-October 1991): 36(5)

CHAPTER 6

ENVIRONMENTAL PHILOSOPHY, ETHICS AND ECONOMICS

(Council on Environmental Quality Assessment) 12, 24
(See also *Health and the Environment: Vegetarianism*) 1297
(Related topics in *The Environmental Movement*) 419
(Related topics in *Animal Rights*) 581

ENVIRONMENTAL PHILOSOPHY,
ETHICS AND ECONOMICS

(Council on Environmental Quality Assessment) 12, 24
(See also *Health and the Environment: Vegetarianism*) 1297
(Related topics in *The Environmental Movement*) 419
(Related topics in *Animal Rights*) 518

SHORT LIST OF SOURCES
FOR GENERAL RESEARCH

Books

Anderson, Terry L.; and Leal, Donald R. *Free market environmentalism*. San Francisco: Pacific Research Institute for Public Policy; Boulder, Colo.: Westview, 1991. 192 p.: ill. (HC110.E5 A6665 1991)
> Focusing on free-market methods of improving environmental quality, this work challenges many traditional environmental policies. Compares a policy of incentives to one of active government participation. Includes examples of successful free-market environmental policy as applied to energy development, land and water management, and outdoor recreation. Also demonstrates the difficulties the free market has had in dealing with ocean management and pollution control. Contains footnotes and a bibliography.

Attfield, Robin. *The ethics of environmental concern*. 2d ed. Athens: University of Georgia Press, 1992. 249 p. (GF80 .A88 1991)
> Originally published in 1983 and reprinted here with a new preface, the author examines traditional attitudes toward nature and the degree to which they affect ecological problems. Analyzes the Judeo-Christian beliefs of man's dominion over Earth, the tradition of stewardship, and the belief in progress. Examines concerns of applied ethics and considers obligations to future generations.

Brennan, Andrew. *Thinking about nature: an investigation of nature, value, and ecology*. Athens: University of Georgia Press, 1988. 235 p.: ill. (QH540.5 .B74 1988)
> "Eco-humanism claims that, among the relevant frameworks that political and ethical thinking ought to use is one deriving from scientific ecology. My business in this book is to consider our attitude to nature, the moral standing, if any, of other natural things apart from ourselves, and the question of whether the

biological sciences--in particular ecology--can give us the insight or information that may help us plot a sensible strategy for our future dealings with nature."

Cairncross, Frances. *Costing the Earth: the challenge for governments, the opportunities for business*. Boston: Harvard Business School Press, 1992. 341 p.: ill. (HC79.E5 C27 1992)
    Maintaining that government and the private sector have an equal responsibility in promoting environmental solutions, this work examines such topics as pollution taxes and permits, conservation incentives, sustainable development, pricing techniques, government subsidies, energy efficiency, the potential profits of environmentally conscious businesses, and international issues. A good overview that bridges the gulf between environmentalism and economics.

Devall, Bill; and Sessions, George. *Deep ecology: living as if nature mattered*. Salt Lake City, Utah.: Peregrine Smith, 1985. 266 p.
    Deep ecology, first named by Arne Naess in 1973, contrasts with reform ecology, resource conservation, preservation, and ecological sensibility in that deep ecology attempts to appreciate the natural world for its inherent values apart from any benefits it might provide humanity. By entering into a harmonious relationship with all beings so as to realize their intrinsic goodness, we become more fully realized individuals within a larger system.

Dower, Nigel, ed. *Ethics and environmental responsibility*. Brookfield, Vt: Gower, 1989. 146 p.
    Topics include the link between the environmentalist perspective and global interdependence; biocentrism versus human-centered environmentalism; obligations to future generations; animal rights; democracy and environmental policy; and nuclear power.

Engel, J. Ronald, ed. *Ethics of environment and development*. Tucson: University of Arizona Press, 1990.
    These essays discuss the complexity of the ethics of sustainability as complicated by the opposing views of science and traditional concepts of nature. Presents the views of leaders from various religions, including Christianity which often perceives environmentalism to be in conflict with Christian theology.

Hargrove, Eugene C., ed. *Religion and environmental crisis*. Athens: University of Georgia Press, 1986. 222 p. (GF80 .R45 1986)
    Examines the moral debate surrounding the relationship between humans and their environment in a historical and metaphorical context, East and West, ancient and modern. Explores ways in which religion can help people live ethically in an endangered world.

Krutch, Joseph Wood. *The voice of the desert*. New York: Morrow, 1954.
    This classic work explains how observing life through nature--the desert in this case--reveals a world far more complex than the sophisticated alienation of a

pessimistic urban philosopher. By questioning the definitions of human consciousness, Krutch lays the groundwork for belief as it emerges from the natural world.

Plant, Judith, ed. *Healing the wounds: the promise of ecofeminism.* Philadelphia, Pa.: New Society, 1989.
Maintaining that the environmental and women's movements have much in common, this work is an excellent introduction to this cutting-edge perspective. Includes an analysis of the women's movement in India and how opposition to clear cutting in Alaska is similar to American witchcraft. Contains a call for the reestablishment of sacred groves where female and male power is balanced and a critique of the "deep ecology" movement.

Rifkin, Jeremy. *Biosphere politics: a new consciousness for a new century.* New York: Crown, 1991. 388 p. (GF21 .R52 1991)
Argues that the enclosure of land during late medieval England was the beginning of the social process that has led to modern market forces. The nation-state depended on the war machine to enclose and protect territory, which became the symbol of security. Our dependence on machines to protect us led to human isolation from the holistic concept of life and nature, and a disassociation from their animal nature. However, biosphere politics will cause the system of global enclosure to break down as nations realize their dependence on the biosphere for survival.

Swanson, Timothy; and Barbier, Edward B., eds. *Economics for the wilds.* Washington, D.C.: Island Press, 1992.
Argues that--in contrast to the view that wilderness should never be used for economic gain--an economic strategy that values the resources of the wilds offers the best long-term security for them. Providing communities with incentives to properly manage their resources will ensure that people respect their natural resources.

General interest periodicals

"Survey of the environment: costing the Earth." *Economist* 312 (September 2-8, 1989): 1-18.
Argues that sensible economics can avoid much environmental damage, but that powerful lobbies skew the system. Tax breaks, subsidies, and trade protection put money into the pockets of vocal, well-organized groups at the expense of consumers. Rich democracies frequently give special treatment to groups such as big farmers or logging companies whose activities cause environmental damage.

Eisenberg, Evan. "The call of the wild: nature's four lessons for ecologists." *New Republic* 202 (April 30, 1990): 30-38.

"For some ecologists, global warming is a spiritual crisis, for others it is a problem in science and public policy. But there are paths between the Deep Ecologists' apocalyptic pessimism and the Planet Managers' dreams of total control. The key is to let nature be our guide.

Guha, Ramachandra. "Toward a cross-cultural environmental ethic." *Alternatives* 15 (Fall 1990): 431-447.
    Examines the U.S. debate on environmental ethics, recasting it as a debate about social utopias; for every theory of human-nature interaction is itself embedded in a larger theory of humans in society. Focuses on three perspectives of the human-nature relationship (agrarianism, wilderness thinking, and scientific industrialism), arguing that each forms part of a larger philosophy of social reconstruction. Compares these environmental perspectives across two cultures-- India and the U.S.

Joseph, Lawrence E. "Britain's whole Earth guru." *New York Times Magazine* (November 23, 1986): 67, 95, 106.
    Discusses James Lovelock's theory of Gaia which proposes that the Earth is a self-controlling whole system in which narrow atmospheric limits are maintained.

Sale, Kirkpatrick. "Ecofeminism--a new perspective." *Nation* 245 (September 26, 1987): 302-305.
    "Two of the most potent and durable ideas of the 1960s--feminism and ecological politics--have begun to come together in a new and fruitful way at last. The resulting hybrid, 'ecofeminism,' has finally taken on a distinct life of its own and appears to be influencing a growing number of groups and movements."

Tokar, Brian. "Exploring the new ecologies: social ecology, deep ecology and the future of green political thought." *Alternatives* 15 (November-December 1988): 31-43.
    Discusses deepening divisions between social ecology, which "emphasizes the embeddedness of human consciousness in nature, a radical ecological critique of hierarchy and domination in society, and the historical unity of ecological and social concerns" and deep ecology, which "purports to speak more directly for the biosphere as a whole and seeks a better relationship between the human species and other forms of life."

Environmental and professional journals

Merchant, Carolyn. "Earthcare: women and the environmental movement." *Environment* 23 (June 1981): 6-13, 38-40.
    Discusses the simultaneous emergence of the feminist and the ecology movements; traces women's involvement with environmental protection; urges further participation in appropriate low-impact technology as essential to sustain future life on Earth.

Sattaur, Omar. "Cuckoo in the nest." *New Scientist* 116 (December 24-31, 1987): 16-18.
"Fifteen years ago, James Lovelock proposed his Gaia hypothesis, which regards the planet Earth as a living organism. Since that time the idea has been making people uncomfortable. The hypothesis speaks for the Earth and not necessarily for the human beings on it."

INTRODUCTORY SOURCES

Books (annotated)

Allison, Lincoln. *Ecology and utility: the philosphical dilemmas of planetary management.* Rutherford, N.J.: Fairleigh Dickinson University Press, 1991. 185 p. (GF21 .A45 1991)
Focuses on the modern environmentalist movement in England and maintains that movement's philosophical underpinnings are drawn from a variety of sources. Argues that utilitarian ideologies are as impractical as radical Green ones. Seeks to bring the Greens and the more moderate utilitarians together.

Attfield, Robin. *The ethics of environmental concern.* 2d ed. Athens: University of Georgia Press, 1992. 249 p. (GF80 .A88 1991)
Originally published in 1983 and reprinted here with a new preface, the author examines traditional attitudes toward nature and the degree to which they affect ecological problems. Analyzes the Judeo-Christian beliefs of man's dominion over Earth, the tradition of stewardship, and the belief in progress. Examines concerns of applied ethics and considers obligations to future generations.

Berry, Wendell. *The unsettling of America: culture and agriculture.* San Francisco: Sierra Club, 1977.
Maintains that ecological responsibility can result only from responsible food production. The author is a farmer, a poet, and a novelist.

Blackstone, William T., ed. *Philosophy and environmental crisis.* Athens: University of Georgia Press, 1985.
Seven philosophers address the value systems behind our actions and attempt to define the social, ethical, political and legal values of an environmental ethic.

Bormann, F. Herbert; and Kellert, Stephen R. *Ecology, Economics, Ethics: The Broken Circle.* New Haven, Ct.: Yale University Press, 1992.
Looking at environmental problems from ecological, economical, and ethical viewpoints, the twelve essays which comprise this volume address such issues as species diversity, agriculture, pollution, market mechanisms, and environmental values. Maintains that modern environmental problems are the result of a focus on short-term gain. Specific, practical proposals are included in each article.

Botkin, Daniel B. *Discordant harmonies: a new ecology for the twenty-first century.* New York: Oxford University Press, 1990.

> Maintains that scientific thought and myth are intrinsically related. For instance nature has been associated with everything from God to machines. But the focus here is the relatively new myths surrounding resource management. In an attempt to correct this, the author maintains we need to use the computer as a metaphor for nature, for it is both simultaneous and variable. Contains notes, a glossary, and an index.

Boyden, Stephen. *Western civilization in biological perspective: patterns in biohistory.* Oxford, England: Clarendon, 1987.

> Examines the evolution of Western society and maintains that man has evolved from a state of nature to a state of culture. Offers an ecological critique of modern society and the effects it has had on the biosphere.

Brennan, Andrew. *Thinking about nature: an investigation of nature, value, and ecology.* Athens: University of Georgia Press, 1988. 235 p.: ill. (QH540.5 .B74 1988)

> "Eco-humanism claims that, among the relevant frameworks that political and ethical thinking ought to use is one deriving from scientific ecology. My business in this book is to consider our attitude to nature, the moral standing, if any, of other natural things apart from ourselves, and the question of whether the biological sciences--in particular ecology--can give us the insight or information that may help us plot a sensible strategy for our future dealings with nature."

Caldwell, Lynton Keith. *Between two worlds: science, the environmental movement, and policy choice.* New York: Cambridge University Press, 1990. 224 p. (HC79.E5 C33 1990)

> Maintaining that the Earth is in a transition period between that of a dying planet and that of a sustained postmodern world, the author examines how science and technology have redefined sustainability. Instead of focusing on the Earth itself, however, this work analyzes the way people understand the Earth, as reflected in global political systems. He describes how environmentalism must be included in political and economic debates. Offers suggestions on how a political framework can effectively manage a postmodern world.

Callicott, J. Baird, ed. *Companion to "A Sand Country Almanac."* Madison: University of Wisconsin Press, 1987.

> These essays, produced for a centenary celebration of Aldo Leopold's birth, explain the significance, sources, and applications of Leopold's land ethic. It is stronger in its humanities--based on an understanding of Leopold's work--than in its scientific explanation of its importance.

Callicott, J. Baird. *In defense of the land ethic.* Albany: State University of New York Press, 1989. 325 p. (GF80 .C35 1989)

> Argues that environmental ethics pits economic determinists against animal liberationists against land ethicists in determining whether the environmental

ethic is based on individualism or holism. Citing Plato, Hume, and other philosophers, Callicott develops the theories first articulated by Aldo Leopold.

Callicott, J. Baird; and Ames, Roger T., eds. *Nature in Asian traditions of thought: essays in environmental philosophy*. Albany: State University of New York Press, 1989. 335 p.
Maintains that traditional Western philosophy can no longer cope with the environmental crisis and examines the proposition that Asian philosophies may be better equipped to meet the challenges of the modern age. Discusses the philosophies of China, Japan, and India, as well as Buddhism.

Cooper, David E.; and Palmer, Joy A., eds. *Environment in question: ethics and global issues*. New York: Routledge, 1992. 256 p.
Assesses environmental issues--such as animal rights and radioactive waste--from an ethical or moral perspective. As a whole, the essays uphold the need for sustainable development and a healthy environment.

Dickens, Peter. *Society and nature: towards a green social theory*. Philadelphia, Pa.: Temple University Press, 1992. 203 p.: ill. (HM206 .D52 1992)
Focusing on the alienation of human beings from the natural world and the place of nature in their "deep mental structures." Uses a Marxist perspective to address a central debate in contemporary social science regarding the role of culture in the environmental movement.

Dower, Nigel, ed. *Ethics and environmental responsibility*. Brookfield, Vt: Gower, 1989. 146 p.
Topics include the link between the environmentalist perspective and global interdependence; biocentrism versus human-centered environmentalism; obligations to future generations; animal rights; democracy and environmental policy; and nuclear power.

Eckersley, Robyn. *Environmentalism and political theory: toward and ecocentric approach*. Albany, N.Y.: State University of New York, 1992. 274 p.
Examines the relationship between traditional and green political thought. Argues for a comprehensive ecocentric political theory. Contains a bibliography.

Ehrlich, Gretel. *Islands, the universe, home*. New York: Viking, 1992.
These essays and meditations explore the link between geography and imagination. The author seeks to understand the human need for a sense of place. Threads run throughout her writing include time and landscape, primal instincts about the Earth, and bondage to the Earth. She meditates on the natural world and in doing so confirms the value and necessity of meditation.

Engel, J. Ronald, ed. *Ethics of environment and development*. Tucson: University of Arizona Press, 1990.
These essays discuss the complexity of the ethics of sustainability as complicated

by the opposing views of science and traditional concepts of nature. Presents the views of leaders from various religions, including Christianity which often perceives environmentalism to be in conflict with Christian theology.

Evernden, Neil. *The natural alien: humankind and environment.* Toronto: University of Toronto Press, 1985.
> Reappraises the international environmental movement and the characteristics inherent in Western industrial societies that defeat environmentalism.

Frome, Michael. *Conscience of a conservationist: selected essays.* Knoxville: University of Tennessee Press, 1989.
> This collection of conservation essays depicts the gradual attitudinal change of the author: from quiet gentility to outright anger. This change of course reflects the growing evidence of environmental degradation which has occurred over the years. Names various groups including the forestry service and individuals for failing in their duties to protect the environment. The author, however, maintains that what natural resources we still have can and must be saved for the creatures of the Earth, including humans.

Gunn, Alastair S.; and Vesilind, Aarne. *Environmental ethics for engineers.* Chelsea, Mich.: Lewis, 1986. 155 p.
> Some of the ideas addressed include the following: there may be values that transcend society with which engineers must learn to cope; that engineers are in a strategic position to understand environmental problems and to influence environmental policy; that the Christian idea of man as master of the universe is in opposition to the environmental idea that man is the steward of nature; that nature has intrinsic value apart from its value to humans; that man might not be as superior as he has often thought; and that competitive enterprise has caused environmental disaster.

Hargrove, Eugene C. *Foundations of environmental ethics.* Englewood Cliffs, N.J.: Prentice-Hall, 1989. 229 p.
> Asks why Western philosophers have been so unreceptive to environmental philosophy. Addresses the Western concepts of property rights and the importance of individualism, and how these concepts are linked to attitudes toward the natural world. For specialists.

Holmes, Rolston. *Environmental ethics: duties to and values in the natural world.* Philadelphia, Pa.: Temple University Press, 1988. 391 p.
> Holmes develops complex arguments that include the following: (1) Although there is a duty within a culture to alleviate suffering, in an ecosystem suffering is permitted but should be no greater than ecological functional suffering. (2) Humans are superior to all nonhumans. (3) All living creatures are intrinsically valuable. (4) Destroying species is like tearing pages out of an unread book. (5) We have a more compelling duty to existing individuals than to future ones. The

latter sections deal with more practical ethical matters, such as the obligation of business to the environment. For specialists.

Krutch, Joseph Wood. *The voice of the desert*. New York: Morrow, 1954.
This classic work explains how observing life through nature--the desert in this case--reveals a world far more complex than the sophisticated alienation of a pessimistic urban philosopher. By questioning the definitions of human consciousness, Krutch lays the groundwork for belief as it emerges from the natural world.

Lane, John. *The living tree: art and the sacred*. Bideford, England: Green Books, 1988.
The author, who is an artist, argues for a better state of social and spiritual ecology. Maintains that while economic individualism, "capitalism," has led to the destruction of the soul and psyche, the individual can be reintegrated through nature. Attacks modern art as anti-ecological and, like William Morris in the nineteenth century, believes that social ills can be healed by returning the spirit to nature for instruction.

Laszlo, Ervin. *The age of bifurcation: understanding the changing world*. New York: Gordon and Breach, 1991. 126 p.
Maintains that the world will undergo a major cultural and environmental transformation, or bifurcation, in the next ten years. Argues that the status quo is simply unsustainable. Contains an index and bibliography.

Lembke, Janet. *Dangerous birds: a naturalist's aviary*. New York: Lysons and Burford, 1992. 192 p. (QL677.75 .L46 1992)
These essays focus on "ecological morality." One discusses the author's realization that a neighbor is eating robins. Another ponders humanity's role in the natural world. A sensitive study.

Naess, Arne. *Ecology, community and lifestyle: outline of ecosophy*. Cambridge: Cambridge University Press, 1989. 223 p.: ill. (GF21 .N34 1989)
In laying out his philosophy of ecology, Naess discusses the concepts behind biospherical egalitarianism, identification, self realization, and nonviolence. Naess's ideas might lay the groundwork for new phiolosophies for the twenty-first century.

Nash, Roderick. *The rights of nature: a history of environmental ethics*. Madison: University of Wisconsin Press, 1989. 290 p. (GF80 .N36 1989)
In tracing the history of environmental ethics, the author finds precursors of the animal rights and deep ecology movements in the natural rights tradition and the work of various little-known individuals. Discusses the development of environmental ethics as a subfield, makes a comparison between the nineteenth-century abolitionist movement and modern radical environmentalism, and discusses civil disobedience and nonviolent resistance. Includes an extensive bibliography.

Norton, Bryan G. *Why preserve natural variety?* Princeton, N.J.: Princeton University Press, 1988. 278 p.

> Examines the philosophy of the "intrinsic value" of nature, in other words, its value in its own right without regard to its value to any other object. Identifies the converse of intrinsic value as "instrumental value," which can be divided into "demand value" (for the need it can satisfy) or a "transformative value" (for the preference it can change, for better or worse). Maintains that intrinsic value philosophy results in the decline of species preservation and that preservation should occur not with individuals but through a wide geographical system.

Oates, David. *Earth rising: ecological belief in an age of science.* Corvallis: Oregon State University Press, 1989. 255 p. (QH540.7 .O27 1989)

> Argues that "the instinctive individualism of the Western mind is undergoing a deep change" which will have profound effects on the West's social organization and on technology that is wreaking havoc on ecosystems. Traces the roots of ecological thought, rejects neo-Darwinism, and develops a philosophy centered around the interrelatedness of dependent organisms.

Odum, Eugene P. *Ecology and our endangered life-support systems.* 2d ed. Sunderland, Mass: Sinauer Associates, 1993.

> Argues that the exploitable pioneer mentality that championed individual rights and independence must give way to a less selfish, cooperative mentality based on interdependent societal rights. Sees cities as the dynamic center of society but also as parasites that depend on the land, and they therefore force themselves to the center of any land plan uses and ecological policies.

Oelschlaeger, Max. *The idea of wilderness: from prehistory to the age of ecology.* New Haven, Ct.: Yale University Press, 1991. 477 p. (GF21 .O34 1991)

> Chronicles the concept of nature through Western history, explaining how humans slowly stopped perceiving nature as something to depend upon and began to see it as something to exploit. Summarizes the philosophical challenges which must be met if environmental destruction is to be avoided.

Oelschlaeger, Max, ed. *Wilderness condition: essays on environment and civilization.* San Francisco, Calif.: Sierra Club, 1992. 345 p.

> These essays discuss the meanings and limitations of words like "nature" and "wilderness." Compares Chinese nature philosophies with Western ones. Contains extensive footnotes and bibliograhic references.

Palm, Risa I. *Natural hazards: an integrative framework for research and planning.* Baltimore, Md.: Johns Hopkins University Press, 1990. 184 p. (GB5014 .P35 1990)

> This general text focuses on linking individual responses to environmental risks to macro-level responses of society. Reviews numerous different perspectives used by geographers, though the author seems to be most sympathetic to a structuralist, Marxist viewpoint. More a work of social geography and political science.

Peccei, Aurelio; and Ikeda, Daisaku. *Before it's too late.* Tokyo: Kodansha International, 1984. 154 p.

> Peccei, former head of Fiat and Olivetti, and Ikeda, a Buddist monk, exchange ideas about man's place in nature. Peccei argues that people are trying to reshape the Earth as if they alone were supposed to inhabit it, whereas the planet is beautiful and generous because of its many forms of life. Man's desire for power is the fundamental evil that estranges him from nature and religion and is the most important single force that can reestablish man's peace and harmony with nature.

Rifkin, Jeremy. *Biosphere politics: a new consciousness for a new century.* New York: Crown, 1991. 388 p. (GF21 .R52 1991)

> Argues that the enclosure of land during late medieval England was the beginning of the social process that has led to modern market forces. The nation-state depended on the war machine to enclose and protect territory, which became the symbol of security. Our dependence on machines to protect us led to human isolation from the holistic concept of life and nature, and a disassociation from their animal nature. However, biosphere politics will cause the system of global enclosure to break down as nations realize their dependence on the biosphere for survival.

Roszak, Theodore. *The voice of the Earth.* New York: Simon and Schuster, 1992. 367 p.: ill. (BD581 .R69 1992)

> Exploring the correlation between environmental degradation and the uneasy state of the human psyche, this work maintains that the former has been caused by an outdated view of the world as mindless matter rather than an interrelated web of open, integrated systems. Discusses the anthropic principle, Deep Ecology, the Gaia hypothesis, and systems theory.

Scherer, Donald, ed. *Upstream/downstream: issues in environmental ethics.* Philadelphia, Pa.: Temple University Press, 1990. 242 p. (GF80 .U67 1990)

> Focusing on the environmental ethics of one entity polluting the resources of another, this work covers numerous related topics, including the rights of future generations, public policy and global warming, scientific predictions, international environmental law, compensation for requiring environmental safety modifications, and the use of a moral "as well as cost-benefit" analysis.

Schumm, Stanley A. *To interpret the Earth: ten ways to be wrong.* New York: Cambridge University Press, 1991. 133 p.: ill. (QE40 .S38 1991)

> This definitive work analyzes the philosophical and intellectual problems involved in interpreting the Earth, with particular emphasis given to discussing and challenging the scientific methods of the field scientist. The four chapters are the results of lectures originally delivered in 1982 and cover the topics of diagnosing the Earth, the scientific method, problems of explanation and extrapolation, and scientific approaches and solutions.

Sheldrake, Rupert. *The rebirth of nature: the greening of science and God*. London: Century, 1990. 215 p.: ill. (BL65.N35 S44 1990)

Traces the history of biology and the Mother Nature metaphor. Restates the Gaia hypothesis as being the purposeful regulation of conditions of the Earth by organisms. Discusses morphic fields and morphic resonance.

Snyder, Gary. *The practice of the wild*. Berkeley, Calif.: North Point, 1990.

As a writer of the late Beat Generation, the author discusses the spiritual aspects of environmental awareness as he examines man's relationship to nature.

Stone, Christopher D. *Earth and other ethics: the case for moral pluralism*. New York: Harper and Row, 1987. 280 p.

Argues that the further development of enviornmental law is dependent on important developments in moral philosophy. Until we can decide why we should preserve a natural object in a specific condition--the ethical rationale for preservation--we cannot set the boundaries for the legal rights of nonhuman entities, such as trees and rivers.

Taylor, Paul W. *Respect for nature: a theory of environmental ethics*. Princeton, N.J.: Princeton University Press, 1986. 329 p. (GF80 .T39 1986)

Argues that a biocentric ethical system implies different and broader obligations than a human-centered system and maintains that moral principles in human ethics must be symmetrical with environmental ethics. However, although all living things have inherent worth, not all are subject to the attribution of "primary moral rights"; thus, some living things have a higher priority than others.

Thomas, Keith. *Man and the natural world: a history of modern sensibility*. New York: Pantheon Books, 1984. 427 p.

Chronicles the history of humanity's attitudes toward nature. Discusses how Tudor and Stewart England believed that nature was a resource for man to exploit. Traces the movement toward a more sensible attitude toward nature.

VanDeVeer, Donald; and Pierce, Christine, eds. *People, penguins, and plastic trees*. Belmont, Calif.: Wadsworth, 1986. 268 p. (QH75 .P46 1986)

Explores man's duty to animals and different views on the morality of animals. Examines nonanimal species and the wilderness, including conservation and preservation thinking. Discusses the foundations of environmental ethics, focusing on the conflicts between traditional and holistic approaches. Concludes with an examination of the relationship between economics, ecology, and ethics.

Winner, Langdon. *The whale and the reactor: a search for limits in an age of high technology*. Chicago: University of Chicago Press, 1986. 200 p. (T14 .W54 1986)

Argues that technology cannot be understood without placing it in a social and political context, which can provide perspective to the issues of justice, democracy, power, and freedom. Maintains that technologies not only facilitate human actions but reshape human activities and their meaning. Consequently, we must constantly evaluate the moral limits of a modern technology whose limits seem boundless and uncontrollable.

Young, John. *Post environmentalism*. London: Belhaven, 1990. 230 p.
Examines environmentalist writings from the 1960s onward, links environmentalism to ecology and other political and moral issues, and tries to forge a new environmental philosophy, which the author terms post environmentalism. Maintains that this "new global morality" will bring together everyone from animal rights activists to those who advocate zero population growth.

Young, John. *Sustaining the Earth*. Cambridge, Mass.: Harvard University Press, 1990. 225 p. (QH75 .Y68 1990)
Analyzes many of the current environmental issues and offers reassuring evidence that the situation is not without hope. Discusses such issues as the sustainability of traditional, non-Western cultures; the sources for an environmental ethic within Western culture; and the link between sustainable development and equality. Believes that environmentalism will become a plank of all party platforms.

Books (unannotated)

Ahrens, John H. *Preparing for the future: an essay on the rights of future generations.* Bowling Green, Ohio: Bowling Green State University, Social Philosophy & Policy Center, 1983. 44 p. (GF80 .A36 1983)

Allen, T. F. H. *Hierarchy: perspectives for ecological complexity*. Chicago: University of Chicago Press, 1982. 310 p.: ill. (QH541 .A45 1982)

Allen, T. F. H. *Toward a unified ecology*. New York: Columbia University Press, 1992.

Animal rights, environmental ethics debate: the environmental perspective. Albany: State University of New York Press, 1992. 273 p. (HV4711 .A575 1992)

Bennett, Jane. *Unthinking faith and enlightenment: nature and the state in a post-Hegelian era*. New York: New York University Press, 1987. 166 p. (B802 .B445 1987)

Beyond spaceship Earth: environmental ethics and the solar system. San Francisco: Sierra Club Books, 1986. 336 p., 4 p. of plates: ill. (some color) (QB501.5 .B49 1986)

Bookchin, Murray. *The philosophy of social ecology: essays on dialectical naturalism*. New York: Black Rose Books, 1990. 198 p. (GF80 .B66 1990)

Bowers, C. A. *Education, cultural myths, and the ecological crisis: toward deep changes*. Albany, N.Y.: State University of New York Press, 1992.

Daniels, Neil M. *The morality maze: an introduction to moral ecology.* Buffalo, N.Y.: Prometheus Books, 1991. 227 p. (BJ1012 .D28 1992)

Drengson, Alan R. *Beyond environmental crisis: from technocrat to planetary person.* New York: P. Lang, 1989. 259 p. (GF21 .D73 1989)

Dubos, Rene J. *The world of Rene Dubos: a collection from his writings.* New York: Henry Holt, 1990. 418 p. (QR6 .D8325 1990)

Dwelling, place, and environment: towards a phenomenology of person and world. New York: Columbia University Press, 1989. 310 p.: ill. (GF21 .D83 1989)

Earthbound: new introductory essays in environmental ethics. Philadelphia, Pa.: Temple University Press, 1984. 371 p. (GF80 .E16 1984b)

Easwaran, Eknath. *The compassionate universe.* Petaluma, Calif.: Nilgiri Press, 1989. 188 p. (GF80 .E18 1989)

Ecosystem health: new goals for environmental management. Washington, D.C.: Island Press, 1992.

Environment in question: ethics and global issues. New York: Routledge, 1992. 256 p.: ill., maps (GF80 .E57 1991)

Environmental ethics: philosophical and policy perspectives. Burnaby, Canada: Institute for the Humanities/SFU Publications, 1986. 199 p. (HC79.E5 E5774 1986)

Environmental philosophy. Canberra, Australia: Department of Philosophy, Research School of Social Sciences, Australian National University, 1982. 385 p. (QH541.145 .E58 1982)

Environmental philosophy: a collection of readings. New York: University of Queensland Press, 1983. 303 p.: ill. (GF80 .E59 1983)

Ethical issues of population aid: culture, economics, and international assistance. New York: Irvington Publishers, 1981. 360 p.

Ethics and environmental responsibility. Brookfield, Vt.: Avebury, 1989. 146 p. (GF80 .E83 1989)

Ethics and the environment. Englewood Cliffs, N.J.: Prentice-Hall, 1983. 236 p. (HC79.E5 E75 1983)

Ethics and the environment. Lanham, Md. (Long Island Philosophical Society): University Press of America, 1992.

Ethics of environment and development: global challenge, international response. Tucson: University of Arizona Press, 1990. 264 p.: ill. (HD75.6 .E84 1991)

Finding home: writing on nature and culture from Orion magazine. Boston, Mass.: Beacon Press, 1992.

Flynn, Eileen P. *Cradled in human hands: a textbook on environmental responsibility.* Kansas City, Mo.: Sheed & Ward, 1991. 155 p. (GF80 .F57 1991)

Frankel, Boris. *The post-industrial utopians.* Cambridge, England: Polity, 1987. Distributed by B. Blackwell. 303 p.

Fuller, Robert C. *Ecology of care: an interdisciplinary analysis of the self and moral obligation.* Louisville, Ky.: Westminster/John Knox Press, 1992. 121 p. (BJ1475 .F85 1992)

Goldsmith, Edward. *The way: an ecological world-view.* Boston: Shambhala, 1993. Distributed by Random House.

Grundmann, Reiner. *Marxism and ecology.* New York: Oxford University Press, 1991. 324 p.: ill. (HB97.5 .G752 1991)

Hargrove, Eugene C. *Foundations of environmental ethics.* Englewood Cliffs, N.J.: Prentice Hall, 1989. 229 p. (GF80 .H37 1989)

Heidegger and ecological thinking. Kirksville, Mo.: Thomas Jefferson University Press, 1991.

Johnson, Lawrence E. *A morally deep world: an essay on moral significance and environmental ethics.* New York: Cambridge University Press, 1991. 301 p. (GF80 .J64 1991)

Jonesburg, Harry. *The waste streams of ignorance.* Dayton, Ohio: Les Livres, 1992. 199 p. (GF21 .J66 1992)

Land, the city, and the human spirit: America the beautiful--an assessment. Austin, Tex.: Lyndon Baines Johnson Library, Lyndon B. Johnson School of Public Affairs, Center for the Study of American Architecture, 1985. 146 p.: ill. (HN65 .L316 1985)

Learning to listen to the land. Washington, D.C.: Island Press, 1991. 282 p. (GF49 .L43 1991)

Lee, Keekok. *Social philosophy and ecological scarcity.* New York: Routledge, 1989. 425 p. (HD75.6 .L44 1989)

Macy, Joanna. *World as lover, world as self.* Berkeley, Calif.: Parallax Press, 1991. 251 p.: ill. (BQ4570.S6 M33 1991)

Martin, Calvin. *In the spirit of the Earth: rethinking history and time.* Baltimore, Md.: Johns Hopkins University Press, 1992. 157 p. (GN388 .M37 1992)

Maybury-Lewis, David. *Millennium: tribal wisdom and the modern world.* New York: Viking, 1992.

McCloskey, Henry J. *Ecological ethics and politics.* Totowa, N.J.: Rowman and Littlefield, 1983. 167 p. (GF80 .M37 1983)

McHarg, Ian L. *Design with nature.* New York: Wiley, 1992. 197 p.: (some color) ill. (HC110.E5 M33 1992)

Mellos, Koula. *Perspectives on ecology: a critical essay.* New York: St. Martin's, 1988. 178 p.: ill. (QH540.5 .M45 1988)

Merchant, Carolyn. *Ecological revolutions: nature, gender, and science in New England.* Chapel Hill: University of North Carolina Press, 1989. 379 p.: ill., maps (GF504.N45 M47 1989)

Merchant, Carolyn. *Radical ecology: the search for a livable world.* New York: Routledge, 1992.

Meyer, Art. *Earth-keepers: environmental perspectives on hunger, poverty, and injustice.* Scottdale, Pa.: Herald, 1991. 264 p. (GF80 .M48 1991)

Nature in Asian traditions of thought: essays in environmental philosophy. Albany: State University of New York Press, 1989. 335 p. (QH540.7 .N37 1989)

Norton, William. *Explorations in the understanding of landscape: a cultural geography.* New York: Greenwood, 1989. 201 p. (GF90 .N66 1989)

Ornstein, Robert E. *New world new mind: moving toward conscious evolution.* New York: Simon & Schuster, 1990, 1989. 302 p.

Passmore, John Arthur. *Man's responsibility for nature: ecological problems and Western traditions.* 2d ed. London: Duckworth, 1980. 227 p. (GF75 .P34 1980)

Pepper, David. *The roots of modern environmentalism.* Dover, N.H.: Croom Helm, 1984. 246 p.: ill. (QH540.5 .P47 1984)

Philosophy and the ecological problems of civilisation. Moscow: Progress Publishers, 1983. 410 p. (GF49 .P47 1983)

Plant, Christopher. *Turtle talk: voices for a sustainable future*. Philadelphia, Pa.: New Society Publishers, 1990. 133 p.: ill. (QH540.5 .P53 1990)

Responsibilities to future generations: environmental ethics. Buffalo, N.Y.: Prometheus Books, 1981. 319 p.: ill. (GF80 .R47)

Rifkin, Jeremy. *Biosphere politics: a cultural odyssey from the middle ages to the new age*. San Francisco, Calif.: HarperSanFrancisco, 1992.

Riker, John H. *Human excellence and an ecological conception of the psyche*. Albany: State University of New York Press, 1991. 239 p. (BJ1533.E82 R55 1991)

Rolston, Holmes. *Environmental ethics: duties to and values in the natural world*. Philadelphia, Pa.: Temple University Press, 1988. 391 p.: ill. (GF80 .R64 1988)

Rolston, Holmes. *Philosophy gone wild: environmental ethics*. Buffalo, N.Y.: Prometheus Books, 1989. 269 p.: ill. (QH540.5 .R649 1989)

Rowthorn, Anne W. *Caring for creation: toward an ethic of responsibility*. Wilton, Conn.: Morehouse, 1989. 163 p. (BT695.5 .R68 1989)

Russell, Peter. *The global brain: speculations on the evolutionary leap to planetary consciousness*. Los Angeles: J. P. Tarcher, 1983. Distributed by Houghton Mifflin. 251 p.: ill. (QH371 .R78 1983)

Sagoff, Mark. *The economy of the Earth: philosophy, law, and the environment*. New York: Cambridge University Press, 1988. 271 p. (HC110.E5 S34 1988)

Sale, Kirkpatrick. *Dwellers in the land: the bioregional vision*. San Francisco: Sierra Club Books, 1985. 217 p. (QH540.5 .S25 1985)

Schaefer, Paul. *Defending the wilderness: the Adirondack writings of Paul Schaefer*. Syracuse, N.Y.: Syracuse University Press, 1989. 250 p.: ill. (QH76.5.N7 S33 1989)

Science Action Coalition. *Environmental ethics: choices for concerned citizens*. Garden City, N.Y.: Anchor Press, 1980. 309 p. (HC79.E5 S3 1980)

Shrader-Frechette, K. S. *Environmental ethics*. Pacific Grove, Calif.: Boxwood Press, 1981. 358 p. (GF80 .S47 1981)

Spretnak, Charlene. *The spiritual dimension of green politics*. Sante Fe, N.M.: Bear, 1986. 95 p. (JN3971.A98 G72376 1986)

Stikker, Allerd. *The transformation factor: towards an ecological consciousness*. Rockport, Mass.: Element Books, 1991.

Taylor, Bob Pepperman. *Our limits transgressed: environmental political thought in America*. Lawrence: University Press of Kansas, 1992. 184 p. (HC110.E5 T37 1992)

Titmuss, Christopher *Spirit for change: voices of hope for a better world*. Alameda, Calif.: Hunter House, 1992.

Tuan, Yi-fu. *Morality and imagination: paradoxes of progress*. Madison, Wis.: University of Wisconsin Press, 1989. 209 p. (BJ1031 .T77 1989)

Ulanowicz, Robert E. *Growth and development: ecosystems phenomenology*. New York: Springer-Verlag, 1986. 203 p.: ill. (QH540.5 .U43 1986)

Wenz, Peter S. *Environmental justice*. Albany: State University of New York Press, 1988. 368 p. (HB523 .W46 1988)

Wright, Will. *Wild knowledge: science, language, and social life in a fragile environment*. Minneapolis: University of Minnesota Press, 1992. 236 p. (GF21 .W75 1992)

General interest periodicals (annotated)

Bookchin, Murray. "What is social ecology?" *Alternatives* 12 (Spring-Summer 1985): 62-68.
> Murray Bookchin has been active on the radical ecology scene for many decades. He has written numerous books and articles during that time. He founded and presently directs the Institute for Social Ecology in Vermont. Included is a description by John Ely of Bookchin's book "Ecology of Freedom".

Diamond, Jared. "The worst mistake in the history of the human race." *Discover* 8 (May 1987): 64-66.
> Maintains that the adoption of agriculture by our hunter-gatherer ancestors was a great mistake: "forced to choose between limiting population or trying to increase food production, we chose the latter and ended up with starvation, warfare, and tyranny."

Eisenberg, Evan. "The call of the wild: nature's four lessons for ecologists." *New Republic* 202 (April 30, 1990): 30-38.
> "For some ecologists, global warming is a spiritual crisis, for others it is a problem in science and public policy. But there are paths between the Deep Ecologists' apocalyptic pessimism and the Planet Managers' dreams of total control. The key is to let nature be our guide. Four organic principles show the way."

"Environmental ethics." *Alternatives* 12 (Winter 1985): 3-41.
> Contents include: "Moral concern and the ecosphere," by J. Livingston; "Moral

concern and animals," by B. Rollin; "The Canadian harp seal hunt: a moral issue," by L. Sumner; "Animal suffering and the ethics of eating meat," by R. Carter; "The paradox of environmental ethics," by D. Torgerson; and "Beyond the domination of nature: moral foundations of a conserver society," by L. Rubinoff.

Guha, Ramachandra. "Toward a cross-cultural environmental ethic." *Alternatives* 15 (Fall 1990): 431-447.
> Examines the U.S. debate on environmental ethics, recasting it as a debate about social utopias; for every theory of human-nature interaction is itself embedded in a larger theory of humans in society. Focuses on three perspectives of the human-nature relationship (agrarianism, wilderness thinking, and scientific industrialism), arguing that each forms part of a larger philosophy of social reconstruction.

Kohak, Erazim. "The relevance of Tolstoy: or Europe after Chernobyl." *Dissent* 34 (Winter 1987): 5-9.
> After viewing a modern version of the Faustian story, the author challenges the assumption that consumption is central to being human. The assumption is "reinforced daily by what the Americans boast about and the Soviets wish for. Yet that assumption has led us to a dead end. Perhaps we need to listen to a different America, that of Thoreau and Emerson, and to a different Russia, that of Tolstoy. We need to hear a different drummer."

Regan, Donald H. "Why we should preserve nature." *QQ* 2 (Fall 1982): 6-8.
> A condensed adaptation of "Duties of preservation." Presents a philosophical theory, which the author's book proposes, to govern our behavior towards endangered species and our ecosystem generally.

Turner, Frederick. "A field guide to the synthetic landscape: toward a new environmental ethic." *Harper's Magazine* 276 (April 1988): 49-55.
> Describes Greene Prairie "planted forty years ago by the ecological restorationist Henry Greene on forty acres of degraded Wisconsin farmland. It is part of the University of Wisconsin's arboretum, which all told has more than a hundred acres of thriving prairie."

General interest periodicals (unannotated)

Bookchin, Murray. "Freedom and necessity in nature: a problem in ecological ethics." *Alternatives* 13 (November 1986): 28(11)

Diamond, Jared. "The golden age that never was." (man and nature) *Discover* 9,12 (December 1988): 70(8)

Jager, Ronald. "The higher ecology: a well, a wall, and the interconnectedness of things." *Harper's Magazine* 270 (May 1985): 73(2)

Leopold, Aldo. "The land ethic." *Wilderness* 48 (Spring 1985): 4(11)

Livingston, John A. "Moral concern and the ecosphere." *Alternatives* 12 (Winter 1985): 3(5)

McKibben, Bill. "The mountain hedonist." (Gary Snyder's philosophy teaches harmony with nature) *The New York Review of Books* 38,7 (April 11, 1991): 29(4)

Rennie, J. C. "Ethical choice in food systems; balancing responsibility." *Vital Speeches of the Day* 56,5 (December 15, 1989): 143(5)

Rubinoff, Lionel. "Beyond the domination of nature: moral foundations of a conserver society." *Alternatives* 12 (Winter 1985): 37(12)

Russell, Nicholas. "Before the wheel." (pre-industrial era economic systems as models for sustainable systems) *History Today* 40 (March 1990): 5(3)

Sale, Kirkpatrick. "Bioregionalism - a sense of place." (new environmentalist theory) *Nation* 241 (October 12, 1985): 336(4)

Schullery, Paul. "Mother's mess; there's no such thing as a balance in nature. It's all up for grabs." *Backpacker* 16,5 (September 1988): 24(2)

Sears, Paul B. "'Coming to terms with environment about us.'" (ecologist discusses the future) *Science Digest* 87 (May 1980): 48(6)

Stegner, Wallace. "It all began with conservation; our first response to the New World's wild Eden was to break it--but seeds of a land ethic were planted early on." (20: A Special Issue on the Environment) *Smithsonian* 21,1 (April 1990): 34(10)

Taylor, Duncan M. "Disagreeing on the basics: environmental debates reflect competing world views." *Alternatives* 18,3 (January-February 1992): 26(8)

Torgerson, Douglas. "The paradox of environmental ethics." *Alternatives* 12 (Winter 1985): 26(11)

Waskow, Arthur. "From compassion to Jubilee." (ecology and social meaning) *Tikkun* 5,2 (March-April 1990): 78(4)

Weeden, Robert B. "An exchange of sacred gifts." (thoughts about ecology) *Alternatives* 16,1 (March-April 1989): 40(10)

Whitacare, David. "Environmentalists emerging; a philosophy of entitlement." *Vital Speeches of the Day* 55,2 (November 1, 1988): 44(4)

Environmental and professional journals (annotated)

Alexander, Nancy C. "Healing community: restoring creation." *Earth Ethics* 2 (Spring 1991): 1, 3-5.
> "Concern about environmental degradation must be linked to care about human communities if a sustainable future is to be achieved. This article shows the connection between those two concerns in the context of the global issues facing the 1992 United Nations Conference on Environment and Development."

"Ethics and the land." *American Land Forum Magazine* 6 (Summer 1986): 17-27.
> An edited version of a panel discussion convened by the American Land Resources Association, March 4, 1986 to address the questions: "Have we developed an American land ethic? What are the pressing issues today?" Includes "Aldo Leopold's Legacy" by C. E. Little.

Goodin, Robert E. "International ethics and the environmental crisis." *Ethics & International Affairs* 4 (1990): 91-105.
> "What is striking about the environmental crisis as it is currently understood is how genuinely global it is, in contrast to traditional environmental problems. The problems at the forefront of present environmentalist discussions are problems like the degradation of the ozone layer and the greenhouse effect." Aims "to use philosophical insights to assist us in deciding the appropriate structure of an international regime for resolving the full range of environmental problems that we now know we face."

Marchetti, Cesare. "Environmental problems and technological opportunities." *Technological Forecasting and Social Change* 30 (August 1986): 1-4.
> Argues that environmental problems are closely linked to religion; and while fairly simple solutions exist for the large ecological problems generated by the intrusion of humanity into the ecosphere, these solutions would require long-term cultural changes.

Rolston, Holmes, III. "Engineers, butterflies, worldviews." *Environmental Professional* 9,4 (1987): 295-301.
> "Natural systems, characterized by speciation, are engineering projects worthy of admiring respect--in the sense that they represent inventive, ingenious, trial and error solutions to problems in survival. Butterflies demonstrate engineering principles. Ecosystems are prolific and satisfactory communities in an objective sense. Culture superimposed on wild nature ought to seek an optimally satisfactory development that maximizes cultural values with minimal loss of natural values. Current environmental policy, though seemingly prohibitive, can liberate environmental professionals from narrow economic constraints and permit them to operate within this more comprehensive worldview. In symbolic as well as specific terms, engineers can and should count butterflies."

539

Environmental and professional journals (unannotated)

Slobodkin, L.B. "Intellectual problems of applied ecology." *BioScience* 38,5 (May 1988): 337(6)

Windle, Phyllis. "The ecology of grief." (care for the environment should be meaningful to people even on an emotional level) *BioScience* 42,5 (May 1992): 363(4)

Law journals

Bern, Martin. "Government regulation and the development of environmental ethics under the Clean Air Act." *Ecology Law Quarterly* 17,3 (1990): 539-580.
> "Describes the moral and ethical elements Congress considered and enacted in the Clean Air Act of 1970"; explains how an approach to environmental regulation that focuses on economic rationality and technological dependence retards the development of environmental ethics related to individual behavior; proposes a theoretical framework for the development of environmental ethics by drawing on utility-based and biocentric perspectives; identifies attributes of American culture that could develop these ethics by reinforcing individual freedom and environmentally sensitive behavior; applies this theory to air quality problems in the Los Angeles Basin.

Bibliographies

Anglemyer, Mary. *A search for environmental ethics: an initial bibliography*. Washington, D.C.: Smithsonian Institution Press, 1980. 119 p. (Z7405.N38 A53)

Anglemyer, Mary. *The natural environment: an annotated bibliography on attitudes and values*. Washington, D.C.: Smithsonian Institution Press, 1984. 268 p. (Z7405.N38 A52)

Davis, Donald Edward. *Ecophilosophy: a field guide to the literature*. San Pedro: R & E Miles, 1989. 137 p.: ill. (Z5322.E2 D38 1989)
> Defined as that which seeks the re-unification of humans and nature, ecophilosophy has become a subject of enough importance to justify this annotated bibliography of 334 sources, of which 280 are books. Most of the works were published in the last ten years. Contains an index.

Elbers, Joan S. *Changing wilderness values, 1930-1990: an annotated bibliography*. New York: Greenwood Press, 1991. 138 p. (Z5861 .E4 1991)

Videos

Voices of the land. Produced by C. McLeod. (Bullfrog Films. Oley, PA 19547). 21 min.

Two spiritual shaman--Red Ute, a Colorado Native American, and Dr. Emmett Alulu, a Hawaiian--speak of the spirituality of the land in their religions. Ecologist Dave Forman then emerges to explain their feelings in eco-philosophical terms. Thoughtful, sometimes brooding, this film is more personal testimony than compelling ecological evidence for preserving the integrity of spiritual lands.

GAIA HYPOTHESIS

Books (annotated)

Joseph, Lawrence E. *Gaia: the growth of an idea*. New York: St. Martin's Press, 1990. 276 p. (QH331 .J74 1990)
> Traces the history of James Lovelock's Gaia hypothesis, which maintained that the Earth is one entity. Follows the careers of those who founded the movement as well as the theoretical evidence for the hypothesis. Discusses a five-day conference on the subject, points out scientific spin-offs, and gives a history of the Earth from a Gaian perspective.

Lovelock, James. *The ages of Gaia: a biography of our living Earth*. New York: Norton, 1988. 252 p.: ill. (QH331 .L688 1988)
> Explains the Gaia hypothesis, which maintains the Earth is one single entity, with all of its individual plants, animals, and ecosystems closely interrelated.

Schneider, Stephen H.; and Schneider, Penelope J., eds. *Scientists on Gaia*. Boston: MIT Press, 1992.
> The Gaia hypothesis postulates that Earth's physical and biological processes are linked and that life is an active determinant of its own environment. The sum total of living organisms is highly adaptable to changing conditions and has been able to counteract physical abuse to Earth. These forty-four essays addresses the many implications of Gaia and argue that the Earth is better able to decide its own remedies than humans. From tectonics to the greenhouse effect, these scientists offer a challenge to environmental theory.

Books (unannotated)

Allaby, Michael. *A guide to Gaia: a survey of the new science of our living Earth*. New York: E. P. Dutton, 1990, 1989. 181 p. (QH331 .A349 1990)

Burger, Julian. *The Gaia atlas of first peoples: a future for the indigenous world*. New York: Doubleday, 1990. 191 p.: ill. (some color) (GN380 .B85 1990)

Derrick, Dan. *Master SimCity/SimEarth: city & planet design strategies.* Carmel, Ind.: SAMS, 1991. 508 p.: ill. (HT166 .D385 1991)

Devereux, Paul. *Earthmind: communicating with the living world of Gaia.* Rochester, Vt.: Destiny Books, 1992.

Ekins, Paul. *The Gaia atlas of green economics.* New York: Anchor Books, 1992.

Gaia 2: emergence. Hudson, N.Y.: Lindisfarne Press, 1991. 272 p.: ill. (QH331 .G2 1991)

Gaia, a way of knowing: political implications of the new biology. Great Barrington, Mass.: Lindisfarne Press, 1987. Distributed by Inner Traditions International, Rochester, Vt. 217 p.: ill. (QH331 .G22 1987)

Lovelock, J. E. *Gaia: a new look at life on Earth.* New York: Oxford University Press, 1987. 157 p.: ill.

Lovelock, J. E. *Healing Gaia: practical medicine for the planet.* New York: Harmony Books, 1991. 192 p.: color ill. (QH343.4 .L69 1991)

Miller, Alan S. *Gaia connections: an introduction to ecology, ecoethics, and economics.* Savage, Md.: Rowman & Littlefield, 1991. 301 p.: ill. (GF80 .M53 1990)

Myers, Norman. *Gaia, an atlas of planet management.* Garden City, N.Y.: Anchor Press/Doubleday, 1984. 272 p.: color ill. (HC79.E5 M94 1984)

Myers, Norman. *The Gaia atlas of future worlds: challenge and opportunity in an age of change.* London: Robertson McCarta, 1990. 190 p.: ill. (some color) maps.

Pedler, Kit. *The quest for Gaia: a book of changes.* London: Paladin, 1991. 222 p.: ill.

Sahtouris, Elisabet. *Gaia: the human journey from chaos to cosmos.* New York: Pocket Books, 1989. 252 p. (QH331 .S224 1989)

Scientists on Gaia. Cambridge, Mass.: MIT Press, 1991. 433 p.: ill. (QH331 .S375 1991)

Wilson, Johnny L. *The SimEarth bible.* Berkeley, Calif.: Osborne McGraw-Hill, 1991. 185 p.: ill. (QB631 .W55 1991)

Zoeteman, Kees. *Gaiasophy: the wisdom of the living Earth: an approach to ecology.* Hudson, N.Y.: Lindisfarne Press, 1991. 374 p.: ill., maps (GF80 .Z64 1991)

General interest periodicals (annotated)

Joseph, Lawrence E. "Britain's whole Earth guru." *New York Times Magazine*

(November 23, 1986): 67, 95, 106.
> Discusses James Lovelock's theory of Gaia which proposes that the Earth is a self-controlling whole system in which narrow atmospheric limits are maintained.

General interest periodicals (unannotated)

Bosveld, Jane. "Life according to Gaia." *Omni* 14,1 (October 1991): 66(4)

Elmer-Dewitt, Philip. "The day I played God: creating a new world is complicated--and risky." (SimEarth - The Living Planet, a computer game that shows how life evolved) *Time* 136,27 (December 24 1990): 74(2)

Merchant, Carolyn. "Gaia's last gasp." (ecology and social meaning) *Tikkun* 5,2 (March-April 1990): 66(4)

Ponte, Lowell. "The man who discovered Mother Nature." (James Lovelock) *Reader's Digest* 139,834 (October 1991): 141(4)

"The veiled goddess." (Gaia) *Economist* 317,7686 (December 22 1990): 101(7)

Environmental and professional journals (unannotated)

Goldsmith, Edward. "Gaia and evolution." *Ecologist* 19 (July-August 1989): 145-153.
> "The 'survival of the fittest' maxim of Darwinism is widely used to justify the disastrous process of unrestrained technological progress and economic development. However, if the world is seen as a single self-regulating system, then progress through competition becomes fundamentally anti-evolutionary. Co-operation is the true evolutionary strategy."

Hughes, Charles J.; and Abram, David. "Gaia." *Ecologist* 15,3 (1985): 92-103.
> "Gaia: a natural scientist's ethic for the future," by Hughes and "The perceptual implications of Gaia" by Abram, discuss Jim Lovelock's thesis that this planet can best be described as a coherent, living entity.

Lovelock, James. "Gaia: the world as living organism." *New Scientist* 112 (December 18, 1986): 25-28.
> "As organisms grow, they benefit the environment as well as themselves. 'Geophysiological' systems thus emerge from the activity of individual organisms."

Lovelock, James. "Hands up for the Gaia hypothesis." *Nature* 344 (March 8, 1990): 100-102.
> "The concept of Gaia, self-regulating Earth, excites both admiration and obloquy. Its inventor describes the genesis and evolution of the hypothesis."

Sattaur, Omar. "Cuckoo in the nest." *New Scientist* 116 (December 24-31, 1987): 16-18. "Fifteen years ago, James Lovelock proposed his Gaia hypothesis, which regards the planet Earth as a living organism. Since that time the idea has been making people uncomfortable. The hypothesis speaks for the Earth and not necessarily for the human beings on it."

Weiner, Jonathan. "In Gaia's garden." *Sciences* 26 (July-August 1986): 2-5. James Lovelock's "Gaia hypothesis" is that "the planetary kingdoms of animals and plants and bacteria are working together to keep life liveable--life shapes Earth to its own ends." Weiner involves H. D. Block's "mechanical biology" computer theory to support this concept.

Environmental and professional journals (unannotated)

Beardsley, Tim. "Gaia; the smile remains, but the lady vanishes." *Scientific American* 261,6 (December 1989): 35(2)

"Gaia hypothesis." (proposition that in some ways the Earth behaves like a living system) *Mother Earth News* (May-June 1986): 30(2)

Krapfnel, Paul. "Becoming a part of Gaia." (holistic view of man and Earth) *Coevolution Quarterly* (Fall 1984): 4(7)

Lovelock, James. "Planetary medicine." (Earth's environmental crisis) *American Health* 8,2 (March 1989): 86(3)

Lovelock, James. "The independent practice of science." *Coevolution Quarterly* (Spring 1980): 22(8)

Mann, Charles. "Lynn Margulis: science's unruly Earth Mother." (includes related information on the Gaia hypothesis) *Science* 252,5004 (April 19 1991): 378(4)

Sagan, Dorion; and Margulis, Lynn. "Gaia and the evolution of machines." *Whole Earth Review* (Summer 1987): 15(7)

Schneider, Stephen H.. "Debating Gaia." (does life control the planet's atmosphere?) (includes related article) *Environment* 32,4 (May 1990): 4(9)

DEEP ECOLOGY

Books (annotated)

Andrews, Valerie. *A passion for this Earth: exploring a new partnership of man, woman,*

and nature. San Francisco: HarperSanFrancisco, 1992. (BF353.5.N37 A53 1992)
>   Using thinkers such as Joseph Campbell, Thomas Berry, Carl Jung, and Annie Dillard as guideposts, explores the origins of the modern attitude toward nature and its consequences. Examines the feminine principle and its inherent closeness to the Earth, as well as the ways in which modern society has suppressed it.

Devall, Bill; and Sessions, George. *Deep ecology: living as if nature mattered.* Salt Lake City, Utah.: Peregrine Smith, 1985. 266 p.
>   Deep ecology, first named by Arne Naess in 1973, contrasts with reform ecology, resource conservation, preservation, and ecological sensibility in that deep ecology attempts to appreciate the natural world for its inherent values apart from any benefits it might provide humanity. By entering into a harmonious relationship with all beings so as to realize their intrinsic goodness, we become more fully realized individuals within a larger system.

Fox, Warwick. *Toward a transpersonal ecology: developing new foundations for environmentalism.* Boston: Shambhala, 1990. 380 p. (GF21.F68 1990)
>   Maintains that the environmental philosophy of deep ecology is not distinguished by its tenant that humans must go through a deep self-examination to discover that all life is valuable. Rather deep ecology is unique because it advocates transcending the self and realizing the connectedness of all living things.

Kealey, Daniel A. *Revisioning environmental ethics.* Albany: State University of New York Press, 1990. 136 p. (GF80 .K43 1990)
>   Maintains that the environmental problems facing the modern world are the result of a prevalent Cartesian mental-rational consciousness structure. The solution to these problems, therefore, lies in replacing this consciousness structure with a more "magical," integral consciousness structure, namely "deep ecology."

Books (unannotated)

Bradford, George. *How deep is deep ecology? with an essay-review on woman's freedom.* Ojai, Calif.: Times Change Press, 1989. 86 p.: ill. (GF21 .B64 1989)

Devall, Bill. *Deep ecology.* Salt Lake City, Utah: G. M. Smith, 1985. 266 p. (GF75 .D49)

Devall, Bill. *Simple in means, rich in ends: practicing deep ecology.* Salt Lake City, Utah: Peregrine Smith Books, 1988. 224 p. (GF75 .D5 1988)

McLaughlin, Andrew. *Regarding nature: industrialism and deep ecology.* Albany: State University of New York Press, 1992.

Wisdom in the open air: the Norwegian roots of deep ecology. Minneapolis: University of Minnesota Press, 1992.

General interest periodicals (annotated)

Tokar, Brian. "Exploring the new ecologies: social ecology, deep ecology and the future of green political thought." *Alternatives* 15 (November-December 1988): 31-43.
  Discusses deepening divisions between social ecology, which "emphasizes the embeddedness of human consciousness in nature, a radical ecological critique of hierarchy and domination in society, and the historical unity of ecological and social concerns" and deep ecology, which "purports to speak more directly for the biosphere as a whole and seeks a better relationship between the human species and other forms of life."

General interest periodicals (unannotated)

Elder, P. S. "Is deep ecology the way?" *Alternatives* 15,2 (April-May 1988): 70(4)

Sale, Kirkpatrick. "The cutting edge: deep ecology and its critics." *Nation* 246,19 (May 14 1988): 670(5)

Environmental and professional journals (annotated)

Fox, Warwick. "Deep ecology: a new philosophy of our time?" *Ecologist* 14,5-6 (1984): 194-204.
  "Deep ecology strives to be non-anthropocentric by viewing humans as just one constituency among others in the biotic community" as opposed to "shallow ecology which views humans as the source of all value and ascribes only instrumental (or use) value to the nonhuman world."

Hinchman, Lewis P.; and Hinchman, Sandra K. "Deep ecology and the revival of natural right." *Western Political Quarterly* 42 (September 1989): 201-228.
  "The writings of Deep Ecologists reveal the outlines of a political theory at odds with the main currents of contemporary thought. This theory revives, in a novel form, an essential element of Aristotelian natural right: that what is 'right by nature' involves the full unfolding of an entity within its proper context. But this context proves to be undamaged nature as a whole (not just the polls), and the full unfolding of human beings presupposes that of other beings as well. Confusions arise, however, when Deep Ecologists try to infer from natural paradigms, such as undisturbed ecosystems, the proper order for human society."

ECOFEMINISM

Books (annotated)

Anderson, Lorraine, ed. *Sisters of the Earth*. New York: Vintage, 1992.
These essays are grouped thematically around kinship, pleasure, wildness, solace, plants and animals, rape, and healing. Suggests that women relate to nature with humility and caring while Western males have treated nature aggressively. Many of the contributors write about their emotional reactions to experiences with nature.

Biehl, Janet. *Rethinking ecofeminist politics*. Boston: Southend Press, 1991. 181 p.
The author addresses the ideas of ecofeminism from a nonecofeminist perspective with the aim of promoting "social ecology." Attacks other ecofeminists, whom she stereotypes as apolitical, antirational, home and nature loving.

Caldecott, Leonie; and Leland, Stephanie. *Reclaim the Earth: women speak out for life on Earth*. London: The Women's Press, 1983.
Although the essays do not deal directly with environmental issues, they do reveal the deep-rooted socioenvironmental concerns of women that are largely ignored by the environmental movement. The editors believe that these "women's issues" must be elevated to global issues in a movement to break down the concept of minority or special interests. The contributors offer deeply personal stories and in some instances readers may wonder what environmental point is being made, which itself may be the point.

Collard, Andree. *Rape of the wild: man's violence against animals and the Earth*. Bloomington: Indiana University Press, 1990. 208 p.
This ecofeminist perspective maintains that men have violated animals in the Earth in much the same way that they have violated women. Includes an examination of matriarchal Old Stone Age society and how this society worshiped the Mother Earth.

Dankelman, Irene; and Davidson, Joan. *Women and environment in the Third World*. London: Earthscan, 1988. 210 p.
The authors argue that women often are given the poorest lands to farm and as a result are most adversely affected by the loss of natural resources. Thus, women are more concerned about preserving resources and could accomplish sustainable development if given the resources. The authors do not base their opinions on scientific research, but they do make their point that women are often ignored when they should be recognized as valuable resources.

Diamond, Irene; and Orenstein, Gloria Fenman, eds. *Reweaving the world: the emergence of ecofeminism*. San Francisco: Sierra, 1990.
A collection of essays written by leaders in the movement that challenges the idea that a comprehensive theory will lead to coherent practice among ecofeminists. The essays cover a range of subjects from the nature/culture dualism to stereotyping ecofeminists.

Gray, Elizabeth Dodson. *Green paradise lost*. Santa Monica, Calif.: Roundtable, 1979.
Gray was one of the first feminists to attack the idea that "stewardship" of nature is paternalistic, and she argues instead for connections to the natural world, espcially to understand human dependence on it. Argues against the assumption that dominance in any form is a necessary part of nature.

Merchant, Carolyn. *The death of nature: women, ecology, and the scientific revolution*. New York: Harper and Row, 1980.
Presents the connections between the exploitation of nature, commercial expansion, and the subjugation of women during the sixteenth and seventeenth centuries. Argues that the vision of nature as the nuturing mother was replaced by industrial society and women as symbols of nature were devalued and exploited.

Plant, Judith, ed. *Healing the wounds: the promise of ecofeminism*. Philadelphia, Pa.: New Society, 1989.
Argues that the environmental and women's movements have much in common. Analyzes of the women's movement in India and how opposition to clear cutting in Alaska is similar to witchcraft. Calls for the establishment of sacred groves where female and male power is balanced; critiques the deep ecology movement.

Shiva, Vandana. *Staying alive: women, ecology and survival in India*. London: Zed, 1988.
Argues that much of the Indian knowledge of ecology lies with the women and relates their efforts to combat development projects such as forestry and dams.

Tempest, Terry. *Refuge: an unnatural history of family and place*. New York: Pantheon, 1992.
Tempest, a naturalist who lives in Salt Lake City, juxtaposes two natural disasters: the inexplicable rising of the Great Salt Lake and her mother's slow death from cancer. Chronicles the devastation of flooding on the surrounding communities and the imbalance of the family ecosystem. She learns that her family was exposed to radiation from an above-ground atomic bomb test when she was an infant and the ensuing insensitive statement from the Atomic Energy Commission that the desert was virtually uninhabited.

Wajcman, Judy. *Feminism confronts technology*. Cambridge: Policy Press, 1991.
Examines male bias in the development and definition of technology, especially in the industrialization of the home and the creation of the housewife. Questions whether technology itself is inherently patriarchal and what more women scientists would contribute.

Books (unannotated)

Adams, Carol J. *The sexual politics of meat: a feminist-vegetarian critical theory*. New York: Continuum, 1990. 256 p.: ill. (HV4708 .A25 1990)

Bradford, George. *How deep is deep ecology? with an essay-review on woman's freedom.* Ojai, Calif.: Times Change Press, 1989. 86 p.: ill. (GF21 .B64 1989)

International Pacific Policy Congress (1991: Port-Vila, Vanuatu) Women's voices on the Pacific. Washington, D.C.: Maisonneuve Press, 1991. 156 p.: ill. maps (HQ1236.5.P16 I57 1991)

Women and the environment: a reader: crisis and development in the Third World. New York: Monthly Review, 1991. 205 p. (HQ1240.5.D44 W64 1991)

Women's issues in water and sanitation: attempts to address an age-old challenge. Ottawa, Canada: International Development Research Centre, 1985. 104 p.: ill. (TD201 .W66 1985)

General interest periodicals (annotated)

Sale, Kirkpatrick. "Ecofeminism--a new perspective." *Nation* 245 (September 26, 1987): 302-305.
> "Two of the most potent and durable ideas of the 1960s--feminism and ecological politics--have begun to come together in a new and fruitful way at last. The resulting hybrid, 'ecofeminism,' has finally taken on a distinct life of its own and appears to be influencing a growing number of groups and movements."

General interest periodicals (unannotated)

Devall, Bill. "Ecofeminism and deep ecology--a personal response." *Canadian Dimension* 25,4 (June 1991): 25(3)

Hynes, H. Patricia. "Beyond global housekeeping." (role of women in the environmental protection movement) *Ms.* 1,1 (July-August 1990): 91(3)

Jones, Patti. "Women are waking up to their planet." *Glamour* 88,5 (May 1990): 270(6)

Kelly, Petra. "Beyond the greens." (ecofeminism) *Ms.* 11,3 (November-December 1991): 70(2)

Larsen, Elizabeth. "Granola boys, eco-dudes, and me." (a woman's involvement in the environmental movement) *Ms.* 2,1 (July-August 1991): 96(2)

Environmental and professional journals (annotated)

Kolinsky, Eva. "The West German greens--a women's party?" *Parliamentary Affairs* 41

(January 1988): 129-148.

"Focuses on the significance of the women's issue for the Green Party and explains why the women's issue rose to priority status and which role it is likely to play in the future.

Merchant, Carolyn. "Earthcare: women and the environmental movement." *Environment* 23 (June 1981): 6-13, 38-40.

Discusses the simultaneous emergence of the feminist and the ecology movements; traces women's involvement with environmental protection; urges further participation in low-impact technology as essential to sustain future life on Earth.

Conference proceedings

Women and the environmental crisis: forum '85: a report of the proceedings of the workshops on women environment and development, July 10 to 20, 1985, Nairobi, Kenya. Nairobi, Kenya: Environment Liaison Centre, 1986. 109 p.: ill. (HQ1240.5.D44 W65 1986)

ENVIRONMENTALISM AND RELIGION

Books (annotated)

Berry, Thomas. *The dream of Earth.* San Francisco: Sierra Club Books, 1988. 247 p. (GF80 .B47 1988)

Argues that orthodox Christianity is opposed to deep ecology because of the overemphasis on human redemption and the transcendence of God. Berry believes that the universe and the planet Earth "need to be experienced as the primary mode of the divine presence, just as it is a primary educator, primary healer, primary commercial establishment, and primary lawgiver for all that exists within this life community." Berry sees the "sense of communion at the heart of reality" as the "central force bringing the ecological age into existence."

Dwivedi, O. P., ed. *World religions and the environment.* New Delhi, India: Gilanjali Publishing House, 1989. 461 p.

Discusses the different ways the world's eight major religions (Judaism, Christianity, Buddhism, Islam, Zoroastrianism, Hinduism, Jainism, and Sikhism) have dealt with nature. Presents the sources and, in an appendix, translations of relevant scripture. Maintains that the world's major religions should work together to forge an environmental ethic.

Engel, J. Ronald, ed. *Ethics of environment and development.* Tucson: University of Arizona Press, 1990.

These essays discuss the complexity of the ethics of sustainability as complicated by the opposing views of science and traditional concepts of nature. Presents the views of leaders from various religions, including Christianity which often perceives environmentalism to be in conflict with Christian theology.

Hargrove, Eugene C., ed. *Religion and environmental crisis*. Athens: University of Georgia Press, 1986. 222 p. (GF80 .R45 1986)
Examines the moral debate surrounding the relationship between humans and their environment in a historical and metaphorical context, East and West, ancient and modern.

McGaa, Ed. *Mother Earth spirituality: Native American paths to healing ourselves and our world*. New York: Harper, 1990.
Maintains that native American spirituality offers a paradigm for a different view of the Earth, away from the modern mechanistic perspective. Presents both the vision and ceremonies of native American reverence for the Earth and proposes that if society accepted this reverence it could reverse environmental destruction.

McGaa, Ed. *Rainbow Tribe: ordinary people journeying on the red road*. San Francisco: HarperSanFrancisco, 1992. 240 p.
Focuses on the Rainbow Tribe, non-native Americans following the ways of native Americans. Maintains that religions more centered on the Earth are the solution to our impending ecological crisis.

Spretnak, Charlene. *The spiritual dimension of green politics*. Santa Fe, N.M.: Bear, 1987. 95 p.
Maintains that Green Politics rejects the Western values of humanism, modernism, and patriarchy. Argues that the Judeo-Christian heritage is based upon pagan ties to nature--such as the way holy days tend to correspond to the lunar cycle--and that modern religion should embrace these older, more environmentally reverent aspects of their traditions.

Books (unannotated)

Abraham, Ralph. *Trialogues at the edge of the West: chaos, creativity, and the resacralization of the world*. Santa Fe, N.M.: Bear, 1992.

Bowman, Douglas C. *Beyond the modern mind: the spiritual and ethical challenge of the environmental crisis*. New York: Pilgrim Press, 1990. 125 p. (BT695.5 .B695 1990)

Covenant for a new creation: ethics, religion, and public policy. Maryknoll, N.Y.: Orbis Books, Graduate Theological Union, 1991. 293 p.; ill. (BT695.5 .C68 1991)

Nollman, Jim. *Spiritual ecology: a guide for reconnecting with nature*. New York: Bantam, 1990. 227 p. (QH540.5 .N65 1990)

Tree of life: Buddhism and protection of nature. N.p.: Buddhist Perception of Nature, 1987. 99 p.: (some color) ill. (QH75 .T74 1987)

Turner, Frederick. *Rebirth of value: meditations on beauty, ecology, religion, and education.* Albany: State University of New York Press, 1991. 188 p. (B831.2 .T87 1991)

General interest periodicals (annotated)

Lilburne, Geoffrey R. "Theology and land-use: moving toward a Christian land-care ethic." *Vital Speeches of the Day* 53 (December 15, 1986): 139-143.
> "The present rural crisis comes at a time when ecologists are speaking of the fragility of our global ecosystem and the possibility of permanent damage to its structure. In turn, these issues are related to the changing patterns of land use and land rights, not only in industrialized society but throughout the Third World. How we relate to the land is ultimately a religious issue."

Rasmussen, Larry. "Toward an Earth charter." *Christian Century* 108 (October 23, 1991): 964-967.
> "Outlines several theological models as possible responses to the eco-crisis."

Santmire, H. Paul. "The liberation of nature: Lynn White's challenge anew." *Christian Century* 102 (May 22, 1985): 530-534.
> The author discusses White's 1966 address "The Historical Roots of Our Ecologic Crisis," which took Christian theology to task for helping to cause and continuing to fuel our global environmental crisis. He concludes that "thanks largely to Lynn White, the liberation of nature is now before us as a theological theme."

General interest periodicals (unannotated)

Cato, Phillip C. "Management of the biosphere: ethical and theological issues." *Vital speeches of the day* 56,2 (November 1 1989): 53(3)

Christiansen, Drew. "Christian theology and ecological responsibility." (Cover Story) *America* 166,18 (May 23 1992): 448(4)

Crawford, Mark. "Religious groups join animal patent battle." *Science* 237 (July 31 1987): 480(2)

Ebenreck, Sara. "An Earth-care ethics." *Catholic World* 233,1396 (July-August 1990): 153(5)

Environmental crisis: put Christian virtues at your disposal. (interview with Frederick W. Krueger) *U.S. Catholic* 54,10 (October 1989): 25(6)

Fehren, Henry. "Where were you?" (preserving God's creation) *U.S. Catholic* 55,4 (April 1990): 39(3)

Frame, Randy. "Christianity and ecology: a better mix than before". *Christianity Today* 34,7(April 23 1990): 38(2)

Himes, Michael J.; and Himes, Kenneth R. "The sacrament of creation: toward an environmental theology." *Commonweal* 117,2 (January 26 1990): 42(8)

Jones, Arthur. "Green asceticism making a virtue of necessity: psychologist promotes search for self-discovery." *National Catholic Reporter* 28,20 (March 20 1992): 19(2)

Jones, Arthur. "Pained cry of the planet exclaimed as world Christians gather in Seoul." (World Conference of Churches ecology conference) *National Catholic Reporter* 26,22 (March 23 1990): 1(3)

Limburg, James. "The way of an eagle in the sky: reflections on the Bible and the care of the Earth." *The Catholic World* 233,1396 (July-August 1990): 148(5)

Stone, Pat. "Christian ecology: a growing force in the environmental movement." *Utne Reader* 36 (November-December 1989): 78(2)

Train, Russell E. "Caring for creation: religion and ecology." *Vital Speeches of the Day* 56,21 (August 15 1990): 664(3)

Environmental and professional journals (annotated)

"New green gospel." *Environmental Action* 22 (January-February 1991): 23-30.
"A wide range of religious institutions are now pulling the environment to the forefront of their ministries." Partial contents include "Preaching for the planet," by Barbara Ruben; "Forging a new land ethic from the Bible," by Ellen Bernstein; and "Theology for a small planet," by Timothy C. Weiskel.

Environmental and professional journals (unannotated)

Meeker, Joseph W. "The Assisi connection." (Christianity and environmental protection) *Wilderness* 51,180 (Spring 1988): 61(3)

Videos

Spirit and nature. Produced by Public Affairs Television. Mystic Fire Video (Department PR, Box 9323, South Burlington, VT 05407). 1991.

This video, hosted by Bill Moyers, is the result of a conference at Middlebury College which addressed the connection between the world's major religions and the environment. The conference included the Dalai Lama, an Islamic scholar, a rabbi, a Protestant theologian, and a native American elder. Moyers intersperses interviews with footage of rituals and illustrative performances.

ENVIRONMENTAL ECONOMICS

(See also the *Environmental Movement: Business*) 481
(Related topics in *Global Environmental Policy*) 255

Books (annotated)

Anderson, Terry L.; and Leal, Donald R. *Free market environmentalism.* San Francisco: Pacific Research Institute for Public Policy; Boulder, Colo.: Westview, 1991. 192 p.: ill. (HC110.E5 A6665 1991)
> Focusing on free-market methods of improving environmental quality, this work challenges many traditional environmental policies. Compares a policy of incentives to one of active government participation. Includes examples of successful free-market environmental policy as applied to energy development, land and water management, and outdoor recreation. Also demonstrates the difficulties the free market has had in dealing with ocean management and pollution control. Contains footnotes and a bibliography.

Blinder, Alan S. *Cleaning up the environment. In his hard heads, soft hearts: tough-minded economics for a just society.* Reading, Mass.: Addison-Wesley, 1987. p. 136-159.
> Points out that current U.S. policies make environmental protection far too costly. "America can achieve its present levels of air and water quality at far lower cost," economists insist. The nation is, in effect, shopping for cleaner air and water in a high-priced store when a discount house is just around the corner. Besides, if we shopped in the discount store, we would probably buy a higher-quality environment than we do now."

Bormann, F. Herbert; and Kellert, Stephen R. *Ecology, economics, ethics: the broken circle.* New Haven, Conn.: Yale University Press, 1991. 233 p.: ill. (QH541 .E31934 1991)
> Looking at environmental problems from ecological, economical, and ethical viewpoints, the twelve essays which comprise this volume address such issues as species diversity, agriculture, pollution, market mechanisms, and environmental values. Maintains that modern environmental problems are the result of a focus on short-term gain. Specific, practical proposals are included in each article. Contains a good bibliography.

Bromley, Daniel W. *Environment and the economy: property rights and public policy.* Oxford: Blackwell, 1991. 247 p. (HC79.E5 B75 1991)

Argues that the economics of natural resources must be viewed with respect to different types of property rights.

Costanza, Robert, ed. *Ecological economics: the science and management of sustainability.* New York: Columbia University Press, 1991. 525 p.

The composite field of ecological economics is beginning to offer new answers for established environmental issues. These thirty-two essays, compiled from a 1990 workshop, form a path-breaking collection that aims to develop an ecological economic world view, which will help to construct better methods of accounting, modeling, and analysis; to suggest institutional changes; and to examine case studies.

Daly, Herman E. *Steady-state economics.* 2d ed. Washington, D.C.: Island Press, 1991. 302 p.: ill. (HD82 .D31415 1991)

This is the revised edition of Daly's 1977 book that introduced the then-radical views on the economics of sustainability, which is now considered the standard work on the subject. New material includes the debate over growth and the environment.

Economic instruments for environmental protection. Paris: Organisation for Economic Co-operation and Development, 1989. Distributed by OECD Publications and Information Centre, Washington, D.C. 131 p. (HC79.P55 E26 1989)

Assesses current economic methods of enforcing environmental protection and evaluates the effectiveness and potential for future use. Examples are taken from fourteen countries and include taxes on effluents, charges on the user and product, tax relief for antipollution measures, and the trading of pollution rights. Along with these measures, the problems incurred in enforcing them are also discussed. The implications of making the polluter pay are also examined.

Ekins, Paul; Hillman, Mayer; and Hutchinson, Robert. *The Gaia atlas of green economics.* New York: Anchor, 1992. 191 p.

Maintains that green economics is the solution for the current global economic crisis and that the economy is intrinsically interconnected to other aspects of society. Defines green economics and offers a new definition of wealth based upon responsible use of resources. Offers a vision of sustainable development and discusses current efforts toward this end. Contains extensive references.

Environmental investments: the cost of a clean environment. Washington, D.C.: Island Press, 1991.

Describes the $115 billion a year commitment Americans have made to protecting the environment and evaluates the effectiveness of those efforts in protecting and restoring the nation's air, water, and land.

Folke, C.; and Kaberger, T., eds. *Linking the natural environment and the economy.* Boston: Kluwer Academic, 1991.

> This work aims to narrow the gap between environmentalists and economists by demonstrating that natural resources can no longer be consumed unthinkingly. Analyzes specific links between the two areas through institutions, environmental effects, and energy flows. Examines the role of environmental resources in economics through examples such as wetlands, landscape change, coastal, and marine ecosystems. Analyzes anthropogenic effects on the environment in developing countries.

Horowitz, Daniel. *The morality of spending: attitudes toward the consumer society in America, 1875-1940.* Baltimore, Md.: Johns Hopkins University Press, 1985. 254 p.

> Environmental historians will find here a synopsis of ideas about the ethics of spending in past eras, some of which apply to modern consumerism. Crises, such as the world wars, the Depression, and world oil shortages, momentarily remind society that communal and nonmaterial pleasures are more valuable than individually purchasable ones, which in turn gives way to greater spending.

Kane, Hal; and Starke, Linda. *Time for a change.* Washington, D.C.: Island Press, 1991.

> Argues that looking at environmental problems individually, rather than in relation to each other, is not the way to develop policy or succeed in solving problems. Outlines the need to see the full cost of exploiting resources and provides models for sustainable life styles and policies.

Kula, E. *Economics of natural resources and the environment.* New York: Chapman and Hall, 1992.

> A good introduction to the basics of environmental economics.

Krabbe, J. J.; and Heijman, W. J. M, eds. *National income and nature: externalities, growth, and steady state.* Boston: Kluwer, 1992.

> Includes a methodology for correcting national income statistics for environmental damage.

MacNeil, Jim; Winsemius, Pieter; and Yakushiji, Taizo. *Beyond interdependence: the meshing of the world's economy and the Earth's ecology.* New York: Oxford University Press, 1991. 159 p.: ill. (HD75.6 .M33 1991)

> This short work demonstrates the pro-environment actions large industrialized nations can take which will simultaneously bring about microeconomic efficiency and international competitiveness. The authors aptly demonstrate the importance of addressing environmental issues to world security, as well as illustrating the vital role developing countries play in those issues. Suggestions for economic suggestions and ways to encourage international cooperation are included.

McNeely. Jeffrey A. *Economics and biological diversity.* Gland, Switzerland: IUCN, 1988. 236 p.

Reviews economic incentives to promote conservation, argues the values and benefits of biological diversity, and outlines the mechanisms for funding conservation. Presents twenty case studies in Third World and developed countries and outlines the programs that work well, work indifferently, and could work a lot better. Argues that a strict cost-benefit analysis cannot be applied in the absence of social equity considerations.

Mitchell, Robert Cameron; and Carson, Richard T. *Using surveys to value public goods: the contingent valuation method.* Washington, D.C.: Resources for the Future, 1989. 463 p.
"Argues that at this time the contingent valuation method offers the most promising approach for determining public willingness to pay for many public goods."

Peet, John. *Energy and the ecological economics of sustainability.* Washington, D.C.: Island Press, 1991.
Argues that in order to come to terms with the present crisis we must understand its roots in the classical economics upon which modern society is based and must initiate a fundamental revision in setting economic priorities. Only when people view themselves in the context of the global ecosystem can they accept the physical limits set by the natural world.

Perrings, Charles. *Economy and environment: a theoretical essay on the interdependence of economic and environmental systems.* New York: Cambridge University Press, 1987. 179 p. (HD75.6 .P47 1987)
Modern economics maintains "the sovereignty of the individual, the sanctity of private property, and the domination of the present. Its effect is to justify the abdication of our collective responsibility for the outcome of our actions. By exploring the time behavior of a jointly determined economy-environment system, this essay tackles the theoretical implications of the environmental blindness that underpins the formal foundations of the market solution."

Raufer, Roger K. *Acid rain and emissions trading: implementing a market approach to pollution control.* Totowa, N.J.: Rowman & Littlefield, 1987. 161 p.: ill. (TD196.A25 R37 1987)
Raufer rejects both the emissions free and BTU tax theories for reducing acid rain pollution in favor of "emission trading," which would allow low polluting companies to sell some of their pollution allotment to high polluting companies. Environmental advocates are often caught up in the idea of punishing pollutors, which the tax theories would do. Raufer, on the other hand, proposes giving companies positive incentives to reduce pollution and to help them recoup the cost by making a profit on the pollution they do not create.

Sagoff, Mark. *The economy of the Earth: philosophy, law, and the environment.* Cambridge: Cambridge University Press, 1988. 275.
Written for the specialist, this book challenges the traditional economic approach

to setting environmental goals through cost-benefit analysis, trade-offs, and efficiency.

Schumacher, E. F. *Small is beautiful: economics as if people mattered*. New York: Harper, 1973.
> Challenges economic materialism as being the foundation of environmental destruction. Discusses the spiritual and moral consequences of environmental irresponsibility. Provides suggestions for alternative ways of living which has resulted in the theory of sustainable development.

Singh, Narindar. *Economics and the crisis of ecology*. 3d rev. ed. London: Bellew, 1989.
> Argues that ecological destruction is the inevitable result of the military/industrial complex of both Eastern and Western powers and that only their decline can foster ecological progress.

Swanson, Timothy; and Barbier, Edward B., eds. *Economics for the wilds*. Washington, D.C.: Island Press, 1992.
> Argues that--in contrast to the view that wilderness should never be used for economic gain--an economic strategy that values the resources of the wilds offers the best long-term security for them.

Winpenny, J. T. *Values for the environment*. London: H.M.S.O., 1991.
> Designed to be of use for policymakers, this work begins with a summary of important habitats such as rain forests and watersheds. Analyzes practical applications of economics to such things as desertification and biodiversity.

Young, Michael D. *Sustainable investment and resource use: equity, environmental integrity, and economic efficiency*. Park Ridge, N.J.: Parthenon, 1992. 176 p.: ill. (HD75.6 .Y68 1992)
> A good introduction to environmental economics. Discusses sustainable development.

Books (unannotated)

Allaby, Michael. *The politics of self-sufficiency*. New York: Oxford University Press, 1980. 242 p. (HD9980.5 .A44)

Analysing the options: cost-benefit analysis in differing economic systems. Nairobi: United Nations Environment Programme, 1982. 151 p.: ill. (HC79.E5 A492 1982)

Boulding, Kenneth Ewart. *Towards a new economics: critical essays on ecology, distribution, and other themes*. Brookfield, Vt., Edward Elgar, 1992. 344 p. (HB171 .B642 1992)

Brown, Lester Russell. *Saving the planet: how to shape an environmentally sustainable global economy*. New York: W. W. Norton, 1991. 224 p. (HD75.6 .B76 1991)

Ciriacy-Wantrup, Siegfried V. *Natural resource economics: selected papers*. Boulder, Colo.: Westview, 1985. 321 p.: ill. (HC55 .C52 1985)

Common, Michael S. *Environmental and resource economics*. New York: Longman, 1988.

Conrad, Jon M. *Natural resource economics: notes and problems*. New York: Cambridge University Press, 1987. 231 p.: ill. (HC59 .C693 1987)

Cooper, Charles. *Economic evaluation and the environment: a methodological discussion with particular reference to developing countries*. London: Hodder and Stoughton, 1981. 161 p. (HC79.E5 C654)

Daly, Herman E. *For the common good: redirecting the economy toward community, the environment, and a sustainable future*. Boston: Beacon Press, 1989. 482 p.: ill. (HD75.6 .D35 1989)

Dietz, Frank J., et al., eds. *Environmental policy and the economy*. New York: Elsevier Science, 1991. 331 p. (HD2129.Z8 D45 1991)

Downing, Paul B. *Environmental economics and policy*. Boston: Little, Brown, 1984. 334 p.: ill. (HC79.E5 D69 1984)

Ecological economics: a practical programme for global reform. Atlantic Highlands, N.J.: Zed Books, 1992.

Ecological economics: the science and management of sustainability. New York: Columbia University Press, 1991. 525 p.: ill. map. (HD75.6 .E29 1991)

Economic-ecological modeling. New York: North-Holland, 1987. Distributed by Elsevier Science Publishing. 329 p.: ill. (HC79.E5 E267 1987)

Economic policy towards the environment. Cambridge, England: Blackwell, 1991. 326 p.: ill. (HC79.E5 E273 1991)

Economic valuation techniques for the environment: a case study workbook. Baltimore, Md.: Johns Hopkins University Press, 1986. 203 p.: ill. (HC79.E5 E274 1986)

Economics of the environment. Brookfield, Vt: Elgar Publishing, 1992.

Economics, growth, and sustainable environments: essays in memory of Richard Lecomber. New York: St. Martin's Press, 1988. 205 p.: ill. (HC79.E5 E2774 1988)

Economics of ecosystem management. Boston: W. Junk, 1985. Distributed by Kluwer Boston, 1985. 244 p.: ill. (HC79.E5 E278 1985)

559

Economics of environment. New Delhi: Lancer International in association with India International Centre, 1987. 76 p.: ill. (HC440.E5 E36 1987)

Economics of the environment: selected readings. 3d ed. New York: W. W. Norton, 1993.

Edwards, F., ed. *Environmental auditing: the challenge of the 1990s.* Calgary, Alberta, Canada: University of Calgary Press, 1992. 346 p.

Eggert, Jim. *Meadowlark economics: perspectives on ecology, work, and learning.* Armonk, N.Y.: M. E. Sharpe, 1992. 127 p.: ill. (HD75.6 .E35 1992)

Ekins, Paul. *The Gaia atlas of green economics.* New York: Anchor Books, 1992. 191 p.: (some color) ill. color maps. (HC79.E5 E363 1992)

Energy, economics, and the environment: conflicting views of an essential interrelationship. Boulder, Colo.: Westview, 1981. 200 p.: ill. (TJ163.2 .E474)

Environmental regulation and the U.S. economy. Baltimore, Md.: (Resources for the Future): Johns Hopkins University Press, 1981. 163 p.: ill. (HC110.E5 E4989 1981)

Faber, Malte Michael. *Entropy, environment, and resources: an essay in physio-economics.* New York: Springer-Verlag, 1987. 205 p.: ill. (HC79.E5 F2313 1987)

Fisher, Anthony C. *Resource and environmental economics.* New York: Cambridge University Press, 1981. 284 p.: ill. (HC59 .F558)

Forsund, Finn R. *Environmental economics and management: pollution and natural resources.* New York: Croom Helm, 1988. 307 p.: ill. (HC79.E5 F64 1988)

Freeman, A. Myrick. *The economics of environmental policy.* Malabar, Fla.: R. E. Krieger, 1984. 184 p.: ill. (HC110.E5 F69 1984)

Freeman, Christopher. *The economics of hope: essays on technical change, economic growth, and the environment.* New York: Pinter Publishers, 1992. Distributed by St. Martin's Press. 249 p.: ill. (HD74.5 .F74 1992)

Hafkamp, Wilhelmus A. *Economic-environmental modeling in a national-regional system: an operational approach with multi-layer projection.* New York: North-Holland, 1984. Distributed by Elsevier Science Publishing. 235 p.: ill. (HC79.E5 H3 1984)

Handbook of natural resource and energy economics. New York: North-Holland, 1985. Distributed by Elsevier Science Publishing, 1985: ill. (HD9502.A2 H257 1985)

Hartwick, John M. *The economics of natural resource use.* New York: Harper & Row, 1986. 530 p.: ill. (HC59 .H3558 1986)

560

Hyman, Eric. *Combining facts and values in environmental impact assessment: theories and techniques.* Boulder, Colo.: Westview, 1988. 304 p.: ill. (HD75.6 .H95 1988)

Jacobs, Michael. *The green economy: environment, sustainable development, and the politics of the future.* Concord, Mass.: Pluto Press, 1991. 312 p.: ill. (HD75.6 .J33 1991)

Kassiola, Joel Jay. *The death of industrial civilization: the limits to economic growth and the repoliticization of advanced industrial society.* Albany: State University of New York Press, 1990. 297 p. (HD75.6 .K38 1990)

Kelman, Steven. *What price incentives? economists and the environment.* Boston, Mass.: Auburn House, 1981. 170 p. (HC79.E5 K4 1981)

Lowe, J. F. *The economics of environmental management.* Deddington, Oxford: P. Allan, 1980. 344 p.: ill.

Lowe, J. F. *Total environmental control: the economics of cross-media pollution transfers.* New York: Pergamon Press, 1982. 126 p.: ill. (HC79.E5 L68 1982)

Luten, Daniel B. *Progress against growth: Daniel B. Luten on the American landscape.* New York: Guilford Press, 1986. 366 p.: ill. (HC110.E5 L88 1986)

Makhijani, Arjun. *From global capitalism to economic justice: an inquiry into the elimination of systemic poverty, violence, and environmental destruction in the world economy.* New York: Apex Press, 1992.

Making pollution prevention pay: ecology with economy as policy. New York: Pergamon Press, 1982. 156 p.: ill. (HD69.P6 M28 1982)

Martinez Alier, Juan. *Ecological economics: energy, environment, and society.* New York: Basil Blackwell, 1987. 286 p. (HD85.S7 M363513 1987)

Mills, Edwin S. *The economics of environmental quality.* 2d ed. New York: W. W. Norton, 1986. 368 p.: ill. (HC110.E5 M54 1986)

National income and nature: externalities, growth, and steady state. Boston: Kluwer Academic Publishers, 1992. 232 p.: ill. (HD75.6 .N38 1991)

Neher, Philip A. *Natural resource economics: conservation and exploitation.* New York: Cambridge University Press, 1990. 360 p.: ill. (HC59 .N382 1990)

Nichols, Albert L. *Targeting economic incentives for environmental protection.* Cambridge, Mass.: MIT Press, 1984. 189 p.: ill. (HC79.E5 N47 1984)

Pearce, David W. *Blueprint for a green economy.* London: Earthscan, 1989. 192 p.

Pearce, David W. *Economics of natural resources and the environment.* New York: Harvester Wheatsheaf, 1990. 378 p.: ill.

Randall, Alan. *Resource economics: an economic approach to natural resource and environmental policy.* 2d ed. New York: Wiley, 1987. 434 p. (HC59 .R256 1987)

Renner, Michael. *Jobs in a sustainable economy.* Washington, D.C.: Worldwatch Institute, 1991. 58 p. (HC106.8 .R463 1991)

Robertson, James. *Future wealth: a new economics for the 21st century.* New York: Bootstrap Press, 1990.

Seneca, Joseph J. *Environmental economics.* 3d ed. Englewood Cliffs, N.J.: Prentice-Hall, 1984. 349 p.: ill. (HC110.E5 S45 1984)

Siebert, Horst. *Economics of the environment: theory and policy.* Rev. ed. New York: Springer-Verlag, 1992.

Structural adjustment and the environment. Boulder, Colo.: Westview, 1992.

Tietenberg, Thomas H. *Environmental and natural resource economics.* 3d ed. New York: HarperCollins, 1992. 678 p.: ill. (HC79.E5 T525 1992)

Tisdell, Clement Allan. *Economics of environmental conservation: economics for environmental & ecological management.* New York: Elsevier, 1991. 233 p.: ill. (HD75.6 .T563 1991)

Tisdell, Clement Allan. *Environmental economics: policies for environmental management and sustainable development.* Brookfield, Vt.: E. Elgar, 1992.

Valuation methods and policy making in environmental economics. New York: Elsevier, 1989. 259 p.: ill. (HC79.E5 V25 1989)

Valuing the Earth: economics, ecology, ethics. Cambridge, Mass.: MIT Press, 1993.

Wykle, Lucinda. *Worker empowerment in a changing economy: jobs, military production, and the environment.* New York: Apex Press, 1991. 84 p.: ill. (HD5724 .W95 1991)

Zsolnai, Laszlo. *Making a meta-economics.* Budapest: Karl Marx University of Economic Sciences, Department of Sociology, 1986. 88 p.: ill. (HD75.6 .Z77 1986)

General interest periodicals (annotated)

Brooks, David; Smith, Fred L.; and Kushner, Kathy H. "A sustainable environmentalism: saving the Earth from its friends." *National Review* 42 (April 1, 1990): 28-33, 59.

"Does protecting the environment require collectivism and an end to economic growth? David Brooks examines the full agenda of the environmental activists. Fred L. Smith and Kathy H. Kushner present a conservative alternative: restoring property rights and individual responsibilities so that free markets will protect natural resources."

Daly, Herman E. "Ultimate confusion: the economics of Julian Simon." *Futures* 17 (October 1985): 446-450.
"Argues that there are profound mistakes and exaggerations in Simon's influential and popular ideas. It discusses Simon's denial of resource finitude and his views that neither ecology nor entropy exists."

Pearce, David. "Resource scarcity and economic growth in poor developing countries: missing links in Simon's cornucopian philosophy." *Futures* 17 (October 1985): 440-445.
"Underlying Julian Simon's cornucopian philosophy is a model of economic-demographic interaction which derives normative views in favour of moderate population growth. This article criticizes Simon's unsatisfatory treatment of discounting in this procedure, and more significantly, his denial of mankind--ecosystems interactions."

Richardson, John M., Jr. "'The Resourceful Earth': optimism and confrontation." *Futures* 17 (October 1985): 464-474.
"*The Resourceful Earth: A Response to Global 2000,* edited by Julian Simon and the late Herman Kahn, is a recent contribution to the continuing debate over 'optimistic' v, 'pessimistic' futures and how projections about the future should be made. This article is concerned more with issues of methodology, fact and philosophy raised by *The Resourceful Earth* than with assessing its rightness or wrongness. The intention of the article is to move beyond the debate not to perpetuate it."

"Survey of the environment: costing the Earth." *Economist* 312 (September 2-8, 1989): 1-18.
Argues that sensible economics can avoid much environmental damage, but that powerful lobbies skew the system. Tax breaks, subsidies, and trade protection put money into the pockets of vocal, well-organized groups at the expense of consumers. Rich democracies frequently give special treatment to groups such as big farmers or logging companies whose activities cause environmental damage.

General interest periodicals (unannotated)

Bayon, Ricardo. "Ecology vs. economics: time to reconsider." (interview with Konrad Von Moltke) *World Press Review* 37,8 (August 1990): 14(2)

Cook, James. "The ghosts of Christmas yet to come." (world economic growth and the environment) *Forbes* 149,13 (June 22, 1992): 92(4)

Dreisbach, Robert. "Must we relearn how to use surplus?" (economies based on the production of non-essentials) *Alternatives* 17,2 (July-August 1990): 65(4)

"Greening economics." (survey of the environment; costing the Earth) *Economist* 312,7618 (September 2, 1989): S5(2)

"Growth can be green." (how to be clean and prosperous too) (editorial) *Economist* 312,7617 (August 26, 1989): 12(2)

MacNeill, Jim; Winsemius, Pieter; and Yakushiji, Taizo. "The West's shadow ecologies." (ecology and the world economy) *New Perspectives Quarterly: NPQ* 8,4 (Fall 1991): 28(4)

Magnusson, Paul; and Hong, Peter. "Save the dolphins - or free trade? A recent GATT ruling challenges U.S. environmental laws." *Business Week* 3252 (February 17, 1992): 130D(2)

McGuire, Richard. "Economic priorities and the environmentalist: the dimensions of our dilemma." *Vital Speeches of the Day* 58,11 (March 15, 1992): 329(5)

Repetto, Robert. "Nature's resources as productive assets." *Challenge* 32,5 (September-October 1989): 16(5)

Simon, Julian L. "Now (I think) I understand the ecologists better." (an economist's view of ecologists) *Futurist* 21 (September-October 1987): 18(2)

"Sootbusters." (European Community carbon dioxide tax) (editorial) *Economist* 321,7728 (October 12, 1991): 19(2)

Weiner, Edith; and Brown, Arnold. "Exonomics: what economics fails to explain." *Futurist* 24,4 (July-August 1990): 36(4)

Environmental and professional journals (annotated)

Burnett, John. "Ecology, economics and the environment." *Royal Bank of Scotland Review* 167 (September 1990): 3-14.
> Examines the effects of man's actions--frequently agricultural or technological--on the environment, i.e., the ecological consequences. "These effects, although ecological and environmental in nature, are ultimately economic in their consequences. But economic theory and planning has rarely included such depreciation in its calculations, certainly not in a country's gross national product (GNP); neither has the question of replacing or restoring such assets always been addressed, nor the cost of such restoration, if indeed they can be restored at all."

"Greener budget." *Economics* 314 (January 27, 1990): 61-63.
"Taxing polluters could be an efficient way for the government to raise revenue while helping to clean up the environment."

Hahn, Robert W. "Economic prescriptions for environmental problems: how the patient followed the doctor's orders." *Journal of Economic Perspectives* 3 (Spring 1989): 95-114.
"Chronicles the experience with both marketable permits and emissions charges. It also provides a selective analysis of a variety of applications in Europe and the United States and shows how the actual use of these tools tends to depart from the role which economists have conceived for them." Shows that regulatory systems involving multiple policy instruments are the rule rather than the exception, and that the level of oversight can affect the implementation of policies.

Repetto, Robert. "Accounting for environmental assets." *Scientific American* 266 (June 1992): 94-98, 100.
Evaluates the lack of economic values on natural resources and concludes that "the basic measuring instrument must be recalibrated if policymakers are to recognize and be held accountable for the wholesale disruption of natural systems now under way."

Takekawa, John Y.; and Garton, Edward O. "How much is an evening grosbeak worth?" *Journal of Forestry* 82 (July 1984): 426-428.
Estimates "the economic value of bird predation on western spruce budworm on two stands in north-central Washington by substituting the cost to spray with insecticides to produce the same mortality rate as birds cause. It would cost at least $1,820 per square km per year over a 100-year rotation." Evaluates the cost-effectiveness of biological control with evening grosbeaks.

Environmental and professional journals (unannotated)

Corcoran, Elizabeth; and Wallich, Paul. "Green economists." *Scientific American* 262,5 (May 1990): 86(2)

Greanville, David P. "Turning inward: the promise of the steady-state economy." *The Animals' Agenda* 9,2 (February 1989): 40(3)

Hahn, Robert W.; and Stavins, Robert N. "Economic incentives for environmental protection: integrating theory and practice." *American Economic Review* 82,2 (May 1992): 464(5)

Holden, Constance. "Multidisciplinary look at a finite world." (ecological economics) *Science* 249,4964 (July 6 1990): 18(2)

Oates, Wallace E.; Portney, Paul R.; and McGartland, Albert M. "The net benefits of incentive-based regulation: a case study of environmental standard setting." *American Economic Review* 79,5-6 (December 1989): 1233(10)

Sloan, J. J. "Making a start on avoiding calamity." (economic collapse) *Countryside & Small Stock Journal* 73,4 (July-August 1989): 54(2)

Law journals

Stewart, Richard B. "Controlling environmental risks through economic incentives." *Columbia Journal of Environmental Law* 13,2 (1988): 153-169.
> "Economic incentive systems, including pollution charges, transferable pollution permits, and waste deposit and refund programs, use of market principles to achieve environmental goals while avoiding many of the dysfunctions of centralized regulation." This article considers objections and obstacles to expanded use of economic incentives, discusses how to design and implement the incentives, and argues that we cannot continue to rely mostly on centralized commands to achieve environmental goals.

Reports

Dasgupta, Partha. *The environment as a commodity.* Helsinki, Finland: World Institute for Development Economics Research of the United Nations University, 1990. 50 p. (HC59 .D363 1990)

Environmental policy in a market economy: selected papers from the "Congress Environmental Policy in a Market Economy," Wageningen, Netherlands, 8-11 September 1987. Wageningen, Netherlands: Pudoc, 1988. 205 p.: ill. (HC79.E5 E5787 1988)

"Forum: U.S. productivity growth and living standards." *Congressional Research Service Review* 10 (June 1989): 1-23.
> Includes "Productivity and environmental regulation," by John L. Moore.

Hahn, Robert W.; and Stavins, Robert N. *Incentive-based environmental regulation: a new era from an old idea?* Cambridge, Mass.: Energy and Environmental Policy Center, John F. Kennedy School of Government, Harvard University, 1990. 40 p. (Discussion paper no. E-90-13)
> "Begins with a brief overview of conventional and alternative approaches to environmental regulation, and a review of previous U.S. experience with incentive-based policies. We chronicle how a shift in attitudes among influential interest groups is leading to new consideration of market-based proposals at the federal level, and we seek to explain why these changes are occurring."

Jorgenson, Dale W.; Slesnick, Daniel T.; and Wilcoxen, Peter J. *Carbon taxes and economic welfare*. Washington, D.C.: Brookings Institution, 1992. (papers on economic activity: microeconomics): 393-454.

First, describes "the inter-temporal general equilibrium model of the U.S. economy employed in our evaluation of the effect of carbon taxes." Outlines the framework for measuring the welfare of individual households and combining these into an overall measure of social welfare. Analyzes the effects of taxes required to hold U.S. carbon dioxide emissions constant at 1990 levels. Compares the growth of the U.S. economy with these taxes to a base case with no controls on emissions. Evaluates the distributional effect of carbon taxes, disaggregating the overall economic effect of the taxes to the level of individual households.

Macro-economic impact of environmental expenditure. Paris: Organisation for Economic Co-operation and Development, 1985. Distributed by OECD Publications and Information Centre, Washington, D.C. 120 p. (HC79.E5 M322 1985)

Nicolaisen, Jon; and Hoeller, Peter. *Economics and the environment: a survey of issues and policy options*. Paris: Organisation for Economic Co-operation and Development, 1990. 76 p. (Working papers no. 82)

Examines overuse of environmental resources in a market economy, and governmental policies designed to offset environmental degradation. Offers suggestions for preserving or expanding wealth.

Norton, Bryan G. *On the inherent danger of undervaluing species*. College Park, Md.: Center for Philosophy and Public Policy, 1983. 28 p. (Working paper: PS-3)

Argues that one need not advocate intrinsic value for any particular species because "natural diversity, in and of itself, has utilitarian value ... which does not depend upon discovering some economic or industrial use for it."

Pearce, David W. *Environmental policy benefits: monetary valuation*. Paris: Organisation for Economic Co-operation and Development, 1989. Distributed by OECD Publications and Information Centre, Washington, D.C. 83 p.: ill. (HC79.E5 P385 1989)

Protecting the environment: a free market strategy. Washington, D.C.: Heritage Foundation, 1986. 88 p. (HC110.E5 P77 1986)

Rao, Vaman. *A taxonomy of the techniques of economic impact analysis: final report*. Springfield, Ill. (325 W. Adams, Room 300, Springfield 62704-1892): Illinois Department of Energy and Natural Resources, 1989. 91 p. (HC107.I33 E557 1989)

Stavins, Robert N.; Wirth, Timothy E.; and Heinz, John. *Project 88--round II: incentives for action: designing market-based environmental strategies; a public policy study sponsored by Senator Timothy E. Wirth, Colorado, and Senator John Heinz, Pennsylvania*. New York: Carnegie Corporation, 1991. 95 p.

"Focuses on design issues associated with incentive-based policies for three problem areas of particular importance: global climate change due to the greenhouse effect; generation and disposal of solid and hazardous waste; and management of natural resources."

Stokoe, Peter. *Integrating economics and EIA: institutional design and analytical tools: a background paper prepared for the Canadian Environmental Assessment Research Council.* Hull, Quebec: Canadian Environmental Assessment Research Council, 1991. 44, 44 p.: ill. (HD75.6 .S75 1991)

Toman, Michael A.; and Crosson, Pierre. *Economics and 'sustainability:' balancing tradeoffs and imperatives.* Washington, D.C.: Resources for the Future, 1991. 37 p. (ENR-91-05)
> "Seeks to clarify the disparities between economists and ecologists in the interpretation of sustainability, and to begin identifying some common ground between the two disciplines. Our hope is that by at least providing some common vocabulary, unnecessary divisiveness can be avoided."

Weidenbaum, Murray. *Protecting the environment: harnessing the power of the marketplace.* St. Louis, Mo.: Center for the Study of American Business, Washington University, 1989. 19 p. (Formal publication no. 92)
> "If the government were to levy a fee on the amount of pollutants discharged, that would provide an incentive to reduce the actual generation of wastes. Some companies would find it cheaper to change their production processes than to pay the tax. Recycling and reuse systems would be encouraged."

Federal government reports

Environmental investments: the cost of a clean environment; report of the Administrator of the Environmental Protection Agency to the Congress of the United States. Washington, D.C.: Environmental Protection Agency, 1990. 501 p. (EPA-230-11-90-083, November 1990)
> Presents data on environmental pollution control costs during the period 1972-1987, projects these costs for each subsequent year to the year 2000 under a number of assumptions, and breaks them down in a variety of ways. These ways include differentiating between capital, operating, and annualized costs, as well as the medium where the pollution is controlled, the sector (e.g., public, private) from which the control is funded, new versus existing regulations, whether the control is primarily a result of a Federal mandate or the result of local initiative, and to the extent permitted by the data, by pollutant controlled."

Conference proceedings

Economy & ecology: the economics of environmental protection: a symposium. Edmonton, Alberta, Canada: Canadian Society of Environmental Biologists, 1985. 224 p.: ill.

Environment and economics: results of the International Conference on Environment and Economics, 18th-21st June, 1984. Paris: Organisation for Economic Co-operation and Development, 1985. Distributed by OECD Publications and Information Centre, Washington, D.C. 247 p.: ill. (HC79.E5 I532 1984)

Environment and economy, partners for the future: a conference on sustainable development: the proceedings. Winnipeg, Canada: Sustainable Development Coordination Unit, Executive Council, Province of Manitoba, 1989. 445 p. (HD75.6 .E55 1989)

Harvard University's John F. Kennedy School of Government/Project 88 Conference: harnessing market forces to protect the environment; summary of proceedings. Cambridge, Mass.: Energy and Environmental Policy Center, John F. Kennedy School of Government, Harvard University, 1989. 50 p. (M-89-02)
> A central theme of Project 88 is "that inefficient natural resource use and environmental degradation could be reduced by ensuring that consumers and producers face the true costs of their decisions--not just their direct costs, but the full social costs of their actions." Conference participants included: Robert Stavins, C. Boyden Gray, Frederic Krupp, Joseph Kalt, Dale Bumpers, William K. Reilly, Henry Waxman, Senators Heinz and Wirth, and Dean Allison.

Valuing the environment: economic approaches to environmental evaluation: proceedings of a workshop held at Ludgrove Hall, Middlesex Polytechnic, on 13 and 14 June 1990. New York: Belhaven Press, 1992.

Bibliographies

Kazis, Richard. "Jobs and the environment: a selected and annotated bibliography." Chicago: *Council of Planning Librarians*, 1983. 13 p. (CPL bibliography no. 116)
> "Surveys the literature on the economic impact of environmental regulations--the impact on jobs, inflation, productivity and innovation, and evidence on the costs and benefits of environmental protection. Focuses on job blackmail--what it is, how it works, why it is effective, and how its use has affected relations among unions and environmental groups. Surveys materials on the history of the environmental and occupational health and safety movements. Examines the current economic crisis and surveys reform proposals which address the nation's related failures to provide sufficient jobs and to protect environmental quality."

McCarl, Henry N. *Bibliography on environmental economics, 1980-1990.* Monticello, Ill.: Vance Bibliographies, 1990. 91 p. (Z5863.P6 M353 1990)

SUSTAINABLE DEVELOPMENT
███████████████████

(Related topics in *Global Environmental Policy*) 255

Books (annotated)

Holmberg, Johan. *Making development sustainable.* Washington, D.C.: Island Press, 1992. 362 p.: ill., maps. (HD75.6 .M35 1992)
> Begins with an analysis of political institutions and the economic principles upon which public policy is based, then follows with studies about how to increase sustainability in agriculture, urban development, industry, forestry, and energy use, and ends with a discussion of the potential for financing new policy initiatves to reduce population growth and consumption. A separate chapter is devoted to an examination of successful development in African drylands.

Jacobs, Michael. *The green economy: environment, sustainable development, and the politics of the future.* Concord, Mass.: Pluto, 1991. 312 p.: ill. (HD75.6 .J33 1991)
> Summarizes different perspectives on ecologically sustainable economic development. Analyzes the weaknesses of the traditional, neo-Classic viewpoint. Maintains that common property resources should not be left to market manipulation because future generations, who have a vested interest in these resources, do not have a voice in how they are handled today. Present policy must take into account such factors as wealth distribution, politics, and economic power.

Schmidheiny, Stephan. *Changing course: a global business perspective on development and the environment.* Cambridge, Mass.: MIT Press, 1992. 374 p. (HD75.6 .S35 1992)
> Maintains that business must change its goals and assumptions if it is to play an active role in sustainable development, and presents thirty-seven case studies of businesses that participate in sustainable development.

Sustainable development of the biosphere. New Rochelle, N.Y.: Cambridge University Press, 1986. 491 p.: color ill. (HD75.6 .S87 1986)
> Contains papers presented at a meeting "to synthesize in policy terms our understanding of global ecological and geophysical systems as they are linked with industrial and resource development activities; to characterize the issues of global environmental change in terms of their ability to inhibit or promote regional development; to explore institutional and organizational designs for more effective international research, policymaking, and management, concerning interactions between environment and regional development."

Thibodeau, Francis R.; and Field, Hermann H. *Sustaining tomorrow: a strategy for world conservation and development.* Hanover, N.H.: University Press of New England, 1984. 186 p.
> This book is a product of the World Conservation Strategy of the International Union for the Conservation of Nature and Natural Resources in cooperation with the United Nations Environment Programme and World Wildlife Fund. Outlines views of eighteen experts on how to balance environmental protection with equitable economic development.

World development report 1992: development and the environment. New York: Oxford University Press, 1992. 308 p.

The fifteenth in this annual series emphasizes the need to integrate environmental considerations into development policymaking.

Books (unannotated)

Barbier, Edward. *Economics, natural-resource scarcity and development: conventional and alternative views.* London: Earthscan, 1989. 223 p.: ill. (HC59 .B373 1989)

Bojo, Jan. *Environment and development: an economic approach.* 2d rev. ed. Boston: Kluwer Academic, 1992. 211 p.: ill. (HD75.6 .B65 1992)

Common property resources: ecology and community-based sustainable development. New York: Belhaven, 1989. 302 p.: ill. (HD75.6 .C647 1989)

Court, Thijs de la. *Beyond Brundtland: green development in the 1990s.* Atlantic Highlands, N.J.: Zed, 1990. 139 p.: ill. (HD75.6 .C69513 1990)

Davis, John. *Greening business: managing for sustainable development.* Cambridge, Mass.: Blackwell, 1991. 215 p.: ill. (HD75.6 .D38 1990)

Dube, Shyama Charan. *Modernization and development: the search for alternative paradigms.* Atlantic Highlands, N.J.: Zed, 1988. 144 p. (HD87 .D8 1988)

Earth and us: population, resources, environment, development. Boston: Butterworth-Heinemann, 1991. 107 p. (HD75.6 .P66 1990)

Ecodevelopment: concepts, projects, strategies. New York: Pergamon, 1984. 247 p.: ill. (HD75.6 .E26 1984)

Ecological economics: the science and management of sustainability. New York: Columbia University Press, 1991. 525 p.: ill. map. (HD75.6 .E29 1991)

Economics for the wilds: wildlife, diversity, and development. Washington, D.C.: Island Press, 1992. 226 p.: ill. maps. (QL82 .E26 1992)

Economy and ecology: towards sustainable development. Norwell, Mass.: Kluwer Academic, 1989. 348 p.: ill. (HD75.6 .E33 1989)

Ethics of environment and development: global challenge, international response. Tucson: University of Arizona Press, 1990. 264 p.: ill. (HD75.6 .E84 1991)

Global development and the environment: perspectives on sustainability. Washington, D.C.: Resources for the Future, 1992. 91 p.: ill. (HD75.6 .G54 1992)

Global resources: opposing viewpoints. San Diego, Calif.: Greenhaven, 1991. 288 p.: ill. (HD75.6 .G56 1991)

Goldsmith, Edward. *The great U-turn: de-industrializing society.* Hartland, England: Green, 1988. 217 p.: ill. (HC21 .G65 1988)

Gourlay, K. A. *World of waste: dilemmas of industrial development.* Atlantic Highlands, N.J.: Zed, 1992. 247 p. (TD897.5 .G68 1992)

Grassroots environmental action: people's participation in sustainable development. New York: Routledge, Chapman & Hall, 1992

Henderson, Hazel. *Paradigms in progress: life beyond economics.* Indianapolis, Ind.: Knowledge Systems, 1991. 293 p.: ill. (HD75.6 .H46 1991)

In search of indicators of sustainable development. Boston: Kluwer Academic, 1991. 126 p.: ill. map. (HC329.5.E5 I56 1991)

Indiresan, P. V. *Managing development: decentralisation, geographical socialism, and urban replication.* Newbury Park, Calif.: Sage, 1990. 284 p.: ill. (HD75.6 .I53 1990)

Jacobs, Michael. *The green economy: environment, sustainable development, and the politics of the future.* Concord, Mass.: Pluto, 1991. 312 p.: ill. (HD75.6 .J33 1991)

Kane, Hal. *Time for change: a new approach to environment and development.* Washington, D.C.: Island Press, 1992. 141 p.: ill. (HD75.6 .K36 1992)

Khozin, Grigorii Sergeevich. *Talking about the future: can we develop without disaster?* Moscow: Progress, 1988. 144 p.: color ill. (HD75.6 .K5 1988)

Making development sustainable: redefining institutions, policy, and economics. Washington, D.C.: Island Press, 1992.

Mikesell, Raymond Frech. *Economic development and the environment: a comparison of sustainable development with conventional development economics.* New York: Mansell, 1992.

Modelling for population and sustainable development. New York (International Social Science Council with the co-operation of UNESCO and the Institute for Environmental Studies, Free University, Amsterdam): Routledge, Chapman & Hall, 1991. 261 p.: ill. (HB849.51 .M645 1990)

Simonis, Udo Ernst. *Beyond growth: elements of sustainable development.* Berlin: Edition Sigma, 1990. 151 p.: ill. (HD75.6 .S57 1990)

Sustainable development. Portland, Oreg.: F. Cass in association with the European Association of Development Research and Training Institutes (EADI), Geneva, 1991. 132 p.: ill. (HD75.6 .S86 1991)

Sustainable environmental management: principles and practice. Boulder, Colo.: Westview, 1988. 292 p.: ill. (HD75.6 .S88 1988)

Tisdell, Clement Allan. *Environmental economics: policies for environmental management and sustainable development*. Brookfield, Vt.: E. Elgar, 1992.

Tisdell, Clement Allan. *Natural resources, growth, and development: economics, ecology and resource-scarcity*. New York: Praeger, 1990. 186 p.: ill. (HD75.6 .T565 1990)

Young, Michael Denis. *Sustainable investment and resource use: equity, environmental integrity, and economic efficiency*. Park Ridge, N.J.: Parthenon, 1992. 176 p.: ill. (HD75.6 .Y68 1992)

General interest periodicals (annotated)

Gardner, Julia; and Roseland, Mark. "Thinking globally." *Alternatives* 16 (October-November 1989): 26-48.
> Argues that "successful pursuit of sustainable development depends upon interdependent principles aimed at meeting human needs, maintaining ecological integrity, attaining social self-determination, and establishing social equality." Discusses the strategies and actions that have emerged from conventional interpretations of sustainable development and the paradigms that have the potential to redirect humans toward sustainable development with equity.

Repetto, Robert. "Environmental productivity and why it is so important." *Challenge* 33 (September-October 1990): 33-38.
> "Although the economic costs of pollution are known to be significant, and although natural systems that provide materials and energy or serve as a repository for wastes are increasingly stressed, environmental dimensions of productivity change remain completely unexplored. This article represents a first exploratory step in that direction, taking as an object of study the electric power industry in the United States."

General interest periodicals (unannotated)

Allman, William F. "The preservation paradox." (peaceful coexistence between conservation and economic development) *U.S. News & World Report* 104,16 (April 25, 1988): 53(2).

Batisse, Michel. "A partnership with nature." *UNESCO Courier* (November 1991): 14(2).

Batisse, Michel. "Reviewing the accounts." *UNESCO Courier* (January 1991): 45(3).

di Castri, Francesco. "Time to act." *UNESCO Courier* (November 1991): 40(3).

Gibson, Robert. "On having your cake and eating it too." *Alternatives* 16,3 (October--November 1989): 67(3).

Hodel, Donald Paul. "A sensible balance." *Reader's Digest* 133,796 (August 1988): 117(4).

Kneen, Brewster. "The contradiction of sustainable development." *Canadian Dimension* 23,1 (January-February 1989): 12(4).

Kuhn, Thomas R. "Striving for balanced growth: preserving the environment while fueling the economy." *Vital Speeches of the Day* 55,20 (August 1, 1989): 637(4).

McGuire, Richard "Economic priorities and the environmentalist: the dimensions of our dilemma." (speech by commissioner of the agriculture and markets of the State of New York) *Vital Speeches of the Day* 58,11 (March 15, 1992): 329(5).

Orton, David. "A no-growth economy." *Canadian Dimension* 24,4 (June 1990): 30(2).

Pearce, Fred. "Mega-projects create some giant fears: is development always good for the Earth?" *World Press Review* 37,9 (September 1990): 38(2).

Serafin, Rafal. "Sustainable development of the biosphere." (book reviews) *Alternatives* 15,1 (December-January 1987): 50(3).

von Droste, Bernd. "No room in the Ark." (effect of economic growth on animal species) *UNESCO Courier* (November 1991): 36(4).

Environmental and professional journals (annotated)

Allenby, Braden R. "Achieving sustainable development through industrial ecology." *International Environmental Affairs* 4 (Winter 1992): 56-68.
> "Our species will reach sustainable levels, of course. It is our choice as to whether this occurs through natural population control mechanisms (starvation, epidemics) and social collapse, or planned evolution toward sustainable development. If we choose the latter course, we must begin now to explore and understand the industrial ecology metasystem."

Barbier, Edward B. "The concept of sustainable economic development." *Environmental Conservation* 14 (Summer 1987): 101-110.

"Attempts to refine and substantiate the concept of sustainable economic development." Sustainable economic development argues that "real" improvements cannot occur unless "the strategies which are being formulated and implemented are ecologically sustainable over the long term, are consistent with social values and institutions, and encourage 'grassroots' participation in the development process."

Doyle, Derek. "Sustainable development: growth without losing ground." *Journal of Soil and Water Conservation* 46 (January-February 1991): 8-13.

Discusses sustainable development from the viewpoint of the natural resource manager, recounting lessons from the past, examining global forces of change, sampling issues of contention in North America, and recommending management strategies.

"Environment and development." *Resources* 106 (Winter 1992): whole issue (38 p.)

Contents include: "The difficulty in defining sustainability," by Michael A. Toman; "The role of natural assets in economic development," by Raymond J. Knopp; "Sustainable agriculture," by Pierre R. Crosson; "Climate variability and development," by Peter M. Morrisette and Norman J. Rosenberg; and "Using benefit-cost analysis to prioritize environmental problems," by Alan J. Krupnick.

Grue, Harald. "The Brundtland report." *Ceres* 21 (July-August 1988): 28-30.

Havard Altern interviews the Permanent Secretary of Norway's Ministry of Agriculture. The discussion centers on sustainable development and its impact in Norway on agricultural policy and environmental protection--three themes detailed in the World Commission on Environment and Development's Brundtland Report.

Hill, David R. "Sustainability, Victor Gruen, and the cellular metropolis." *Journal of the American Planning Association* 58 (Summer 1992): 312-326.

This well-known practitioner of planning and architecture "also contributed significantly to the theory of good urban form. He argued that a world system of medium-sized, dense, urbane, cellular metropolises would improve considerably both human happiness and global ecological sustainablity. This article first places Gruen's prototype in the contemporary literature on global sustainability, and outlines his life and work as a paractitioner/theorist."

McNeely, Jeffrey A. "How conservation strategies contribute to sustainable development." *Environmental Conservation* 17 (Spring 1990): 9-13.

The challenge facing nations today "is no longer deciding whether conservation is a good idea, but rather how it can be implemented in the national interest and within the means available in each country." What is the national interest, and why are insufficient means made available for conserving biological resources?

The chief conservation officer of the IUCN "suggests that the answers to both those questions come from the field of economics, and outlines how conservation strategies can draw on economics to support government policies which will promote forms of development that are sustainable in the long run."

Myers, Norman; and Brundtland, Gro Harlem. "Making the world work for people." *International Wildlife* 19 (November-December 1989): 12-15.
" 'Sustainable development' is the new buzzword for saving the earth, but what exactly does it mean?" Interviews Gro Harlem Brundtland, head of the U.N.'s World Commission on Environment and Development, about its report entitled "Our Common Future."

Orians, Gordon H. "Ecological concepts of sustainability." *Environment* 32 (November 1990): 10-15, 34-39.
"Perhaps a protocol could be developed for identifying and ranking VECs (Valued Ecosystem Components) given different time and spatial scales, so that the approximate limits to sustainable development imposed by a VEC could be determined. With such information, it may be possible to formalize the means by which conflicts regarding different VECs might be resolved and weighted values might be assigned to each VEC."

Ponting, Clive. "Historical perspectives on sustainable development." *Environment* 32 (November 1990): 4-9, 31-33.
Examines the histories of Mesopotamia, the Mediterranean, and the Maya as they relate to the concept of sustainable development.

Rees, William E. "The ecology of sustainable development." *Ecologist* 20 (January-February 1990): 18-23.
"As the term 'sustainable development' has been embraced by the political mainstream, so it has been stripped of its original concern with ensuring future ecological stability. It is no longer a challenge to the conventional economic paradigm but rather has become another excuse for continued economic growth. True sustainability demands a radically different economics which fully recognizes the processes and limits of the biosphere."

Sachs, Wolfgang. "Environment and development: the story of a dangerous liaison." *Ecologist* 21 (November-December 1991): 252-257.
Maintains that states and development organizations have interpreted environmental destruction as the result of poverty and inefficiency. While "economic growth, more efficient production methods and better monitoring of the biosphere the only answers to the problem. Thus the fundamental issues which society must address--how much it should produce and consume--are ignored."

Simon, David. "Sustainable development: theoretical construct or attainable goal?" *Environmental Conservation* 16 (Spring 1989): 41-48.

Reviews "current thinking on sustainability, and then raises some pertinent questions about how readily such concepts can be operationalized in the context of the prevailing world order."

Watkins, G. C. "The role of energy economists in promoting sustainable energy development." *Energy Policy* 20 (June 1992): 575-580.

"The role of energy in pursuit of policies seeking sustainable development is crucial. Correspondingly, the work of energy economists will be affected in many traditional areas of analysis and will require enhanced scope and new expertise. This would lead to a better understanding of the place of natural resources in the production process, better delineation of trade offs between avoidance of ecological degradation and economic stagnation, and more interdisciplinary feedback."

Wheeler, Joseph C. "The interwoven strands of development." *OECD Observer* 167 (December 1990-January 1991): 31-33.

Traces the issues of population growth and environmental degredation and their importance to both rich and poor countries; discusses their importance to the crafting of UNCED.

Environmental and professional journals (unannotated)

Burton, Ian. "Our common future; the World Commission on Environment and Development." *Environment* 29 (June 1987): 25(5).

Carey, John. "Will saving people save our planet?" *International Wildlife* 22,3 (May-June 1992): 12(11).

Catton, William R., Jr. "The world's most polymorphic species; carrying capacity transgressed two ways." (putting ecological issues on the public policy agenda) *BioScience* 37 (June 1987): 413(7).

MacNeil, Jim. "Strategies for sustainable economic development." *Scientific American* 261,3 (September 1989): 154(10).

Tester, Frank. "As if people matter more: rethinking impact assessment and international development." *Canadian Dimension* 23,1 (January-February 1989): 16(2).

Law journals

Muldoon, Paul R. "The international law of ecodevelopment: emerging norms for development assistance agencies." *Texas International Law Journal* 22 (Winter 1987): 1-52.

"The international law of ecodevelopment merges environmental and development concerns to create a framework for the rational use, allocation, and management of natural resources and the conservation and rehabilitation of local, regional, and global environments." Examines the emergence and progression of new international legal principles and their impacts upon development agencies.

Reports

Colby, Michael E. *The evolution of paradigms of environmental management in development.* Washington, D.C.: Strategic Planning and Review Department, World Bank (1818 H Street NW, Washington 20433), 1989. 37 p.: ill. (HD75.6 .C64 1989)

Environmental accounting for sustainable development. Washington, D.C.: World Bank, 1989. 100 p.
> Based on a UNEP-World Bank Symposium. Contains: "Environmental and resource accounting: an overview," by Salah El Serafy and Ernest Lutz; "The proper calculation of income from depletable natural resources," by Salah El Serafy; "Measuring pollution within the framework of the national accounts," by Derek W. Blades; "Environmental accounting in development policy: the French experience," by Jacques Theys; "Environmental and nonmarket accounting in developing countries," by Henry M. Peskin; "A proposed environmental accounts framework," by Henry M. Peskin; and "Environmental accounting and the system of national accounts," by Peter Bartelmus.

Rees, William E. *Sustainable development and the biosphere: concepts and principles.* Chambersburg, Pa. (American Teilhard Association for the Future of Man): Anima, 1990. 28 p. (HD75.6 .R44 1990)

Toman, Michael A.; and Crosson, Pierre. *Economics and "sustainability": balancing tradeoffs and imperatives.* Washington, D.C.: Resources for the Future, 1991. 37 p. (ENR-91-05)
> Seeks "to clarify the disparities between economists and ecologists in the interpretation of sustainability, and to begin identifying some common ground between the two disciplines." Hopes that by providing some common vocabulary, unnecessary divisiveness can be avoided.

Federal government reports

Managing natural resources for sustainable development: special report on the environment. Washington, D.C.: Agency for International Development, 1987. 48 p. (HC59 .E55 1987)
> Contents: "Meeting today's challenge;" "Biological diversity;" "Environmental health and safety;" "Coastal zones;" "Forestry and fragile lands;" "Haiti agroforestry;" and "The challenge ahead."

Conference proceedings

Environment and development: building sustainable societies: lectures from the 1987 summer forum at the University of Wisconsin-Madison. Madison, Wis.: Institute for Environmental Studies, University of Wisconsin-Madison: Copies from Institute for Environmental Studies, Office of Publications, Information, and Outreach, 1988. 121 p.: ill. (HD75.6 .E54 1988)

Environment and economy, partners for the future: a conference on sustainable development: the proceedings. Winnipeg, Manitoba, Canada: Sustainable Development Coordination Unit, Executive Council, Province of Manitoba, 1989. 445 p. (HD75.6 .E55 1989)

Ministerial Seminar EC-EFTA Environment--sustainable development, Copenhagen, 14-15 November, 1989. Copenhagen: Ministry of the Environment, 1989. 47 p. (HD75.6 .M56 1989)

Power of convening: collaborative policy forums for sustainable development: proceedings of an international workshop held at Claremont, California, October 5-7, 1989. Sacramento, Calif.: California Institute of Public Affairs in cooperation with IUCN-the World Conservation Union and the Center for Politics and Policy of the Claremont Graduate School, 1990. 112 p.: ill. (HD75.6 .P68 1990)

Bibliographies

Merideth, Robert W. *Global sustainability: a selected, annotated bibliography.* Madison, Wis.: Institute for Environmental Studies, University of Wisconsin-Madison (550 N. Park Street, 15 Science Hall, Madison 53706), 1991. 46 p. (Z7164.E2 M473 1991)

CHAPTER 7

ANIMAL RIGHTS

continued

ANIMAL RIGHTS

(See also *Endangered Species: Captive Breeding*) 2719
(See also *Health and the Environment: Vegetarianism*) 1297

SHORT LIST OF SOURCES
FOR GENERAL RESEARCH

Books

Baird, Robert M.; and Rosenbaum, Stuart E. eds. *Animal experimentation: the moral issues.* Buffalo, N.Y.: Prometheus Books, 1991. 182 p. (HV4915.A64 1991)
This collection of previously published materials brings together the full range of opinions on animal rights, as well as summarizing the various ways humans use animals. Essays from both ends of the spectrum, written by prominent supporters and opponents of animal rights, are skillfully juxtaposed with critiques that challenge their arguments. Some intermediate positions are also presented.

Clark, Stephen R. L. *The nature of the beast: are animals moral?* New York: Oxford University Press, 1982. 127 p.
Reflects and speculates on a variety of concepts relating to morality, including intelligence and language, freedom and necessity, me and mine, altruism, parenthood, territory and dominance. Defines animal morality as "the impulses and inhibitions, learned and instinctual, that seem to play the part in beasts that morals play in us." Those are "the morals of Nature." Concludes that animals are ethical in so far as they respond to particular situations in a way that a good man would respect, but that animals are not moral because they do not construct intellectual systems by which to judge their behavior.

Cook, Lori. *A shopper's guide to cruelty-free products.* New York: Bantam, 1991. 262 p. (TX336 .C66 1991)
Maintains that no practical benefits come from animal testing and that results of tests on animals are usually inconclusive. Rates various products according to the amount of animal testing that went into their production. The majority of the book is an alphabetical list of companies that do not test their products on animals. Offers alternatives to animal testing, such as not using unproven ingredients, using statistical tests, and using organ cultures.

Fox, Michael. *Inhumane society: the American way of exploiting animals*. New York: St. Martin's, 1990. 268 p.

> Asserts that this country lacks ethical awareness of animals. Discusses the treatment of farm animals, including the amputation of pigs' tales and the debeaking and declawing of poultry. Argues that pets should be chosen from animal shelters rather than custom bred. Offers many suggestions for saving the animals and, in so doing, the environment.

Hargrove, Eugene C., ed. *Animal rights, environmental ethics debate: the environmental perspective*. Albany: State University of New York Press, 1992. 273 p. (HV4711 .A575 1992)

> Traces the animal rights/environmental ethics dispute in an attempt to place the basic ideas in a historical context. All sides of the issue are presented by the numerous contributors who advocate different perspectives.

Johnson, Andrew. *Factory farming*. Cambridge, England: Blackwell, 1991. 272 p.: ill. (SF140.L58 J64 1991)

> Presents balanced reasons why factory farming is unethical and unnecessary, as well as environmentally unsound. Discusses various systems of intensive farming in terms of animal welfare and the environment. Includes a chapter on fish farming. Relates progressive steps that have been taken against factory farming in other countries. Points out that most meat contains animal dung. Contains notes, an index, and helpful appendices.

General interest periodicals

Cowley, Geoffrey. "Of pain and progress." *Newsweek* 112 (December 26, 1988): 50-55, 57, 59.

> "A growing social movement raises a thorny ethical question: do the practical benefits of animal experimentation outweigh the moral costs?"

Elshtain, Jean Bethke. "Why worry about the animals?" *Progressive* 54 (March 1990): 17-23.

> Discusses the public's growing awareness of animal abuse in science/research labs, war research experiments, factory farms, and household products testing. Examines tactics, such as civil disobedience and violation of laws, that animal-welfare activists are taking to protect the animals.

White, Robert J. "The facts about animal research." *Reader's Digest* 132 (March 1988): 127-132.

> " 'I am convinced that most Americans are unaware of the devasting effect that animal-rights extremists are having on lifesaving medical research,' warns this eminent scientist and neurosurgeon."

Wright, Robert. "Are animals people too?" *New Republic* 202 (March 12, 1990): 20-23, 26-27.
  Notes that the animal rights movement has grown over the past decade. Reasons that if animals are sentient beings capable of experiencing pleasure and pain than inflecting pain on them raises numerous moral questions.

Zak, Steven. "Ethics and animals." *Atlantic Monthly* 263 (March 1989): 69-74.
  "Our mores and laws now grant that it's wrong to torture animals to no purpose. But, an animal-rights activist argues, members of other species have a far greater moral claim on us than that."

Environmental and professional journals

"Suffer the animals." *Environmental Action* 21 (May-June 1990): 23-30.
  Explores the shared concerns and points of conflict of environmentalists and animal protectionists regarding animal-derived toxicity testing and "factory farming."

Connif, Richard. "Fuzzy-wuzzy thinking about animal rights." *Audubon* 92 (November 1990): 120-122, 124, 126-133.
  "The animal rights movement has elevated ignorance about the natural world almost to the level of a philosophical principle." Argues that nature exhibits no ethical prohibition against violence and killing.

Fox, Michael W. "Livestock animals . . . a question of rights: the plowboy interview with Dr. Michael Fox." *Mother Earth News* 79 (January-February 1983): 16-20, 22.
  An interview with the director of the Institute for the Study of Animal Problems presents his views on modern livestock rearing practices, many of which he describes as inhumane.

Gavaghan, Helen. "Animal experiments the American way." *New Scientist* 134 (May 16, 1992): 32-37.
  Contends that the U.S. Animal Welfare Act offers very few rules, regulations or standards concerning the actual practice of research using animals, and those it does can sometimes be ignored in the interests of science.

Hampson, Judith. "The secret world of animal experiments." *New Scientist* 134 (April 11, 1992): 24-30.
  Urges public accountability for animal experiments. An accompanying article suggests that scientists should give details of any adverse effects on animals so that the same mistakes can be avoided by others.

Junkin, Elizabeth Darby. "Solomon's child." *Buzzworm: The Environmental Journal* 1 (Spring 1989): 20-29.

Examines controversy over the use of chimpanzies for AIDS research. Late in 1988, the Fish and Wildlife Service reclassified wild populations from "threatened" to "endangered" while leaving captive animals in the threatened category and legally available for research.

Sleeper, Barbara. "The high price of fur." *Animals* 121 (November-December 1988): 14-19.
"The fur industry argues that if we can kill animals to eat meat, conduct experimental research, and wear leather clothes, shoes, and handbags, then why not kill for fur as well? Animal-rights activists argue that the methods used to trap animals in the wild are cruel and barbaric and that the fur farms are run like animal concentration camps. The deaths of these animals, they say, serve no other purpose than to supply an unnecessary luxury."

Solomon, Robert C. "Has not an animal organs, dimensions, senses, affections, passions?" *Psychology Today* 16 (March 1982): 36, 39-41, 43-45.
Describes the views of scientists on animal consciousness, animal suffering, and feeling.

Videos

In defense of animals: a portrait of Peter Singer. Directed and produced by Julie Akeret. Distributed by Bullfrog Films, Oley, Pa. 1989. 28 min.
Addresses the issue of animal rights, makes a case for vegetarianism, points out some questionable government requirements regarding animal testing, and seems to advocate its total elimination. Contains interviews from people of all ages and graphic footage of animal testing. Discusses animal rights in anthropocentric terms.

INTRODUCTORY SOURCES

Books (annotated)

Budiansky, Stephen. *The covenant of the wild: why animals choose domestication*. New York: William Morrow, 1992.
Maintains that the relationship between humans and animals is more symbiotic than one of a master and servant. Looks to relationships between species in the natural world as parallel examples to relationships between humans and domesticated animals. Argues that the horrors of laboratory experiments are often emphasized by people who are unaware of some of nature's horrors.

Hemmer, Helmut. *Domestication: the decline of environmental appreciation*. New York: Cambridge University Press, 1990.

> Focuses on the differences between domestic and wild animals. Maintains that domestic animals are the evolutionary product of less aggressive and less environmentally sensitive wild animals. Thus, they are the product of regressive evolution.

Leahy, Michael P. T. *Against liberation: putting animals in perspective*. New York: Routledge, 1991. 273 p. (HV4708 .L43 1991)

> Analyzes recently published works supporting animal rights, and claims that animal-rights advocates make two mistakes. They improperly give animals human characteristics, and they misuse terminology when they describe how animals behave. Asserts that animal-rights advocates misinterpret animal behavior by thinking of the way animals respond to stimuli as "understanding." Contends that understanding and self-awareness are dependent upon language, and it is therefore man's possession of language that distinguishes him from other animals. Since animals cannot share understanding, they should not been given the same rights as humans. Anticipates potential arguments against this argument and applies actual issues, including experimentation, killing for food and fur, hunting for sport, and zoos. Claims to be offering a concise philosophical position against animal rights.

Magel, Charles R. *Keyguide to information sources in animal rights*. Jefferson, N.C.: McFarland, 1989. 267 p. (Z7164.C45 M36 1989)

> Aims to guide readers to the thought, literature, and organizations of the animal rights movement. Begins with an introduction that discusses the history of the movement. Information is divided into six bibliographic essays which cover the topics of philosophy, medicine and science, education, law, religion, and vegetarianism. Also includes a separate annotated bibliography of 335 major works as well as a bibliography for works that are mentioned in the essays but not included in the annotated section. Contains a list of almost two hundred organizations, both in the U.S. and abroad.

Midgley, Mary. *Animals and why they matter*. Athens: University of Georgia Press, 1985.

> Addresses the anthropomorphizing of animals, vegetarianism, the limits of animal welfare, humanitarian reform and attitude toward other species, and the parallels between animal rights and women's rights.

Newkirk, Ingrid. *Free the animals!: the untold story of the Animal Liberation Front and its founder, Valerie*. Chicago: Noble Press, 1992.

> Focuses on the Animal Liberation Front (ALF) and its anonymous founder. The author has first-hand knowledge of the founder and of how this secret organization operates. ALF is known for such activities as rescuing animals from research laboratories.

Preece, Rod; and Chamberlain, Lorna. *Animal welfare and human values.* Waterloo, Ontario, Canada: Wilfrid Laurier University Press, 1993. 280 p.
> Discusses the full range of animal welfare issues, from fur and trapping, experimentation, and animals raised for food, to animals in entertainment. Argue for a "trans-species democracy" in which humans' intrinsic primal link to the animal kingdom is given greater appreciation.

Rowan, Andrew N., ed. *Animals and people sharing the world.* Hanover, N.H.: University Press of New England, 1992.
> According to the contributors, the human-animal bond extends far back into human history, but only recently have scholars turned their attention to the interaction between humans and animals. Various chapters include pet-keeping, emotions toward pets, the need to care for animals, expressions of alter egos, and the role of small and large animals in our lives.

Books (unannotated)

Adams, Carol J. *The sexual politics of meat: a feminist-vegetarian critical theory.* New York: Continuum, 1990. 256 p.: ill. (HV4708 .A25 1990)

Animal rights and welfare. New York: H. W. Wilson, 1991. 168 p. (HV4708 .A55 1991)

Animal rights: opposing viewpoints. San Diego, Calif.: Greenhaven Press, 1989. 235 p.: ill. (HV4711 .A58 1989)

Animal welfare and the law. New York: Cambridge University Press, 1989. 283 p. (KD3424.A75 A53 1990)

Animals and people sharing the world. Hanover, N.H.: University Press of New England, 1988. 192 p.: ill. (SF411.4 .A56 1988)

Animals and their legal rights: a survey of American laws from 1641 to 1990. 4th ed. Washington, D.C. (P.O. Box 3650, Washington 20007): Animal Welfare Institute, 1990. 441 p. (KF3841 .L4 1990)

Beyond the bars: the zoo dilemma. Rochester, Vt.: Thorsons, 1987. 208 p.: ill. (HV4708 .B49 1987)

Boyd, B. R. *The new abolitionists: animal rights and human liberation: an essay.* Rev. ed. San Francisco, Calif.: Taterhill Press, 1987. 24 p.: ill. (HV4708 .B69 1987)

Bresler, Fenton S. *Beastly law.* North Pomfret, Vt.: David and Charles, 1986. 126 p.: ill. (K183 .B74 1986)

Carman, Russ. *The illusions of animal rights*. Iola, Wis.: Krause Publications, 1990. 160 p. (HV4708 .C37 1990)

Collard, Andree. *Rape of the wild: man's violence against animals and the Earth*. Bloomington: Indiana University Press, 1989, 1988. 187 p. (GF50 .C63 1989)

Dawkins, Marian Stamp. *Animal suffering: the science of animal welfare*. New York: Chapman and Hall, 1980. 149 p. (HV4708 .D38)

Dolan, Edward F. *Animal rights*. New York: F. Watts, 1986. 144 p.: ill. (KF3841.Z9 D65 1986)

Domestication, conservation, and use of animal resources. New York: Elsevier Science, 1983. 357 p.: ill. (SF61 .D65 1983)

Duffy, Maureen. *Men and beasts: an animal rights handbook*. London: Paladin, 1984. 160 p. (HV4805.A3 D84 1984)

Extended circle: a commonplace book of animal rights. New York: Paragon House, 1989. 436 p. (PN6084.A57 E98 1989)

Favre, David S. *Animal law*. Westport, Conn.: Quorum Books, 1983. 253 p. (KF390.5.A5 F38 1983)

Fox, Michael W. *Animals have rights, too*. New York: Continuum, 1991. 174 p.: ill. (HV4708 .F718 1991)

Frey, Raymond Gillespie. *Interests and rights: the case against animals*. New York: Oxford University Press, 1980. 176 p. (HV4708 .F76)

Hannah, Harold W. *Legal briefs from the Journal of the American Veterinary Medical Association*. Schaumburg, Ill.: American Veterinary Medical Association, 1986. 256 p. (KF3835.A75 H36 1986)

Hollands, Clive. *Compassion is the bugler: the struggle for animal rights*. Edinburgh: Macdonald, 1980. 201 p.: plates. ill. (HV4805.A3 H64)

In defence of animals. New York: Blackwell, 1985. 224 p. (HV4711 .I6 1985)

Leiber, Justin. *Can animals and machines be persons?: a dialogue*. Indianapolis, Ind.: Hackett, 1985. 76 p. (BD331 .L44 1985)

Lynge, Finn. *Arctic wars, animal rights, endangered peoples*. Hanover, N.H.: Dartmouth College: University Press of New England, 1992.

Managing the behaviour of animals. New York: Chapman and Hall, 1990. 257 p.: ill. (QL751 .M2198 1990)

Mighetto, Lisa. *Wild animals and American environmental ethics.* Tucson: University of Arizona Press, 1991. 177 p.: ill. (QL85 .M54 1991)

Moretti, Daniel S. *Animal rights and the law.* New York: Oceana Publications, 1984. 147 p. (KF3841.Z95 M67 1984)

Morris, Desmond. *The animal contract: sharing the planet.* New York: Warner Books, 1990. 169 p. (QL85 .M67 1991)

O'Barry, Richard. *Behind the dolphin smile.* Chapel Hill, N.C.: Algonquin Books; Dallas, Tex.: Taylor Publishing, 1988. 252 p.: plates. (some color) ill. (QL795.D7 O23 1988)

O'Connor, Karen. *Sharing the kingdom: animals and their rights.* New York: Dodd, Mead, 1984. 144 p.: ill. (HV4764 .O38 1984)

Owen, Marna A. *Animal rights—yes or no?* Minneapolis, Minn.: Lerner, 1992.

Parker, Ian S. C. *Oh quagga!: thoughts on people, pets, loving animals, shooting them, and conservation.* Nairobi, Kenya: I. Parker, 1983. 131 p.: ill. (QL85 .P37 1983)

Regan, Tom. *All that dwell therein: animal rights and environmental ethics.* Berkeley: University of California Press, 1982. 249 p. (HV4711 .R36 1982)

Regan, Tom. *The case for animal rights.* Berkeley: University of California Press, 1983. 425 p. (HV4708 .R43 1983)

Regan, Tom. *The thee generation: reflections on the coming revolution.* Philadelphia, Pa.: Temple University Press, 1991. 180 p. (BJ1469 .R43 1991)

Rothschild, Miriam. *Animals and man.* New York: Oxford University Press, 1986. 98 p. (HV4711 .R58 1986)

Ruesch, Hans. *Slaughter of the innocent.* New York: Civitas Publications, 1983. 446 p.

Salt, Henry Stephens. *Animals' rights: considered in relation to social progress.* Clark Summit, Pa.: Society for Animal Rights, 1980. 240 p. (HV4708 .S17 1980)

Scientific perspectives on animal welfare. New York: Academic Press, 1982. 131 p. (HV4704 .S34 1982)

Silverman, B. P. Robert Stephen. *Defending animals' rights is the right thing to do.* New York: Shapolsky Publishers, 1991.

Singer, Peter. *Animal liberation.* 2d ed. New York: New York Review of Books, 1990. Distributed by Random House. 320 p.: ill. (HV4708 .S56 1990)

Spiegel, Marjorie. *The dreaded comparison: human and animal slavery.* 2d ed. New York: Mirror Books, 1989. 108 p.: ill. (HV4708 .S63 1989)

Steffens, Bradley. *Animal rights: distinguishing between fact and opinion.* San Diego, Calif.: Greenhaven, 1989. 32 p.: ill. (HV4711 .A57 1989)

Sweeney, Noel. *Animals and cruelty and law.* Bristol: Alibi, 1990. 119 p. (KD667.A5 S94 1990)

Tester, Keith. *Animals and society: the humanity of animal rights.* New York: Routledge, 1991.

Townend, Christine Elizabeth. *In defence of living things.* Sydney, Australia: Wentworth, 1980. 125 p.: ill. (HV4890.A3 T68)

Turner, James. *Reckoning with the beast: animals, pain, and humanity in the Victorian mind.* Baltimore, Md.: Johns Hopkins University Press, 1980. 190 p.: ill. (HV4708 .T87)

General interest periodicals (annotated)

Brownlee, Shannon. "First it was 'save the whales', now it's 'free the dolphins'." *Discover* 7 (December 1986): 70-72.
> Opposes the position which declares that capitivity endangers local populations and that treatment of captive marine mammals is shockingly bad.

Budiansky, Stephen. "The ancient contract." *U.S. News & World Report* 106 (March 20, 1989): 75-79.
> Discusses the relationships between man and animals, reflecting on the evolutionary reasons which favored domestication. Examines the positions of animal rights activists as well as the view which considers animals to be incapable of feeling.

Curtis, Patricia. "If a gibbon bangs his head, it's best to bang yours, too." *Smithsonian* 14 (March 1984): 119-120, 122, 124.
> Describes the work of Shirley McGreal, who founded the International Primate Protection League, a volunteer organzation which works to improve the conditions of primates in zoos, laboratories, and the wild.

Revkin, Andrew C. "Water hazard." *Discover* 9 (September 1988): 72-76, 78.
> "As Florida's populations of people and alligators boom, the reptile is invading backyards and golf courses and prompting thousands to call for gator aid."

Describes the daily tasks of Lieutenant Dick Lawrence, wildlife officer for the Florida Game and Fresh Water Fish Commission, who responds to complaints from the public regarding alligators, and with a team of professional trappers, removes these animals, which are later killed.

Smith, Tony. "Dietary reform is crucial." *Alternatives* 14 (August-September 1987): 92-94.
Urges the adoption of vegetarianism for ethical, economic, and health reasons.

Solomon, Robert C. "Has not an animal organs, dimensions, senses, affections, passions?" *Psychology Today* 16 (March 1982): 36, 39-41, 43-45.
Describes the views of scientists on animal consciousness, animal suffering and feeling.

Wright, Robert. "Are animals people too?" *New Republic* 202 (March 12, 1990): 20-23, 26-27.
Notes that the animal rights movement has grown over the past decade. Reasons that if animals are sentient beings capable of experiencing pleasure and pain then inflecting pain on them raises numerous moral questions.

General interest periodicals (unannotated)

Adler, Jerry. "Emptying the cages: does the animal kingdom need a bill of rights?" *Newsweek* 111,21 (May 23, 1988): 59(2).

Binding, Paul. "Animal, mineral, or vegetable?" (European Community regulation of animal rights) *New Statesman Society* 4,169 (September 20, 1991): 18(2).

Boehler, Karen. "Hamming it up with animals in space." (role of animals in the space program) *Ad Astra* 3,4 (May 1991): 40(5).

Cole, John R. "Animal rights and wrongs." *The Humanist* 50,4 (July-August 1990): 12(4).

Collard, Sneed B., III. "Refocusing animal rights." *The Humanist* 50,4 (July-August 1990): 10(3).

Feder, Kenneth L.; and Park, Michael Alan. "Animal rights: an evolutionary perspective." *The Humanist* 50,4 (July-August 1990): 5(4).

Freeman, Alan; and Mensch, Betty. "Scratching the belly of the beast." (treatment of pets; experimentation on animals) *Tikkun* 4,5 (September-October 1989): 34(5).

Fischer, Mary A. "Pet peeves: publisher Gil Michaels is fighting for real-live creature comforts." (publisher of "The Animals' Voice") *Los Angeles* 33,9 (September 1988): 60(5).

Gilsdorf, Ethan. "Animal rights, wronged." *Harper's Magazine* 283,1699 (December 1991): 4(2).

Hearne, Vicki. "What's wrong with animal rights: of hounds, horses, and Jeffersonian happiness." *Harper's Magazine* 283,1696 (September 1991): 59(6).

Kanner, Bernice. "A dog's life: love and death at the ASPCA." (American Society for the Prevention of Cruelty to Animals) *New York* 25,17 (April 27, 1992): 48(6).

Minetree, Harry; and Guernsey, Diane. "Animal rights--and wrongs." *Town and Country* 142,5096 (May 1988): 158(9).

Obee, Bruce. "The great killer whale debate: should captive orcas be set free?" *Canadian Geographic* 112,1 (January-February 1992): 20(12).

Ripstein, Arthur. "Animal rights and wrongs." *Canadian Dimension* 22,5 (July-August 1988): 8(3).

Rosenberger, Jack. "Whose life is it, anyway?" *New York* 23,2 (January 15, 1990): 30(2).

Singer, Peter; Lynch, Berkley A.; and Sperling, Susan. "Unkind to animals." *The New York Review of Books* 36,6 (April 13, 1989): 52(2).

Steinbach, Alice. "Whose life is more important: an animal's or a child's?" *Glamour* 88,1 (January 1990): 140(7).

Walker, Alice. "Am I blue? thoughts on animal feelings, human rights, and justice for all." *Utne Reader* 31 (January-February 1989): 98(3).

Environmental and professional journals (annotated)

Bateson, Patrick. "Do animals feel pain?" *New Scientist* 134 (April 25, 1992): 30, 32-33.
    Maintains that it is possible to tell when an animal suffers, but that the process is complicated.

Connif, Richard. "Fuzzy-wuzzy thinking about animal rights." *Audubon* 92 (November 1990): 120-122, 124, 126-133.
    "The animal rights movement has elevated ignorance about the natural world almost to the level of a philosophical principle." Argues that nature exhibits no ethical prohibition against violence and killing.

Leepson, Marc. "Animal rights." Congressional Quarterly's *Editorial Research Reports* 2,5 (1980)
> Considers the concept of "natural rights" for animals, examines the role of animals in research, and the feasibility of research without animals. Also discusses "factory farming" and motives for vegetarianism.

Environmental and professional journals (unannotated)

Cairns, John, Jr.; and Weis, Judith S. "The myth of the most sensitive species: multispecies testing can provide valuable evidence for protecting the environment." (includes related article on single species testing) *BioScience* 36 (November 1986): 670(3).

Cooke, Patrick. "A rat is a pig is a dog is a boy: the debate over animal rights is full of equations that don't add up." *In Health* 5,4 (July-August 1991): 58(7).

Forbes, Dana. "Where animals come first: the Black Beauty Ranch." *Animals' Agenda* 9,7 (September 1989): 26(4)

Skolnick, Andrew A. "Killing people to save animals and the environment by screaming wolf." *JAMA: the Journal of the American Medical Association* 266,16 (October 23, 1991): 2186(3).

Williams, Ted. "Driving out the dread serpent." (Opp, Alabama Rattlesnake Rodeo) *Audubon* 92,5 (September 1990): 26(6).

Wolkomir, Richard; and Wolkomir, Joyce. "Caught in the crossfire." (animal welfare in war zones) *International Wildlife* 22,1 (January-February 1992): 4(8).

Law journals

Daniel, Michelle D. "Air transportation of animals: passengers or property?" *Journal of Air Law and Commerce* 51 (Winter 1986): 497-529.
> Considers the special problems inherent in air transportation of animals and examines the animal welfare laws which govern treatment of animals during transportation and property laws which govern compensation to the owner in the case of injury or death of animals.

DeCapo, Thomas A. "Challenging objectionable animal treatment with the shareholder proxy proposal rule." *University of Illinois Law Review* 1 (1988): 119-149.
> Contends that when a publicly owned corporation is assocated with objectional animal treatment, the shareholder proxy proposal rule of federal securities law can provide a vehicle for challenging that treatment.

Linder, Douglas O. " 'Are all species created equal?' and other questions shaping wildlife law." *Harvard Environmental Law Review* 12 (1988): 157-200.
> Traces the history of animal welfare law and considers how the environmental and animal rights movements are likely to influence the development of this body of law.

Winters, Mary A. "Cetacean rights under human laws." *San Diego Law Review* 21 (July-August 1984): 911-940.
> Argues that by giving cetaceans rights and allowing people to sue on their behalf, the goal of effective protection will be more readily achieved.

Reports

Animals as sentinels of environmental health hazards. Committee on Animals as Monitors of Environmental Hazards, National Research Council. Washington, D.C.: National Academy Press, 1991. 160 p.: ill. (RA1199.4.A54N38 1991)
> Explains how animals could be used for determining ecological dangers, threats to human health, and as an early warning system for risk assessment and management. Contains data and references. No index or glossary.

Kuker-Reines, Brandon. *Environmental experiments on animals.* Boston, Mass. (1 Bulfinch Place, Boston 02114): New England Anti-Vivisection Society, 1984. 41 p. (QP45 .K85 1984)

Responsible animal regulation: a discussion of animal regulation and control problems prepared for city and county officials, humane societies, and legislative bodies. Washington, D.C. (2100 L Street NW, Washington 20037): Humane Society of the United States, 1986. 13 p. (HV4764 .R47 1986)

Bibliographies

Magel, Charles R. *A bibliography on animal rights and related matters.* Washington, D.C.: University Press of America, 1981. 602 p. (Z7164.C45 M35)

Nordquist, Joan. *Animal rights: a bibliography.* Santa Cruz, Calif.: Reference and Research Services, 1991. 68 p. (Z7164.C45 N67 1991)

Palen, Roberta. *Animal control: a preliminary bibliography.* Monticello, Ill.: Vance Bibliographies, 1980. 21 p. (Z7164.C45 P34)

Reference

Directory of animal care and control agencies. 2d ed. Denver, Colo. (9725 E. Hampden

Avenue, Denver 80231): American Humane Association, 1981. 196 p. (HV4702 .D57 1981)

Pocket guide to the humane control of wildlife in cities and towns. Washington, D.C. (2100 L Street NW, Washington 20037): Humane Society of the U.S., 1990. 112 p.: ill. (SB603.3 .P63 1990)

Pratt, Dallas. *Animal films for humane education.* New York: Argus Archives, 1986. 284 p. (HV4708 .P73 1986)

Books for young adults

Arnold, Caroline. *Pets without homes.* New York: Clarion, 1983. 46 p.: ill. (HV4708 .A76 1983)

Barton, Miles. *Animal rights.* New York: Gloucester, 1987. 32 p.: (some color) ill. (HV4708 .B37 1987)

Bloyd, Sunni. *Animal rights.* San Diego, Calif.: Lucent Books, 1990. 127 p.: ill. (HV4708 .B58 1990)

Burroughs, Nigel. *Nature's chicken: a book for animal lovers.* Summertown, Tenn.: Book Publishing, 1992.

Curtis, Patricia. *Animals and the new zoos.* New York: Lodestar, 1991. 60 p.: (some color) ill. (QL77.5 .C87 1991)

Curtis, Patricia. *The animal shelter.* New York: E. P. Dutton, 1984. 163 p.: ill. (HV4708 .C87 1984)

Evans, Rose. *Friends of all creatures.* San Francisco, Calif.: Sea Fog Press, 1984. 122 p.: ill. (HV4708 .E94 1984)

Goodman, Billy. *A kid's guide to how to save the animals.* New York: Avon, 1991. 136 p.: ill. (QL83 .G66 1991)

Greene, Carol. *Caring for our animals.* Hillside, N.J.: Enslow, 1991. 32 p.: ill. (QL49 .G75 1991)

Guernsey, JoAnn Bren. *Animal rights.* New York: Crestwood House, 1990. 48 p.: color ill. (HV4764 .G84 1990)

Hoff, Syd. *The man who loved animals.* New York: Coward, McCann, and Geoghegan, 1982. 48 p.: ill. (HV4764 .H63 1982)

Kelty, Jean McClure. *If you have a duck--: adventures to help children create a humane world.* Rev. ed. Youngstown, Ohio: George Whittell Memorial Press, 1982. 104 p.: ill.

Kerven, Rosalind. *Equal rights for animals.* New York: F. Watts, 1993.

Landsman, Sandy. *Castaways on Chimp Island.* New York: Atheneum, 1986. 202 p. (PZ7.L23175 Cas 1986)

Lavie, Arlette. *Animal rights.* New York: Child's Play International, 1991.

Lee, Gregory. *Animal rights.* Vero Beach, Fla.: Rourke, 1991. 64 p.: ill. (HV4764 .L44 1991)

Loeper, John J. *Crusade for kindness: Henry Bergh and the ASPCA.* New York: Atheneum, 1991. 103 p.: ill. (HV4764 .L64 1991)

Newkirk, Ingrid. *Kids can save the animals!: 101 easy things to do.* New York: Warner, 1991. 234 p.: ill. (HV4708 .N494 1991)

Poynter, Margaret. *Too few happy endings: the dilemma of the humane societies.* New York: Atheneum, 1981. 130 p. (HV4702 .P69)

Poynter, Margaret. *What's one more?* New York: Atheneum, 1985. 117 p.: ill. (HV4708 .P69 1985)

Pringle, Laurence P. *The animal rights controversy.* San Diego, Calif.: Harcourt Brace, 1989. 103 p.: ill. (HV4764 .P75 1989)

Scott, Elaine. *Safe in the spotlight: the Dawn Animal Agency and the Sanctuary for Animals.* New York: Morrow Junior Books, 1991. 77 p.: ill. (HV4766.N732 S367 1991)

Shorto, Russell. *Careers for animal lovers.* Brookfield, Conn.: Millbrook, 1992. 64 p.: (some color) ill. (HV4764 .S53 1992)

Twinn, Michael. *Who cares about animal rights?* New York: Child's Play International, 1992.

Ziefert, Harriet. *Bob and Shirley: a tale of two lobsters.* New York: HarperCollins, 1991. 32 p.: color ill. (PZ7.Z487 Bo 1991)

Videos

The price they pay. Produced by Heidi Ann Perkins. (Varied Directions, 69 Elm Street, Department BL, Camden, Maine 04843.) 1991. 12 min.

This graphic, often difficult to watch, video asks if fur clothing derived from animal pelts is worth the pain that the animals suffer. Examples of animal suffering include leg-hold traps, which are legal in all but three states and cause an animal intense pain but do not immediately kill them, and suffocation and electrocution techniques, which are employed on animal farms. Suggests using imitation animal products.

A question of respect. Produced by Charles Passerman for American Society for the Prevention of Cruelty to Animals. Directed by Kirk Wolfinger. (Varied Directions, 69 Elm Street, Department BL, Camden, ME 04843). 1989. 11 min.

MOVEMENT AND LEADERS

(See also *Environmental Movement: History*) 447
(See also *Environmental Philosophy: Ecofeminism*) 547

Books (unannotated)

Jasper, James M. *The animal rights crusade: the growth of a moral protest.* New York: Maxwell Macmillan International, 1992. 214 p. (HV4764 .J37 1992)

Kalechofsky, Roberta. *Autobiography of a revolutionary: essays on animal and human rights.* Marblehead, Mass.: Micah Publications, 1991. 189 p. (HV4918 .K35 1991)

Regan, Tom. *The struggle for animal rights.* Clarks Summit, Pa.: International Society for Animal Rights, 1987. 197 p. (HV4764 .R45 1987)

Ryder, Richard D. *Animal revolution: changing attitudes towards speciesism.* Cambridge, Mass.: B. Blackwell, 1989. 385 p.: ill. (HV4705 .R93 1989)

General interest periodicals (annotated)

Driscoll, Lisa McGurrin. "A corporate spy story." *New England Business* 11 (May 1989): 28, 30-33, 76-78.
   Recounts the attempt by animal rights activist Fran Stephanie Trutt to kill Leon C. Hirsch, chairman of United States Surgical Corporation. Explores the controversy surrounding U.S. Surgical's role in the pipe-bombing episode--was their covert infiltration of the animal rights group instigative?

Elshtain, Jean Bethke. "Why worry about the animals?" *Progressive* 54 (March 1990): 17-23.

Discusses the public's growing awareness of animal abuse in science/research labs, war research experiments, factory farms, and household products testing. Examines tactics, such as civil disobedience, that animal welfare activists are taking to protect animals.

"Just like us?" *Harper's Magazine* 277 (August 1988): 43-52.
Two leading animal rights activists discuss their views with a philosopher and a constitutional scholar.

Oliver, Charles. "Liberation zoology." *Reason* 22 (June 1990): 22-27.
"The new animal activists don't talk of animal welfare; they want animal rights. For them, all sentient beings have equal moral status."

General interest periodicals (unannotated)

"Beastly idea." (criticism of animal rights movement) *National Review* 42,6 (April 1, 1990): 15(2).

Behar, Richard. "Meet the meatless." *Forbes* 143,6 (March 20, 1989): 43(2).

Hughes, Jane. "Reigning cats and dogs." (criticism of animal rights movement) *National Review* 42,14 (July 23, 1990): 35(3).

"Man's mirror." (history of animal rights) *Economist* 321,7733 (November 16, 1991): 21(3).

Pantridge, Margaret. "The improper Bostonian." *Boston* 83,6 (June 1991): 68(5).

"Raider's life."(Animal Liberation Front) *New Statesman Society* 3,118 (September 21, 1990): 12(2).

Rose, Steven. "Proud to be speciest." *New Statesman Society* 4,148 (April 26, 1991): 21(2).

Sager, Mike; and Mahurin, Matt. "Inhuman bondage." (Animal Liberation Front) *Rolling Stone* 522 (March 24, 1988): 86(7).

Singer, Peter. "Ten years of animal liberation." *The New York Review of Books* 31 (January 17, 1985): 46(6).

Tyrrell, R. Emmett, Jr. "Save the chickens." *The American Spectator* 22,2 (February 1989): 10(2).

Wolkomir, Richard. "Pet power on the march." *Reader's Digest* 132,794 (June 1988): 54(3).

Environmental and professional journals (annotated)

Doherty, Jim. "Eradicate, suppress, destroy." *Audubon* 84 (September 1982): 101-107.
Discusses public reactions to the use of 1080 to poison coyotes in sheep country.

"Militant morality." *Hastings Center Report* 19 (November-December 1989): 23-45.
This collection of essays on civil disobedience includes "The fiery fight for animal rights," by Christine M. Jackson; "Grassroots opposition to animal exploitation," by Steve Siegel; and "To do or not to do?," by Peter Singer.

Rowan, Andrew. "The development of the animal protection movement." *Journal of NIH Research* 1 (November-December 1989): 97-100.
Presents a history of the forces behind the current animal protection movement and assesses the impact of Peter Singer's book *Animal Liberation,* which was published in 1975.

Starr, Douglas. "Equal rights." *Audubon* 86 (November 1984): 30-32, 34, 35.
Describes the activities of animal rights activists.

Environmental and professional journals (unannotated)

Barker, Leigh. "Violence, infiltration, and sabotage: what we can learn from other movements." *Animals' Agenda* 9,7 (July-August 1989): 26(4)

Bartlett, Kim. "Of meat and men: a conversation with Carol Adams." (feminist, animal rights activist) *Animals' Agenda* 10,8 (October 1990): 12(4)

Beedy, Kevin J. "The politics of animal rights." *Animals' Agenda* 10,2 (March 1990): 17(5)

Bring, Ellen. "Joyce Tischler: legal activist." (executive director of the Animal Legal Defense Fund) (interview) *Animals' Agenda* 11,6 (July-August 1991): 40(4)

Burgos, Javier. "Animal rights: the suicide of a movement." *Animals' Agenda* 9,1 (January 1989): 39(2)

Cave, George; and Stuchell, Dana. "The path to anti-vivisection." *Animals' Agenda* 9,1 (January 1989): 39(3)

Chui, Glennda. "Activists beset UC, Stanford labs." (University of California; biology labs under attack by neighborhood activists and animal rightists) *Science* 239,4845 (March 11, 1988): 1229(4)

Clifton, Merritt. "A late April Fool? Or something worse?" (Sydney Singer, husband of Good Shepherd Foundation head Tanja Keogh-Singer, masquerades as animal rights advocate) *Animals' Agenda* 11,5 (June 1991): 34(2)

Clifton, Merritt. "Activists take the fifth in grand jury probe." (probe of Animal Liberation Front; includes article on other court cases involving animal rights activists) *Animals' Agenda* 10,6 (July-August 1990): 36(2)

Clifton, Merritt. "Birdshooters, police attack activists at Hegins." (Hegins Pa. pigeon shoot) *Animals' Agenda* 10,9 (November 1990): 40(2)

Clifton, Merritt. "Bombs and bombast." (Fran Trutt arrested while planting pipe bomb at U.S. Surgical Corp.) *Animals' Agenda* 9,2 (February 1989): 28(2)

Clifton, Merritt. "Earth First! Founder busted in possible set-up. (environmentalist Dave Foreman) *Animals' Agenda* 9,7 (September 1989): 20(2)

Clifton, Merritt. "Hegins, 1991: 400 birds freed, 91 arrests." (Labor Day pigeon shoot in Pennsylvania) *Animals' Agenda* 11,9 (November 1991): 34(2)

Clifton, Merritt. "HSUS in hot water again." (Humane Society of the United States) *Animals' Agenda* 11,4 (May 1991): 33(2)

Clifton, Merritt. "Out of the cage: the movement in transition." *Animals' Agenda* 10,1 (January-February 1990): 26(5)

Clifton, Merritt. "Public opinion." (attitudes towards animal rights) *Animals' Agenda* 10,4 (May 1990): 35(2)

Clifton, Merritt. "Reshuffle at HSUS." (Humane Society of the U.S.) *Animals' Agenda* 11,10 (December 1991): 33(2)

Clifton, Merritt. "Smears, suits, and espionage." (increased confrontation between animal researchers and animal rights activists) *Animals' Agenda* 10,4 (May 1990): 37(2)

Clifton, Merritt. "Thousands march for animals." (animal rights demonstration in Washington D.C.) *Animals' Agenda* 10,7 (September 1990): 38(2)

Clifton, Merritt. "Tony LaRussa: going to bat for the animals." (column) *Animals' Agenda* 10,2 (March 1990): 10(2)

Cornell, Marly. "Sue Coe: rebel with many causes." (animal rights activist) (interview) *Animals' Agenda* 9,2 (February 1989): 7(3)

Davis, Karen. "Re-searching the heart: an interview with Eldon Kienholz. (animal rights advocate) (interview) *Animals' Agenda* 11,3 (April 1991): 12(3)

"DJ of animal rights: Shelton Walden." *Animals' Agenda* 9,10 (November 1989): 10(2)

Greanville, David P. "Ted Kirkpatrick: metal with a conscience." (heavy metal rocker an animal rights advocate) *Animals' Agenda* 11,1 (January-February 1991): 46(2)

Greanville, David Patrice. "Frontline reports." (animal rights hits big-time media) *Animals' Agenda* 9,4 (April 1989): 44(3)

Greanville, David Patrice. "Environmentalists and animal rightists--the new odd couple?" *Animals' Agenda* 9,8 (October 1989): 22(3)

Hogshire, Jim. "Sam LaBudde: Earth activist." *Animals' Agenda* 9,8 (October 1989): 8(2)

Iacobbo, Karen L.T. "Ann Cottrell Free: poet/journalist/animal advocate extraordinaire." *Animals' Agenda* 11,1 (January-February 1991): 38(2)

Iacobbo, Karen; and Iacobbo, Michael. "Sandy Larson: humane educator." (education coordinator, New England Anti-Vivisection Society) (Profile) *Animals' Agenda* 10,4 (May 1990): 9(2)

Iacobbo, Karen; and Iacobbo, Michael. "Ken Shapiro and PsyETA." (founder of Psychologists for the Ethical Treatment of Animals) *Animals' Agenda* 9,3 (March 1989): 8(2)

Janson, Thor. "Johnny Appleseed: pioneer of animal rights." *Animals' Agenda* 10,10 (December 1990): 40(2)

Knox, Margaret L.; and Montgomery, Sy. "The rights stuff." (animal rights and the environmentalist movement) (includes related article on Jane Goodall's work with chimpanzees) *Buzzworm: The Environmental Journal* 3,3 (May-June 1991): 30(8).

Krawiec, Richard. "Dealing with the media: advice from a journalist." (tips for animal rights activists) *Animals' Agenda* 10,9 (November 1990): 16(2)

Maggitti, Phil. "Annette Lantos: bring concern for animals into Congress." (wife of California Congressman Tom Lantos) *Animals' Agenda* 11,1 (Jan-February 1991): 40(2)

Maggitti, Phil. "Marian Probst: in the shadow of the curmudgeon." (animal rights activist) *Animals' Agenda* 9,6 (June 1989): 7(2)

Maggitti, Phil. "Marv Levy: Buffalo coach tackles the issues head on." (Buffalo Bills coach and animal rights activist) *Animals' Agenda* 9,10 (November 1989): 8(3)

Maggitti, Phil. "Priscilla Feral: in for the long haul." (Friends of Animals' president) *Animals' Agenda* 10,1 (January-February 1990): 11(3)

Maggitti, Phil. "Susan Rich: compassion begins in the home." (People for the Ethical Treatment of Animals volunteer) (Profile) *Animals' Agenda* 10,4 (May 1990): 8(2)

Maggitti, Phil. "The opposition in motion." (counteroffensives against the anti-fur movement) *Animals' Agenda* 10,5 (June 1990): 17(5)

Maggitti, Phil; and Ross, Kenneth D. "Veterinarians: for or against animal rights?" (includes related article) *Animals' Agenda* 9,2 (February 1989): 12(11)

Mason, Jim. "Becky Sandstedt: stockyard activist." (Minneapolis animal rights advocate) *Animals' Agenda* 11,3 (April 1991): 20(2)

Moran, Victoria. "De-stressing the activist." (compassionate living) (column) *Animals' Agenda* 9,4 (April 1989): 51(2)

Moran, Victoria. "Gary Null: guru of health and wholeness." (talk show host, animal rights activist) (column) *Animals' Agenda* 9,11 (December 1989): 8(3)

Moran, Victoria. "Reaching rural America." *Animals' Agenda* 10,1 (January-February 1990): 50(2)

Newkirk, Ingrid. "Total victory, like checkmate, cannot be achieved in one move." *Animals' Agenda* 12,1 (Jan-February 1992): 43(3)

Pacelle, Wayne. "An interview with Luke Dommer." (founder of Committee to Abolish Sport Hunting) *Animals' Agenda* 9,1 (January 1989): 6(4)

Pacelle, Wayne. "Animal rights in Bloom: an interview with Berke Breathed." *Animals' Agenda* 9,7 (July-August 1989): 7(6)

Pacelle, Wayne. "Game commission arms Florida youths." (Florida Game and Fresh Water Fish Commission) *Animals' Agenda* 9,1 (January 1989): 24(2)

Pavlova, T. "Eastern Europe: the animal defense movement in the Soviet Union." *Animals' Agenda* 10,8 (October 1990): 38(4)

Plous, S. "Toward more effective activism: advice from a psychologist." (tips for animal rights activists) *Animals' Agenda* 9,11 (December 1989): 24(3)

Regan, Tom. "Reader beware: when it comes to a declaration, what is meant may not be what is said." *Animals' Agenda* 11,8 (October 1991): 24(3)

Regan, Tom; and Francione, Gary. "A movement's means create its ends." *Animals' Agenda* 12,1 (January-February 1992): 40(4)

Schwartz, Sheila. "Humane education: collaborating for pro-animal school programming." *Animals' Agenda* 9,8 (October 1989): 12(2)

Sommer, Mark. "Animal rhythms: rockin' for animal rights." *Animals' Agenda* 9,7 (July-August 1989): 14(5)

Sommer, Mark. "Grace Slick: animal magnetism." *Animals' Agenda* 9,7 (September 1991): 44(2)

Troiano, Linda. "Owl connections." (new grassroots power in animal rights movement) *American Health* 8,2 (March 1989): 90(2).

Weil, Zoe. "Teaching animal issues." (includes bibliography) *Animals' Agenda* 12,2 (March 1992): 20(4)

Wise, Steven. "Legal advice for avoiding lawsuits." (tips for animal rights activists) *Animals' Agenda* 10,6 (July-August 1990): 24(3)

"Who gets the money?" (how dollars donated to animal rights causes are spent) *Animals' Agenda* 11,6 (July-August 1991): 34(2)

Reports

Dakers, Sonya. *Animal rights campaigns: their impact in Canada, 18 January 1988, reviewed 21 March 1991.* Ottawa: Research Branch, Library of Parliament, 1991. 19 p. (Current issue review 88-4 E)
  "Describes how the anti-sealing campaign adversely affected the livelihoods of sealers in Labrador and Newfoundland and of the Inuit in northern Canada, who hunted only adult seals. It then outlines how a similar campaign might affect trapping and fur ranching in this country. Also touched on are two other activities (animal research and factory farming) that have been attracting sporadic but well-publicized attention from various animal rights groups and seem likely to be the next targets of animal rights campaigns in this country."

Books for young adults

Curtis, Patricia. *Animal rights: stories of people who defend the rights of animals.* New York: Four Winds Press, 1980. 148 p.: ill. (HV4725.U33 C87)

Videos

In defense of animals: a portrait of Peter Singer. Directed and produced by Julie Akeret.

Distributed by Bullfrog Films (Oley, PA, 1989). 28 min.

> Addresses the issue of animal rights, makes a case for vegetarianism, points out some questionable government requirements regarding animal testing, and seems to advocate its total elimination. Contains interviews from people of all ages and graphic footage of animal testing. Discusses animal rights in anthropocentric terms.

ETHICAL AND RELIGIOUS CONSIDERATIONS

Books (annotated)

Clark, Stephen R. L. *The nature of the beast: are animals moral?* New York: Oxford University Press, 1982. 127 p.

> Reflects and speculates on a variety of concepts relating to morality, including intelligence and language, freedom and necessity, me and mine, altruism, parenthood, territory and dominance. Defines animal morality as "the impulses and inhibitions, learned and instinctual, that seem to play the part in beasts that morals play in us." Those are "the morals of Nature." Concludes that animals are ethical in so far as they respond to particular situations in a way that a good man would respect, but that animals are not moral because they do not construct intellectual systems by which to judge their behavior.

Dombrowski, Daniel A. *Hartshorne and the metaphysics of animal rights.* Albany: State University of New York Press, 1988. 159 p.

> Examines egalitarianism in considering the value of animals. During a time when some animal rights advocates hold that the value of any animal is equal to the value of any other, maintains that it is not a question of whether animals have rights but which rights do they have? Concludes that the human treatment of animals is a moral issue and that Western countries should become more vegetarian. Not only are the methods for raising and slaughtering animals inhumane, raising animals results in the destruction of habitats for wild animals and thus has a doubly cruel result.

Hargrove, Eugene C., ed. *Animal rights, environmental ethics debate: the environmental perspective.* Albany: State University of New York Press, 1992. 273 p. (HV4711 .A575 1992)

> Traces the animal rights/environmental ethics dispute in an attempt to place the basic ideas in a historical context. All sides of the issue are presented by the numerous contributors who advocate different perspectives.

McDaniel, Jay B. *Of God and pelicans.* Louisville, Ky.: Westminister Press, 1989. 168 p.

> Draws on the currents of animal rights, feminist philosophy, and Buddism to

propose an understanding of God through the needs of nonhuman creatures and an awareness of the environmental crisis. Defines his theology as "relational pantheism" in which "there are multiple creative powers of which God is the primordial but not exclusive instance." Defends individualistic animal rights and laments the pain endured by all animals while acknowledging that the value of plants and animals has some relation to their function in the ecosystem.

Mighetto, Lia. *Wild animals and the American environmental ethics.* Tucson: University of Arizona Press, 1991. 177 p.: ill. (QL85 .M54 1991)
Divides conservationists by the amount of pain they will tolerate for animals. Begins by showing how the Forest Reserve Act (1891) was passed to preserve large tracts of land for hunting. The Society for the Prevention of Cruelty to Animals was founded to combat any pain inflected on animals by humans. Some environmentalists accept predation and death as a natural part of ecology; that man is the predator doesn't make any difference as long as the slaughter of animals for a mass market is not the issue. For others, any killing is unconsciona- ble; hopes that conservationists will gravitate toward the latter position.

Rachels, James. *Created from animals.* New York: Oxford University Press, 1990.
Maintains that Darwinism is incompatible with the idea that humans have a special worth through their relationship with a deity. Asserts that the moral separation between humans and other species relies solely on differences between the two. Upholds that these differences are matters of degree instead of kind. Makes a case against factory farming and some animal testing.

Radner, Daisie; and Radner, Michael. *Animal consciousness.* Buffalo, N.Y.: Prometheus, 1989.
Aims to destroy the principal barrier to the study of animal consciousness--the cartesian theory that humans are composed of a body and a soul and animals only a body. Analyzes attempts to communicate with nonhuman primates in terms of animal consciousness. Maintains that if one accepts the theory of evolution, one must accept that animal consciousness is possible.

Regan, Tom, ed. *Animal sacrifices: religious perspectives on the use of animals in science.* Philadelphia, Pa.: Temple University Press, 1986. 270 p.
Contains papers from the religious community that treat the theology and religious stance of Christianity, Judaism, Islam, Hinduism, Jainism, Buddhism, and Confucianism on animal rights. Christianity has had the least interest in animal rights because of the belief dervied from the Greeks that humans alone are rational beings with moral values and are the only species capable of redemption. In Islam, the Koran specifically provides a moral basis for the responsibility people have to animals. In Hinduism a person should not harm animals because it would impede spiritual growth. Jainists believe that experimentation on animals disrupts their spiritual development.

Rodd, Rosemary. *Biology, ethics, and animals.* New York: Oxford University Press, 1990. 272 p. (HV4708 .R63 1990)

> Synthesizes the biology and philosophy of animal rights and examines the theories and known facts surrounding the way decisions are made to treat different species. Issues such as taxonomy, anatomical structure, evolutionary kinship, and animal communication are examined. Concludes that the distinction between animals and humans is intrinsically vague, both conceptually and biologically. Cites evidence that animals are conscious and that some possess the very human traits of belief and emotion. Attempts to strike a moderate balance between human needs the need to reduce the costs to animals. Maintains that animals' roles as companions and workers should be reevaluated and that animal consciousness makes killing animals intrinsically harmful, regardless of whether or not pain is inflicted.

Rollin, Bernard E. *The unheeded cry: animal consciousness, animal pain and science.* New York: Oxford University Press, 1989. 308 p. (HV4915 .R65 1989)

> A historical outline of moral attitudes toward animals asks whether or not animals feel pain. Maintains that the nineteenth-century Darwin-Romanes view that animals indeed feel pain was not disproven, but simply lost in the flux of changing perspectives, which included a combination of empiricism, positivism, and behaviorism. Asserts that the link between behaviorism and determinism resulted in a reductionist foundation which did not uphold that animals experienced pain and thus the belief that science should operate outside the realm of common-sense and human values.

Ryder, Richard D. *Animal revolution: changing attitudes towards speciesism.* London: B. Blackwell, 1989.

> Speciesism refers to discrimination by man against other species. Traces the history of this idea from the ancient Egyptians to the present. Critical of the Royal Society for the Protection of Cruelty to Animals, with which the author was involved. Addresses wildlife conservation, animal experimentation, and farming. Maintains that speciesism is a wrong on par with racism.

Serpell, James. *In the company of animals: a study of human-animal relationships.* London: B. Blackwell, 1986.

> Traces the development of the modern attitude that livestock are simply food-producing machines. Examines history "all the way back to the Romans," philosophy, and animal behavior. Juxtaposes the modern economist who sees animals in terms of potential profit and the modern pet-owner who is completely dedicated to his or her animal. Examines the behavior of domesticated animals. Explains how people distance themselves from animals that they kill or mistreat.

Sperling, Susan. *Animal liberators: research and morality.* Berkeley: University of California Press, 1988.

> Attempts to explain why the animal rights movement has emerged in some

Western countries in the twentieth century. Maintains that a fundamental difference exists between those who simply push for more humane treatment of animals and those who see animal experimentation as a sign of technology's manipulation of society and its morals. Compares the animal rights movement of the late 1980s in the United States with that of an antivivisection protest of late nineteenth-century Britain.

Wenzel, George. *Animal rights, human rights: ecology, economy, and ideology in the Canadian arctic.* Toronto: University of Toronto Press, 1991. 224 p.
Examines the effects a strict animal-rights stance can have on human societies. Looks to the movement to ban seal hunting in Canada; demonstrates how cultural prejudices and questionable ecological aims have skewed the agenda of the animal-rights movement. Maintains that banning seal hunting in Canada could be detrimental to the Inuit, a culture believed to be ecologically sound.

Books (unannotated)

Animal rights, environmental ethics debate: the environmental perspective. Albany: State University of New York Press, 1992. 273 p. (HV4711 .A575 1992)

Animal rights and human obligations. 2d ed. Englewood Cliffs, N.J.: Prentice-Hall, 1989. 280 p. (HV4711 .A56 1989)

Brown, Les. *Cruelty to animals: the moral debt.* Houndmills, Basingstoke, Hampshire: Macmillan, 1988. 240 p. (HV4708 .B76 1988)

Carruthers, Peter. *The animals issue: moral theory in practice.* New York: Cambridge University Press, 1992.

Clark, Stephen R. L. *The moral status of animals.* New York: Oxford University Press, 1984. 221 p.

Fox, Michael W. *Returning to Eden: animal rights and human responsibility.* New York: Viking, 1980. 281 p. (HV4708 .F72 1980)

Johnson, Lawrence E. *A morally deep world: an essay on moral significance and environmental ethics.* New York: Cambridge University Press, 1991. 301 p. (GF80 .J64 1991)

Judaism and animal rights: classical and contemporary responses. Marblehead, Mass.: Mican, 1992.

Linzey, Andrew. *Christianity and the rights of animals.* New York: Crossroad, 1987. 197 p. (HV4708 .L564 1987)

Regenstein, Lewis. *Replenish the Earth: a history of organized religions' treatment of animals and nature--including the Bible's message of conservation and kindness toward animals.* New York: Crossroad, 1991. 304 p. (BL435 .R44 1991)

Rollin, Bernard E. *Animal rights and human morality.* Buffalo, N.Y.: Prometheus, 1981. 182 p. (HV4708 .R64)

Sapontzis, Steve F. *Morals, reason, and animals.* Philadelphia, Pa.: Temple University Press, 1987. 302 p. (HV4708 .S23 1987)

Tannenbaum, Jerrold. *Veterinary ethics.* Baltimore, Md.: Williams and Wilkins, 1989. 358 p. (SF756.39 .T36 1989)

Wynne-Tyson, Jon. *The extended circle: a dictionary of humane thought.* Fontwell, Sussex, England: Centaur, 1985. 436 p.: ill. (PN6084.A57 W9 1985)

General interest periodicals (annotated)

"Environmental ethics." *Alternatives* 12 (Winter 1985): 3-41.
   Includes "Moral concern and animals," by B. Rollin; "The Canadian harp seal hunt: a moral issue," by L. Sumner; and "Animal suffering and the ethics of eating meat," by R. Carter.

Zak, Steven. "Ethics and animals." *Atlantic Monthly* 263 (March 1989): 69-74.
   Argues that people have a far greater moral responsibility to animals than just to stop torturing them for no purpose.

General interest periodicals (unannotated)

Linzey, Andrew. "The theological basis of animal rights." (Cover Story) *Christian Century* 108,28 (October 9, 1991): 906(4).

Stafford, Tim. "Animal lib: despite silliness and fanaticism on both sides, the animal-rights debate remains an inherently religious issue." *Christianity Today* 34,9 (June 18, 1990): 18(5).

Wall, James M. "An uncaged vision of nonhuman creation." (animal rights and theology) (editorial) *Christian Century* 106,31 (October 25, 1989): 947(2).

Environmental and professional journals (unannotated)

Bartlett, Kim. "A conversation with Andrew Linzey: on Christianity and animals."

(Anglican minister and animal rights advocate) *Animals' Agenda* 9,4 (April 1989): 7(19)

Gaffney, James. "Animals and ethics: a Catholic blind spot." *America* 163,12 (October 27, 1990): 297(3).

Hogshire, Jim. "Animals and Islam." *Animals' Agenda* 11,8 (October 1991): 10(5)

Mangan, Dennis. "Arthur Schopenhauer: philosopher of compassion (1788-1860)." *Animals' Agenda* 11,10 (December 1991): 22(2)

Moran, Gabriel. "Dominion over the Earth: does ethics include all creatures?" *Commonweal* 114,21 (December 4, 1987): 697(5).

Rosen, Steven. "Ahimsa: animals and the East." (respect for animals is a cornerstone of eastern religions; includes bibliography) (Cover Story) *Animals' Agenda* 10,8 (October 1990): 20(6)

Singer, Sydney. "The neediest of all animals." (medical ethics, healing and animal rights) (column) *Animals' Agenda* 10,5 (June 1990): 50(2)

"World Council of Churches report." (respect for animals advocated) *Animals' Agenda* 9,4 (April 1989): 10(5)

Reports

Fox, Michael W. *St. Francis of Assisi, animals, and nature*. Washington, D.C. (2100 L Street NW, Washington 20037): Center for Respect of Life and Environment, 1989. 34 p.: ill. (BT695.5 .F69 1989)

Hyland, J. R. *The slaughter of terrified beasts: a Biblical basis for the humane treatment of animals*. Sarasota, Fla.: Viatoris Ministries, 1988. 74 p. (BS680.A5 H95 1988)

Mistichelli, Judith Adams. *Ethical issues surrounding cross-species organ transplantation*. Washington, D.C.: Joseph and Rose Kennedy Institute of Ethics, Georgetown University, 1985. 19 p.
  Presents a current overview of bioethical issues related to cross-species transplants. Includes a bibliography of major news accounts, editorials, and legislation and regulations concerning organ transplants, human experimentation, and animal experimentation/animal rights.

Bibliographies

Clingerman, Karen J. *Ethical and moral issues relating to animals, January 1979-March 1990*. Beltsville, Md.: National Agricultural Library, 1990. 33 p. (Series QB 90-48)

EXPLOITATION AND ABUSE OF ANIMALS
███████████████████████████████████

Books (annotated)

Fox, Michael. *Inhumane society: the American way of exploiting animals.* New York: St. Martin's, 1990. 268 p.

> Asserts that this country lacks ethical awareness of animals. Discusses the treatment of farm animals, including the amputation of pigs' tales and the debeaking and declawing of poultry. Argues that pets should be chosen from animal shelters rather than custom bred.

General interest periodicals (annotated)

Linden, Eugene. "An uneasy dip with the dolphins." *Time* 134 (November 27, 1989): 80-81.

> "For some conservationists, 'dolphin-fondling' programs (as they are dismissively called) are just one more way in which humans deprive highly intelligent animals of their freedom and put them at risk of disease or mishandling for the entertainment of customers and the enrichment of owners."

General interest periodicals (unannotated)

Shahin, Jim. "Pet project." (abuse investigators) *Texas Monthly* 20,5 (May 1992): 66(5).

Environmental and professional journals (unannotated)

Barnard, Neal D. "The psychology of abuse." *Animals' Agenda* 11,5 (June 1991): 50(2)

Carbonne, Susan. "Acting and animals: Earl Holliman." (actor, animal rights activist) *Animals' Agenda* 9,7 (September 1989): 7(2)

Clifton, Merritt. "Wildlife photographers: will they ever get the picture? (photographers endanger animals) *Animals' Agenda* 11,1 (Jan-February 1991): 12(5)

Clifton, Merritt. "Hearings on military animal abuse?" *Animals' Agenda* 12,1 (Jan-February 1992): 34(2)

Clifton, Merritt. "Working to prevent abuse." (United States Department of Agriculture fights animal abuse) *Animals' Agenda* 11,3 (April 1991): 22(2)

Clifton, Merritt. "More no-kill shelter scandals." *Animals' Agenda* 10,9 (November 1990): 37(2)

Clifton, Merritt. "Connecticut Humane plans no changes." (Connecticut Humane Society) *Animals' Agenda* 9,7 (July-August 1989): 22(2)

Clifton, Merritt. "Wanted: wildlife--dead or alive." (includes related articles on illegal wildlife traffic and the United States Fish and Wildlife Service) *Animals' Agenda* 11,5 (June 1991): 12(11)

Clifton, Merritt. "Animals' voice silenced?" *Animals' Agenda* 9,7 (September 1991): 39(2)

Colbourn, Keith; and O'Barry, Richard. "Dolphins in captivity: wasted lives, wasted minds." *Animals' Agenda* 9,3 (March 1989): 12(8)

Davis, Karen. "Speaking for Dr. Frankenstein's creatures today." (column) *Animals' Agenda* 10,2 (March 1990): 48(2)

Greanville, David P. "Spain--conservation makes some strides." *Animals' Agenda* 11,2 (March 1991): 25(2)

Greanville, David P. "Spain--the heart also rises." *Animals' Agenda* 9,7 (September 1991): 31(2)

Greanville, David Patrice. "Cruelty to animals a cultural trait?" (Columbia) *Animals' Agenda* 9,6 (June 1989): 30(2)

Greanville, David P. "Petitions on bullfighting and fur labelling considered." (Western Europe) *Animals' Agenda* 9,2 (February 1989): 33(2)

Greanville, David P. "Spain: animal via crucis continues." (country's long record of animal mistreatment) *Animals' Agenda* 10,10 (December 1990): 25(2)

Greanville, David Patrice. "Holocaust at the animal shelter." *Animals' Agenda* 10,1 (January-February 1990): 44(3)

Jennings, Gary. "The fur, feathers, and scales of justice." (a brief history of animal prosecution and punishment) *Animals' Agenda* 10,10 (December 1990): 30(4)

Kay, David E. "No kill animal shelters:" do it right--or don't do it. (includes guidelines on running a shelter) *Animals' Agenda* 11,8 (October 1991): 41(3)

King, Marcia. "Killing horses to collect insurance." *Animals' Agenda* 10,9 (November 1990): 24(2)

King, Marcia. "Throwaway animals." (includes related articles on stray animals) (Cover Story) *Animals' Agenda* 11,4 (May 1991): 12(9)

King, Marcia. "The American Horse Protection Association: defending equines." *Animals' Agenda* 11,10 (December 1991): 24(2)

King, Marcia. "The horse show." *Animals' Agenda* 11,10 (December 1991): 16(2)

Maggitti, Phil. "Where the unicorn is king: a look at the circus." (mistreatment of circus animals) *Animals' Agenda* 9,10 (November 1989): 22(7)

Maggitti, Phil. "If you can grow it, you can show it - a guide to animal fancies in America." *Animals' Agenda* 11,10 (December 1991): 12(7)

Maggitti, Phil. "They shoot up horses, don't they?" (Cover Story) *Animals' Agenda* 10,9 (November 1990): 18(7)

Maggitti, Phil. "Nancy Burnet: on a wing and a prayer and perseverance." (founder of United Activists for Animal Rights and the Coalition to Protect Animals in Entertainment) *Animals' Agenda* 9,3 (March 1989): 6(2)

Maggitti, Phil. "Dog and cat collectors." (people who "collect" hundreds of dogs and cats under abusive conditions) *Animals' Agenda* 10,1 (Jan-February 1990): 20(5)

Maggitti, Phil. "Rescuing animals on the front lines." *Animals' Agenda* 10,8 (October 1990): 34(4)

Mason, Jim. "A trip to the world's largest exotic animal auction." *Animals' Agenda* 9,6 (June 1989): 46(3)

Meade, Bill; and Bevan, Laura. "Cockfighting: cruelty not courage." (includes related articles) *Animals' Agenda* 9,2 (February 1989): 44(3)

Merbarti, Don. "The dog show." *Animals' Agenda* 11,10 (December 1991): 18(2)

Mills, Eric. "Rodeo: American tragedy or legalized cruelty? Some would call it both." *Animals' Agenda* 10,2 (March 1990): 24(6)

"No-kill shelters in trouble again." (animal shelters) *Animals' Agenda* 10,7 (September 1990): 33(2)

Stopping the war on animals." (congressional inquiry into military experiments on animals) *Animals' Agenda* 10,8 (October 1990): 29(2)

LABORATORY ANIMALS

Books (annotated)

Baird, Robert M.; and Rosenbaum, Stuart E. eds. *Animal experimentation: the moral issues*. Buffalo, N.Y.: Prometheus, 1991. 182 p. (HV4915 .A64 1991)
This collection of previously published materials brings together the full range of opinions on animal rights, as well as summarizing the various ways humans use animals. Essays from both ends of the spectrum, written by prominent supporters and opponents of animal rights, are skillfully juxtaposed with critiques that challenge their arguments. Some intermediate positions are also presented.

Dickinson, Lynda. *Victims of vanity: animal testing of cosmetics and household products*. New York: Summerhill/Sterling, 1990. Distributed by Sterling Publishing, New York. 160 p.
Describes various experiments on animals and discusses alternative methods for testing consumer products. Descriptions rouse sympathy for test animals and outrage against companies that use them. Lists cruelty-free companies and animal welfare groups.

Education and training in the care and use of laboratory animals: a guide for developing institutional programs. Washington, D.C.: National Academy Press, 1991. 139 p. (SF406 .E38 1990)
Designed to help in planning educational programs for animal care workers. Provides annotated outlines for short courses on caring for animals. These courses were created with the objective of helping institutions meet recent federal regulations which require educational programs for personnel involved with animals in teaching and research. Individual sections address such issues as animal surgery, management of pain, and techniques which are specific to certain species. Other outlines offer information on other important issues, such as legal regulations, ethics, euthanasia, drugs, surgery, and animal care facilities.

Fox, Michael W. *Laboratory animal husbandry: ethology, welfare, and experimental variables*. Albany: State University of New York Press, 1986. 267 p.: ill. (SF406 .F69 1986)
Points out several overlooked factors which can have an effect on laboratory animal health and well-being "and subsequently on experiment results." For example the amount of human handling can affect plasma cortiscosterone, prolactin, and growth hormone levels. Also maintains that experiments all to often do not consider social factors in their experiments, such as the effects of having an animal that is highly interactive with other members of its species isolated. Maintains that taking more care to ensure the welfare of laboratory animals is not only morally right, it is good science.

Langley, Gill, ed. *Animal experimentation: the consensus changes*. New York: Chapman

and Hall, 1989. 268 p. (HV4711 .A55 1989)
> Introduces the complex philosophical and ethical issues surrounding animal rights and animal experimentation. Topics covered include the existence of animal suffering, practical alternatives to making experimentation more humane, the responsibilities of researchers and teachers dealing with the issue, and current animal rights legislation. Although a variety of perspectives are presented in these essays, the consensus is that traditional attitudes toward animal experimentation must change.

Lembeck, Fred, ed. *Scientific alternatives to animal experiments*. Translated by Jacqui Welch. Chichester, West Sussex, England, 1989. Distributed by Chapman and Hall, New York. 247 p.: ill. (QL51 .A4813 1989)
> The thirty-seven essays which comprise this work have two broad aims, and they are divided accordingly: to justify the use of animals in biomedical experimentation and to explain the methods used in these experiments, including alternative methods and their limitations. Maintains that recent medical advances have required the use of animals in experimentation and provides historical examples.

Philips, M. T.; and Sechzer, J. A. *Animal research and ethical conflict*. New York: Springer-Verlag, 1989.
> Analyzes the literature published in the U.S. between 1966 and 1885 on laboratory animal welfare and maintains that institutional animal care and use committees, which are responsible for animal care, need to balance scientific progress with the need for humane treatment of animals. Over half of the book is reference material. Includes a summary of international legislation.

Books (unannotated)

Alternatives to animal use in research, testing, and education. New York: M. Dekker, 1988. 441 p.: ill. (QL51.2.U6 A48 1988)

Animal experimentation and animal rights. Phoenix, Ariz.: Oryx Press, 1987. 75 p. (Z7164.C45 A55 1987)

Animal experimentation: the moral issues. Buffalo, N.Y.: Prometheus, 1991. 182 p. (HV4915 .A64 1991)

Animal models for oral drug delivery in man: in situ and in vivo approaches. Washington, D.C.: American Pharmaceutical Association, Academy of Pharmaceutical Sciences, 1983. 179 p.

Animal research, animal rights, animal legislation. Champaign, Ill. (309 W. Clark Street, Champaign 61820): Society for the Study of Reproduction, 1990. 45 p. (KF3841.Z9 A55 1990)

Animal sacrifices: religious perspectives on the use of animals in science. Philadelphia, Pa.: Temple University Press, 1986. 270 p. (BL439.5 .A55 1986)

Animals and alternatives in toxicity testing. New York: Academic Press, 1983. 550 p.: ill. (RA1199 .A538 1983)

Animals in biomedical research: replacement, reduction, and refinement: present possibilities and future prospects. New York: Elsevier Science Publishing, 1991. 288 p.: ill. (R853.A53 A54 1991)

Animals in research: new perspectives in animal experimentation. New York: J. Wiley, 1981. 373 p.: ill. (QL55 .A56)

Animals in toxicological research. New York: Raven Press, 1982. 214 p.: ill. (RA1199.A54)

Behringer, Marjorie P. *Techniques and materials in biology: care and use of living animals, plants, and microorganisms.* 2d ed. Malabar, Fla.: R. E. Krieger, 1989.

Covino, Joseph. *Lab animal abuse: vivisection exposed!* Berkeley, Calif.: New Humanity Press, 1990. 532 p. (HV4915 .C68 1990)

Dalton, John Call. *John Call Dalton on experimental method.* New York: Arno Press, 1980. 71, 108 p. (QL55 .D34 1980)

Experimental animals in biomedical research. Boca Raton, Fla.: CRC, 1990. : ill. (R853.A53 E86 1990)

Fox, Michael Allen. *The case for animal experimentation: an evolutionary and ethical perspective.* Berkeley: University of California Press, 1986. 262 p. (HV4915 .F67 1986)

Guide to the care and use of experimental animals. 2 vols. Ottawa: Canadian Council on Animal Care, 1980-1984. (QL55 .G75 1980)

Hendriksen, Coenraad F. M. *Laboratory animals in vaccine production and control: replacement, reduction, and refinement.* Boston: Kluwer Academic, 1988. 158 p.: ill. (QR189 .H47 1988)

Importance of animal experimentation for safety and biomedical research. Boston: Kluwer Academic, 1990. 246 p.: ill. (R853.A53 I46 1990)

Inevitable bond: examining scientist-animal interactions. New York: Cambridge University Press, 1992. 399 p.: ill. (QL55 .I44 1992)

Laboratory animals: an introduction for new experimenters. New York: Wiley, 1987. 342 p. ill. (QL55 .L274 1987)

Lives in the balance: the ethics of using animals in biomedical research: the report of a Working Party of the Institute of Medical Ethics. New York: Oxford University Press, 1991. 352 p.: ill. (R853.A53 E74 1991)

Phillips, Mary T. *Animal research and ethical conflict: an analysis of the scientific literature, 1966-1986.* New York: Springer-Verlag, 1989. 251 p.: ill. (HV4915 .P55 1989)

Pratt, Dallas. *Alternatives to pain in experiments on animals.* New York: Argus Archives, 1980. 283 p.: ill. (HV4930 .P68)

Recognition and alleviation of pain and distress in laboratory animals. Washington, D.C.: National Academy, 1992. 137 p.: ill. (SF996.5 R43 1992)

Ronner, Peter M. *Too true to be good: animal experiments in drug research.* Basel, Switzerland: Pharma Information, 1982. 53 p. (RM301.25 .R66 1982)

Rowan, Andrew N. *Of mice, models, and men: a critical evaluation of animal research.* Albany: State University of New York Press, 1984. 323 p. (HV4764 .R69 1984)

Science, medicine, and animals. Washington, D.C.: National Academy, 1991. 30 p.: ill. (HV4915 .S35 1991)

Scientific alternatives to animal experiments. Chichester, West Sussex, England: E. Horwood, 1989. Distributed by Chapman and Hall. 247 p.: ill. (QL51 .A4813 1989)

Sharpe, Robert. *The cruel deception: the use of animals in medical research.* Wellingborough, Northamptonshire, England: Thorsons, 1988. Distributed by Sterling Publishing, New York. 288 p.: plates. ill. (HV4915 .S497 1988)

Sperling, Susan. *Animal liberators: research and morality.* Berkeley: University of California Press, 1988. 247 p.: plates. ill. (HV4764 .S68 1988)

Through the looking glass: issues of psychological well-being in captive nonhuman primates. Washington, D.C.: American Psychological Association, 1991. Distributed by APA Order Department, Hyattsville, Md. 285 p.: ill. (SF407.P7 T49 1991)

UFAW handbook on the care and management of laboratory animals. 6th ed. New York: Churchill Livingstone, 1986.

Use of laboratory animals in biomedical and behavioral research. Washington, D.C.: National Academy Press, 1988. 102 p. (QL55 .U74 1988)

Vivisection in historical perspective. New York: Croom Helm, 1987. 373 p. (HV4915 .V58 1987)

General interest periodicals (annotated)

Angier, Natalie. "The electronic guinea pig." *Discover* 4 (September 1983): 76-78, 80.
    Describes new computer programs to reduce the need for dissections.

Boyce, John R.; and Lutes, Christopher. "Animal rights: how much pain is a cure worth?" *Christianity Today* 29 (September 6, 1985): 35-38.
    Argues that alternatives to animal experimentation should be used whenever possible, but that animals simply must be used in many cases so that cures may be found.

Cowley, Geoffrey. "Of pain and progress." *Newsweek* 112 (December 26, 1988): 50-55, 57, 59.
    "A growing social movement raises a thorny ethical question: do the practical benefits of animal experimentation outweigh the moral costs?"

Goodall, Jane. "A plea for the chimps." *New York Times Magazine* (May 17, 1987): 108-110, 118-119.
    Describes the cognitive abilities and emotional needs of chimpanzees. Reflects that the best way to improve the quality of life for laboratory chimpanzees is to "increase the number of carefully trained caretakers, and to select people who have compassion for their charges."

Hubbell, John G. "The 'animal rights' war on medicine." *Reader's Digest* 136 (June 1990): 70-76.
    Proclaims that animal research "is an unqualified success story." Notes that animal research has led to important vaccines, effective treatment for diabetes, and advances in organ transplants. Characterizes animal rights activists as "extremists" trying to stop vital research.

White, Robert J. "The facts about animal research." *Reader's Digest* 132 (March 1988): 127-132.
    " 'I am convinced that most Americans are unaware of the devasting effect that animal-rights extremists are having on lifesaving medical research,' warns this eminent scientist and neurosurgeon."

General interest periodicals (unannotated)

"Alternatives to animals." (using cells instead of laboratory animals in product research) *Economist* 313,7631 (December 2, 1989): 97(2).

"Animal rights and wrongs." (Department of Agriculture makes animal rights rules) *Economist* 313,7630 (November 25, 1989): 28(2).

Biermann, Karl. "Why animal experimentation should continue." *The Humanist* 50,4 (July-August 1990): 8(3).

Binding, Paul. "Prisoners of science." (animal rights and research) *New Statesman Society* 4,181 (December 13, 1991): 21(2).

Carey, John. "Will relief for lab animals spell pain for consumers?" (also includes a related article on the role of animals in lab testing) *Business Week* 3131 (October 30, 1989): 43(2).

Greene, Katherine; and Greene, Richard. "Necessary evil?" (animals in medical research; includes related articles) *Redbook* 173,5 (September 1989): 160(5).

Hanson, Gayle. "Tactics turn rabid in dissection war." (animal rights) *Insight* 7,38 (September 23, 1991): 18(4).

Hoyt, John A.; Newkirk, Ingrid E.; and McCabe, Katie. "Who will live, who will die?" (exchange on laboratory animal abuse; includes related article on John A. McArdle's dismissal) *The Washingtonian* 22 (October 1986): 105(4).

Karpati, Ron. "A scientist: 'I am the enemy'; a medical researcher takes on his accusers--the animal-rights movement." *Newsweek* 114,25 (December 18, 1989): 12(2).

MacFarlane, Andrew. "The truth about animal research." (includes related article on major contributions of animal research) *Reader's Digest* 136,816 (April 1990): 89(5).

Relin, David Oliver. "Welcome to the monkey house." (animal research laboratory) (includes related article on animal rights) *Science World* 48,14 (April 3, 1992): 6(8).

Siegel, Steve. "Animal research is unnecessary and dangerous to human health." *Utne Reader* 35 (September-October 1989): 47(3).

Singer, Peter. "Animal liberators: research and morality." *The New York Review of Books* 36,1 (February 2, 1989): 36(3).

Singer, Peter. "Use of laboratory animals in biomedical and behavioral research." *The New York Review of Books* 36,1 (February 2, 1989): 36(3).

Wall, James M. "Animals in research: a conflict in caring." (editorial) *The Christian Century* 105,29 (October 12, 1988): 883(2).

Weil, Robert. "Inhuman bondage: the untold story of a midnight raid that doomed a lab and sent reverberations through the scientific community." (Animal Liberation Front raid on University of Pennsylvania's Experimental Head Injury Laboratory) *Omni* 9 (November 1986): 64(9).

Environmental and professional journals (annotated)

"Alternatives to animal testing." *Chemical Times and Trends* 10 (July 1987): 7-25, 38-40, 57, 59.
>   Contains the essays "Balancing animal rights and consumer safety: progress to date," by Herbert N. Prince, Daniel L. Prince, and Richard N. Prince; "EPA testing aim: reducing use of animals," by Theodore Farber; "The low volume eye irritation test: a case study in progress toward validation," by John F. Griffith; and "Toward zero-based use of animals," by Henry Spira.

"Animal experimentation: issues for the 1980s." *Science Technology and Human Values* 9 (Spring 1984): 40-50.
>   Outlines areas of disagreement between vivisectionists and the bio-medical community. Describes legislation designed to limit experimentation on animals and argues that some use of animals is essential to research and education.

"Animal experiments: the great debate." *New Scientist* 133 (April 4, 1992): 25-30, 32-35.
>   Discussion of animal experiments by scientists involved: "The researchers' dilemma," by Lynda Birke and Mike Michael; "Views from behind the barricade," by Lynda Birke and Mike Michael; and "Trapped in a guilt cage," by Arnold Arluke.

"Animals in research." *Journal of Medicine and Philosophy* 13 (May 1988): whole issue (121-221 p.)
>   Contains essays on a variety of subjects, including "Standards for animal research: looking at the middle," by Rebecca Dresser; "Institutional animal care and use committees: a new set of clothes for the emperor?" by Lawrence Finsen; "Animal welfare and animal rights," by L. W. Sumner; "On justifying the exploitation of animals in research," by S. F. Sapontzis; and "The question is not 'Can they talk?' " by Gene Namkoong and Tom Regan.

"Animals, science, and ethics." *Hastings Center Report* 20 (May-June 1990, suppl.): 1-32.
>   Discusses the ethical basis for animal testing and reviews the policy issues involved. Contains a bibliography.

Bateson, Patrick. "When to experiment on animals." *New Scientist* 109 (February 20, 1986): 30-32.
>   Urges regulation that will take into account both the needs of science and the welfare of the animals used in research.

Baum, Rudy M. "Biomedical researchers work to counter animal rights agenda." *Chemical and Engineering News* 68 (May 7, 1990): 9-16, 21-24.
>   Reports that many biomedical researchers, who dismissed animal rights advocates a decade ago, now view them as a serious threat.

Birke, Lynda. "Better homes for laboratory animals." *New Scientist* 120 (December 3, 1988): 50-55.
> Asserts that animal welfare and the usefulness of animals to science are better served by studying the biological needs of animals.

Burghardt, Gordon M.; and Herzog, Harold A., Jr. "Beyond conspecifics: is Brer Rabbit our brother?" *BioScience* 30 (November 1980): 763-768.
> Discusses the ethics of vivisection, emphasizing the impossibility of deriving a consistent, universal set of principals to guide our dealings with members of other species.

Cherfas, Jeremy. "Chimps in the laboratory: an endangered species." *New Scientist* 109 (March 27, 1986): 37-41.
> Claims that chimpanzees are valuable in medical research because they are so like humans. But chimps are becoming more scarce because their natural habitat is being destroyed, and they do not breed well in captivity.

Cohn, Jeffrey P. "The animal research ruckus." *Government Executive* 23 (October 1991): 14- 17.
> "When animal-rights activists went wild over the government's treatment of monkeys, they created a host of problems for federal research and regulatory agencies."

Coile, D. Caroline; and Miller, Neal E. "How radical animal activists try to mislead humane people." *American Psychologist* 39 (June 1984): 700-701.
> Defends the use of animal experimentation in experimental psychology and comments that the radical animal activists are in fact diverting energy and funds away from areas where abuse is common: "10 million pet dogs and cats are abandoned each year to die of starvation, disease, and accidents."

Donnelley, Strachan. "Speculative philosophy, the troubled middle, and the ethics of animal experimentation." *Hastings Center Report* 19 (March-April 1989): 15-21.
> Maintains that judicious animal research can be ethically justified.

Gavaghan, Helen. "Animal experiments the American way." *New Scientist* 134 (May 16, 1992): 32-37.
> Contends that the U.S. Animal Welfare Act offers very few rules, regulations, or standards concerning the actual practice of research using animals, and those it does provide can sometimes be ignored in the interests of science.

Hampson, Judith. "The secret world of animal experiments." *New Scientist* 134 (April 11, 1992): 24-30.
> Urges public accountability for animal experiments. An accompanying article suggests that scientists should give details of any adverse effects on animals so that the same mistakes can be avoided by others.

Holden, Constance. "A pivotal year for lab animal welfare." *Science* 232 (April 11, 1986): 147-150.

> Forecasts a reduction in laboratory animal use and a continuation of conflict between the scientific community and the "radical fringe of the animal welfare movement."

Howard-Jones, Norman. "A CIOMS ethical code for animal experimentation." *WHO Chronicle* 39,2 (1985): 51-56.

> Early in 1985 the Council for International Organizations of Medical Sciences (CIOMS) published International Guiding Principles for Biomedical Research Involving Animals. This was the culmination of a three-year programme initiated in 1982 with the encouragement of the WHO Advisory Committee on Medical Research and the active collaboration of expert staff members of WHO.

Iglehart, John K. "The use of animals in research." *New England Journal of Medicine* 313 (August 8, 1985): 395-400.

> Reviews recent developments in the struggle between the medical-research community and animal-welfare advocates over the use of animals in research.

Junkin, Elizabeth Darby. "Solomon's child." *Buzzworm: The Environmental Journal* 1 (Spring 1989): 20-29.

> Examines controversy over the use of chimpanzies for AIDS research. Late in 1988, the Fish and Wildlife Service reclassified wild populations from "threatened" to "endangered" while leaving captive animals in the threatened category and legally available for research.

Langley, Gill. "Redundancy for the laboratory guinea pig." *New Scientist* 102 (May 3, 1984): 12-16.

> Foresees new methods of testing the toxicity of drugs and chemicals replacing the use of laboratory animals.

Lays, Julie. "Do rabbits have rights?" *State Legislatures* 13 (August 1987): 18-21.

> "As public outcry against the use of animals in research and testing grows, state legislators are being asked to take a more active role in regulating animal research within their states."

Loeb, Jerod M.; Hendee, William R.; and Smith, Steven J. "Human vs. animal rights: in defense of animal research." *JAMA: Journal of the American Medical Association* 262 (November 17, 1989): 2716-2720.

> A condensation of an AMA analysis of the use of animals and research and the consequences of banning such research.

Matlack, Carol. "Animal-rights furor." *National Journal* 23 (September 7, 1991): 2143-2146.

> "Opponents of animal experimentation are raising big money and making head-

lines. But medical research, pharmacy, and cosmetic interests are winning the battle over federal policy."

Moss, Thomas H. "The modern politics of laboratory animal use." *Science Technology and Human Values* 9 (Spring 1984): 51-56.
"Because the laboratory animals issue is one among relatively few areas in which the public expresses concern about the procedures, ethics, and approaches of scientific research . . . many scientists now recognize that the scientific community must ensure that the issue is handled sensitively and credibly."

Newman, Alan. "Research versus animal rights: is there a middle ground?" *American Scientist* 77 (March-April 1989): 135-137.
Seeks to define a middle ground between the militant animal rights position of Peter Singer and Ingrid Newkirk and the equally fervent pro-animal research position of some scientists.

Overcast, Thomas D.; and Sales, Bruce D. "Regulation of animal experimentation." *JAMA: Journal of the American Medical Association* 254 (October 11, 1985): 1944-1949.
Examines the nature of the controversy between animal activists and the scientific community over the need for and desirability of additional regulation of animal experimentation.

Pardes, Herbert; West, Anne; and Pincus, Harold Alan. "Physicians and the animal-rights movement." *New England Journal of Medicine* 324 (June 6, 1991): 1640-1643.
Contends that the current animal-rights movement threatens the future of health science and calls on physicians to inform the public about the need for animal experimentation.

Pavelock, Lisa A. "Towards legal rights for laboratory animals?" *Journal of Legislation* 10 (Winter 1983): 198-212.
Reviews philosophical views concerning the protection of animals and analyzes legislation proposed in the 97th Congress that sought to reconcile the concerns and interests of both scientists and animal activists.

Ritvo, Harriet. "Toward a more peaceable kingdom." *Technology Review* 95 (April 1992): 55-61.
"The mounting controversy over animal rights raises questions that challenge age-old assumptions about scientific research."

Stevens, Christine. "Animal torture in corporate dungeons." *Business and Society Review* 49 (Spring 1984): 39-43.
Advocates passage of the Impounded Standards for Laboratory Animals Act, adequate funding for USDA's veterinary services, and international adoption of the OECD Guidelines for Testing Chemicals.

Stevens, Christine. "Mistreatment of laboratory animals endangers biomedical research." *Nature* 311 (September 27, 1984): 295-297.
Urges the passage of pending legislation to improve the care of lab animals, avoid needless duplication of experiments and unnecessary pain.

Tannenbaum, Jerrold; and Rowan, Andrew N. "Rethinking the morality of animal research." *Hastings Center Report* 15 (October 1985): 32-43.
Argues that the debate on animal research has entered a new phase in reevaluating the moral status of animals and examining the biological and philosophical meaning of animal pain and suffering.

Weiss, Rick. "Test tube toxicology." *Science News* 133 (January 16, 1988): 42-45.
Describes new methods to test the safety of household chemical, cosmetic and therapeutic products which will provide alternatives to using laboratory animals.

Environmental and professional journals (unannotated)

"Animals in research." *JAMA: the Journal of the American Medical Association* 261,24 (June 23, 1989): 3602(5).

Barnard, Neal D. "Animal experiments in stroke research." (column) *Animals' Agenda* 10,10 (December 1990): 50(2)

Bernard, Neal D. "Barbara Orlans, Ph.D.: on research reforms and regulations." (animal experimentation; includes related article) (interview) *Animals' Agenda* 9,5 (May 1989): 7(5)

Barnard, Neal D. "Biological warfare experiments: cruel, impractical, and dangerous." *Animals' Agenda* 11,2 (March 1991): 50(2)

Barnard, Neal D. "Cleaning up U.S. Surgical." (animal rights activists attempt to end animal experimentation at medical supply company) *Animals' Agenda* 9,7 (September 1989): 44(2)

Barnard, Neal D. "Dog days in medical school." (ridding medical schools of animal experimentation) *Animals' Agenda* 9,4 (April 1989): 42(2)

Barnard, Neal D. "Longer life expectancy: who gets the credit?" (lack of importance of animal experimentation) (column) *Animals' Agenda* 9,5 (May 1989): 45(2)

Barnard, Neal D. "The Nazi experiments." (column) *Animals' Agenda* 10,3 (April 1990): 8(2)

Barnard, Neal D. "Product testing on animals." *Animals' Agenda* 11,3 (April 1991): 50(2)

Barnard, Neal D. "Studying the AIDS epidemic." *Animals' Agenda* 9,7 (July-August 1989): 43(2)

Breo, Dennis L. "Animal rights vs research: A question of the nation's scientific literacy." *JAMA: the Journal of the American Medical Association* 264,19 (November 21, 1990): 2564(2).

Budkie, Michael A. "Researching your local research facility." (resources available to animal rights activists) *Animals' Agenda* 10,8 (October 1990): 16(3)

Chui, Glennda. "Activists beset UC, Stanford labs." (includes related article) *Science* 239,4845 (March 11, 1988): 1229(4).

Clifton, Merritt. "Flattery will get them nowhere." (vivisectors sponsored by American Medical Association testify against animal defenders in Washington) *Animals' Agenda* 10,10 (December 1990): 46(2)

Clifton, Merritt. "Industry responds to product testing campaigns." (animal testing in the manufacture of cosmetics) *Animals' Agenda* 9,8 (October 1989): 40(3)

Clifton, Merritt. "More fraud in animal-based research." *Animals' Agenda* 11,5 (June 1991): 33(2)

Clifton, Merritt. "Shake-up at Sleepy Hollow." (corporate acquisition raises furor about use of animals in cosmetics testing) *Animals' Agenda* 9,11 (December 1989): 34(2)

Clifton, Merritt. "World Laboratory Animal Liberation Week." *Animals' Agenda* 10,6 (July-August 1990): 38(2)

Cohen, Murry J.; et al. "Use of animals in medical education." (Letter to the Editor; includes reply) *JAMA: the Journal of the American Medical Association* 266,24 (December 25, 1991): 3421(3).

Cohn, Jeffrey P.; and Root, Michael. "Beyond the lab rat: what biomedicine is learning from unconventional organisms." (includes related article) *BioScience* 39,8 (September 1989): 518(5).

Comfort, Nathaniel C. "Can you love animals and kill them?" *Utne Reader* 35 (September-October 1989): 46(4).

Davis, Karen. "What's wrong with pain anyway?" *Animals' Agenda* 9,2 (February 1989): 50(2)

Erickson, Deborah. "Blood feud: researchers begin fighting back against animal-rights activists." *Scientific American* 262,6 (June 1990): 17(2).

Francione, Gary L. "The importance of access to animal care committees: a primer for activists." (Animal Welfare Act requires animal experimenters to establish Institutional Animal Care and Use committees, but are they effective?) *Animals' Agenda* 10,7 (September 1990): 44(4)

Frazier, Claude A. "Lesson from a cat: a physician speaks out on animal experimentation." *Animals' Agenda* 10,4 (May 1990): 44(2)

Gill, Thomas J., III; et al. "The rat as an experimental animal." *Science* 245,4915 (July 21, 1989): 269(8).

Goldberg, Alan M.; and Frazier, John M. "Alternatives to animals in toxicity testing." *Scientific American* 261,2 (August 1989): 24(7).

Goldsmith, Marsha F. "At Southwest Foundation, scientists run show--but research animals are essential stars." *JAMA: the Journal of the American Medical Association* 262,19 (November 17, 1989): 2648(3).

Greanville, David P. "Soviet Union: struggle against biomedical research gains momentum." *Animals' Agenda* 11,5 (June 1991): 25(2)

Grey, Marilyn. "Animal testing and the law." *Animals' Agenda* 11,4 (May 1991): 39(2)

Holden, Constance. "A preemptive strike for animal research." (New York University defends research) *Science* 244,4903 (April 28, 1989): 415(2).

Holden, Constance. "Animal regulations: so far, so good: most institutions are reasonably content with new NIH standards but there is concern about forthcoming Agriculture regulations." *Science* 238,4829 (November 13, 1987): 880(3).

Holden, Constance. "Billion dollar price tag for new animal rules." *Science* 242,4879 (November 4, 1988): 662(2).

Holden, Constance. "Compromise in sight on animal regulations: rules governing dogs and primates are being revised for the third time; less financial distress for scientists predicted." *Science* 245,4914 (July 14, 1989): 124(2).

Holden, Constance. "Experts ponder simian well-being." (regulations governing the psychological well-being of primates) *Science* 241,4874 (September 30, 1988): 1753(3).

Holden, Constance. "Monkey saga continues." (animal rights case in Louisiana) *Science* 247,4941 (January 26, 1990): 406(2).

Holden, Constance. "Universities fight animal activists." *Science* 243,4887 (January 6, 1989): 17(3).

Kaplan, John. "The use of animals in research." *Science* 242,4880 (November 11, 1988): 839(2).

Kaufman, Stephen; and Barnard, Neal D. "How useful are animal models?" *Animals' Agenda* 10,7 (September 1990): 18(2)

Maggitti, Phil. "Lawrence Carter: the prodigal poster child." (Health Care Consumer Network director, cerebral palsy victim opposed to animal medical research) *Animals' Agenda* 11,6 (July-August 1991): 45(2)

Norman, Colin. "Cat study halted amid protests." (barbiturate addiction study) *Science* 242,4881 (November 18, 1988): 1001(2).

Palca, Joseph. "Import rules threaten research on primates." *Science* 248,4959 (June 1, 1990): 1071(3).

Reitman, Judith; and Clifton, Merrit. "Addicted by Yale." (animal experimentation at Yale University; includes related article) *Animals' Agenda* 9,5 (May 1989): 20(4)

Ritvo, Harriet. "Plus ca change: antivivisection then and now." *BioScience* 34 (November 1984): 626(8). (QH1.A277)

Rosenberger, Jack. "Hiding behind dead babies." (justification of animal laboratory research) *Animals' Agenda* 10,4 (May 1990): 36(2)

Rowan, Andrew N. "Why scientists should seek alternatives to animal use." *Technology Review* 89 (May-June 1986): 22(2).

Shapiro, Kenneth; and Carr, John. "The politics of psychology." (members of American Psychological Association's Committee on Animal Research and Ethics conduct painful experiments on animals) *Animals' Agenda* 11,4 (May 1991): 42(2)

Shulman, Seth. "Who needs mousetraps?" (alternatives to using laboratory rats in research) *Technology Review* 89 (November-December 1986): 11(2).

Smith, Steven J. "Animals in research." *JAMA: the Journal of the American Medical Association* 259,13 (April 1, 1988): 2007(2).

Snow, Bonnie. "Online searching for alternatives to animal testing." *Online* 14,4 (July 1990): 94(4).

Thomas, James A.; Hamm, Thomas E., Jr.; and Perkins, Pamela L. "Animal research at Stanford University." *The New England Journal of Medicine* 318,24 (June 16, 1988): 1630(3).

Thompson, Richard C. "Reducing the need for animal testing." *FDA Consumer* 22,1 (February 1988): 15(3).

"Three R's." (alternatives to current state of animal experimentation) *Scientific American* 254 (April 1986): 68(2).

"TSU shuts down cat lab at Cornell." (Trans-Species Unlimited, animal rights movement) *Animals' Agenda* 9,3 (March 1989): 22(2)

Ulrich, Roger E. "Animal research: a psychological ritual." (animal experimentation in psychological research has no demonstrable value in helping humans) *Animals' Agenda* 11,4 (May 1991): 40(4)

"Use of animals in medical education." *JAMA: the Journal of the American Medical Association* 266,6 (August 14, 1991): 836(2).

Vaughan, Christopher. "Animal research: ten years under siege." *BioScience* 38,1 (January 1988): 10(4).

Law journals

Brody, Mimi. "Animal research: a call for legislative reform requiring ethical merit review." *Harvard Environmental Law Review* 13,2 (1989): 423-484.
> Proposes an "ethical merit review" process to determine whether a research project's potential justifies its cost in terms of animal suffering.

Dresser, Rebecca. "Research on animals: values, politics, and regulatory reform." *Southern California Law Review* 58 (July 1985): 1147-1201.
> Examines the problem of regulating research on animals and proposes legal reform that balances the interest of lab animals and those who benefit from animal research.

Garvin, Larry T. "Constitutional limits on the regulation of laboratory animal research." *Yale Law Journal* 98 (December 1988): 369-388.
> Legislative successes of animal rights activists have been mostly on the state and local levels. Of special note are two measures recently enacted in Massachusetts.

Goodkin, Susan L. "The evolution of animal rights." *Columbia Human Rights Law Review* 18 (Spring 1987): 259-288.
> Argues that laboratory animals have the right to be free from pain inflicted for research purposes. Indicates that because such research has obvious human benefits it is the most difficult issue facing advocates of animal rights.

McDonald, Karen L. "Creating a private cause of action against abusive animal research." *University of Pennsylvania Law Review* 134 (January 1986): 399-432.

Concludes that "court actions by animal rights groups may serve to mobilize public opinion sufficiently to motivate legislators to enact laws recognizing legal rights for animals."

Spira, Henry. "Animal rights and toxicology: a quiet but profound revolution." *Food Drug Cosmetic Law Journal* 46 (January 1991): 89-95.
    Describes ideas being developed and implemented to eliminate unnecessary suffering of laboratory animals.

Reports

Biohazard: the silent threat from biomedical research and the creation of AIDS: a report by the National Anti-Vivisection Society. London: National Anti-Vivisection Society, 1987. 80 p.: ill. (RC607.A26 N385 1987)

Explanatory report on the European Convention for the Protection of Vertebrate Animals Used for Experimental and Other Scientific Purposes. Strasbourg, Austria: Council of Europe, 1986. Distributed by Manhattan Publishing, Croton, N.Y. 75 p.: ill. (KJC6237.A41986 A34 1986)

Interdisciplinary principles and guidelines for the use of animals in research, testing, and education. New York: New York Academy of Sciences, 1988. 13 p. (HV4930 .N48 1988)

Nethery, Lauren B. *Animals in product development and safety testing: a survey*. Washington, D.C. (2100 L Street NW, Washington 20037): Institute for the Study of Animal Problems, 1985. 52 p. (RA1199 .N466 1985)

Statistics of experiments on living animals: Great Britain, 1982. London: H.M.S.O., 1983. 35 p.: ill. (QL55 .S726 1983)

Use of animals in biomedical research: the challenge and response. Washington, D.C.: American Medical Association, 1989. 25 p. (AMA white paper)
    The AMA supports "the humane use of animals for biomedical research. Research involving animals is essential to improving the health and well-being of the American people, and the American Medical Association actively opposes any legislation, regulation, or social action that inappropriately limits such research." Describes "the importance of animals to past, present and future biomedical research and medical practice. The history and philosophies of the animal rights movement are considered together with arguments raised by animal welfare groups and responses by the scientific community."

Stephens, Martin L. *Alternatives to current uses of animals in research, safety testing, and education: a layman's guide*. Washington, D.C. (2100 L Street NW, Washington 20037): Humane Society of the United States, 1986. 86 p.: ill. (RB125 .S74 1986)

Welsh, Heidi J. *Animal testing and consumer products.* Washington, D.C.: Investor Responsibility Research Center, 1990. (HV4930 .W45 1990)

Federal government reports

Alternatives to animal use in research, testing, and education. Washington, D.C.: (Office of Technology Assessment), G.P.O., 1986. 441 p. (OTA-BA-273). Summary, 49 p. (OTA-BA-274)

> Analyzes the scientific, regulatory, economic, legal, and ethical considerations involved in alternative technologies in biomedical and behavioral research, toxicity testing, and education. Included is a detailed examination of federal, state, and institutional regulation of animal use, and a review of recent developments in ten other countries.

National Chimpanzee Breeding Program: report of the Ad Hoc Task Force. Washington, D.C.: Department of Health and Human Services 1982 11 p.

> Outlines a coordinated plan to ensure a self-sustaining, breeding population both to provide animals for research and to safeguard an endangered species.

Conference proceedings

Centre for Medicines Research, Surrey, England: 1984 Workshop: Long-term animal studies: their predictive value for man. Boston, Mass.: MTP Press, 1986. 156 p.: ill. (RS189 .C46 1984)

National Symposium on Imperatives in Research Animal Use: Scientific Needs and Animal Welfare. Bethesda, Md.: National Institutes of Health, 1984. 345 p.

Science and animals: addressing contemporary issues: from a conference held by the Scientists Center for Animal Welfare in Washington, D.C. on June 22-25, 1988. Bethesda, Md. (4805 St. Elmo Avenue, Bethesda 20814): Scientists Center for Animal Welfare, 1989. 149 p.: ill. (HV4704 .S33 1989)

Bibliographies

Murphy, Richard A. *Annotated bibliography on laboratory animal welfare.* Bethesda, Md. (4805 St. Elmo Avenue, Bethesda 20814): Scientists Center for Animal Welfare, 1991. 91 p. (Z7164.C45 M87 1991)

FARM ANIMALS

(See also *Food Safety: Animal Hormones and Medicines*) 1174

Books (annotated)

Coats, C. David. *Old MacDonald's factory farm: the myth of the traditional farm and the shocking truth about animal suffering in today's agribusiness*. New York: Continuum, 1989. 186 p.: ill. (SF140.L58 C63 1989)
>   Maintains that animals grown for food, clothing, or cosmetics are cruelly handled, poorly housed in overcrowded pens or cages, genetically manipulated, forced to consume food and drugs, and suffer other mistreatment. Includes a discussion of food ingredient labelling and an appendix of animal welfare organizations. Contains a bibliography.

Johnson, Andrew. *Factory farming*. Cambridge, England: Blackwell, 1991. 272 p.: ill. (SF140.L58 J64 1991)
>   Presents balanced reasons why factory farming is unethical and unnecessary, as well as environmentally unsound. Discusses various systems of intensive farming in terms of animal welfare and the environment. Includes a chapter on fish farming. Relates progressive steps that have been taken against factory farming in other countries. Points out that most meat contains animal dung. Contains notes, an index, and helpful appendices.

Books (unannotated)

Ethical, ethological, and legal aspects of intensive farm animal management. Boston: Birkhauser Verlag, 1987.

Fox, Michael W. *Agricide: the hidden crisis that affects us all*. New York: Schocken Books, 1986. 194 p.: plates. ill. (SF51 .F69 1986)

Fox, Michael W. *Farm animals: husbandry, behavior, and veterinary practice: viewpoints of a critic*. Baltimore, Md.: University Park Press, 1984. 285 p.: ill. (SF61 .F63 1984)

Gold, Mark. *Assault and battery: what factory farming means for humans and animals*. London: Pluto Press, 1983. 172 p.: ill. (HV4805.A3 G65 1983)

Mason, Jim. *Animal factories*. Rev. ed. New York: Harmony Books, 1990. 240 p.: ill. (SF140.L58 M37 1990)

General interest periodicals (annotated)

Japenga, Ann. "Livestock liberation." *Harrowsmith Country Life* 4 (November-December 1989): 34, 36-43, 96.
>   "The issue of farm-animal welfare wil come up again and again in the next

decade as a growing number of activists across the country continue to press for improvements in the way livestock is treated."

Krohe, James, Jr. "Anatomy of a food business." *Across the Board* 19 (May 1982): 44-49. "Hogs are produced in factories housing as many as 15,000 head. Critics say the system of confinement reduces the animals to 'biomachines.' "

General interest periodicals (unannotated)

Feder, Barnaby J. "Pressuring Perdue." (Henry Spira) *The New York Times Magazine* 138 (November 26, 1989): 32.

McGuire, Richard. "Agriculture and animal rights: ethical agendas and agriculture. *Vital Speeches of the Day* 55,24 (October 1 1989): 766(3)

Nimble, J. B. "Anxious beef." *Alternatives* 16,2 (June-July 1989): 29(2).

Environmental and professional journals (annotated)

Appleby, Michael. "Frustration on the factory farm." *New Scientist* 129 (March 30, 1991): 34, 36.
"Animals kept in close confinement pace their cages or bite their bars because of frustration, not boredom. Changes in housing can reduce the problem."

Birchall, Annabelle. "The rough road to slaughter." *New Scientist* 128 (November 24, 1990): 33-36, 38.
"Farm animals often spend their last days travelling long distances cramped into lorries, hungry and thirsty. But there are ways to transport animals in far less discomfort."

Dawkins, Marian Stamp; and Nicol, Christine. "No room for manoeuvre." *New Scientist* 123 (September 1989): 44-46.
"Farm animals could benefit from housing standards agreed in Brussels. But the Eurocrats have given each battery hen floor space smaller than a telephone directory."

Fox, Michael W. "Livestock animals . . . a question of rights: the plowboy interview with Dr. Michael Fox." *Mother Earth News* 79 (January-February 1983): 16-20, 22.
An interview with the director of the Institute for the Study of Animal Problems presents his views on modern livestock rearing practices, many of which he describes as inhumane.

Fox, Michael W. "The dairy cow debacle: the government mandates face branding." *Humane Society News* 31 (Summer 1986): 4-9.

Reports on the Humane Society's opposition to hot-iron branding mandated by the USDA on cows to be slaughtered in the whole-herd buy-out program. USDA has amended its instructions to allow freeze-branding, a more humane branding, but has not withdrawn the hot-iron branding policy.

Klintberg, Patricia. "The real deal about veal." *Farm Journal* 114 (November 1990): 16-17.
"Despite a wealth of experience establishing the benefits of confinement raising to both calves and producers, consumers are being led on a guilt trip about veal that may be irreversible."

Mason, Jim. "Is factory farming really cheaper?" *New Scientist* 105 (March 28, 1985): 12-15.
"Modern farmers keep their animals in strict confinement and under intensive management. Their justification is that these practices result in cheaper food, but the savings are exaggerated and hidden costs undermine the argument."

Mench, Joy A.; and Van Tienhoven, Ari. "Farm animal welfare." *American Scientist* 74 (November-December 1986): 598-603.
"This article describes techniques such as confinement of hens, sows, and veal calves, and discusses research on animals' stress responses and behavioral preferences."

Russell, John; and Ainsworth, Earl. "What farmers should know about animal welfare." *Farm Journal* 105 (October 1981): 21-23; 64.
"Deliberate cruelty to animals ought never be tolerated . . . but this is a far cry from insisting that animals have 'inalienable rights' to roam free and that farmers violate these 'rights' when they confine animals for the efficient production of food."

"Suffer the animals." *Environmental Action* 21 (May-June 1990): 23-30.
Explores the concerns and points of conflict of environmentalists and animal protectionists regarding animal-derived toxicity testing and factory farming.

Webster, John. "Sense and sensibility down on the farm." *New Scientist* 119 (July 21, 1988): 41-44.
"Biotechnology seems to provide us with new ways of producing bigger and better livestock. But are these techniques worth the trouble, and what is the cost to the animals?"

Environmental and professional journals (unannotated)

Davis, Karen. "Cry fowl!" (animal cruelty in chicken processing plants) *Animals' Agenda* 11,3 (April 1991): 46(2)

Freese, Betsy. "New role for boars." (possible elimination of the practice of castrating male meat-hogs; includes related article warning of possible attacks by animal rights activists) *Successful Farming* 89,10 (October 1991): 32(2).

Greanville, David P. "Horsemeat consumption hits slump." *Animals' Agenda* 9,11 (December 1989): 27(2)

Greanville, David Patrice. "Animals in the electronic eye." *Animals' Agenda* 9,5 (May 1989): 46(3)

Mason, Jim. "America's other drug problem: down on the factory pharmacy." (use of drugs and hormones in husbandry; includes related article on livestock diseases and specific drugs used to treat them) *Animals' Agenda* 10,6 (July-August 1990): 47(3)

Mason, Jim. "Taking stock: from farm to slaughter." *Animals' Agenda* 11,3 (April 1991): 16(8)

Moran, Victoria; and Clifton, Merritt. "They used to call them slaughterhouses." (includes related articles) *Animals' Agenda* 11,3 (April 1991): 40(5)

Sommer, Mark. "Farm animal abuse goes to the ballot." (Humane Farming Initiative loses in Massachusetts) *Animals' Agenda* 9,2 (February 1989): 26(2)

Spencer, Earl F. "A farmer's thoughts on animal rights." *Countryside and Small Stock Journal* 72,2 (March-April 1988): 62(2).

Reports

Durning, Alan B.; and Brough, Holly B. *Taking stock: animal farming and the environment.* Washington, D.C.: Worldwatch Institute, 1991. 62 p. (Worldwatch paper 103)
> "Livestock create an array of problems . . . because human institutions have driven some forms of animal farming out of alignment with the ecosystems in which they operate. Many governments--including those of China, the European Community, and the United States--subsidize ecologically harmful methods of growing feed crops and raising animals." Maintains that reform is necessary, including less meat consumption, ecological taxes, etc.

Fox, Michael W. *Factory farming.* Washington, D.C.: (2100 L Street NW, Washington 20037): Humane Society of the U.S., 1980. 37 p.
> Describes intensive farming techniques which involve the total confinement of certain domestic livestock. Suggests the need for more humane consideration and treatment both for their own sake and for economic reasons.

Fox, Michael W. *Farm animal welfare and the human diet*. Washington, D.C. (2100 L. Street NW, Washington 20037): Humane Society of the United States, 1983. 22 p.: plates: ill. (HV4764 .F69 1983)

Fox, Michael W. *The hidden costs of beef*. Washington, D.C. (2100 L Street NW, Washington 20037): Humane Society of the U.S., 1989. 45 p.: ill. (HD9433.U4 F65 1989)

Guither, Harold D.; and Curtis, Stanley E. *Changing attitudes toward animal welfare and animal rights: the meaning for the U.S. food system*. East Lansing: Michigan State University Cooperative Extension Service, 1984. 6 p.
    Analyzes animal welfare and rights issues related to the production of domestic food animals. Discusses economic effects of welfare measures.

Scientific aspects of the welfare of food animals. Ames, Iowa: Council for Agricultural Science and Technology, 1981. 54 p. (Report no. 91)
    Assesses whether modern scientific methods of food-animal production are compatible with the welfare of animals. Surveys current methods of managing poultry, swine, cattle, sheep, and research animals.

HUNTING

(See also *Endangered Species*) 2615
(See also *Wildlife Conservation: Waterfowl*) 2585
(See also *Wildlife Conservation: Hunting*) 2598

Books (unannotated)

Ammon, William H. *The Christian hunter's survival guide*. Old Tappan, N.J.: F. H. Revell, 1989. 95 p. (BV4597.4 .A55 1989)

Baker, Ron. *The American hunting myth*. New York: Vantage Press, 1985. 287 p. (SK361 .B35 1985)

Carman, Russ. *The illusions of animal rights*. Iola, Wis.: Krause Publications, 1990. 160 p. (HV4708 .C37 1990)

Huntington, Henry P. *Wildlife management and subsistence hunting in Alaska*. Seattle: University of Washington Press, 1992.

Marks, Stuart A. *Southern hunting in black and white: nature, history, and ritual in a Carolina community*. Princeton, N.J.: Princeton University Press, 1991. 327 p.: ill., map (SK113 .M37 1991)

Thomas, Richard H. *The politics of hunting.* Brookfield, Vt.: Gower, 1983. 313 p. (SK185 .T48 1983)

Whisker, James B. *The right to hunt.* Croton-on-Hudson, N.Y.: North River Press, 1981. 173 p. (SK21 .W46)

Wallace, Ronald L. *The tribal self: an anthropologist reflects on hunting, brain, and behavior.* Lanham, Md.: University Press of America, 1991. 149 p. (BF701 .W34 1991)

General interest periodicals (annotated)

Reiger, George. "And still they go a-hunting." *New York Times Magazine,* June 11, 1989: 68, 81-82.
"Despite the outcry from animal-rights groups, big-game hunting is on the increase. The thrill is in the chase."

Reisner, Marc. "Bad news, bears." *California Magazine* 12 (March 1987): 71-73, 110-112, 128.
"California's last black bears are being slaughtered by hunters who kill them, skin them, remove the gall bladder and leave the rest behind. On the black market, the galls are as precious as cocaine." Koreans believe bear gall can cure anything, and powdered bear gall may be worth $5,000 a pound.

Satchell, Michael. "The American hunter under fire." *U.S. News & World Report* 108 (February 5, 1990): 30-31, 33-37.
Focuses on hunting in the U.S. and current efforts by several national organizations to launch an anti-hunting campaign and to promote legislation banning hunts. Also discusses the effect of hunting on U.S. wildlife.

Williams, Joy. "The killing game." *Esquire* 114 (October 1990): 113-114, 116, 118, 121, 124.
"In the hunting magazines, hunters freely admit the pleasure of killing to one another. But in public, most hunters are becoming a little wary about raving on as to how much fun it is to kill things. Hunters are increasingly relying upon their spokesmen and supporters, state and federal game managers and wildlife officials, to employ the drone of a solemn bureaucratic language and toss around a lot of questionable statistics to assure the nonhunting public (93 percent) that there's nothing to worry about."

Williams, Ted. "Should they shoot Bambi?" *Boston Magazine* 76 (November 1984): 178, 246-248, 250-251.
Describes controversy over hunting of the overpopulous deer herd at Crane Memorial Reservation in Ipswich. Although Massachusetts wildlife officials favor the hunt, animal-rights group Friends of Animals opposes it and may succeed in banning the hunt again this year. Meanwhile, deer starve.

General interest periodicals (unannotated)

Almy, Gerald. "First blood: a little 3-point buck taught me the meaning of death." *Sports Afield* 196 (November 1986): 132(2)

Atwill, Lionel. "Anti-Hunting 101: as the hunting debate heats up, a class at West Virginia University takes a scholarly look at these often emotional issues." *Field & Stream* 96,4 (August 1991): 46(3)

Atwill, Lionel. "What is the UBNJ? And how is it converting anti-hunters. (United Bowhunters of New Jersey) *Field & Stream* 96,7 (November 1991): 41(3)

Barsness, John. "For the asking." (asking a stranger for permission to hunt) *Field & Stream* 90 (January 1986): 26(2)

Binding, Paul. "Unsporting lives." (fox hunting) *New Statesman Society* 4,175 (November 1 1991): 14(2)

Bodio, Stephen J. "Why we should hunt." (good hunting as a morally, ecologically, and esthetically sound pursuit) *Outdoor Life* 175 (February 1985): 58(4)

Bouwman, Fred. "Reading, writing, and anti-hunting." (wildlife education) *Outdoor Life* 175 (April 1985): 62(4)

Brister, Bob. "Cutting our losses: an animal that is shot badly and escapes will eventually perish. What dies with the creature is a bit more of the public's tolerance for the sport of hunting." *Field & Stream* 94,8 (December 1989): 74(3)

Conley, Clare. "The tag name." (H. Cotler Co.'s clothing tags and its anti-hunting campaign; includes related information on harassment of hunters) *Outdoor Life* 186,5 (November 1990): 71(3)

Conner, Beverly; and Leahy, John. "Hunter harassment: plaguing our American heritage." (includes related articles on hunting laws) *Outdoor Life* 186,4 (October 1990): 82(17).

Conniff, Richard. "Fuzzy-wuzzy thinking about animal rights." *Outdoor Life* 187,2 (February 1991): 68(6)

DeJong, Cornell. "Beyond the guns and the game--hunters and hunting in New York: a question of baiting." (baiting deer) *Field & Stream* 93,5 (September 1988): 56(5)

Forbes, Dana. "Liberating the killing field." (site of the Fred Coleman Memorial Pigeon Shoot in Hegins, Pennsylvania) *MS.* 2,4 (January-February 1992): 84(2)

Garfield, Bob. "My son the hunter: what's a nice Jewish boy doing in this sport?" *Sports Illustrated* 69,21 (November 14 1988): 100B(3)

Gresham, Tom. "Handling hunter harassment." *Sports Afield* 208,3 (September 1992): 60(3)

Hill, Gene. "Ethics: for sportsmen, right or wrong is often a matter of conscience, not law." *Field & Stream* 92,10 (February 1988): 13(2)

Hill, Gene. "What if?" (psychology of hunters) (column) *Field & Stream* 91 (January 1987): 32(2)

Howard, Walter E. "Animal rights vs. hunters: wildlife must be protected from nature, not just man." *Outdoor Life* 187,4 (April 1991): 110(2)

Kerasote, Ted. "Hunting the hardest question." (ethical values in hunting) *Sports Afield* 199,4 (April 1988): 17(2)

Lea, Sydney. "The death of a hunting dog." (respect for hunting) *Sports illustrated* 75,24 (December 2, 1991): 5(2)

Lewis, Steven. "The hated season." (deer hunting) *The New York Times Magazine* 138 (November 26, 1989): 24

Lopez, Barry. "The mind of the hunter." *Harper's* 272 (March 1986): 30(2)

Magnuson, Jon. "Reflections of an Oregon bow hunter." (Cover Story) *The Christian Century* 108,9 (March 13, 1991): 292(4).

Mann, E. B. "What's in a name?" (anti-hunting groups) *Field & Stream* 88 (April 1984): 22(2)

Matthews, Jim. "The domino effect: this is one dangerous set of dominoes." Anti-hunters have already toppled the first one and banned mountain lion hunting in California. *Outdoor Life* 187,3 (March 1991): 61(6)

McIntyre, Thomas. "Debunking animal rightist 'truths.' " *Sports Afield* 208,6 (December 1992): 22(2)

McIntyre, Thomas. "Hunter equals conservationist: it's time to wake up and realize that antihunters and animal rightists are driving a wedge between our sport and our caring for the environment." *Sports Afield* 205,6 (June 1991): 17(2)

McIntyre, Thomas. "Shooting Bambi's mom." (deer hunting) *Sports Afield* 206,4 (October 1991): 35(3).

McIntyre, Thomas. "Teach your children well." (children's views of hunting) *Sports Afield* 207,4 (April 1992): 18(3)

McIntyre, Thomas. "The public war on hunters." (antihunting sentiments in the media) *Sports Afield* 204,6 (December 1990): 20(3)

McIntyre, Thomas. "This land was your land." (access to public lands) *Sports Afield* 205,2 (February 1991): 24(2)

Nelson, Sarah Jane. "Posted: no trespassing: a rising tide of rural growth leaves hunters and property owners with a world neither is accustomed to, or likes." *Country Journal* 17,6 (December 1990): 62(4)

Pacelle, Wayne. "An interview with Luke Dommer." (founder of Committee to Abolish Sport Hunting) *Animals' Agenda* 9,1 (January 1989): 6(4)

Petzal, David E. "Hunting as war." (high-tech equipment versus skill) *Field & Stream* 96,7 (November 1991): 16(2)

Petzal, David E. "Joy Williams hates you." (response to anti-hunting article in Esquire) *Field & Stream* 95,8 (December 1990): 16(2)

Reiger, George. "A limiting mentality." (dove shooting; lack of appreciation for game) *Field & Stream* 94,8 (December 1989): 14(2)

Reiger, George. "Passing along the tradition." (teaching hunting ethics) *Field & Stream* 95,12 (April 1991): 12(2)

Reiger, George. "When ethics should be law." (hunting laws) (column) *Field & Stream* 93,8 (December 1988): 13(2)

Reynolds, Brad. "Eskimo hunters of the Bering Sea." *National Geographic* 165 (June 1984): 814(21)

Rutledge, Archibald. "Why I taught my boys to be hunters." *Outdoor Life* 190,6 (December 1992): 69(5)

Satchell, Michael. "The American hunter under fire." (animal rights activists) *U.S. News & World Report* 108,5 (February 5 1990): 30(6)

Shedd, Warner. "The giant stirs." (sportsmen versus anti-hunting groups; includes related article on hunter harassment laws) *Outdoor Life* 188,4 (October 1991): 80(4)

Sisson, Dan. "Grandpa and the kid." (dealing with emotional responses to hunting) *Field & Stream* 96,6 (October 1991): 62(3)

Whisler, Norman J. "Carnivorous anti-hunters." *Guns & Ammo* 28 (September 1984): 37(4)

Williamson, Lonnie. "The swat team: the anti-hunters are swarming, but sportsmen are fighting back with a new organization that'll help take away some of the sting." (hunters form United Conservation Alliance to fight animal rights groups) *Outdoor Life* 187,6 (June 1991): 38(3).

Williamson, Lonnie. "Know thy enemy: the best defense is often a good defense. Learning about animal rightists could very well be the key to stopping them." (column) *Outdoor Life* 187,2 (February 1991): 49(3)

Woodward, Bob. "Trigger happy." (acceptance of hunters by non-hunters) *Backpacker* 15 (January 1987): 8(2)

Environmental and professional journals (annotated)

Caro, Tim. "Big-game hunters are not biologists." *New Scientist* 14 (December 13, 1984): 12-15.
> "If hunters admitted that they hunted for personal gratification and not because they were managing animal populations we might dismiss them less readily and consider their effects on economics and conservation a little more seriously."

Decker, Daniel J.; and Brown, Tommy L. "How animal rightists view the 'wildlife management--hunting system.' " *Wildlife Society Bulletin* 15 (Winter 1987): 599-602.
> "The animal rightist does not believe animals should be killed for recreational purposes. Consequently some wildlife management professionals believe that the animal rights movement is one of the greatest threats to wildlife conservation faced by the profession."

Heberlein, Thomas. "Stalking the predator: a profile of the American hunter." *Environment* 29 (September 1987): 6-11, 30-33.
> "The American hunter has a major influence on the country's environment. Through his excursions into fields, grazing lands, and forests, he takes his toll of game. Perhaps more important, as a politically active citizen, he influences state programs for wildlife management and the efforts of landowners and government agencies to preserve and maintain habitat."

"Hunting and fishing in southeast Alaska." *Alaska Review of Social and Economic Conditions* 28 (June 1991): whole issue (24 p.)
> "Findings of the Tongass Resource Use Cooperative Study (TRUCS), a 1988 survey carried out jointly by the (University of Alaska Anchorage) Institute of Social and Economic Research, the Alaska Department of Fish and Game, and the U.S. Forest Service. The study documents hunting and fishing for household use

in all permanent southeast Alaska communities except the largest, Juneau and Ketchikan."

Knox, Margaret L. "In the heat of the hunt." *Sierra* 75 (November-December 1990): 48, 50-59.
"Those who shoot animals--for food or for sport--find themselves under the gun . yet more and more these days, hunting is an exercise in scientific and bureaucratic wildlife management." Four sidebar articles are included: "Why I hunt," by Dan Sisson; "Why I don't hunt," by Steve Ruggeri; "Why we hunt," by Humberto Fontova; and "Iguana hunt," by David Sobel.

LaBudde, Sam. "Hunting the good life: killing the walrus." *Earth Island Journal* 4 (Fall 1989): 28-30.
"When walrus were hunted with traditional hide boats and harpoons, a hunter would do well to take several in the course of a year. Today, automatic weapons, speedboats and chain saws (to remove the heads from the body) have allowed the traditional Spring hunts to devolve into ruthless quests for ivory--the international black market's 'white gold.' "

Laycock, George. "How to kill a wolf." *Audubon* 92 (November 1990): 44, 46, 48.
"Across much of Alaska there is no limit on the number of wolves one can kill with the aid of an airplane--for the seven-member Alaska Board of Game, which sets the state's game laws, has refused to call off the aerial attack on wolves, regardless of the fact that most Alaskans seem to condemn this brand of shooting as contrary to all rules of fair chase. Pressure from a select minority of perhaps seventy-five private plane owners keeps the practice alive."

Lord, Miles W. "In favor of the wolf." *Defenders* 59 (March-April 1984): 10-15.
Reprints the decision of Minneapolis Federal District Judge Lord, which declared that the 1983 fish and wildlife regulations permitting sport taking of wolves in parts of Minnesota violated the Endangered Species Act.

Sangiacomo, Michael; and Robb, Donna. "Pain and loathing in Hegins." *Animals* 125 (November-December 1992): 22-25.
About 6000 pigeons were killed at the 59th Annual Labor Day Pigeon Shoot in Hegins, Pennsylvania. Approximately 2000 people from all over the country came to stage a massive protest at the killing fields and end these pigeon shoots. "Judging by the 12,000 people who came to the shoot, a strong case can be made that the protest has become the event's main attraction."

Smith, Bradley F. "Hunter education: improving program implementation." *Journal of Environmental Education* 16 (Fall 1984): 24-28.
"Attempts to identify and analyze areas that prevent hunter education programs from being implemented to their fullest."

Thomas, Richard H. "Hunting as a political issue." *Parliamentary Affairs* 39 (January 1986): 19-30.
   Outlines the history of organized opposition to hunting in England and concludes that changing attitudes towards farming and hunting, "increasing concern for the environment and for animals, and the improved skills of both sets of lobbyists, suggest that hunting is unlikely to survive in its present form into the 21st century."

Environmental and professional journals (unannotated)

Clifton, Merritt; and Barnes, Donald J. "Pennsylvania pigeon blow-away won't go away." (animal rights activists protest 54th annual Labor Day Pigeon Shoot; includes related article) *Animals' Agenda* 9,11 (December 1989): 31(2)

Gaddis, Mike. "Taking a life." (rationalization of hunting animals) *Audubon* 92,6 (November 1990): 108(4)

Greanville, David P. "Hunting referendum fails." (Italy) *Animals' Agenda* 10,9 (November 1990): 43(2)

Knox, Margret L. "In the heat of the hunt." (society's attitudes about hunting) *Sierra* 75,6 (November-December 1990): 48(8)

Maggitti, Phil. "Heidi Prescott: hunt saboteur." *Animals' Agenda* 11,9 (November 1991): 8(2)

"No quiet on the hunting front." (anti-hunting demonstrations) *Animals' Agenda* 10,7 (September 1990): 31(2)

Pacelle, Wayne. "Game commission arms Florida youths." *Animals' Agenda* 9,1 (January 1989): 24(2)

Rainer, Robert. "Challenging hunters in New Brunswick: Bruce Cummings." *Animals' Agenda* 9,7 (September 1989): 8(2)

Tennesen, Michael; Mills, Judy; and Weintraub, David. "Poaching, ancient traditions, and the law." (includes related article on medicinal value of bear gall bladders to Chinese culture) *Audubon* 93,4 (July-August 1991): 90(8).

Law journals

Axline, Michael D. "Constitutional law--the end of a wildlife era: *Hughes v. Oklahoma.*" *Oregon Law Review* 60,4 (1981): 413-430.

Concludes that the Hughes decision threatens the high degree of autonomy which states have enjoyed in the management of wildlife within their borders. "The future of state wildlife now turns on the detail with which states document the costs imposed by nonresident hunters."

Weyhrauch, Bruce B. "Waterfowl and lead shot." *Environmental Law* 16 (Summer 1986): 883-934.
　　"Traces piecemeal federal and state attempts to regulate hunters' use of lead shot. Concludes that Congress should repeal the Stevens Amendment, which requires state approval of nontoxic shot zones; states should act aggressively to eliminate lead shot; and private hunting and conservation organizations should educate the public about the toxicity of lead shot and the suitability of steel shot as a safe and effective substitute."

FUR TRAPPING AND TRADE

Books (annotated)

Nilsson, Greta; et al. *Facts about furs.* Washington, D.C.: Animal Welfare Institute, 1980. 258 p.
　　Partial contents include: "The fur trade, a short history"; "Animals killed for the fur trade"; "From fur bearer to fur wearer: the agony of the transition"; "Legislation regulating the taking of furbearers"; and "The fur trade and endangered species."

Books (unannotated)

Bonner, W. Nigel. *Seals and man: a study of interactions.* Seattle: Washington Sea Grant Publication, 1982. Distributed by the University of Washington Press. 170 p.

Bright, Michael. *Killing for luxury.* New York: Gloucester Press, 1992.

Geary, Steven M. *Fur trapping in North America.* Rev. ed. Piscataway, N.J.: Winchester, 1984. 154 p.: ill. (SK283.6.N7 G43 1984)

Skinned. North Falmouth, Mass.: International Wildlife Coalition, 1988. 242 p. (HV4758 .S55 1988)

General interest periodicals (annotated)

Gilbert, Bil. "This cat is shy and set in its ways, but a master of its profession."

Smithsonian 18 (February 1988): 78-86.
> Article on the natural history of the Canadian lynx observes that "the marked increase in trapping raises a number of questions about the status of the lynx throughout Alaska."

Williams, Ted. "The far-from-tender trap." *Boston Magazine* 75 (November 1983): 243-244, 246-249.
> "There's a perfectly reasonable argument against trapping wild animals. Unfortunately, the people doing the arguing aren't reasonable."

General interest periodicals (unannotated)

"Alaska's trappers." *Alaska* 49 (May 1983): 56(2)

Barber, John. "The persecution of the fur trade." (Canada) *Reader's Digest* 139,832 (August 1991): 37(5).

Beck, Melinda. "The growing furor over fur: a brutal status symbol?" *Newsweek* 112,26 (December 26, 1988): 52(2).

Bifoss, Francis J. "Trapper's diary." *Alaska* 50 (February 1984): 18(5)

Brower, Montgomery. "From the wilderness to Saks Fifth Avenue, the fur trail leads from the hunt to riches." (Russian fur trade) *People* 27 (April 6 1987): 98(4)

Buss, Michael. "Reprieve for the fur-trade canoe." *Beaver* (Summer 1984): 30(4)

Carroll, Christine. "Spot market: bobcats on Texas ranches may end up as fur coats in West Germany." *Texas Monthly* 17,2 (February 1989): 104(4)

"Champions of synthetic fiber, angry human stars make the fur fly at a benefit for animal rights." *People* 31,9 (March 6 1989): 266(2)

Deutschman, Alan. "Fur enough." (selling fur operations) *Fortune* 121,4 (February 12 1990): 133(2)

Dyson, John. "The fur trade: must it die of shame?" *Reader's Digest* 127 (December 1985): 49(6)

Fellingham, Christine. "Are fur coats becoming extinct?" (includes related article on popularity of fake furs) *Glamour* 90,12 (December 1992): 183(2)

Fendi, Paola; et al. "Battle-weary designers take a stand on fashion's oldest fabric. (fur designers express their views) *Savvy* 11,10 (October 1990): 52(2)

Fenge, Terry. "The animal rights movement: a case of evangelical imperialism. (Native Trappers and the Fur Trade) *Alternatives* 15,3 (September-October 1988): 69(3).

"Fur this winter: sales are cool, the issue is hot." *Glamour* 88,10 (November 1990): 193(2)

Gold, Philip. "Rights clad in fur coats and put under the scalpel." *Insight* 5,19 (May 8 1989): 15(3)

Hough, Robert. "Dressed to kill." (animal-rights activists challenged by pro-fur groups) *Saturday Night* 107,4 (May 1992): 20(8)

"Hunt for illegal skins and furs: the U.S. gets tougher with retailers to protect endangered species." *Business Week* (March 7 1983): 70(2)

"Indigenous Survival International." *Alaska* 52 (September 1986): 69(3)

Kasindorf, Jeanie. "The fur flies." *New York* 23,2 (January 15 1990): 26(8)

Lee, John Alan. "Seals, wolves and words: loaded language in environmental controversy." *Alternatives* 15,4 (November-December 1988): 20(10)

Lynden, Patricia. "The fur dilemma: both sides of the controversy." (includes related articles) *Connoisseur* 219,934 (November 1989): 112(10).

McQuade, Walter. "The fur trade." (illustration) *Fortune* 107 (February 7 1983): 76(10)

Neuhaus, Cable. "What's up, Doc? A real-life Elmer Fudd tries to turn rabbits into riches." (food and furs) *People* 19 (April 4 1983): 119(2)

Okun, Stacey. "Which side are you on?" *Savvy* 11,10 (October 1990): 48(5)

Park, Ed. "A coyote coat for Lue." *Outdoor Life* 179 (February 1987): 58(7)

Payne, Michael; and Thomas, Gregory. "Literacy, literature, and libraries in the fur trade." *Beaver* (Spring 1983): 44(10)

Plummer, William. "Trendy Aspen warms up for a showdown in the bitter cold war over furs." (pro- and antifur activists) *People* 33,4 (January 29, 1990): 40(3)

Shapiro, Harriet. "Even in hard times an ill wind blows furrier Ernest Graf some good." *People* 19 (February 7, 1983): 86(2)

"Stoles could be a steal this year, if you're thinking mink. *Money* 18,1 (January 1989): 18(2)

Struzik, Ed. "Animal rights vs. native rights: are opponents of the fur trade fighting speciesism, or practicing cultural genocide?" *Nature Canada* 17,3 (Summer 1988): 26(7).

Symonds, William C. "Now, the trapper is an endangered species." *Business Week* 3212 (May 6 1991): 24A(2)

Taylor, Zack. "Land of the beaver." (northern fur traders) *Sports Afield* 194 (September 1985): 90(4)

Trachtenberg, Jeffrey A. "Make mine mink." (fur trade survey; includes related article) *Forbes* 135 (May 6, 1985): 62(3)

Trachtenberg, Jeffrey A. "R.I.P. Fred the furrier." (Fred Schwartz and Fur Vault ads) *Forbes* 143,4 (February 20, 1989): 122(2)

Trachtenberg, Jeffrey A. "What's a shopper to do?" (buying furs) *Forbes* 134 (December 3, 1984): 248(3)

Environmental and professional journals (annotated)

Buyukmichi, Ned; and Jessup, David A. "Steel leghold traps." *California Veterinarian* 37 (January 1983): 36-37, 39-42.
    Two articles on trapping: the title article by Buyukmichi condemns the use of steel leghold traps and "Trapping: a wildlife veterinarian's view" by Jessup, presents the trappers' arguments.

Clark, Bill. "Leg-hold trapping." *Defenders* 55 (December 1980): 388-391.
    Presents evidence from several countries that refutes American trappers' arguments against banning the leg-hold trap.

Gentile, John R. "The evolution of antitrapping sentiment in the United States: a review and commentary." *Wildlife Society Bulletin* 15 (Winter 1987): 490-503.
    "Antitrapping efforts are considered a serious threat, not only by trappers, but also by hunters and fishermen. In response, those favoring the use of animals have formed a powerful protrapping alliance. By contrast, those opposed to trapping have had difficulty consolidating their efforts. The 99.98% failure rate of antitrapping legislation can be attributed more to the inability of antitrappers to unite than to a lack of popular support for their goal."

Laycock, George. "The legacy of Gerasim Pribilof." *Audubon* 88 (January 1986): 95-100, 102.
    Outlines the history of the fur seal industry in the Aleutian Islands and regards as misguided the attempts of animal welfare groups to scuttle the northern fur seal treaty in the belief that they are protecting the seals. Other Bering Sea wildlife are also showing population declines.

Mitchell, John G. "The trapping question: soft skins and sprung steel." *Audubon* 84 (July 1982): 65-88.
> Discusses traps and trapping in Canada and the U.S. and considers the morality of the fur trade.

Moody, Jim. "The snag in park trapping." *National Parks* 58 (March-April 1984): 16-20.
> "In a controversial move, Interior has blocked Park Service rules in order to allow trapping at eleven national parks."

Newman, Peter C. "Canada's fur-trading empire." *National Geographic* 172 (August 1987): 192-228.
> Describes Hudson's Bay Company over its three centuries of fur trade.

Nilsson, Greta. "Bringing back the river otter." *Defenders* 60 (May-June 1985): 5-9.
> "Although trapping and habitat change have extirpated otters in parts of the nation, ten states have reintroduced them, and 21 protect them against trappers." Urges additional protection from trapping, especially for females nursing young.

Sleeper, Barbara. "The high price of fur." *Animals* 121 (November-December 1988): 14-19.
> "The fur industry argues that if we can kill animals to eat meat, conduct experimental research, and wear leather clothes, shoes, and handbags, then why not kill for fur as well? Animal-rights activists argue that the methods used to trap animals in the wild are cruel and barbaric. The deaths of these animals, they say, serve no other purpose than to supply an unnecessary luxury."

Turbak, Gary. "A tale of two cats." *International Wildlife* 15 (March-April 1985): 4-10.
> Because they are in such sharp competitions for food, bobcats and lynx occupy different ranges. Both are trapped for their fur--bobcat in the lower 48 and lynx in Canada and Alaska. "It is unlikely that harvesting the animals could endanger them, however."

Williams, Ted. "Small cats: forgotten, exploited." *Audubon* 87 (November 1985): 34, 36, 38-41.
> "The sudden pressure on small cats has resulted from a shift in the world's fur trade away from big cats. The U.S. still trades heavily in bobcat and Canada lynx. The poor example it has set for the world by first failing to strictly regulate bobcat harvest and then by institutionalizing that failure with recent amendment of its Endangered Species Act has obscured the fact that the act itself is a very good one."

Woodsum, Karen. "Crisis in wolf country." *Defenders* 59 (January-February 1984): 18-19, 22-27.
> "Minnesota wants to revive sport trapping of wolves, posing a threat to the last major wolf population south of Canada."

Environmental and professional journals (unannotated)

Bushnell, Candace. "Grand illusions: as the controversy over wearing fur rages on, there are--at last--some stunning fakes." *Health* 21,9 (September 1989): 72(7)

Clifton, Merritt. "Anti-fur crusade continues." (Tree-Species Unlimited continues efforts against fur coat industry *Animals' Agenda* 10,1 (January-February 1990): 36(2)

Clifton, Merritt. "Falling profits for the fur industry: furriers feel the pain." (includes related articles) *Animals' Agenda* 9,11 (December 1989): 15(3)

Clifton, Merritt. "Fur takes a hiding." (anti-fur demonstrations) *Animals' Agenda* 9,5 (May 1989): 20(2)

Clifton, Merritt. "Fur farms: where the sun doesn't shine." *Animals' Agenda* 11,9 (November 1991): 12(4)

Clifton, Merritt. "Furriers on the defensive." *Animals' Agenda* 10,3 (April 1990): 39(2)

Clifton, Merritt. "Furriers losing their skins." *Animals' Agenda* 10,2 (March 1990): 39(2)

Clifton, Merritt. "Knocking fur to the canvas." (includes article about celebrities and fur) *Animals' Agenda* 10,9 (November 1990): 32(3)

Clifton, Merritt. "Public opinion." (attitudes towards animal rights) *Animals' Agenda* 10,4 (May 1990): 35(2)

Dafoe, Christopher. "Canada and the fur trade." (history of the Hudson's Bay Co.) *Beaver* 71,2 (April-May 1991): 4(2)

Daniloff, Ruth. "Russian Sable--'soft gold': conservation efforts brought this reclusive animal back from the brink of extinction--and helped create a Soviet monopoly on its fur." *International Wildlife* 16 (July-August 1986): 20(4)

Duckworth, Harry W. "The last coureurs de bois." (fur trading history in Canada) *Beaver* (Spring 1984): 4(9)

"Fur industry seeking friends." (form umbrella group Putting People First to rally support against animal rights activists) *Animals' Agenda* 10,8 (October 1990): 28(2)

Greanville, David P. "Petitions on bullfighting and fur labelling considered." (Western Europe) *Animals' Agenda* 9,2 (February 1989): 33(2)

Jackson, John. "Inland from the Bay." (Hudson's Bay Company Archives map details the fur trade in Northwestern Canada) *Beaver* 72,1 (February-March 1992): 37(6)

Jotham, Neal R. "Is trapping cruel?" *Nature Canada* 12 (April-June 1983): 16(4)

Kantor, Michael. "Fur flies in pelt dispute." (way of life for many native Canadians) *Sierra* 73,4 (July-August 1988): 25(3).

Maggitti, Phil. "The opposition in motion." (counteroffensives against the anti-fur movement) *Animals' Agenda* 10,5 (June 1990): 17(5)

Moran, Victoria. "When the fur wearer isn't the furbearer." (managing face-to-face confrontations with fur wearers) *Animals' Agenda* 9,7 (September 1989): 50(2)

Newman, Peter C. "The beaver and the bay." (history of the Hudson's Bay Company) *Canadian Geographic* 109,4 (August-September 1989): 56(9)

Newman, Peter C.; and Fleming, Kevin. "Three centuries of the Hudson's Bay Company: Canada's fur trading empire." *National Geographic* 172 (August 1987): 192(38)

Ray, Arthur. "Rivals for fur." (excerpts from *The Canadian Fur Trade in the Industrial Age*) *Beaver* 70,2 (April-May 1990): 30(14)

Roach, Thomas R. "The saga of northern radio." (fur-trade radio network of northern Canada) *Beaver* (Summer 1984): 19(5)

Smith, Tony. "Is furwearing an act of philanthropy?" (Native trappers and the fur trade) *Alternatives* 15,3 (September-October 1988): 66(3)

Stardom, Eleanor. "Twilight of the fur trade." (history of the Hudson Bay Company, 1870-1884) *Beaver* 71,4 (August-September 1991): 6(13)

Williams, Glyndwr. "The Hudson's Bay Company and the fur trade: 1670-1870" *Beaver* (Autumn 1983): 4(72)

Law journals

Chaimov, Gregory A.; and Durr, James E. "*Defenders of Wildlife, Inc. v. Endangered Species Scientific Authority*: the court as biologist." *Environmental Law* 12 (Spring 1982): 773-810.
     Discusses the implications of this case in which the court invalidated endangered species scientific authority guidelines permitting the annual export of bobcat pelts.

Reports

Changing U.S. trapping policy: a handbook for activists. Washington, D.C.: Defenders of

Wildlife, 1984. 56 p.

"Defenders of Wildlife has felt for some time that there is a need to collect in one volume, the relevant information about trapping, its role in predator and disease control, and past efforts to reform trapping. This publication therefore attempts to compile a broad base of information useful to that end and to achieving changes in trapping policies."

Pryor, Vivian. *Statement of the National Wildlife Federation before the Subcommittee on Health and the Environment of the House Energy and Commerce Committee on H.R. 1797.* Washington, D.C.: National Wildlife Federation, 1984. 8 p.

Contends that "the modern leghold trap does not have steel-jawed teeth. It is practical, efficient and safe to use and poses less danger to humans, pets, and livestock than killing traps. Wildlife research, control of wildlife damage, and harvest of renewable wildlife resources are all frequently accomplished most efficiently and effectively by trapping."

Federal government reports

North Pacific fur seals: current problems and opportunities concerning conservation and management; a special report to the President and the Congress. Washington, D.C.: U.S. National Advisory Committee on Oceans and Atmosphere, 1985. 84 p.

Reviews the principal issues involved in U.S. consideration of extending or rejecting the 1984 Protocol of the Interim Convention on Conservation of North Pacific Fur Seals. "On balance, NACOA believes that it is in the national interest to extend the Convention. Although NACOA finds problems with the Convention and its implementation by the United States, we believe that corrective measures are called for, not abandonment of the convention."

CITIZEN PARTICIPATION

Books (annotated)

Cook, Lori. *A shopper's guide to cruelty-free products.* New York: Bantam, 1991. 262 p. (TX336 .C66 1991)

Maintains that no practical benefits come from animal testing and that results of tests on animals are usually inconclusive. Rates various products according to the amount of animal testing that went into their production. The majority of the book is an alphabetical list of companies that do not test their products on animals. Offers alternatives to animal testing, such as not using unproven ingredients, using statistical tests, and using organ cultures.

Fraser, Laura; et al. *The animal rights handbook: everyday ways to save animal lives.* Los

Angeles: Living Planet Press, 1990. Distributed by Publishers Group West, Emeryville, Calif. 113 p.: ill. (HV4764 .A65 1990)

A comprehensive guide to animal rights. Covers topics such as pets, wildlife, education, clothing, food products, and laboratory animals. Urges readers to forgo leather and fur, to give up eating meat, to boycott circuses and rodeos, and to use only cruelty-free cosmetics. Lists numerous resources, including relevant publications, organizations, and businesses.

Sequoia, Anna. *67 ways to save the animals*. New York: Harper and Row, 1990. 128 p. (HV4764 .S47 1990)

Offers information and suggestions for how to curb the inhumane treatment of domestic, wild, and agricultural animals. Suggestions--which range from boycotting greyhound racetracks to refusing to wear fur--are followed by background material, relevant addresses, and possible tactics. Concludes with a selection of addresses of animal rights organizations, listings of cruelty-free products, and ways to join adopt-an-animal programs.

Books (unannotated)

Crail, Ted. *The animal activist's handbook: in defense of an animal-wonderful world*. Sacramento, Calif.: Animal Protection Institute of America, 1980. 93 p.: ill. (HV4764 .C7)

Shopping guide for caring consumers: a guide to products that are not tested on animals. Washington, D.C.: People for the Ethical Treatment of Animals, 1992.

Reports

How to organize a humane society. Washington, D.C. (2100 L Street NW, Washington 20037): Humane Society of the United States, 1985. 12 p. (HV4763 .H68 1985)